Power Electronics and Motor Drives

Power Electronics and Motor Drives

Advances and Trends

Bimal K. Bose

Condra Chair of Excellence in Power Electronics/Emeritus
The University of Tennessee
Knoxville, Tennessee

ELSEVIER

AMSTERDAM • BOSTON • HEIDELBERG • LONDON
NEW YORK • OXFORD • PARIS • SAN DIEGO
SAN FRANCISCO • SINGAPORE • SYDNEY • TOKYO

Academic Press is an imprint of Elsevier

Academic Press is an imprint of Elsevier
30 Corporate Drive, Suite 400, Burlington, MA 01803, USA
525 B Street, Suite 1900, San Diego, California 92101-4495, USA
84 Theobald's Road, London WC1X 8RR, UK

This book is printed on acid-free paper. ∞

Library of Congress Cataloging-in-Publication Data
Application submitted

British Library Cataloguing-in-Publication Data
A catalogue record for this book is available from the British Library.

ISBN 13: 978-0-12-088405-6
ISBN 10: 0-12-088405-4

For information on all Academic Press publications
visit our Web site at www.books.elsevier.com

Printed in the United States of America
Transferred to digital printing in 2015

Working together to grow
libraries in developing countries

www.elsevier.com | www.bookaid.org | www.sabre.org

ELSEVIER   BOOK AID International   Sabre Foundation

CONTENTS

Please note that the previous printing included a CD-ROM.

The material is now only available on the companion website:
http://booksite.elsevier.com/9780120884056/

ABOUT THE AUTHOR

Dr. Bimal K. Bose (*Life Fellow, IEEE*) has held the Condra Chair of Excellence in Power Electronics at the University of Tennessee, Knoxville, since 1987. Prior to this, he was a research engineer at General Electric Corporate Research and Development (now GE Global Research Center) in Schenectady, New York (1976–1987), faculty member at Rensselaer Polytechnic Institute, Troy, New York (1971–1976), and faculty member of Bengal Engineering and Science University (formerly Bengal Engineering College) for 11 years. He has done extensive research in power electronics and motor drive areas, including converters, PWM techniques, microcomputer/DSP control, motor drives, and application of expert systems, fuzzy logic, and neural networks to power electronic systems. He has authored or edited seven books, published more than 190 papers, and holds 21 U.S. patents. He has given invited presentations, tutorials, and keynote addresses throughout the world. He is a recipient of a number of awards and honors that include the IEEE Power Electronics Society William E. Newell Award (2005), IEEE Millennium Medal (2000), IEEE Meritorious Achievement Award in Continuing Education (1997), IEEE Lamme Gold Medal (1996), IEEE Industrial Electronics Society Eugene Mittelmann Award for lifetime achievement in power electronics (1994), IEEE Region 3 Outstanding Engineer Award (1994), IEEE Industry Applications Society Outstanding Achievement Award (1993), General Electric Silver Patent Medal (1986) and Publication Award (1987), and the Calcutta University Mouat Gold Medal (1970).

PREFACE

I am presenting this novel book on advances and trends in power electronics and motor drives to the professional community with the expectation that it will be given the same wide and enthusiastic acceptance by practicing engineers, R&D professionals, university professors, and even graduate students that my other books in this area have. Unlike the traditional books available in the area of power electronics, this book has a unique presentation format that makes it convenient for group presentations that use Microsoft's PowerPoint software. In fact, a disk is included that has a PowerPoint file on it that is ready for presentation with the core figures. Presentations can also be organized using just selected portions of the book.

As you know, power electronics and motor drive technology is very complex and multidisciplinary, and it has gone through a dynamic evolution in recent years. Power electronics engineers and researchers are having a lot of difficulty keeping pace with the rapid advancements in this technology. This book can be looked on as a text for a refresher or continuing education course for those who need a quick review of recent technological advancements. Of course, for completeness of the subject, the core technology is described in each chapter. A special feature of the book is that many examples of recent industrial applications have been included to make the subject interesting. Another novel feature is that a separate chapter has been devoted to the discussion of typical questions and answers.

During the last 40+ years of my career in the industrial and academic environment, I have accumulated vast amounts of experience in the area of power electronics and motor drives. Besides my books, technical publications, and U.S. patents, I have given tutorials, invited presentations, and keynote addresses in different countries around the world at many IEEE as well as non-IEEE conferences. A mission in my life has been to promote power electronics globally. I hope that I have been at least partially successful. I pursued the advancement of power electronics technology aggressively from its beginning and have tried to present my knowledge and experience in the whole subject for the benefit of the professional community. However, the book should not be considered as a first or second course in power electronics. The reader should have a good background in the subject to assimilate the content of the book.

Each page contains one or more figures or a bulleted chart with explanations given below it—just like a tutorial presentation. The bulk of the figures are taken from my personal presentation materials from tutorials, invited seminars, and class notes. A considerable amount of material is also taken from my other publications, including the published books.

Unlike a traditional text, the emphasis is on physical explanation rather than mathematical analysis. Of course, exceptions have been made where it is absolutely necessary. After description of the core material in each chapter, the relevant advances and trends are given from my own experience and perspective. For further digging into the subject, selected references have been included at the end of each chapter. I have not seen a similar book in the literature. With its novel and unique presentation format, I describe it as a 21st-century book on power electronics. If opportunity arises, I will create a complete video course on the entire subject in the near future.

The content of the book has been organized to cover practically the entire field of power electronics. Chapter 1 gives a broad introduction and perspective on importance and applications of the technology. Chapter 2 describes modern power semiconductor devices that are viable in industrial applications. Chapter 3 deals with the classical power electronics, including phase-controlled converters and cycloconverters, which are still very important today. Chapter 4 describes voltage-fed converters, which are the most important type of converter in use today and will remain so tomorrow. The chapter includes a discussion of different PWM techniques, static VAR compensators, and active filters. Chapter 5 describes current-fed converters, which have been used in relatively large power applications. Chapter 6 describes different types of ac machines for variable-frequency drives. Chapter 7 deals with control and estimation techniques for induction motor drives, whereas Chapter 8 deals with control and estimation techniques for synchronous motor drives. Chapter 9 covers simulation and digital control in power electronics, including modern microcomputers and DSPs. The content of this chapter is somewhat new and very important. Chapter 10 describes fuzzy logic principles and their applications, and Chapter 11 provides comprehensive coverage of artificial neural networks and their applications. Finally, Chapter 12 poses some selected questions and their answers which are typical after any tutorial presentation.

This book could not have been possible without active contributions from several of my professional colleagues, graduate students, and visiting scholars in my laboratory. The most important contribution came from Lu Qiwei, a graduate student of China University of Mining and Technology (CUMT), Beijing, China, who devoted a significant amount of time to preparing a large amount of the artwork for this book. Professor Joao Pinto of the Federal University of Mato Grosso do Sul (UFMS) in Brazil made significant contributions to the book in that he prepared the demonstration programs in fuzzy logic and neural network applications. I also acknowledge the help of his graduate students. Dr. Wang Cong of CUMT provided help in preparation of the book. Dr. Kaushik Rajashekara of Rolls-Royce gave me a lot of ideas for the book and worked hard in checking the manuscript. Dr. Hirofumi Akagi of the Tokyo Institute of Technology, Japan, gave me valuable advice. Dr. Marcelo Simoes of the Colorado School of Mines and Ajit Chattopadhyay of Bengal Engineering and Science University, India, also deserve thanks for their help. Finally, I would like to thank my graduate students and visiting scholars for their outstanding work, which made the book possible. Some of them are Drs. Marcelo Simoes; Jason Lai of Virginia Tech; Luiz da Silva of Federal University of Itajuba, Brazil; Gilberto Sousa of Federal University of Espirito Santo, Brazil; Wang Cong; Jin Zhao of Huazhong University of Science and Technology,

China; M. H. Kim of Yeungnam College of Science & Technology, Korea; and Nitin Patel of GM Advanced Technology Vehicles. In my opinion, they are the best scholars in the world—it is often said that great graduate students and visiting scholars make the professor great. I am also thankful to the University of Tennessee for providing me with opportunities to write this book. Finally, I acknowledge the immense patience and sacrifice of my wife Arati during preparation of the book during the past 2 years.

Bimal K. Bose
June 2006

LIST OF VARIABLES AND SYMBOLS

$d^e\text{-}q^e$	Synchronously rotating reference frame direct and quadrature axes
$d^s\text{-}q^e$	Stationary reference frame direct and quadrature axes (also known as $\alpha\text{-}\beta$ axes)
f	Frequency (Hz)
I_d	dc current (A)
I_f	Machine field current
I_L	rms load current
I_m	rms magnetizing current
I_P	rms active current
I_Q	rms reactive current
I_r	Machine rotor rms current (referred to stator)
I_s	rms stator current
$i_{dr}{}^s$	d^s axis rotor current
$i_{ds}{}^s$	d^s axis stator current
i_{dr}	d^e axis rotor current (referred to stator)
i_{qr}	q^e axis rotor current (referred to stator)
i_{qs}	q^e axis stator current
J	Rotor moment of inertia (kg-m^2)
X_r	Rotor reactance (referred to stator) (ohm)
X_s	Synchronous reactance
X_{ds}	d^e axis synchronous reactance
X_{lr}	Rotor leakage reactance (referred to stator)
X_{ls}	Stator leakage reactance
X_{qs}	q^e axis synchronous reactance

α	Firing angle
β	Advance angle
γ	Turn-off angle
δ	Torque or power angle of synchronous machine
θ	Thermal impedance (Ohm); also torque angle
θ_e	Angle of synchronously rotating frame ($\omega_e t$)
θ_r	Rotor angle
θ_{sl}	Slip angle ($\omega_{sl} t$)
μ	Overlap angle
τ	Time constant (s)
L_c	Commutating inductance (H)
L_d	dc link filter inductance
L_m	Magnetizing inductance
L_r	Rotor inductance (referred to stator)
L_s	Stator inductance
L_{lr}	Rotor leakage inductance (referred to stator)
L_{ls}	Stator leakage inductance
L_{dm}	d^e axis magnetizing inductance
L_{qm}	q^e axis magnetizing inductance
m	PWM modulation factor for SPWM ($m = 1.0$ at undermodulation limit, i.e., $m' = 0.785$)
m'	PWM modulation factor, where $m' = 1$ at square wave
p	Number of poles
P	Active power
P_g	Airgap power (W)
P_m	Mechanical output power
Q	Reactive power
R_r	Rotor resistance (referred to stator)
R_s	Stator resistance
S	Slip (per unit)

T	Time period(s); also temperature (°C)
T_e	Developed torque (Nm)
T_L	Load torque
t_{off}	Turn-off time
V_c	Counter emf
V_d	dc voltage
V_I	Inverter dc voltage
V_f	Induced emf
V_m	Peak phase voltage (V)
V_g	rms airgap voltage
V_R	Rectifier dc voltage
v_s	Instantaneous supply voltage
v_d	Instantaneous dc voltage
v_f	Instantaneous field voltage
v_{dr}^{s}	d^s axis rotor voltage (referred to stator)
v_{ds}^{s}	d^s axis stator voltage
v_{dr}	d^e axis rotor voltage (referred to stator)
v_{qr}	q^e axis rotor voltage (referred to stator)
v_{qs}	q^e axis stator voltage
φ	Displacement power factor angle
ψ_a	Armature reaction flux linkage (Weber-turns)
ψ_f	Field flux linkage
ψ_m	Airgap flux linkage
ψ_r	Rotor flux linkage
ψ_s	Stator flux linkage
ψ_{dr}^{s}	d^s axis rotor flux linkage (referred to stator)
ψ_{ds}^{s}	d^s axis rotor flux linkage
ψ_{dr}	d^e axis rotor flux linkage (referred to stator)
ψ_{qr}	q^e axis rotor flux linkage (referred to stator)

ψ_{qs}	q^e axis stator flux linkage
ω_e	Stator or line frequency $(2\pi f)$ (rad/s)
ω_m	Rotor mechanical speed
ω_r	Rotor electrical speed
ω_{sl}	Slip frequency
$\hat{X}$	Peak value of a sinusoidal phasor or sinusoidal space vector magnitude; also estimated parameter, where X is any arbitrary variable
$\bar{X}$	Space vector variable; also designated by the peak value $\hat{X}$ where it is a sinusoid

CHAPTER 1

Introduction and Perspective

FIGURE 1.1   What is power electronics?

CONVERSION AND CONTROL OF ELECTRICAL POWER
BY
POWER SEMICONDUCTOR DEVICES

MODES OF CONVERSION

- RECTIFICATION: AC – to – DC
- INVERSION: DC – to – AC
- CYCLOCONVERSION: AC – to – AC
  (Frequency changer)
- AC CONTROL: AC – to – AC
  (Same frequency)
- DC CONTROL: DC – to – DC

Power electronics deals with conversion and control of electrical power with the help of electronic switching devices. The magnitude of power may vary widely, ranging from a few watts to several gigawatts. Power electronics differs from signal electronics, where the power may be from a few nanowatts to a few watts, and processing of power may be by analog (analog electronics) or digital or switching devices (digital electronics). One advantage of the switching mode of power conversion is its high efficiency, which can be 96% to 99%. High efficiency saves electricity. In addition, power electronic devices are more easily cooled than analog or digital electronics devices. Power electronics is often defined as a hybrid technology that involves the disciplines of power and electronics. The conversion of power may include ac-to-dc, dc-to-ac, ac-to-ac at a different frequency, ac-to-ac at the same frequency, and dc-to-dc (also called *chopper*). Often, a power electronic system requires hybrid conversion, such as ac-to-dc-to-ac, dc-to-ac-to-dc, ac-to-ac-to-ac, etc. Conversion and regulation of voltage, current, or power at the output go together. A power electronics apparatus can also be looked on as a high-efficiency switching mode power amplifier. If charging of a battery is required from an ac source, an ac-to-dc converter along with control of the charging current is needed. If a battery is the power source and the speed of an induction motor is to be controlled, an inverter is needed. If 60-Hz ac is the power source, a frequency converter or ac controller is needed for speed control of the induction motor. A dc-to-dc converter is needed for speed control of a dc motor in a subway or to generate a regulated dc supply from a storage battery. Motor drives are usually included in power electronics because the motors require variable-frequency and/or variable-voltage power supplies with the help of power electronics.

FIGURE 1.2   Features of power electronics.

- HARMONICS AND EMI AT LOAD AND SOURCE SIDE
- NONLINEAR DISCRETE TIME SYSTEM
- COMPLEXITY IN ANALYSIS, MODELING, SIMULATION, DESIGN, AND TESTING
- FAST ADVANCING TECHNOLOGY IN LAST THREE DECADES
- FAST GROWTH IN APPLICATIONS

INDUSTRIAL
COMMERCIAL
RESIDENTIAL
AEROSPACE
MILITARY
UTILITY SYSTEM
TRANSPORTATION

- GLOBAL EXPANSION OF TECHNOLOGY AND APPLICATIONS

Because power electronics equipment is based on nonlinear switching devices, it generates undesirable harmonics in a wide frequency range that flow in the load as well as in supply lines. A fast rate of change in voltage (dv/dt) and current (di/dt) due to switching creates electromagnetic interference (EMI) that couples with sensitive control circuits in its own and neighboring equipment. A switching mode converter with a discrete mode of control constitutes a nonlinear discrete time system and adds complexity to the analysis, mathematical modeling, computer simulation, design, and testing of the equipment. The design and testing phases become especially difficult at high power due to harmonics and EMI problems. In spite of this complexity, power electronics technology has been advancing at a rapid rate during the last three decades. Dramatic cost and size reductions and performance improvements in recent years are promoting extensive application of power electronics in the industrial, commercial, residential, aerospace, military, utility, and transportation environments. Power electronics–based energy and industrial motion control systems are now expanding globally to include the developing countries.

FIGURE 1.3   Why is power electronics important?

- ELECTRICAL POWER CONVERSION AND CONTROL AT HIGH EFFICIENCY
- APPARATUSES HAVE LOW COST, SMALL SIZE, HIGH RELIABILITY, AND LONG LIFE
- VERY IMPORTANT ELEMENT IN MODERN ELECTRICAL POWER PROCESSING AND INDUSTRIAL PROCESS CONTROL
- FAST GROWTH IN GLOBAL ENERGY CONSUMPTION
- ENVIRONMENTAL AND SAFETY PROBLEMS EXPERIENCED BY FOSSIL AND NUCLEAR POWER PLANTS
- INCREASING EMPHASIS ON ENERGY SAVING AND POLLUTION CONTROL FEATURES BY POWER ELECTRONICS
- GROWTH OF ENVIRONMENTALLY CLEAN SOURCES OF POWER THAT ARE POWER ELECTRONICS INTENSIVE (WIND, PHOTOVOLTAIC, AND FUEL CELLS)

Modern solid-state power electronic apparatus is highly efficient compared to the traditional M-G sets, mercury-arc converters, and gas tube electronics. The equipment is static and has a low cost, small size, high reliability, and long life. Power electronics and motion control constitute vital elements in modern industrial process control that result in high productivity and improved product quality. Essentially, the importance of power electronics can be defined as close to that of computers. In a modern automobile plant, for example, power electric–controlled robots are routinely used for assembling, material handling, and painting. In a steel-rolling mill, motor drives with high-speed digital signal processor (DSP) control produce steel sheets in high volume with precise control of widths and thicknesses. Globally, electrical energy consumption is growing by leaps and bounds to improve our standard of living. Most of the world's energy is produced in fossil and nuclear fuel power plants. Fossil fuel plants create environmental pollution problems, whereas nuclear plants have safety problems. Power electronics helps energy conservation by improved efficiency of utilization. This not only provides an economic benefit, but helps solve environmental problems. Currently, there is a growing trend toward using environmentally clean and safe renewable power sources, such as wind and photovoltaics, which are heavily dependent on power electronics. Fuel cell power generation also makes intensive use of power electronics.

FIGURE 1.4 Power electronics applications.

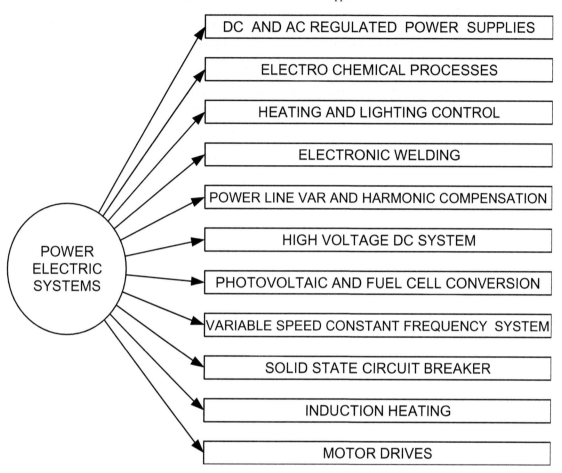

The spectrum of power electronics applications is very wide, and this figure illustrates some key application areas. One end of the spectrum consists of dc and ac regulated power supplies. The dc switching mode power supplies (SMPS) from an ac line or dc source are routinely used in electronics apparatuses, such as a computer, radio, TV, VCR, or DVD player. An example of an ac-regulated supply is an uninterruptible power supply (UPS) system, in which single- or three-phase 60/50-Hz ac can be generated from a battery source. The power supply may also be generated from another ac source where the voltage and frequency may be unregulated. Electrochemical processes, such as electroplating, anodizing, production of chemical gases (hydrogen, oxygen, chlorine, etc.), metal refining, and metal reduction, require dc power that is rectified from ac. Heating control, light dimming control, and electronic welding control are based on power electronics. Modern static VAR compensators (SVC or SVG), based on converters, help improve a system's power factor. They are also key elements for modern flexible ac transmission systems (FACTS). Active harmonic filters (AHFs) are being increasingly used to filter out harmonics generated by traditional diode and thyristor converters. High voltage dc (HVDC) systems are used for long-distance power transmission or to inter-tie two systems with dissimilar frequencies. Here, the line power is rectified to dc and then converted back to ac for transmission. Photovoltaic (PV) arrays and fuel cells generate dc, which is converted to ac for normal consumption or feeding to the grid. A variable-speed constant frequency (VSCF) system converts a variable frequency power from a variable-speed ac generator to a constant frequency, for use in, for example, wind generation systems and aircraft ac power supplies. Solid-state dc and ac circuit breakers and high-frequency induction and dielectric heating equipment are widely used. The dc and ac motor drives possibly constitute the largest area of applications in power electronics.

FIGURE 1.5 Application examples in variable-speed motor drives.

- TRANSPORTATION—EV/HV, SUBWAY, LOCOMOTIVES, ELEVATORS
- HOME APPLIANCES—BLENDERS, MIXERS, DRILLS, WASHING MACHINES
- PAPER AND TEXTILE MILLS
- WIND POWER GENERATION
- AIR CONDITIONERS AND HEAT PUMPS
- ROLLING AND CEMENT MILLS
- MACHINE TOOLS AND ROBOTICS
- PUMPS AND COMPRESSORS
- SHIP PROPULSION
- COMPUTERS AND PERIPHERALS
- SOLID-STATE STARTERS FOR MACHINES

This figure shows some examples of motor drive applications that will be discussed later in detail. A drive can be based on a dc or ac motor. For speed control, a dc motor requires variable dc voltage (or current), whereas an ac motor requires a variable-frequency, variable-voltage (or variable-current) power supply. Although dc drives constitute the bulk of current applications, modern advancements in ac drive technology are promoting their increasing acceptance, leading the dc drives toward obsolescence. Although process control is the main motivation for most of the drives, energy saving is the goal in some applications (e.g., air conditioning and heat pumps). The range of power, speed, and torque varies widely in various applications. Rolling mills and ship propulsion need high power (multi-megawatts); transportation, wind generation, starter-generator, pumps, etc., normally fall into the medium-power range (a few kilowatts to several megawatts), whereas computer and residential applications normally require low power (hundreds of watts to several kilowatts). While the majority of applications require speed control, some applications require position control and torque control. Again, ac motor drives can be based on induction or synchronous motors. Often, solid-state starters are used for soft-starting of ac motors, which normally operate at constant speed. An engineer has to design or select an economical and reliable drive system based on an appropriate machine, converter, and control system. These will be discussed later in detail.

FIGURE 1.6   Power electronics in industrial competitiveness.

- COMMUNICATION AND TRANSPORTATION TECHNOLOGY ADVANCEMENTS HAVE TURNED REMOTE COUNTRIES INTO CLOSE NEIGHBORS—WE NOW LIVE IN GLOBAL VILLAGE
- NATIONS ARE INCREASINGLY MORE INTERDEPENDENT
- FUTURE WARS WILL BE FOUGHT ON AN ECONOMIC FRONT, RATHER THAN A MILITARY FRONT
- INDUSTRIAL AUTOMATION AND GLOBAL COMPETITIVENESS OF NATIONS—KEY TO SURVIVAL AND ECONOMIC PROSPERITY
- POWER ELECTRONICS WITH MOTION CONTROL AND COMPUTERS ARE THE MOST IMPORTANT TECHNOLOGIES FOR INDUSTRIAL AUTOMATION IN 21st CENTURY

The figure highlights the important role of power electronics in terms of the industrial competitiveness of the world in the 21st century. Power electronics with motion control is now an indispensable technology for industrial process control applications. Fortunately, we are now living in an era of industrial renaissance when not only the power electronics and motion control technologies, but also computers, communication, information, and transportation technologies are advancing rapidly. The advancement of these technologies has turned geographically remote countries in the world into close neighbors day by day. We now practically live in a global society, particularly with the recent advancement in Internet communication. The nations of the world have now become increasingly dependent on each other as a result of this closeness. In the new political order of the world in the post-Communism era, the possibility of global war appears remote. In spite of great diversity among nations, we can safely predict that in this century major wars in the world will be fought on an economic front rather than a military front. In the new global market, free from trade barriers, the nations around the world will face fierce industrial competitiveness for survival and improvement of standards of living. In the highly automated industrial environment, where companies struggle to produce high-quality, cost-effective products, it appears that two technologies will be most dominant: computers and power electronics with motion control.

FIGURE 1.7   How can we solve or mitigate environmental problems?

- PROMOTE ALL ENERGY USAGE IN ELECTRICAL FORM
- CENTRALIZE FOSSIL FUEL POWER GENERATION AND APPLY ADVANCED EMISSION STANDARDS
- MOVE TOWARD GREATER USE OF RENEWABLE ENERGY SOURCES: HYDRO, WIND, AND PHOTOVOLTAIC
- REPLACE ICE VEHICLES WITH ELECTRIC AND HYBRID VEHICLES
- CONSERVE ENERGY BY EFFICIENT USE OF ELECTRICITY
- PREVENT ENERGY WASTE

Environmental pollution problems due to burning of fossil fuels (coal, oil, and natural gas) are becoming dominant issues in our society [1, 21]. The pollutant gases, such as CO_2, SO_2, NO_x, HC, O_3, and CO, cause global warming, acid rain, and urban pollution problems. With the rapidly increasing energy consumption trend, pollution is posing a serious threat for the future. The question is how can we solve or mitigate our environmental problems? As a first step, all of our energy consumption should be promoted in electrical form, and then advanced emission control standards can be applied in central fossil fuel power plants. The problems then become easier to handle when compared to distributed consumption of coal, oil, and natural gas. As emission control technologies advance, more and more stringent controls can be enforced in central power stations. The emission problems can be mitigated by emphasizing safe and environmentally clean renewable energy sources, such as hydro, wind, and photovoltaics of which hydro has been practically tapped in full. Urban pollution can be solved by widespread use of EV, HV, trolley buses/trams, and subway transportation. Wind, PV, EV, HV, trolley buses/trams, and subway drives are all heavily dependent on power electronics. Conservation of energy by more efficient use of electricity with the help of power electronics, and thus reduction of fuel consumption, is not only a definite way to reduce environmental pollution, but also to preserve our dwindling fuel resources. Unfortunately, availability of cheap energy promotes wastage. It has been estimated that approximately one-third of the energy generated is simply wasted in the United States [3] because energy is cheap, and consumers are negligenct. In Japan, for example, energy is typically four times more expensive and, therefore, the desire to conserve energy, particularly with power electronics, is far greater.

FIGURE 1.8   Energy saving with power electronics.

- CONTROL OF POWER BY ELECTRONIC SWITCHING IS MORE EFFICIENT THAN OLD RHEOSTATIC CONTROL
- ROUGHLY 60% TO 65% OF GENERATED ENERGY IS CONSUMED IN ELECTRICAL MACHINES, MAINLY PUMPS AND FANS
- VARIABLE-SPEED, FULL-THROTTLE FLOW CONTROL CAN IMPROVE EFFICIENCY BY 30% AT LIGHT LOAD
- LIGHT-LOAD REDUCED-FLUX MACHINE OPERATION CAN FURTHER IMPROVE EFFICIENCY
- VARIABLE-SPEED AIR CONDITIONERS/HEAT PUMPS CAN SAVE ENERGY BY 30%
- 20% OF GENERATED ENERGY IS USED IN LIGHTING
- HIGH-FREQUENCY FLUORESCENT LAMPS ARE TWO TO THREE TIMES MORE EFFICIENT THAN INCANDESCENT LAMPS

Energy saving is one of the most important goals for power electronics applications [1]. Switching mode power control instead of traditional rheostatic control is highly efficient. Rheostatic speed control in a subway dc drive is still used in many parts of the world. According to the Electric Power Research Institute (EPRI) estimates, 60% to 65% of generated electrical energy in the United States is consumed in motor drives of which the major part is used for pump- and fan-type drives. The majority of these pumps and fans work in an industrial environment for control of fluid flow. In such applications, traditionally, the motor runs at constant speed and the flow is controlled by a throttle opening, where a lot of energy is lost due to turbulence. In contrast, variable-speed operation of the motor with the help of power electronics at full throttle opening is highly efficient. Again, most of the machines operate at light load most of the time. Motor efficiency can be improved by reduced flux operation instead of operating with rated flux. Air conditioners and heat pumps are normally controlled by on–off switching of thermostats. Instead, variable-speed load-proportional control can provide energy savings of as much as 30%. Roughly 20% of our generated energy is consumed in lighting. If fluorescent lamps are used instead of incandescent lamps, a substantial amount of energy can be saved. Again, use of high-frequency fluorescent lamps with power electronics–based lamp ballasts can save 20% to 30% in energy consumption. Such lamps have other advantages such as longer lamp life, smooth light, and dimming control capability.

FIGURE 1.9 Electric and hybrid vehicle scenario.

- POWER ELECTRONICS AND DRIVES INTENSIVE— SOMEWHAT MATURE TECHNOLOGY
- LIMITATION OF BATTERY TECHNOLOGY
- EV HAS LIMITED RANGE—SUITABLE FOR SHORT-RANGE AND INDOOR APPLICATIONS
- HYBRID VEHICLE CAN REPLACE ICEV, BUT IS MORE EXPENSIVE
- POSSIBLE STORAGE DEVICES:
    BATTERY
    FLYWHEEL
    ULTRACAPACITOR
- POSSIBLE POWER DEVICES:
    IC ENGINE
    DIESEL ENGINE
    STIRLING ENGINE
    GAS TURBINE
    FUEL CELL
- CURRENTLY, HVs ARE MORE VISIBLE DUE TO RISE OF GASOLINE PRICE

Petroleum conservation and environmental (particularly urban) pollution control have been the main motivations for worldwide R&D activities in EV/HV for more than two decades. The world has limited oil reserves, and at the present consumption rate, it will barely last more than 75 years. Industrial nations are primarily dependent on imported oil. Although EV/HVs have been commercially introduced by a number of auto manufacturers around the world, their acceptance level in the market is currently low mainly due to higher initial costs, periodic battery replacement costs, and the difficulty of roadside servicing. Fortunately, they use power electronics extensively, where the technology is somewhat mature for cost and performance. It is essentially the limitations of battery technology that have inhibited the acceptance of EVs in the market. In spite of prolonged R&D, today's propulsion batteries are too heavy, too expensive, have a low cycle life, and have limited storage capability making them suitable only for short-range driving. Having to replace batteries in EV/HVs every few years is an expensive proposition. In addition, fast and simultaneous charging of a large number of EV/HV batteries on utility distribution systems creates problems. An HV can truly replace an ICEV, but the dual needs of both power and energy sources make it more complex and expensive. Although the battery (Ni-MH, lead-acid, Ni-Cd, Li-ion) is the prime storage device, a flywheel or ultracapacitor could also be considered in HVs for temporary storage.

The IC engine is the traditional power source in HVs, but the diesel engine, Stirling engine, gas turbine, and fuel cell are also potential candidates [7]. Extensive R&D is required in storage and power devices to make EV/HVs more economical and acceptable in the market. (With the current trend toward rising gasoline costs, HVs are becoming more visible in the market.)

FIGURE 1.10 Wind energy scenario.

- MOST ECONOMICAL, ENVIRONMENTALLY CLEAN, AND SAFE "GREEN" POWER
- ENORMOUS WORLD RESOURCES—TAPPING ONLY 10% CAN SUPPLY ELECTRICITY NEED FOR THE ENTIRE WORLD
- COMPETITIVE COST WITH FOSSIL FUEL POWER ($0.05/kWh, $1.00/kW)
- TECHNOLOGY ADVANCEMENT IN POWER ELECTRONICS, VARIABLE-SPEED DRIVES, AND VARIABLE-SPEED WIND TURBINES
- CURRENTLY, GERMANY IS THE WORLD LEADER (4800 MW); NEXT IS UNITED STATES (2600 MW)
- CURRENTLY, 1.0% ELECTRICITY NEED IN UNITED STATES; WILL INCREASE TO 5% BY 2020
- CURRENTLY, 13% ELECTRICITY NEED IN DENMARK; WILL INCREASE TO 40% BY 2030
- PROBLEM OF STATISTICAL AVAILABILITY—NEEDS BACKUP POWER
- KEY ENERGY SOURCE FOR FUTURE HYDROGEN ECONOMY

Wind is a very safe, environmentally clean, and economically renewable energy source. The world has enormous wind energy resources. According to estimates from the European Wind Energy Association, tapping only 10% of viable wind energy can supply the electricity needs of the whole world [12, 13]. Recent technological advances in variable-speed wind turbines, power electronics, and machine drives have made wind energy very competitive—almost equal with fossil fuel power. Wind and PV energy are particularly attractive to the one-third of the world's population that lives outside the electric grid. Among the developing countries, for example, India and China have developed large expansion programs for wind energy. Currently, wind is the fastest growing energy technology in the world. Although Germany is currently the world leader in terms of installed capacity, the United States is next. In fact, the U.S. wind potential is so huge that it can meet more than twice its current electricity needs. North Dakota alone has 2.5 times the potential capacity of Germany. Currently, Denmark is the leader in wind energy utilization in terms of its percentage energy need. One of the drawbacks of wind energy is that its availability is sporadic in nature and requires backup power from fossil or nuclear power plants. The so-called future "hydrogen economy" concept will depend on abundant availability of wind energy that can be converted to electricity and then used to produce hydrogen fuel by electrolysis. Stored hydrogen will then be used extensively as an energy source, particularly for fuel cell vehicles.

FIGURE 1.11   Photovoltaic energy scenario.

- SAFE, RELIABLE, STATIC, AND ENVIRONMENTALLY CLEAN
- DOES NOT REQUIRE REPAIR AND MAINTENANCE
- PV PANELS ARE EXPENSIVE—CURRENTLY AROUND $5.00/W, $0.20/kWh
- SOLAR POWER CONVERSION EFFICIENCY IS AROUND 16%
- APPLICATIONS:
  SPACE POWER
  ROOFTOP INSTALLATIONS
  OFF-GRID REMOTE APPLICATIONS
- SPORADIC AVAILABILITY—REQUIRES BACKUP POWER
- CURRENT INSTALLATION (290 MW):
  JAPAN—45%
  USA—26%
  EUROPE—21%
- TREMENDOUS EMPHASIS ON TECHNOLOGY ADVANCEMENT

Photovoltaic devices, such as silicon (crystalline and amorphous), convert sunlight directly into electricity. They are safe, reliable, static, environmentally clean (green), and do not require any repair and maintenance as do wind power systems. The lifetime of a PV panel is typically 20 years. However, with the current technology, PV is expensive—typically five times more costly than wind power. A solar conversion efficiency of around 16% has been reported with the commonly used thin-film amorphous silicon, although 24% efficiency has been possible for thick crystalline silicon [11, 15]. PV power has been widely used in space applications, where cost is not a primary concern, but terrestrial applications are limited because of its high cost. Interestingly, because of its high energy costs, Japan has the highest PV installations. Like wind power, PV is extremely important for off-grid remote applications. As the price falls, the market is steadily growing. With the current trends in research, its price is expected to fall sharply in the future, thus promoting extensive applications. Unfortunately, its availability, like wind power, is sporadic and thus requires backup sources.

FIGURE 1.12 Fuel cell power scenario.

- HYDROGEN AND OXYGEN COMBINE TO PRODUCE ELECTRICITY AND WATER
- SAFE, STATIC, VERY EFFICIENT AND ENVIRONMENTALLY CLEAN
- FUEL CELL TYPES:
    PROTON EXCHANGE MEMBRANE (PEMFC)
    PHOSPHORIC ACID (PAFC)
    DIRECT METHANOL (DMFC)
    MOLTEN CARBONATE (MCFC)
    SOLID OXIDE (SOFC)
- GENERATE HYDROGEN BY ELECTROLYSIS OR BY REFORMER (FROM GASOLINE, METHANOL)
- BULKY AND VERY EXPENSIVE IN CURRENT STATE OF TECHNOLOGY
- SLOW RESPONSE
- POSSIBLE APPLICATIONS:
    FUEL CELL CAR
    PORTABLE POWER
    BUILDING COGENERATION
    DISTRIBUTED POWER FOR UTILITY
    UPS SYSTEM
- SIGNIFICANT FUTURE PROMISE

A fuel cell is an electrochemical device that operates on the reverse process of electrolysis of water; that is, it combines hydrogen and oxygen to produce electricity and water. It is a safe, static, highly efficient (up to 60%), and environmentally clean source of power [16, 17]. Fuel cell stacks can be considered equivalent to series-connected low-voltage batteries. The dc voltage generated by fuel cells is normally stepped up by power electronics based on a dc-to-dc converter and then converted to ac by an inverter depending on the application. The cells are characterized by high output resistance and sluggish transient response (polarization effect). Fuel cell types are defined by the nature of their electrolytes. Hydrogen or hydrogen-rich gas for fuel cells can be generated, respectively, by electrolysis of water or by hydrocarbon fuels (gasoline, methanol) by means of a reformer. In the latter case, pollutant gas is produced. Hydrogen can be stored in cylinders in either a cryogenically cooled liquefied form or as compressed gas. Fuel cells can be used in transportation and in portable or stationary power sources. In the current state of the technology, fuel cells are bulky and very expensive. Phosphoric acid fuel cells are currently available commercially with a typical cost of $5.50/W. However, with intensive R&D, fuel cells look extremely promising for the future. Automakers in the United States, Europe, and Japan have invested heavily to produce competitive fuel cell cars in the future with a target PEMFC fuel cell cost of $0.05/W.

FIGURE 1.13 Fuel cell EV and the concept of a hydrogen economy.

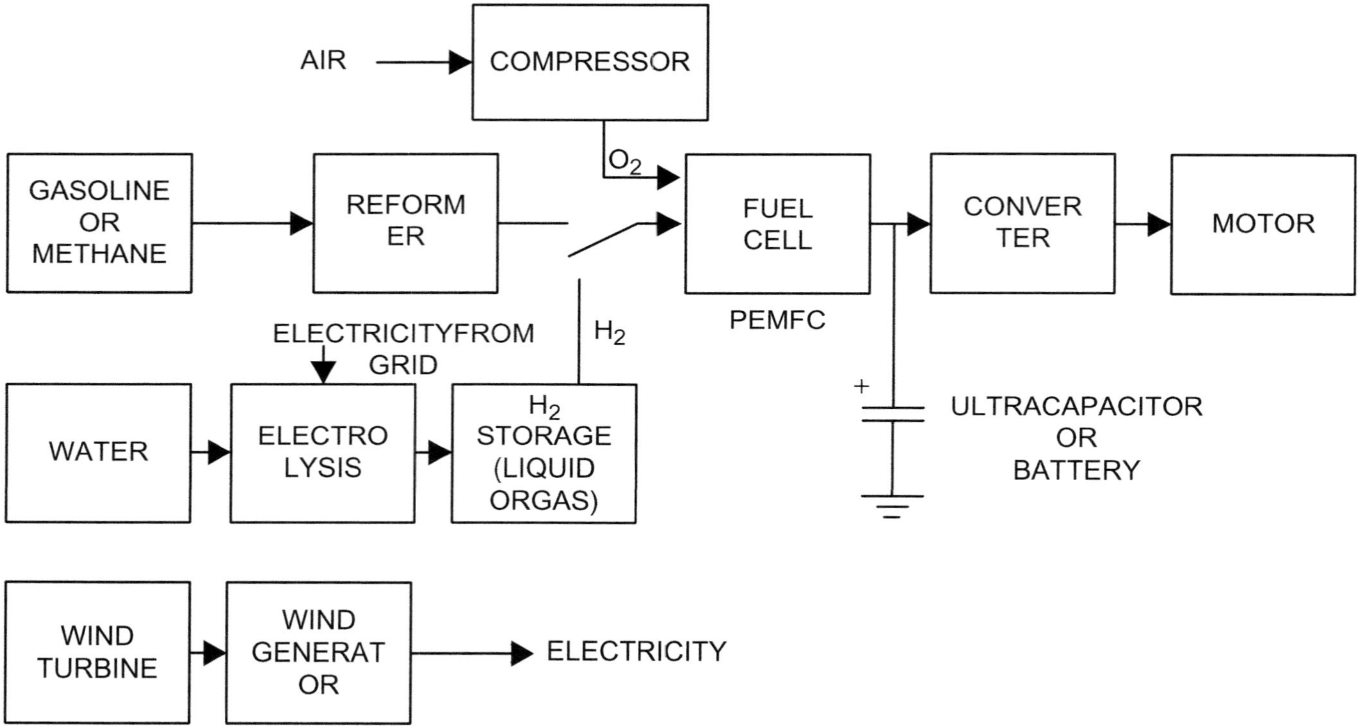

The figure shows the concept for a future fuel cell (FC) vehicle with different possible sources for the H_2 fuel. In an FC vehicle, a PEMFC usually generates the dc power, which is converted to variable-frequency ac for driving the ac motor. Because FCs can not absorb regenerative power, a battery or ultracapacitor storage is needed at the FC terminal. The battery also supplies power during acceleration because of the sluggish response of FCs. The H_2 fuel can be supplied from a tank where it can be stored in liquid or gaseous form. H_2 can be generated by electrolysis of water or H_2-rich gas can be obtained from gasoline/methane by means of a reformer as shown. The electrolysis process requires electricity, which can be supplied from the grid or obtained from a wind generation system as shown. The O_2 for FCs can be obtained from the air by means of a compressor. The problem of air pollution remains with gasoline/methane fuel. The equipment is very expensive, and the need for multiple power conversions makes wind-generated power less efficient. However, H_2 generation is a good way for storing wind energy.

FIGURE 1.14 Power electronics—an interdisciplinary technology.

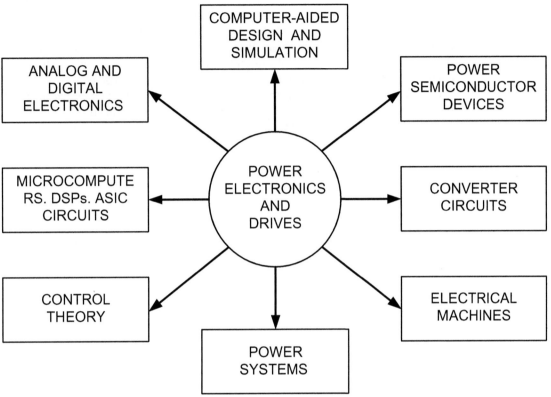

After several decades of evolution, power electronics and motor drives have been established as a complex and multidisciplinary technology. An engineer specializing in this area should have in-depth knowledge of power semiconductor devices, converter circuits, electrical machines, control electronics, microprocessors and DSPs, ASIC chips, control theories, power systems, and computer-aided design and simulation techniques. Knowledge of electromagnetic interference, the passive components (such as inductors, capacitors, and transformers) of such a system, and the accompanying specialized design, fabrication, and testing techniques are equally important. Very recently, the advent of artificial intelligence (AI) techniques, such as expert systems, fuzzy logic, artificial neural networks, and genetic algorithms, have advanced the frontier of the technology. Again, each of these component disciplines is advancing and creating challenges for power electronic engineers. Power semiconductors are extremely delicate and are often defined as the heart of modern power electronics. In-depth knowledge of devices is essential to make the equipment design reliable, efficient, and cost effective. A number of viable converter topologies may exist for a particular conversion function. The selection of optimum topology depends on the power capacity of the equipment, interactions with the load and source, and various trade-off considerations. Electrical machines used in drives, particularly in closed-loop systems, require complex dynamic models. In modern high-performance drives, precise knowledge of machine parameters in the running condition often becomes very difficult to obtain. Power electronic control systems, particularly the drives, are nonlinear, multivariable, discrete time, and often very complex. Therefore, computer-aided design and simulation are often desirable. The complexity of control and signal estimation usually demands the use of microcomputers or DSPs that are going through endless evolution. Because power electronic equipment interfaces with power systems and their applications in utility systems are expanding, an in-depth knowledge of power systems is important.

FIGURE 1.15   Evolution of power electronics.

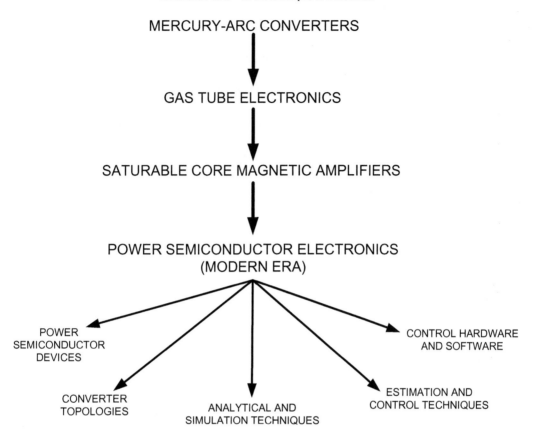

The history of power electronics is around 100 years old. In 1901, Peter Cooper Hewitt of the United States first demonstrated a glass-bulb mercury-arc diode rectifier [20] primarily for supplying power to an arc lamp, which later became the steel tank rectifier. Of course, motor-generator (MG) sets were widely used earlier for power conversion and control. For example, the long-time popularly used Ward-Leonard method of speed control with an MG set was introduced in 1891. Gradually, grid-controlled rectifiers and inverters were introduced. Interestingly, in the New York subway, a mercury-arc rectifier (3000 kW) was first installed in 1930 for dc drives. In 1933, an ignitron rectifier was invented by Slepian, which was another milestone in history. Around the same time (1930s), gas tube electronics using phanotrons and thyratrons were introduced. Gas tube electronics proved unreliable during World War II and, therefore, saturable reactor magnetic amplifiers were introduced and proved to be very rugged and reliable, but bulky. The modern era of solid-state power electronics started with the introduction of the thyristor (or silicon-controlled rectifier). Bell Labs published the historical paper on the PNPN transistor in 1956, and then in 1958, GE commercially introduced the thyristor into the marketplace. Since then, R&D in power electronics has radiated in different directions as shown in the figure. In power devices, R&D continued in different semiconductor materials, processing, fabrication and packaging techniques, device modeling and simulation, characterization, and development of intelligent modules. Starting with diode and thyristor converters, as new devices were introduced, many new converter topologies were invented along with advanced pulse width modulation (PWM) techniques and analytical and simulation methods. Many new control and estimation methods, particularly for drives, were introduced. These include vector or field-oriented control, adaptive and optimal controls, intelligent control, and sensorless control. Control hardware that is based on microprocessors, DSPs, and ASIC chips was introduced, along with software for control and simulation. The advent of powerful personal computers also played an important role in the power electronics evolution.

FIGURE 1.16   Four generations of solid-state power electronics.

- *FIRST GENERATION* (1958–1975) (Thyristor Era)
    - Diode
    - Thyristor
    - Triac

- *SECOND GENERATION* (1975–1985)
    - Power BJT
    - Power MOSFET
    - GTO
    - Microprocessor
    - ASIC
    - PIC
    - Advanced control

- *THIRD GENERATION* (1985–1995)
    - IGBT
    - Intelligent power module (IPM)
    - DSPs
    - Advanced control

- *FOURTH GENERATION* (1995–)
    - IGCT
    - Cool MOS
    - PEBB
    - Sensorless control
    - AI techniques: fuzzy logic, neural networks, genetic algorithm

The evolution of power electronics can be categorized into four generations as indicated in the figure. The first generation spanning around 17 years, when thyristor-type devices dominated, is defined as the thyristor era. In the second generation, lasting about 10 years, self-controlled power devices (BJTs, power MOSFETs, and GTOs) appeared along with power ICs, microprocessors, ASIC chips, and advanced motor controls. In the third generation, the most dominant power device, the IGBT, was introduced and became an important milestone in power electronics history. In addition, SITs, IPMs, and powerful DSPs appeared with further advancements in control. Finally, in the current or fourth generation, new devices, such as IGCTs and cool MOSs appeared. There is currently a definite emphasis on power converters in the power electronic building block (PEBB) or integrated form. Also, sensorless vector control and intelligent control techniques appeared and are now front-line R&D topics.

FIGURE 1.17 Some significant events in the history of power electronics and motor drives.

- 1891 – Ward-Leonard dc motor speed control is introduced
- 1897 – Development of three-phase diode bridge rectifier (Graetz circuit)
- 1901 – Peter Cooper Hewitt demonstrates glass-bulb mercury-arc rectifier
- 1906 – Kramer drive is introduced
- 1907 – Scherbius drive is introduced
- 1926 – Hot cathode thyratron is introduced
- 1930 – New York subway installs grid-controlled mercury-arc rectifier (3 MW) for dc drive
- 1931 – German railways introduce mercury-arc cycloconverters for universal motor traction drive
- 1933 – Slepian invents ignitron rectifier
- 1934 – Thyratron cycloconverter—synchronous motor(400 hp) was installed in Logan power station for ID fan drive (first variable-frequency ac drive)
- 1948 – Transistor is invented at Bell Labs
- 1956 – Silicon power diode is introduced
- 1958 – Commercial thyristor (or SCR) was introduced to the marketplace by GE
- 1971 – Vector or field-oriented control for ac motor is introduced
- 1975 – Giant power BJT is introduced in the market by Toshiba
- 1978 – Power MOSFET is introduced by IR
- 1980 – High-power GTOs are introduced in Japan
- 1981 – Multilevel inverter (diode clamped) is introduced
- 1983 – IGBT is introduced
- 1983 – Space vector PWM is introduced
- 1986 – DTC control is invented for induction motors
- 1987 – Fuzzy logic is first applied to power electronics
- 1991 – Artificial neural network is applied to dc motor drive
- 1996 – Forward blocking IGCT is introduced by ABB

The historical evolution of power electronics and motor drives was marked by many innovations in power devices, converters, PWM techniques, motor drives, control techniques, and applications. Some of the significant events in the history are summarized here with an approximate year. Many of these inventions and their applications will be further described later in the book.

FIGURE 1.18   Where to find information on power electronics.

- Key Books in Power Electronics
    - Bose, Mohan, Rashid, etc.
- IEEE Publications (ieeexplore.ieee.org)
- Individual Author Publication Websites
  (scholar.google.com, scopus.com, etc.)
- Conference Records and Transactions – IAS, IES, PELS, PES, SES,
  APEC, PEDS/PEDES, CIEP – Mexico, etc.
      Proc. of the IEEE
- Conference Records of:
      EPE (Europe)
      ICEM
      IPEC (Japan)
      PCC (Japan)
      PCIM, etc.
- Product Information in Key Power Electronic Company
  Websites:
  ABB, GE, Fuji, Toshiba, Hitachi, Mitsubishi, Siemens, Rockwell, Samsung, etc.

The information on advances and trends in power electronics and drives is scattered widely in recent electrical engineering literature. First, many excellent text and reference books are available, but because the technology is advancing rapidly in recent years, books may not provide the latest information. It is better to start with the books and then fall back into the literature for in-depth information. Again, assimilation of literature information may be difficult without adequate background on the subject. The IEEE transactions publications and the conference records of the IAS (Industry Applications Society), PELS (Power Electronics Society), and IES (Industrial Electronics Society) are very dominant. In addition, the proceedings of the IEEE and other periodically held conference records throughout the world in the English language provide excellent information. Practical product information can be obtained from different company websites (some references are given in Chapter 2). Also, search in general websites (google.com, yahoo.com, etc.) and individual author websites.

Summary

The nature of power electronics and its importance, applications, and historical evolution have been summarized in this introductory chapter. The modern era of power electronics is often defined as an era of "second electronics revolution," whereas the "first electronics revolution" resulted from the development of solid-state electronics. Besides the role of power electronics in industrial automation and energy systems, its role in energy saving and environmental pollution control have been highlighted. The environmentally clean renewable energy systems (wind and photovoltaic), which are being so highly emphasized today, are heavily dependent on power electronics. In addition, power electronics has had a tremendous impact on electric/hybrid vehicles and future fuel cell applications. The continuing technological evolution of power electronics has significantly reduced the cost and improved the performance of power electronic devices, and currently their applications are expanding fast in industrial, commercial, residential, utility, aerospace and military applications. This trend will continue with full momentum in the future.

References

[1] B. K. Bose, "Energy, environment and advances in power electronics," *IEEE Trans. Power Electronics*, vol. 15, pp. 688–701, July 2000.

[2] S. Rahman and A. D. Castro, "Environmental impacts of electricity generation: a global perspective," *IEEE Trans. Energy Conv.*, vol. 10, pp. 307–313, June 1995.

[3] G. R. Davis, "Energy for planet earth," *Scientific American*, pp. 1–10, 1991.

[4] B. K. Bose, "Recent advances in power electronics," *IEEE Trans. Power Elec.*, pp. 2–16, January 1992.

[5] C. C. Chan, "The state of the art of electric and hybrid vehicles," *Proc. IEEE*, vol. 90, no. 2, pp. 247–275, February 2002.

[6] D. Coates, "Advanced battery systems for electric vehicle applications," *American Chemical Society Magazine*, pp. 2.273–2.278, 1993.

[7] B. K. Bose, M. H. Kim, and M. D. Kankam, "Power and energy storage devices for next generation hybrid electric vehicles," *Proc. IEEE 31st Intersoc. Energy Conv. Eng. Conference*, pp. 143–149, August 1996.

[8] B. K. Bose, "Technology advancement and trends in power electronics," *IEEE IECON Conference Record*, pp. 3019–3020, 2003.

[9] W. Sweet, "Power and energy," *IEEE Spectrum*, pp. 62–67, January 1999.

[10] M. Begovic, A. Pregelj, A. Rohatgi, and C. Honsberg, "Green power: status and perspectives," *Proc. IEEE*, vol. 89, pp. 1734–1743, December 2001.

[11] S. R. Bull, "Renewable energy today and tomorrow," *Proc. IEEE*, vol. 89, no. 8, pp. 1216–1226, August 2001.

[12] R. Swisher, C. R. D. Azua, and J. Clendenin, "Strong winds on the horizon: wind power comes of age," *Proc. IEEE*, vol. 89, pp. 1757–1764, December 2001.

[13] N. Hatziargyriou and A. Zervos, "Wind power development in Europe," *Proc. IEEE*, vol. 89, pp. 1765–1782, December 2001.

[14] J. Jaydev, "Harnessing the wind," *IEEE Spectrum*, pp. 78–83, November 1995.

[15] A. Barnet, "Solar electric power for a better tomorrow," *Proc. 25th IEEE Photovoltaic Specialists Conference,* Washington, DC, pp. 1–8, May 1996.

[16] M. W. Ellis, M. R. V. Spakovsky, and D. J. Nelson, "Fuel cell systems: efficient, flexible energy conversion for the 21st century," *Proc. IEEE*, vol. 89, pp. 1808–1818, December 2001.

[17] M. Farooque and H. C. Maru, "Fuel cells—the clean and efficient power generators," *Proc. IEEE*, vol. 89, pp. 1819–1829, December 2001.

[18] R. Anahara, S. Yokokawa, and M. Sakurai, "Present status and future prospects for fuel cell power systems," *Proc. IEEE*, vol. 81, pp. 399–408, March 1993.

[19] K. Rajashekara, "Propulsion system strategies for fuel cell vehicles," *SAE Conference Record,* pp. 56–65, 2000.

[20] C. C. Herskind and M. M. Morack, *A History of Mercury-Arc Rectifiers in North America*, IEEE IA Society, IEEE Service Center, Hoboken, NJ, 1987.

[21] T. Fukao, "Energy, environment and power electronics," *Intl. Power Electronics Conference Record*, Japan, pp. 56–68, April 2005.

CHAPTER 2

Power Semiconductor Devices

FIGURE 2.1   Evolution of power semiconductor devices.

- DIODE (1955)

- THYRISTOR (1958)

- TRIAC (1958)

- GATE TURN-OFF THYRISTOR (GTO) (1980)

- BIPOLAR POWER TRANSISTOR (BPT or BJT) (1975)

- POWER MOSFET (1975)

- INSULATED GATE BIPOLAR TRANSISTOR
  (IGBT) (1985)

- STATIC INDUCTION TRANSISTOR (SIT) (1985)

- INTEGRATED GATE-COMMUTATED
  THYRISTOR (IGCT) (1996)

- SILICON CARBIDE DEVICES

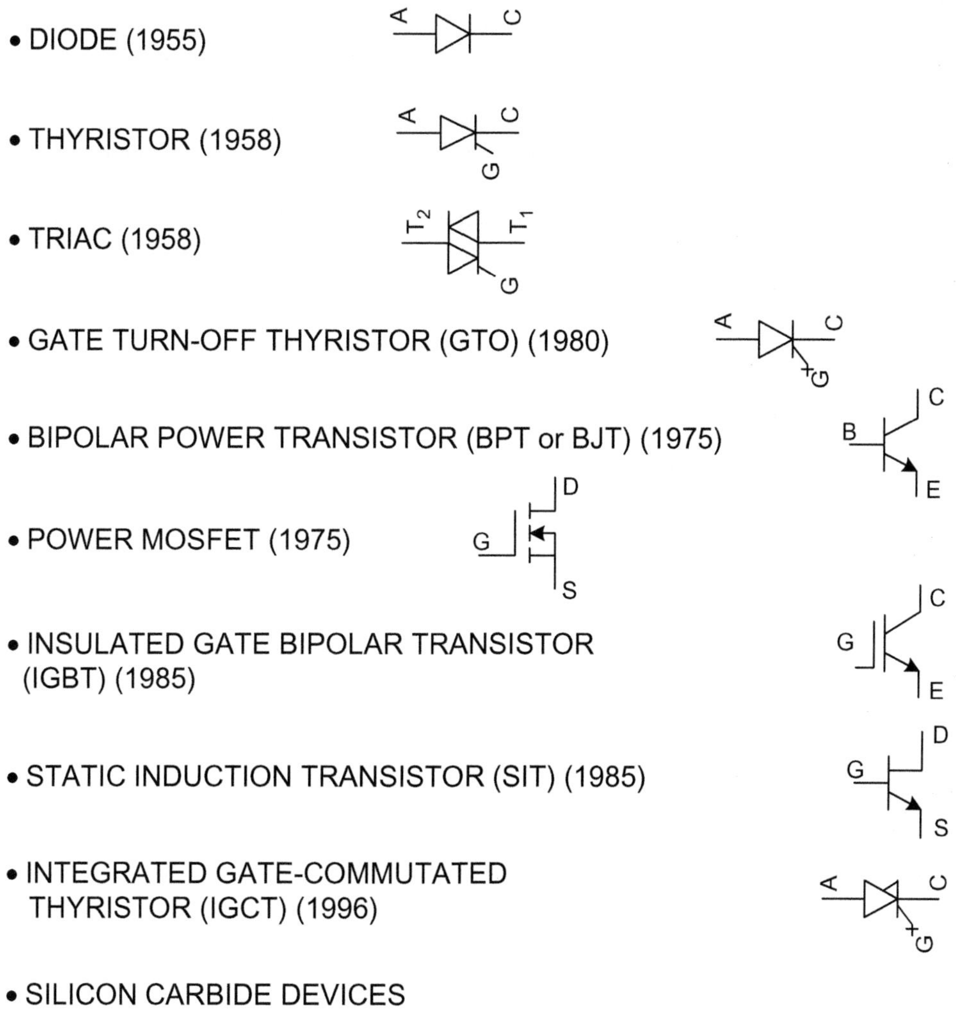

This figure shows the evolution of modern power semiconductor devices with the year of commercial introduction and symbol on the right of each. These power devices constitute the heart of modern power electronics apparatuses and they are almost exclusively based on silicon material. It is no exaggeration to say that today's power electronics evolution has been possible primarily due to device evolution. Again, power semiconductor evolution has closely followed the evolution of microelectronics. Researchers in microelectronics have worked relentlessly to improve semiconductor processing, device fabrication, and packaging, and these efforts have contributed to the successful evolution of so many exotic power devices. Although basically they are on and off switches, practical devices are far more complex, delicate, and "fragile." Thyristors (or silicon-controlled rectifiers) and triacs are essentially the forerunners of modern power semiconductor devices, which operate mainly on a utility system and contribute to power quality and lagging power factor problems. GTOs and IGCTs (or GCTs) are high-power gate-controlled devices that are used in multi-megawatt applications. The BJT, once a very important device, has become obsolete. Power MOSFETs and IGBTs are self-controlled insulated-gate devices that are extremely important today. In particular, the invention of the IGBT in 1983 and its commercial introduction in 1985 have been important milestones in the history of power devices. An SIT is the solid-state equivalent of the vacuum triode; it has a high conduction drop and will not be discussed further. SiC devices with large band-gap semiconducting materials offer extremely high promise for the future. Note that the list does not include devices that did not establish well, such as MOS-controlled thyristors, static induction thyristors, injection-enhanced gate transistors, and MOS turn-off thyristors.

FIGURE 2.2   Power-frequency trends of the devices (from [5]).

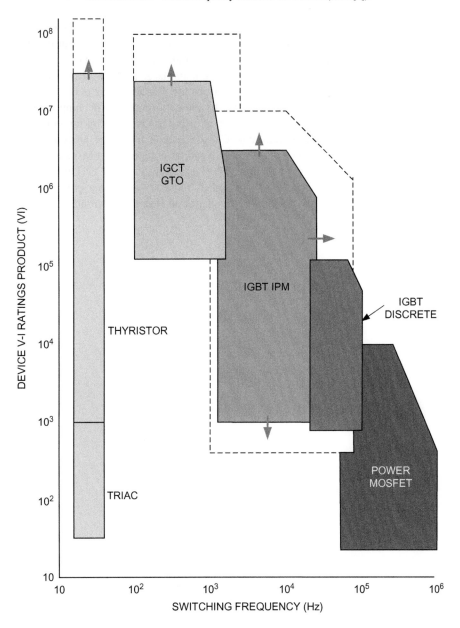

The figure (in log-log scale) shows the power-frequency capability of the current devices and their future trends. The power is given by *V-I* ratings product, that is, the product of the maximum blocking voltage and maximum turn-off current. Note that the BJT is completely removed from the figure. Thyristors and triacs are essentially low-frequency (50/60 Hz) devices, and currently the thyristor has the highest power rating. The future trend, as indicated by the dashed curve, also indicates that is has the highest power rating. High-power thyristors are used in high-voltage dc (HVDC) systems, phase-control type static VAR compensators (SVC), and large ac motor drives. GTOs and IGCTs have a higher frequency range (typically a few hundred hertz to one kilohertz) but their power limits are lower than that of a thyristor. Normally, with a higher power rating, the switching frequency becomes lower, and this is indicated by tapering of the areas at higher frequency. However, an IGCT's switching frequency is somewhat higher than that of a GTO (not shown). IGBT intelligent power modules (IPMs) come next with higher frequency but lower power range. The lower power end of GTO/IGCTs overlaps with IGBTs. The discrete IGBTs have higher frequency and lower power range, as shown. Power MOSFETs have the highest frequency and lowest power range. All of these devices will be described later.

FIGURE 2.3   (a) Diode *V-I* characteristics and (b) turn-off switching characteristics.

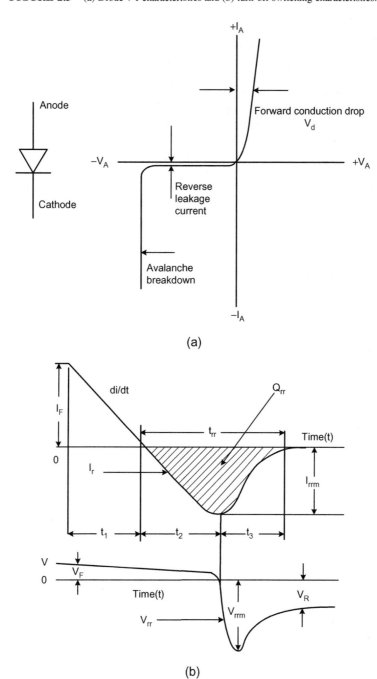

(a)

(b)

The diode is the simplest but most important of the power semiconductor devices. Figure 2.3(a) shows the *V-I* characteristics of a diode. In the forward direction, the device conducts with a small voltage drop (typically 1.0 V), whereas in the reverse direction, it blocks voltage with a small leakage current. With a high enough reverse voltage, it goes through avalanche breakdown and is destroyed. The switching characteristics (turn-off) of the diode are shown in Figure 2.3(b). As reverse voltage is applied in the forward conduction, the current decreases linearly (due to leakage inductance), reverses, and then turns off with a snap. High di/dt at turn-off causes a reverse voltage (V_{rrm}) that is higher than the applied voltage (V_R). The recovery charge Q_{rr} and the corresponding recovery time t_{rr} are important diode parameters. Diodes are classified as slow-recovery diodes, fast-recovery diodes, and Schottky diodes. The former two types have *P-I-N* geometry, whereas the last type has a metal–semiconductor junction. Slow-recovery diodes are used in 50/60-Hz power rectification. Fast-recovery types are used in feedback/freewheeling of converters and snubbers, and Schottky diodes, which provide very fast switching, are used in high-frequency converters.

FIGURE 2.4 (a) Thyristor structure with two-transistor analogy and (b) *V-I* characteristics.

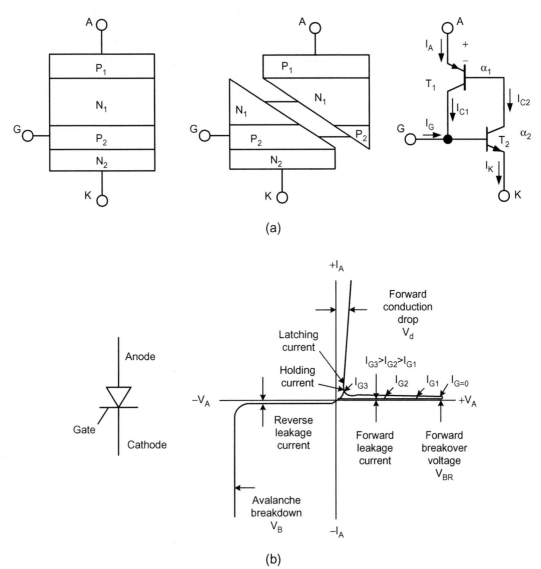

(a)

(b)

A thyristor is basically a three-junction *PNPN* device that can be represented by *PNP* and *NPN* component transistors connected in regenerative feedback mode, as shown in Figure 2.4(a). The device blocks voltage in both the forward and reverse directions. When the anode is positive, the device can be triggered into conduction by a short positive gate current pulse, but once the device is conducting, the gate control is lost. A thyristor can also be turned on by means of excessive anode voltage, high dv/dt, a rise in junction temperature, or light shining on the junctions. The *V-I* characteristics of the device are shown in Figure 2.4(b). With gate current $I_G = 0$, the device blocks voltage in the forward direction, and only small leakage current flows. With $I_G = I_{G3}$, the entire forward blocking voltage is removed, and the device acts as a diode as shown in the figure. The latching current is the minimum anode current required to turn on the device successfully, and holding current is the minimum on-current required to keep the device on without which the device will go to the forward blocking mode.

FIGURE 2.5 Thyristor features.

- SMALL GATE CURRENT PULSE TRIGGERS ON THE DEVICE
- CANNOT BE TURNED OFF BY GATE CURRENT
- SYMMETRIC OR ASYMMETRIC VOLTAGE BLOCKING
- COMMUTATION METHODS:
  - AC LINE
  - LOAD
  - FORCE

- FORCE-COMMUTATED CONVERTERS ARE OBSOLETE
- TRIGGERING POSSIBLE BY:
  - *dv/dt*
  - TEMPERATURE
  - LIGHT

- TURN-ON *di/dt* PROBLEM
- CAN CARRY LARGE TRANSIENT FAULT CURRENT
- FAST FUSE PROTECTION POSSIBLE
- APPLICATIONS WITH PHASE CONTROL:
  - RECTIFIER DC MOTOR DRIVES
  - CYCLOCONVERTER AC MOTOR DRIVES
  - SOLID-STATE INDUCTION MOTOR STARTER,
  - HVDC SYSTEM, ETC.

The figure summarizes the basic characteristics and typical applications of thyristors. Because they have a high turn-on sensitivity, the gate current pulse is short with small energy content. Normally, a train of pulses is fed during on-time so that the device does not turn off inadvertently if anode current falls below the holding current. In utility system applications the device is turned off (or commutated) by a segment of reverse line voltage. It can also be turned off by load-impressed reverse voltage (such as ac machine counter EMF), or by an *R-L-C* circuit transient called forced commutation. For normal phase control application, the device blocks voltage in either direction. Forward-blocking (but reverse-conducting) thyristors, once popular in voltage-fed force-commutated inverters, are now obsolete. The device can carry large short-time current within the limit of junction temperature rise, and a properly coordinated fast fuse can protect the device. High-voltage, high-power devices in utility applications are usually light triggered. Thyristor applications are very wide. Only some sample applications are mentioned.

FIGURE 2.6   (a) Triac symbol with *V-I* characteristics and (b) incandescent light dimmer circuit with triac.

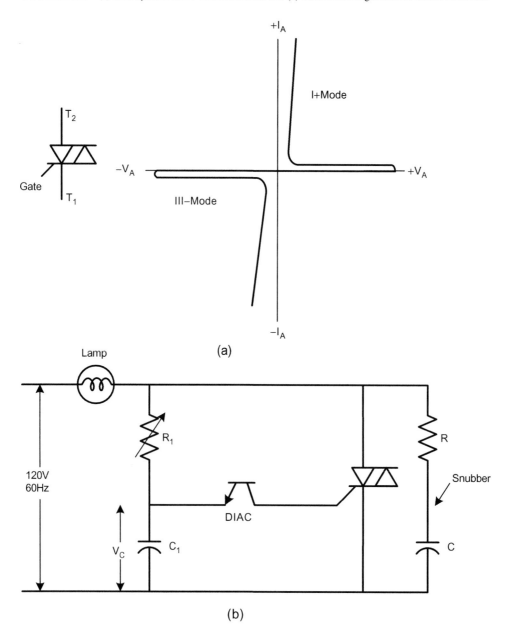

Functionally, a triac can be considered as an integration of a pair of phase-control thyristors connected in inverse-parallel manner. The structure of a triac is more complex than simply a *PNPN* device in parallel with an *NPNP* device. The circuit symbol for a triac and its *V-I* characteristics are given in Figure 2.6(a). In *I+* mode, the terminal T_2 is positive and the device is switched on by a positive gate current pulse. In *III–* mode, the terminal T_1 is positive and the device is turned on by a negative gate pulse. For ac power control, a triac is more economical than an inverse-parallel thyristor combination, but it has a few drawbacks. The gate current sensitivity is somewhat lower and turn-off time is slower than for a thyristor. Besides, the reapplied *dv/dt* capability is poor and, therefore, it is difficult to operate with an inductive load. Triacs are commonly used in incandescent lamp dimming and heating control, where the load is resistive. Figure 2.6(b) shows the popular triac light dimmer circuit. The gate of the triac is supplied from an *R-C* circuit through a diac. A diac is a symmetric voltage-blocking device. The variable resistance R_1 controls the dimming level of the lamp, which is connected in series. When the capacitor voltage V_C in either polarity exceeds the threshold voltage $\pm V_S$ of the diac, a pulse of current triggers the triac at a firing angle (α_f), giving phase-controlled, full-wave current to the lamp. When the triac is on, the *R-C* circuit is shorted and the line voltage is impressed across the lamp. At the end of every half-cycle, the triac turns off and the next half-cycle begins. An *R-C* snubber as shown is important to a triac circuit to reduce reapplied *dv/dt*.

FIGURE 2.7   Triac features.

- INTEGRATES A PAIR OF THYRISTORS
- TURNS ON AT EITHER ANODE VOLTAGE POLARITY BY
  GATE CURRENT
- POOR GATE CURRENT SENSITIVITY
- SYMMETRIC BLOCKING
- LONGER TURN-OFF TIME (USES 60–400 Hz)
- PHASE CONTROL WITH RESISTIVE LOAD
- APPLICATIONS:
    LIGHT DIMMER
    HEATING CONTROL
    SOLID-STATE AC SWITCH, ETC.
    APPLIANCE SPEED CONTROL BY UNIVERSAL MOTOR, ETC.

The integrated construction of a triac gives some control simplicity, but is also accompanied with some disadvantages. The gate current sensitivity of a triac is lower than that of a thyristor, and its turn-off time due to line commutation is longer due to a minority carrier storage effect. For the latter reason, the triac has a low dv/dt rating and should be used mainly with a resistive load. However, it is difficult to avoid some line leakage inductance. The inductance causes snappy turn-off with large $dv/dt,$ which may cause spurious turn-on at the trailing edge. A well-designed RC snubber lowers dv/dt sufficiently to prevent such turn-on. The storage effect also limits the maximum voltage and current ratings of the device (see Figure 2.2). Typical applications for a triac are listed. The ABB (Asea Brown Boveri) Company has recently introduced integrated bidirectional phase-controlled thyristors (BCTs) that have a high power rating (www.abb.com), which basically combines two thyristors in an anti-parallel manner in a single monolithic package. This gives the advantages of reliability, size reduction, and cooling. The BCT can be used in electronic circuit breakers, static VAR compensators, and soft starters of induction motors.

FIGURE 2.8   (a) GTO dc-to-dc converter and (b) switching waveforms.

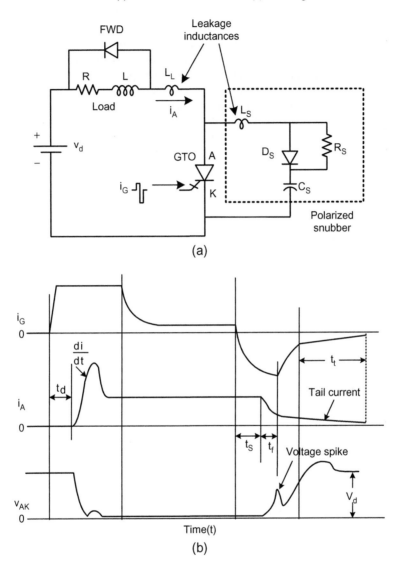

A GTO is a thyristor-like, high-power switching device that can not only be turned on by a positive gate current pulse, but also be turned off by a negative gate current pulse. However, the turn-off current gain is small, typically 4 to 5. This means that a GTO with a 6000-A anode current rating may require –1500-A gate current pulse. High gate current with small energy content can be generated by MOSFETs in parallel in a low inductance path. Figure 2.8(a) shows a GTO-based dc-to-dc converter (called a *chopper*), which is typically used in subway dc motor speed control. The buck or step-down chopper can control the dc voltage at output by duty cycle control operation. The load free-wheeling diode (FWD) circulates the load inductive current when the GTO is off. The circuit is shown with a turn-on snubber inductance (L_L) and turn-off RCD polarized snubber. When the device is turned on by a gate pulse, the turn-on current di/dt is slowed down by L_L, which also reduces turn-on switching loss. The snubber capacitor C_S, which is initially charged to supply voltage, starts discharging to R_S through the snubber leakage inductance L_S and the device. When negative gate current is applied to turn off the device, the anode current begins to fall after storage time (t_s) delay and then falls almost abruptly in fall time (t_f) before going to a long tail current as shown. The inductance L_s contributes to a voltage spike during turn-off as shown. The spike voltage can create a localized junction heating effect, thus causing "second breakdown" failure. A large turn-off snubber with very low L_s is desirable to reduce the turn-off switching losses within the device.

FIGURE 2.9   GTO energy recovery snubbers: (a) passive snubber for chopper and (b) active snubber for half-bridge inverter.

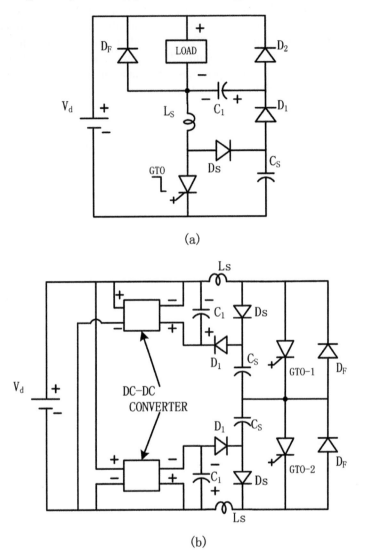

(a)

(b)

As mentioned before, a GTO is a slow switching device, and therefore, its switching loss in the snubber resistance R_S is high. If the converter is operating at a low switching frequency, the snubber loss may not be high and can be tolerated. But, at a high switching frequency, the loss is excessive. Therefore, high-power GTO converters operating in PWM (high-frequency) mode invariably use lossless or energy recovery snubbers (also called regenerative snubbers). Figure 2.9(a) shows a passive snubber energy recovery scheme used in a GTO chopper. When the GTO is turned off, the snubber capacitor C_S charges to the full supply voltage V_d. Later, when the GTO is turned on, the energy stored in C_S is transferred to the capacitor C_1 resonantly through the turn-on snubber inductor L_S and diode D_1. When the GTO is turned off again, the energy in C_1 is absorbed in the load and C_S charges again. Figure 2.9(b) shows an active snubber for a half-bridge converter where the operation principle is the same but the energy in the auxiliary capacitor C_1 is pumped back to the source by a dc-to-dc converter.

FIGURE 2.10   GTO features.

- TURNS ON BY + i_g AND TURNS OFF BY $-i_g$
- TURN-OFF CURRENT GAIN 4–5
- SYMMETRIC OR ASYMMETRIC BLOCKING
- LOW *dv/dt* RATING
- TURN-OFF TAIL CURRENT
- LARGE SNUBBER LOSS
  (CAN USE REGENERATIVE SNUBBER)
- LOW SWITCHING FREQUENCY
  (TYPICALLY 400–1000 Hz)
- APPLICATION IN HIGH POWER:
     CHOPPER DC MOTOR DRIVES
     INVERTER AC MOTOR DRIVES
     STATIC VAR COMPENSATORS, ETC.
     MAGLEV LINEAR SYNCHRONOUS MOTOR DRIVES

The figure summarizes the salient features of GTOs. They are high-power (multi-megawatt) low-switching-frequency devices that can be turned on or off by a low energy gate current pulse, as explained before. However, the turn-off current gain is low. Commercial GTOs are available with asymmetric voltage blocking (for voltage-fed converter) and symmetric voltage blocking (for current-fed converter) capabilities. GTOs are normally used with R-C-D snubbers that give lower efficiency in converters. Of course, regenerative snubbers in PWM converters can give high efficiency. Due to slow switching and high switching losses, GTO switching frequency is lower than that of IGBTs and IGCTs with comparable converter power rating. Typical applications for GTOs are listed in the figure. One very common application is in a multistepped voltage-fed converter on a utility system (see, for example, Figure 4.15), where the switching frequency is same as the line frequency (50/60 Hz). Historically, it is interesting to note that the GTO was invented by GE in 1958, but high-power devices were introduced to the market in the 1980s by several Japanese companies. It can be predicted that with the present trend of device technology, the future of GTOs does not look good.

FIGURE 2.11   (a) Double Darlington transistor and (b) triple Darlington.

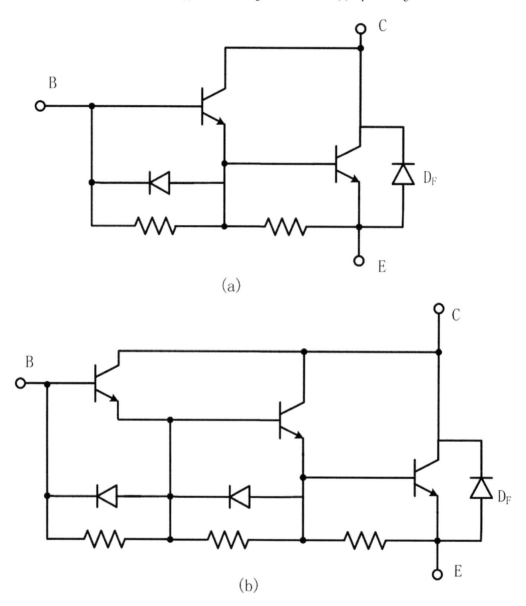

A bipolar junction transistor (BJT or BPT), unlike a thyristor-like device, is a two-junction, self-controlled *PNP* or *NPN* device, in which the collector current is under the control of base drive current. The dc current gain (h_{FE}) of a power transistor is low and varies widely with collector current and collector junction temperature. With a high enough base current, the device turns on or saturates with $V_{CE(S)}$ L1.0 V, when the base current control is lost. At turn-off condition, base current is zero and a small reverse voltage $(-V_{BE})$ is maintained at the base. The current gain of a power transistor can be increased to a high value in a Darlington configuration. Figures 2.11(a) and (b) show, respectively, double Darlington and triple Darlington configurations, which are monolithic, except for the main feedback diode (D_F). However, the disadvantages of Darlington transistors are a higher collector leakage current, higher collector-emitter conduction drop, and somewhat slower switching. These disadvantages are higher in the triple Darlington. The shunt resistors and diodes in a base-emitter circuit reduce collector leakage current and establish base bias voltages, but reduce base current sensitivity. A BJT is a forward-blocking device and is used in a voltage-fed converter with a feedback diode. Although power BJTs have been used extensively in the past, they are practically obsolete now.

FIGURE 2.12   (a) FBSOA of BJT (Powerex KD221K03) (from [9]) and (b) RBSOA.

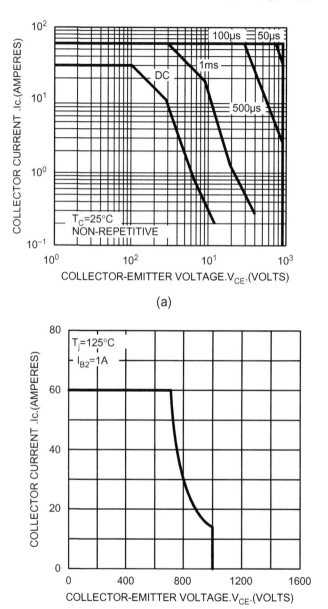

A BJT has an important property known as the *second breakdown effect*. When the device is switched on by the base drive, the collector current tends to crowd near the base-emitter junction periphery, thus constricting the current in a narrow area of the reverse-biased collector junction. This tends to create a hot spot, and because of a negative temperature coefficient effect, the current crowding becomes even worse in a smaller area, causing what is known as the *thermal runaway second breakdown effect*. This regeneration effect collapses the collector voltage, thus destroying the device. The forward bias safe operating area (FBSOA) of a BJT is demonstrated in Figure 2.12(a). In the log-log graph, the dc collector current and peak collector current at saturated condition are indicated on the vertical axis. Similarly, peak collector voltage is indicated on the horizontal axis. The safe active operating area can be exceeded over the dc limit area if the device is switched on a duty-cycle basis. The first negative slope represents the thermal limit junction power dissipation curve, which is dictated by the maximum junction temperature (T_j). The second and third slopes, which operate at higher collector voltage, are limited by the second breakdown effect. The second breakdown effect tends to be more serious during turn-off of an inductive load [see Figure 2.12(b)] when the collector current becomes constricted in a small collector junction area due to base-emitter reverse biasing. These figures indicate that BJTs cannot be operated without snubbers. Well-designed turn-on and turn-off snubbers should shape the load lines so as to limit operation within the SOA envelopes.

FIGURE 2.13 BJT features.

- DARLINGTON TRANSISTOR
    HIGH CURRENT GAIN
    LARGE LEAKAGE CURRENT
    LARGER DROP
- NEEDS NEGATIVE BASE BIAS
- ASYMMETRIC VOLTAGE BLOCKING
- DOMINANT SECOND BREAKDOWN EFFECTS
  (CONTROL BY SNUBBER)
- SWITCHING FREQUENCY OF A FEW KILOHERTZ
- RECENTLY BECAME OBSOLETE
- TYPICAL APPLICATIONS:
    CHOPPER DC MOTOR DRIVE
    INVERTER AC MOTOR DRIVE
    REGULATED DC AND AC SUPPLIES
    UPS SYSTEM

The figure summarizes the basic features of the bipolar junction transistor as discussed earlier. The device is a self-controlled minority-carrier two-junction device. It does not have GTO-like tail current, but the storage of minority carriers can cause a large turn-off delay. The BJT must be operated with snubbers because of second breakdown effect problems. Both turn-on and turn-off snubbers [see Figure 2.8(a)] shape load lines in the active region and divert the device switching loss to the snubber. The typical switching frequency of the device is several kilohertz, but it can be as low as 1.0 kHz in higher converter power ratings. On the other hand, switching frequency is high with a lower power rating as in other devices. Again, historically, the device was invented a long time ago, but Japanese companies brought the power device to market in the mid-1970s. High-power transistors typically have a 1200-V, 400-A rating. Typical applications in voltage-fed converters are indicated in the figure. Although the device has been extensively used in the past, it has recently become obsolete.

FIGURE 2.14   (a) Power MOSFET and (b) *V-I* characteristics (Harris IRF140) (from [10]).

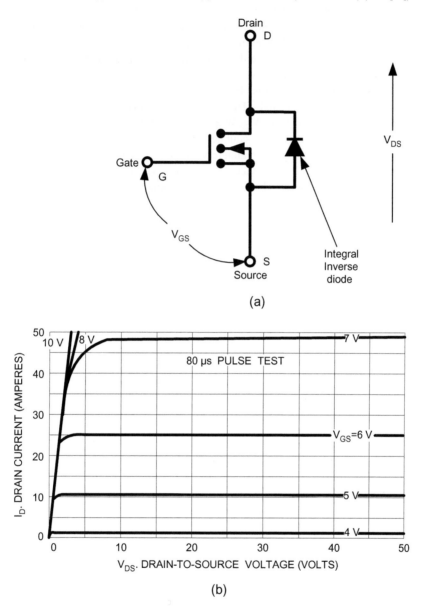

(a)

(b)

A power MOSFET is an extremely important device in low-voltage, high-frequency applications. The circuit symbol of the device is shown in Figure 2.14(a) and its *V-I* characteristics are shown in Figure 2.14(b). Unlike a bipolar transistor, it is a voltage-controlled majority carrier device. With a positive voltage applied to the gate with respect to the source terminal (NMOS), it induces an N-channel and permits electron current to flow from the source to the drain with applied voltage V_{DS}. Because of SiO_2 layer isolation, the gate circuit impedance is extremely high except during switching, when a small capacitor displacement current will occur. The device has an integrated reverse diode that permits free-wheeling current of the same magnitude as that of the main device. However, the body diode is characterized by slow recovery and is often bypassed by an external fast-recovery diode. The static V-I characteristics of the device show a gate circuit threshold voltage (4 V) of V_{GS}. With V_{GS} beyond the threshold value, I_D-V_D characteristics have two distinct regions, a saturated constant resistance ($R_{DS(ON)}$) region and a constant current region. The $R_{DS(ON)}$ region represents a key parameter that determines the conduction voltage drop. For a device, $R_{DS(ON)}$ increases with voltage rating and it has positive temperature coefficient characteristics. Although the conduction loss of a higher voltage MOSFET is high, its switching loss is extremely low. It does not experience minority carrier storage delay as does a BJT.

FIGURE 2.15 Safe operating area (SOA) of power MOSFET (Harris IRF140) (from [10]).

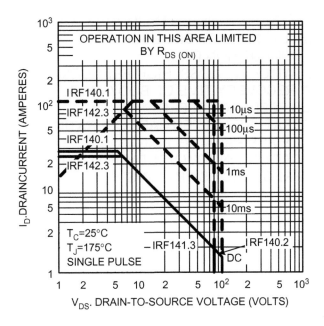

As mentioned earlier, a power MOSFET is a majority carrier device and, therefore, it has a positive temperature coefficient of resistance, and there is no second breakdown problem as with a BJT. If localized and potentially destructive heating occurs within the device, the positive temperature coefficient effect of resistance forces local current concentration to be uniformly distributed across the total area. The safe operating area (SOA) of a power MOSFET is shown in this figure. For example, the particular device (Harris IRF140) has a dc or continuous current rating of 28 A and a peak current rating of 110 A on a pulse basis. Similarly, the maximum drain-to-source voltage ($V_{DS} = 100$ V) is limited by avalanche breakdown. The SOA, shown on a log-log graph, is determined solely by thermal considerations. Within the constraint of the T_J limit of 175°C, the lowest curve (a straight-line negative slope means constant power) corresponds to dc power dissipation. It can be increased to a higher value on a pulsed basis with the absolute limit at 10 μs. The dashed positive slope shown is limited by $R_{DS(ON)}$, which is 0.07 Ohm for this device. Note that there is no secondary slope on SOA curves indicating the absence of second breakdown effect. Higher power rating on lower duty cycle operation is due to lower transient thermal impedance, which will be explained in Figure 2.17.

FIGURE 2.16 Power MOSFET features.

- VOLTAGE-CONTROLLED MAJORITY CARRIER DEVICE
- ASYMMETRIC VOLTAGE BLOCKING
- HIGH CONDUCTION DROP
- LOW-SWITCHING-LOSS, HIGH-FREQUENCY DEVICE
- SLOW RECOVERY TIME OF BODY DIODE
- MILLER FEEDBACK EFFECT
- TEMPERATURE-LIMITED SAFE OPERATING AREA
- EASY DEVICE PARALLELLING
- APPLICATIONS IN LOW-VOLTAGE, LOW-POWER, HIGH-FREQUENCY SWITCHING

    SWITCHING MODE POWER SUPPLIES (SMPS)
    PORTABLE APPLIANCES
    AUTOMOBILE POWER ELECTRONICS, ETC.

The figure summarizes the basic features of power MOSFETs. A MOSFET is an extremely important and popular device in low-voltage, high-switching-frequency, voltage-fed converter applications, such as SMPSs, brushless dc motor drives (BLDM), and portable battery-operated appliances. The device has a low switching loss but high conduction loss, particularly for devices with higher voltage ratings. The device has been used in hundreds of kilohertz (up to several megahertz) switching frequencies. However, the body diode is somewhat slow, which requires it to be bypassed with an external fast-recovery diode. "Trench gate" devices are now available in which the channel conduction resistance ($R_{DS(ON)}$) is reduced so that the conduction drop is lower typically by 20%. In transient operation with high dv/dt in the drain-source circuit, there is some feedback (called *Miller feedback*) to the gate due to stray C_{DG} and C_{GS} capacitors. The devices are easy to parallel for higher current due to a positive temperature efficient of resistance. Recently, Infineon Technology introduced a high-voltage (up to 600 V) COOLMOS, in which the conduction drop is cut down typically to 20% of the conventional device. However, its body diode is very slow.

FIGURE 2.17 Device T_j rise curves with junction power loss waves (from [9]).

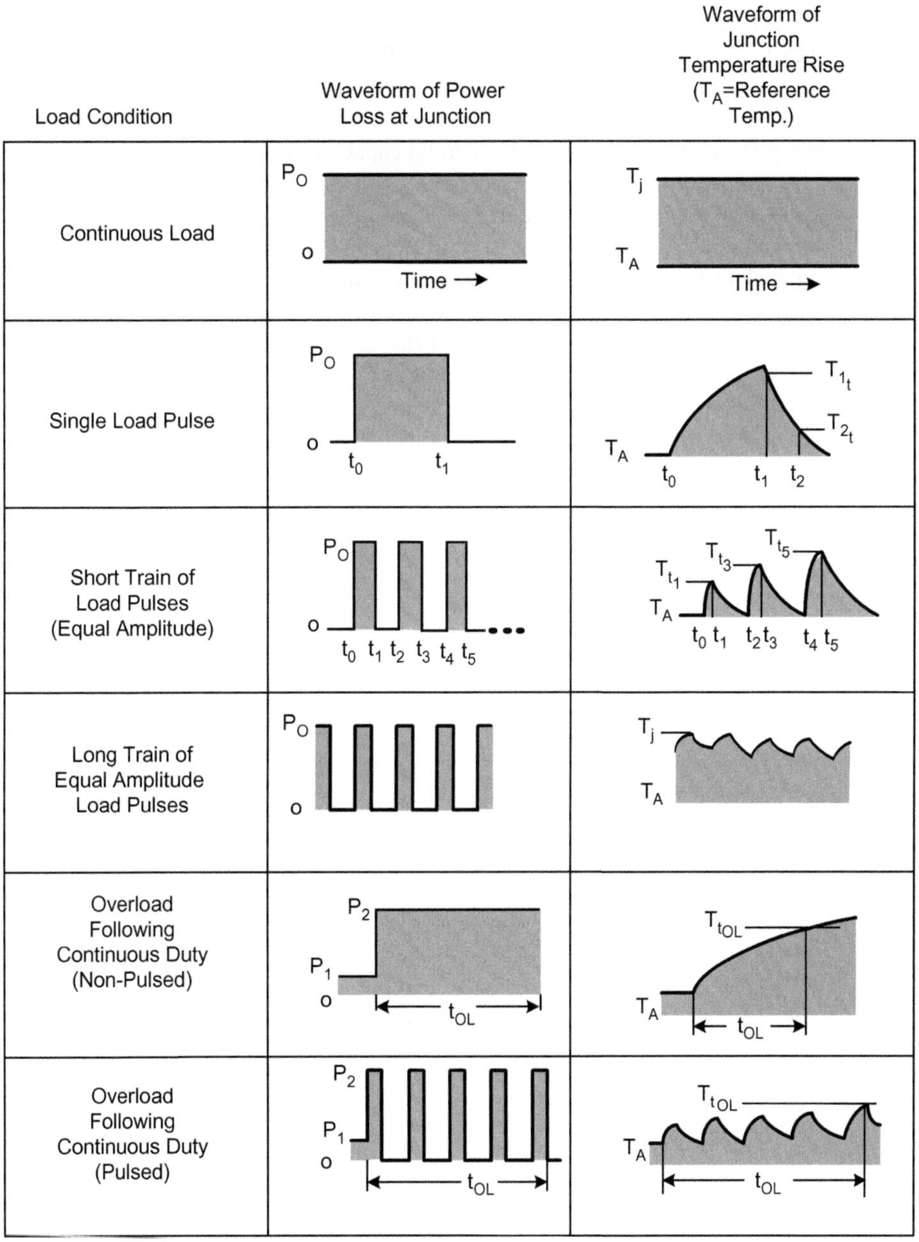

The heat due to power loss in the vicinity of a device junction flows to the case and then to the ambient through the externally mounted heat sink, causing a rise in the junction temperature (T_J). For any device, cooling should be sufficient to limit T_j typically within 125°C. The power loss in the junction (P_0) shown in the figure can be continuous or pulsating on a duty-cycle basis. For continuous P_0, the total thermal resistance (R_{th}) in the heat flow path determines T_J rise $(T_j - T_A = P_0 R_{th})$ over the ambient temperature (T_A). For pulse power dissipation, the thermal capacitance or storage effect should be considered for each medium. The thermal equivalent circuit can be represented by an equivalent R-C circuit, where thermal capacitance C is in parallel with effective R_{th}. The applied current in the circuit is equivalent to P_0 and the junction temperature rise is equivalent to voltage. Because the heating and cooling curves are complementary, a simple superposition principle can be used to calculate the T_J profile as shown in the figure. With a single P_0 pulse, T_J will rise and then fall exponentially as characteristic to a parallel R-C circuit. For the permitted $T_{Jmax} = 125$°C, the heat sink can be designed for the given P_0, or for a given heat sink, P_{0max} can be determined. Using the superposition principle, the T_{Jmax} profile for other load pulses can be determined as shown in the figure. Note that for a fault or overload condition, T_{Jmax} can be exceeded somewhat temporarily. The curves in the figure are valid for all the power devices.

FIGURE 2.18 (a) IGBT symbol with *V-I* characteristics (Powerex CM50TF-28H) (from [9]), (b) saturation characteristics, and (c) transient thermal impedance characteristics.

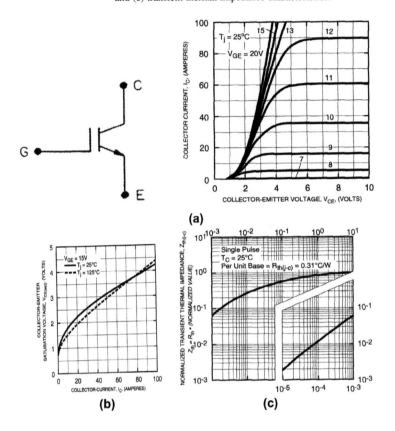

The IGBT is a hybrid device that combines the advantages of a MOSFET's high gate circuit resistance and a bipolar *PNP* transistor's small collector-emitter drop at saturated condition. Currently, it is the most important device for converter applications from several hundred watts up to 1 or 2 MW. The *V-I* characteristics, for example, of the Powerex 1400-V, 50-A (peak) IGBT [9] are shown in Figure 2.18(a) near the saturation region. The device does not conduct any collector current below the threshold voltages of $V_{CE} = 1$ V and $V_{GE} = 7$ V. At the peak rated current (50 A), the conduction drop $V_{CE(sat)}$ is somewhat high (3.1 V) compared to that of a thyristor-type device. The $V_{CE(sat)}$ of the IGBT has negative temperature coefficient characteristics in the operating region shown in Figure 2.18(b), making parallel operation difficult. Figure 2.18(c) shows the transient thermal impedance curves between the device junction and case ($Z_{th(J-C)}$) in two segments in normalized form. The actual value of $Z_{th(J-C)}$ can be obtained by multiplying it with $R_{th(J-C)} = 0.31°/C$, as indicated on the figure. An impedance $Z_{th(C-A)}$ (usually small) for a practical cooling system can be added with it. This figure helps to determine T_{Jmax} for a pulsating type load as discussed in Figure 2.17.

FIGURE 2.19   (a) IGBT half-bridge inverter and (b) switching characteristics of Q_1.

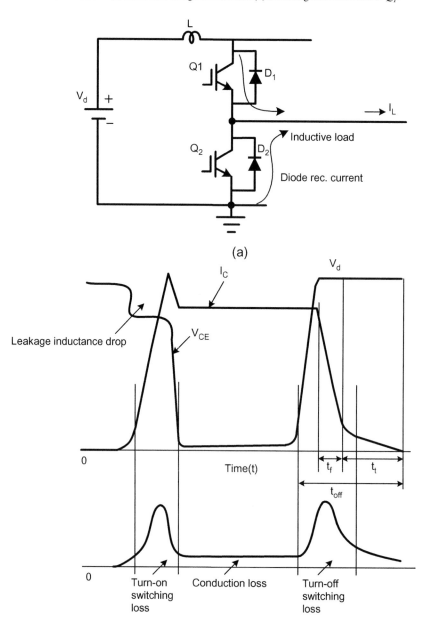

A half-bridge voltage-fed IGBT inverter is shown in Figure 2.19(a), whereas Figure 2.19(b) shows the switching voltage and current waves when no snubbers are used (hard switched). Assume that the load is highly inductive and initially Q_1 is off so that the lower diode D_2 is carrying the full load current I_L. When Q_1 is turned on at $t = 0$, after a short delay time it starts to pick up the load current at full supply voltage (with a small leakage inductance drop) diverting D_2 current. After D_2 forward current goes to zero, current in Q_1 consists of I_L and D_2 reverse recovery current (shown by the hump). When the recovery current is near the peak, the voltage of Q_1 (V_{CE}) falls to zero. When the turn-off gate signal is applied to Q_1, its collector voltage begins to build up with a short delay at full collector current. When full voltage is built up across the device, D_2 begins to pick up the load current. The short fall time (t_f) and relatively long tail time (t_t) of I_C due to minority carrier storage of Q_1 is shown in the figure. The SOA of the device is thermally limited like that of a MOSFET and there is no second breakdown effect. The conduction and switching loss curves, shown at the bottom of Figure 2.19(b), indicate that average switching loss will be high at high switching frequency. Note that the diode recovery current contributes significantly to the turn-on loss. Snubberless operation is possible but will cause high dv/dt and di/dt induced EMI problems. With a snubber, the turn-on di/dt and turn-off dv/dt will be slowed down, causing diversion of switching loss from the device to the snubber.

FIGURE 2.20 IGBT features.

- HYBRID DEVICE—MOS-GATED BJT
- "SMART POWER" CAPABILITY
- ASYMMETRIC VOLTAGE BLOCKING
- SQUARE SOA—SNUBBER OR SNUBBERLESS OPERATION
- LOWER INPUT CAPACITANCE AND IMPROVED MILLER EFFECT
- MODERN INTELLIGENT POWER MODULES (IPMs)
- VERY PROMISING EVOLUTIONARY DEVICE
- APPLICATIONS IN MODERATE TO HIGH POWER:
    - CHOPPER DC MOTOR DRIVE
    - INVERTER AC MOTOR DRIVE
    - REGULATED DC AND AC SUPPLIES
    - UPS SYSTEM
    - STATIC VAR GENERATOR
    - ACTIVE FILTER

The salient features of IGBTs are summarized here. As mentioned earlier, since the device is MOS-gated, the gate drive circuit can be integrated on the same device. Besides the control, some protection features can also be integrated. These integration features are often defined as "smart power." The device has forward-blocking capability like a MOSFET and, therefore, can be used in a voltage-fed converter with a feedback diode. Often the device is used with a series diode for application in current-fed converters where reverse blocking is essential. However, very recently, reverse blocking IGBTs have become available. Although IGBTs have a somewhat higher drop compared to thyristors or GTOs, modern IGBTs are available with trench-gate to reduce the conduction drop. Currently, commercial IGBTs are available with 4.5-kV, 1200-A and 6.5-kV, 700-A ratings; and 10-kV devices are in the test phase. The devices have been applied successfully in series and parallel combinations. IGBT-based converter power level is expanding continuously up to several megawatts. High-power IGBTs typically operate at a 1.0-kHz switching frequency. Simplicity of gate drive, ease of protection, smart power capability, snubberless operation, and higher switching speed make IGBTs very attractive for up to medium-power applications.

FIGURE 2.21 IGBT six-pack intelligent power module (IPM) with gate drive interface logic (from www.pwrx.com).

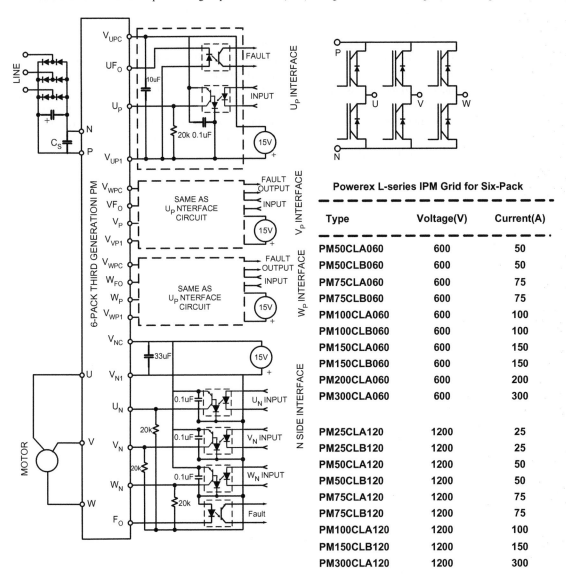

Powerex L-series IPM Grid for Six-Pack

Type	Voltage(V)	Current(A)
PM50CLA060	600	50
PM50CLB060	600	50
PM75CLA060	600	75
PM75CLB060	600	75
PM100CLA060	600	100
PM100CLB060	600	100
PM150CLA060	600	150
PM150CLB060	600	150
PM200CLA060	600	200
PM300CLA060	600	300
PM25CLA120	1200	25
PM25CLB120	1200	25
PM50CLA120	1200	50
PM50CLB120	1200	50
PM75CLA120	1200	75
PM75CLB120	1200	75
PM100CLA120	1200	100
PM150CLB120	1200	150
PM300CLA120	1200	300

IGBT devices are available in discrete form and intelligent power module (IPM) form. The IPM is basically a hybrid of devices with gate drivers and built-in protection features. The advantages of IPMs are lower size and cost, improved reliability, fewer EMI problems, a simple cooling design, and quick converter design and fabrication. Commercial IPMs may be available in various configurations, such as a seven-pack, six-pack, dual-pack, and dc chopper. Figure 2.21 shows, for example, a six-pack Powerex IPM (rectangular block on the left) with peripheral (external) gate drive interface circuits. The six-pack device topology of a three-phase bridge converter with built-in current sensing and feedback diodes is shown on the upper right. Current availability of the modules in various voltage and current ratings is shown on the bottom right. The three-phase induction motor, diode rectifier with dc link filter, and high-frequency capacitor (C_S), as shown, are directly connected to the module. The diode rectifier can be replaced by the PWM rectifier (similar module), if desired. The fault protection (overcurrent, overtemperature, short circuit, and dc undervoltage) signals are available for monitoring. The interface logic circuits (external) are coupled to IPMs with optocoupler isolation. The module typically operates at 20-kHz switching frequency with or without snubbers.

FIGURE 2.22 IGCT with integrated packaging of gate driver (from [15]; photograph used with permission of ABB).

As mentioned earlier, an IGCT is basically a "hard-driven" GTO with unity turn-off current gain. This means that a 3000-A (anode current) device requires a –3000-A gate current to turn it off. The large current pulse should be very narrow with low energy content for fast turn-off. The photograph shows an ABB presspack-type IGCT with a built-in integrated gate drive circuit on the same module. The current pulse with high di/dt (4 kA/µs) is supplied by a large number of MOSFETS operating in parallel in an extremely low leakage inductance path. Skillful design and layout of the control circuit are extremely important.

FIGURE 2.23 IGCT features.

- BASICALLY HARD-DRIVEN GTO
  (TURN-OFF CURRENT GAIN ~1)
- BUILT-IN MONOLITHIC ANTIPARALLEL DIODE
- GATE DRIVER IS BUILT IN TO THE MODULE
- ASYMMETRIC OR SYMMETRIC VOLTAGE BLOCKING
- CONDUCTION DROP–LOWER THAN AN IGBT
- SWITCHING FREQUENCY IS TYPICALLY 1.0 kHz
  (COMPARABLE TO AN IGBT)
- SQUARE SOA
- SNUBBER OR SNUBBERLESS OPERATION
- APPLICATIONS IN HIGH POWER:
  INVERTER MOTOR DRIVES
  DC-LINK HVDC SYSTEM
  STATIC VAR COMPENSATOR
  SHIP PROPULSION

The IGCT features are summarized in the figure. The device is extremely important for high-power (multi-megawatt) present and future applications. It is a serious competitor to the GTO and is expected to oust the latter device from the market in the future. Unlike the GTO, it has a built-in monolithic anti-parallel fast-recovery diode that simplifies the converter design. Currently, IGCTs are available with asymmetric and symmetric blocking capabilities. Otherwise, forward-blocking devices require series diode for reverse blocking. Symmetric blocking devices can be used in current-fed PWM converter applications. IGCT conduction drop is less than that of GTOs and IGBTs, and IGCTs have a higher switching frequency (typically 1.0 kHz) than GTOs. The device has square SOA and, therefore, converters can be built with or without snubber. This is a great advantage over GTOs for which a snubber is mandatory. State-of-the art asymmetric blocking devices are available with 6.5-kV, 3000-A and 4.5-kV, 4000-A ratings; and 10-kV devices are now in the laboratory test phase. Symmetric blocking devices are available at smaller ratings (6000 V, 800 A). IGCT applications are expected to expand quickly in utility applications with multilevel converters.

FIGURE 2.24 Comparison of power MOSFETs, IGBTs, GTOs, and IGCTs.

	Power MOSFET	IGBT	GTO	IGCT
1. Voltage and current ratings (selected device for comparison)	100 V, 28 A* (dc)	1.2 kV, 50 A* (dc)	6 kV, 6000 A* (pk)	4.5 kV, 4000 A* (pK)
2. Present power capability	1.2 kV, 50 A	3.5 kV, 1200 A or higher	6 kV, 6000 A	6.5 kV, 3000 A
3. Voltage blocking	Asymmetric	Asymmetric*	Asymmetric/symmetric Current	Asymmetric/symmetric Current
4. Gating	Voltage	Voltage		
5. Junc. Temp. range	−55 to 175	−20 to 50	−40 to 125	−40 to 125
6. Safe operating area (°C)	Square	Square	2nd breakdown	Square
7. Conduction drop (V) at rated current	2.24	2.65	3.5	2.7
8. Switching frequency	10^6 Hz	1 kHz - 20 kHz	400 Hz	1.0 kHz
9. Turn-off current gain	—	—	4 to 5	1
10. Turn-on di/dt	—	—	500 A/µs	3,000 A/µs
11. Turn-on time	43 ns	0.9 µs	5 µs	2 µs
12. Turn-off time	52 ns	2.4 µs	20a µs	2.5 µs
13. Snubber	Yes or No	Yes or No	Yes (heavy)	Yes or No
14. Protection	Gate control	Gate control	Gate control or very fast fuse	Gate control or very fast fuse
15. Applications	Switching power supply Low-power motor drive	Motor drive UPS, induction heating, etc.	Motor drives SVC, etc. $dv/dt = 1000$ V/µs	Motor drives HVDC, SVC etc. Built-in diode
16. Comments	Body diode can carry full current but sluggish ($t_{rr} = 150$ ns) $I_{pk} = 56$ A	Large power range Very important device currently * Reverse blocking available	High uncontrollable surge current	High uncontrollable surge current $dv/dt = 4000$ V/µs
	*Harris IRF140	*Powerex PM50RVA120 7-pack IPM	*Mitsubishi -FG6000AU-120D	*ABB 5SHY35L4512

FIGURE 2.25 Next-generation power semiconductor materials.

SILICON CARBIDE-DIAMOND
- LARGE BAND GAP
- HIGH CARRIER MOBILITY
- HIGH ELECTRICAL CONDUCTIVITY
- HIGH THERMAL CONDUCTIVITY

RESULT
- HIGH POWER CAPABILITY
- HIGH FREQUENCY
- LOW CONDUCTION DROP
- HIGH JUNCTION TEMPERATURE
- GOOD RADIATION HARDNESS

The power semiconductor devices discussed thus far are exclusively based on silicon material. Silicon has enjoyed a monopoly for a long period of time in both power and microelectronic devices, and this will remain so in the near future. However, new types of materials, such as gallium arsenide, silicon carbide, and diamond (in synthetic thin-film form), show tremendous promise for future generations of devices. SiC (in the near term) and diamond (in the long term) devices are particularly interesting for high-voltage, high-power applications because of their large band gap, high carrier mobility, and high electrical and thermal conductivities compared to silicon material. These properties permit devices with higher power capability, higher switching frequency, lower conduction drop, higher junction temperature, and better radiation hardness. However, processing and fabrication of these materials are difficult and expensive. Most of the power devices based on SiC have been tried successfully in the laboratory. SiC-based power MOSFETs with T_J up to 350°C appear particularly interesting as replacements for medium-power silicon IGBTs in the future. Only SiC-based high-voltage Schottky diodes (300–1700 V, 2–10 A) with close to a 1-V drop and negligible leakage and recovery currents are commercially available now.

FIGURE 2.26 Comparison of total power loss in Si- and SiC-based devices for a half-bridge PWM inverter ($V_d = 400$ V, $I_L = 15$A) (from [16]).

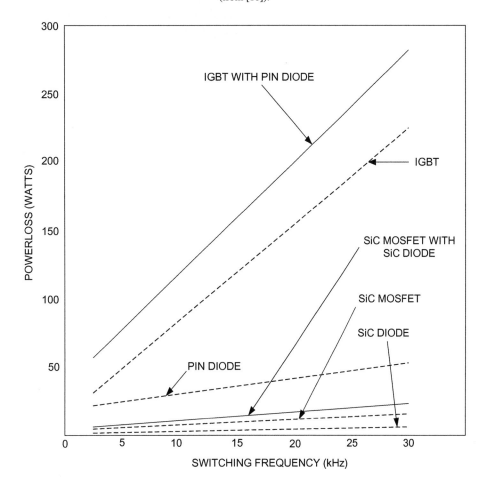

The figure shows a power loss comparison for a single-phase half-bridge PWM snubberless inverter as a function of switching frequency for two cases: (1) using an IGBT with a slow-recovery *PIN* diode and (2) using a SiC MOSFET with a SiC Schottky diode. In both cases, 50% duty-cycle operation is assumed. The loss components in the switching devices and diodes are plotted individually. Figure 2.19, for example, shows a loss curve in an individual IGBT. At low frequency, the conduction loss dominates, and as switching frequency increases, the total power loss increases linearly with frequency, indicating the dominance of switching losses at high frequency. The loss in the IGBT inverter with the *PIN* diode is the highest. It can be shown that turn-on switching loss is higher than that at turn-off because of a large diode recovery current. The SiC MOSFET is extremely fast when switching and has a lower conduction drop. In addition, the SiC diode has a lower drop with negligible recovery current, thus indicating that an inverter with SiC devices will be quite efficient. IGBT and power MOSFET (COOLMOS) inverters with SiC Schottky diodes look very promising for near-term applications.

FIGURE 2.27 Power integrated circuit (PIC) features.

- MONOLITHIC INTEGRATION OF POWER, CONTROL AND PROTECTION ELEMENTS—SMART POWER
- ADVANTAGES OF BETTER COST, SIZE, AND RELIABILITY AND LESS EMI PROBLEMS
- ISOLATION PROBLEM OF LOW- AND HIGH-VOLTAGE DEVICES
- THERMAL MANAGEMENT PROBLEM
- EXAMPLE COMMERCIAL PICs:
  STEPPER MOTOR DRIVE
  BRUSHLESS DC MOTOR (BLDM) DRIVE
  H-BRIDGE INVERTER
  CHOPPER FOR DC MOTOR DRIVE
  GATE DRIVER FOR IGBT
  DC-TO-DC CONVERTER

A power integrated circuit is basically a monolithically integrated power and control circuit, sometimes with protection elements. Figure 2.21, in contrast, illustrates a hybrid integrated circuit. Sometimes, a PIC is defined as an "intelligent power" or "smart power" circuit. It can be differentiated from a high-voltage integrated circuit (HVIC), in which the voltage is high but the current is low so that the power dissipation is low. The advantages of PICs are cost and size reduction, quick design and assembly time, built-in protection, less EMI problems, and overall improvement of converter reliability. In a PIC, high-voltage power devices (i.e., power MOSFET and IGBT) and low-voltage control and protection devices (CMOS, diode, MOSFET, resistor, capacitor, etc.) are integrated on the same chip; therefore, electrical isolation between high-voltage and low-voltage devices becomes a problem. Normally, self, *P-N* junction, and dielectric isolations are used. Again, large dissipated power in a small volume requires a very efficient cooling design. In fact, the problem of cooling limits the PIC power rating to typically below 1.0 kW. Large numbers of PICs are available in the market for various applications as indicated in the figure.

FIGURE 2.28   Power integrated circuit for dc motor drive (Harris HIP4011) (from [18]).

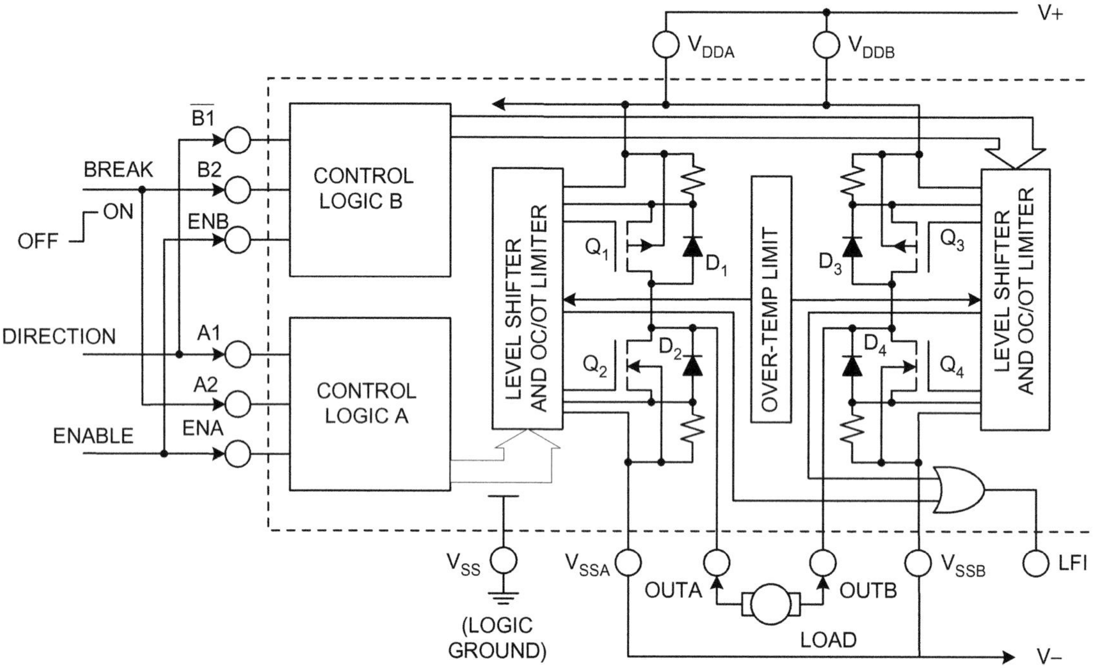

TRUTH TABLE

SWITCH DRIVER A				SWITCH DRIVER B			
INPUTS			OUTPUT	INPUTS			OUTPUT
A1	A2	ENA	OUTA	B1	B2	ENB	OUTB
H	L	H	OH	L	L	H	OH
L	L	H	OL	H	L	H	OL
H	H	H	OL	L	H	H	OL
L	H	H	OL	H	H	H	OL
X	X	L	Z	X	X	L	Z

The figure illustrates the schematic of a commercial low-power PIC for a permanent magnet dc motor drive. A complementary (PMOS and NMOS) power MOSFET H-bridge converter with control and protection is integrated in a chip. The legs can be controlled independently, and the control logic is summarized in the truth table. However, the legs are connected in the figure for simple on–off switching with four-quadrant control. With the Q_1–Q_4 pair on, the motor runs in the forward direction, and with Q_2–Q_3 on, the direction is reverse. During direction reversal, the motor goes through dynamic braking when Q_2–Q_4 are shorted to dissipate the kinetic energy in the armature resistor. The direction and braking are controlled by the input logic signals shown in the figure. The PIC has overcurrent (OC) and overtemperature (OT) limit protections as shown. The current is sensed in the series resistance of each MOSFET as indicated. Because these parameters tend to exceed the safe limit, the gate drives are regulated internally to clamp them to the limit values. It is possible to control the speed of the motor in either direction (4-Q speed control) via PWM control of the legs independently according to the truth table. The PIC can also be used in solenoid driver, relay driver, and stepper motor control.

FIGURE 2.29   Advances in and trends of power semiconductor devices.

- MODERN POWER ELECTRONICS EVOLUTION PRIMARILY FOLLOWED THE POWER DEVICE EVOLUTION, WHICH FOLLOWED THE MICROELECTRONICS EVOLUTION
- GRADUAL OBSOLESCENCE OF PHASE-CONTROL DEVICES (THYRISTOR, TRIAC)
- DOMINANCE OF INSULATED GATE-CONTROLLED DEVICES (IGBT, POWER MOSFET)
- POWER MOSFET WILL REMAIN UNIVERSAL IN LOW-VOLTAGE, HIGH-FREQUENCY APPLICATIONS
- GRADUAL OBSOLESCENCE OF GTOs (LOWER END BY IGBTs AND HIGHER END BY IGCTs)
- REDUCTION OF CONDUCTION DROP IN HIGH-VOLTAGE POWER MOSFET AND IGBT
- SiC-BASED DEVICES WILL BRING RENAISSANCE IN HIGH-POWER ELECTRONICS; DIAMOND DEVICES IN THE LONG RUN

The figure summarizes the advances and trends of modern power semiconductor devices. As mentioned earlier, the modern advancement of power electronics technology has been possible primarily due to advances in power semiconductor devices. Of course, converter topologies, PWM techniques, analytical and simulation methods, CAD tools, control and estimation techniques, and various digital control hardware and software tools also contributed to this advancement. We expect that phase-control devices, like thyristors and triacs, will be obsolete in the future because of line power quality and lagging power factor problems. Of course, in very high power utility applications, thyristors will remain unchallenged in the near future. The insulated-gate devices such as power MOSFETs and IGBTs will be dominant in low- to medium-power applications, as we can see the clear trend now. Power MOSFETs will remain uncontested in low-power, high-frequency applications, and the upper level will be dominated by IGBTs. The GTOs appear to have a bleak future, and the present trend indicates that its lower end will be taken over by IGBTs and higher end by IGCTs. High-voltage insulated-gate devices have the disadvantage of large conduction drop. The trench gate technology is being applied to reduce the conduction drop in power MOSFETs and IGBTs. The large band-gap devices, particularly with silicon carbide, will bring about a renaissance in power electronics in the future because of their high power capability, high temperature, high frequency, and low conduction drop characteristics. Schottky SiC diodes are available now, and power MOSFETs will be available in a few years.

Summary

Important types of modern power semiconductors, such as diodes, thyristors, triacs, GTOs, BJTs, power MOSFETs, IGBTs, SITs, and IGCTs and their electrical characteristics are described. However, devices that have not yet been well established, such as MCTs, SITHs, MTOs, and IEGT, are not covered. Currently, and for the future, power MOSFETs and IGBTs will be extremely important devices, and IGCTs also appear very promising as high-power applications for replacing GTO devices. BJTs have recently become obsolete because the high-power end has been taken over by IGBTs and the lower power end is being covered by power MOSFETs.

The power capabilities of modern devices and their future trends have been summarized. Most of the power devices are available in modular form. In particular, MOS-gated devices with control and protection features are available in the form of IPMs. Important self-controlled devices (power MOSFETs, IGBTs, GTOs, and IGCTs) are compared in a tabular form. The future promise of large band-gap power semiconductor materials (SiC and diamond) are highlighted. A comparison of power loss has been illustrated for a half-bridge PWM inverter using silicon-based IGBTs and SiC-based power MOSFETs.

References

[1] B. J. Baliga, "Trends in power semiconductor devices," *IEEE Trans. on Electron. Devices,* vol. 43, pp. 1717–1732, October 1996.

[2] B. K. Bose, "Power electronics—a technology review," *Proc. IEEE,* vol. 80, pp. 1303–1334, August 1992.

[3] B. J. Baliga, "The future of power semiconductor device technology," *Proc. IEEE,* vol. 89, pp. 822–832, June 2001.

[4] B. K. Bose, *Modern Power Electronics and AC Drives,* Prentice Hall, Upper Saddle River, NJ, 2002.

[5] K. Satoh and M. Yamamoto, "The present state of the art in high power semiconductor devices," *Proc. IEEE,* vol. 89, pp. 813–821, June 2001.

[6] Website product information (some companies):

Powerex	pwrx.com
ABB	abbsem.com
Toshiba	Toshiba.com
Hitachi	Hitachi.co.jp
Fuji	fujielectric.co.jp
Westcode	westcode.com
Mitsubishi	mitsubishielectric.com
Collmer	collmer.com
International Rectifier	irf.com
Harris	harris.com

[7] B. K. Bose, "Power semiconductor devices," Section IV.1, pp. 239–270, in *Modern Electrical Drives,* edited by H. B. Ertan, M. Y. Uctug, R. Colyer, and A. Consoli, Kluwer Academic, The Netherlands, 1994.

[8] N. Mohan, T. M. Undeland, and W. P. Robbins, *Power Electronics,* John Wiley, New York, 1995.

[9] Powerex, *Semiconductor Power Modules, Applications and Technical Data Book,* 1998.

[10] Harris, *Power MOSFETs,* 1994.

[11] B. W. Williams, *Power Electronics,* John Wiley, New York, 1987.

[12] H. Okayama, R. Uchida, M. Koyama, S. Mizoguchi, S. Tamai, H. Ogawa, T. Fujii, and Y. Shimomura, "Large capacity high performance 3-level GTO inverter system for steel rolling mill drives," *IEEE IAS Conference Record,* pp. 174–179, 1996.

[13] Powerex, *IGBT Intelligent Power Modules, Application and Data Book,* 1998.

[14] J. Donlon, J. Achhammer, H. Iwamoto, and M. Iwasaki, "Power modules for appliance motor control," *IEEE IAS Magazine,* pp. 26–34, July/August 2002.

[15] P. K. Steimer, H. E. Gruning, J. Werminger, E. Carroll, S. Klaka, and S. Linder, "IGCT—A new emerging technology for high power low cost inverters," *IEEE IAS Annual Meeting Conference Record,* pp. 1592–1599, 1997.

[16] B. J. Baliga, "Power semiconductor devices for variable frequency drives," *Proc. IEEE,* vol. 82, pp. 1112–1122, August 1994.

[17] T. M. Jahns, "Designing intelligent muscle into industrial motion control," *IEEE Trans. Ind. Electron.,* vol. 37, pp. 329–341, October 1990.

[18] Harris, *Intelligent Power ICs,* 1994.

[19] B. Murari, C. Contiero, R. Gariboldi, S. Sueri, and A. Russo, "Smart power technologies evolution," *IEEE IAS Conference Record,* pp. 10–19, 2000.

CHAPTER 3

Phase-Controlled Converters and Cycloconverters

FIGURE 3.1   General converter characteristics.

- CONVERSION AND CONTROL OF POWER
- MATRIX OF POWER SEMICONDUCTOR SWITCHES
  - NO ENERGY STORAGE
    (INSTANTANEOUS INPUT/OUTPUT POWER
    BALANCE)
  - NONLINEAR DISCRETE TIME TRANSFER
    CHARACTERISTICS
  - GENERATION OF LOAD AND SOURCE HARMONICS
  - EMI GENERATION
- SINGLE-STAGE OR MULTISTAGE HYBRID
- COMMUTATION METHODS
    LINE COMMUTATION
    LOAD COMMUTATION
    FORCED COMMUTATION
    SELF-COMMUTATION
- HARD OR SOFT SWITCHING

A converter basically consists of an array of on–off electronic switches that use power semiconductor devices. If the switches are considered ideal or lossless (zero conduction drop, zero leakage current, and instantaneous turn-on and turn-off times), the instantaneous and average power will balance at input and output of the converter. Switching mode operation makes the converter nonlinear, thus generating source and load harmonics and also EMI problems. The discrete time switching characteristics cause a delay in signal propagation. Of course, a high switching frequency reduces the propagation delay. A converter can be single stage, or multiple conversions may be involved in a cascaded converter system. Several types of commutation (transferring current from the outgoing device to the incoming device) can be used. Thyristor converters are characterized by line (or natural), load, or forced commutation. Line-commutated converters are used extensively in utility systems, and these will be discussed in this chapter. Force-commutated thyristor converters that require auxiliary transient circuits are practically obsolete. Converters that use devices such as power MOSFETs, GTOs, IGBTs, and IGCTs are characterized by self-commutation. Again, a converter can be based on hard switching (as in Figure 2.19) or soft-switching. In a soft-switched converter, dv/dt and di/dt are lowered to reduce EMI problems. These types of converters will be discussed in the next chapter.

FIGURE 3.2 Converter classification.

- AC–to–DC: RECTIFIER
    DIODE
    THYRISTOR PHASE-CONTROLLED
    PWM (VOLTAGE-FED OR CURRENT-FED)
    (HARD- OR SOFT-SWITCHED)
- DC–to–DC
    PWM (BUCK, BOOST, OR BUCK/BOOST)
    RESONANT LINK
    QUASI-RESONANT LINK
- DC–to–AC: INVERTER
    THYRISTOR PHASE-CONTROLLED
    PWM (VOLTAGE-FED OR CURRENT-FED)
    (HARD- OR SOFT-SWITCHED)
- AC–to–AC: AC CONTROLLER (SAME FREQUENCY)
    CYCLOCONVERTER (FREQUENCY CHANGER)
    THYRISTOR PHASE-CONTROLLED
    DC LINK (VOLTAGE-FED OR CURRENT-FED)
    (HARD- OR SOFT-SWITCHED)
    HIGH-FREQUENCY LINK (VOLTAGE-FED OR
    CURRENT-FED)
    MATRIX

The figure summarizes different classes of converters. Power rectification is not only possible by traditional diode and thyristor converters, but also by pulse width modulation (PWM) converters that use self-commutated devices and are based on voltage-fed or current-fed principles. Most of the rectifier topologies (except those with diodes) can be used for inversion also. The dc-to-dc converters can be classified as the buck, boost, buck/boost, flyback, or forward type. The traditional PWM dc-to-dc converters are defined as choppers. PWM converters are usually based on hard switching, but soft switching with resonant (with ac link) or quasi-resonant principles can also be used to reduce the switching loss. An ac-to-ac converter can be in the same frequency (voltage controller) or at different frequency (cycloconverter). A cycloconverter can utilize direct frequency conversion (phase-controlled cycloconverter or matrix converter), or it can accomplish conversion via the dc or high-frequency link.

FIGURE 3.3   Principal topologies of diode rectifiers.

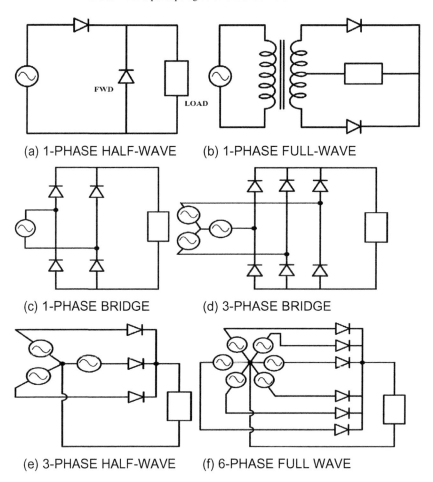

Diode rectifiers are the simplest type of uncontrolled power conversion circuits. Because the device conducts current in one direction, it can be used for power rectification only. The figure shows the principal single-phase and polyphase diode rectifiers. The supply is usually 50/60 Hz, and the load can be *R, RL, RC,* or *RL* with CEMF (such as a dc motor). The slow-recovery *P-I-N* diode is normally used. The recovery current can create a large voltage transient at turn–off, which can be limited by snubber. In designing a diode rectifier, the average current, peak current, and peak reverse voltage of the device are to be limited. Also, adequate heat sink or cooling is required to limit T_J (junction temperature). Often a transformer is used between the supply and rectifier for voltage level conversion, isolation, or tapping. A single-phase half-wave circuit is the simplest. For inductive loads, an FWD in parallel increases the dc output voltage by clamping the negative voltage segment. Single-phase full-wave circuits (two-pulse) can be with the transformer or bridge type. However, the latter type is very common. The half-wave circuits for three-phase (three-pulse) and six phase (six-pulse) rectifiers have poor duty cycles. Three-phase bridge rectifier (six-pulse) use is common. Only a few important diode rectifier topologies will be discussed later in detail. Diode rectifier applications include battery chargers, electrochemical processes, SMPS power supplies, and ac motor drives. More complex 12-pulse (see, for example, Figure 7.23) and 24-pulse rectifiers can be used in high-power applications.

FIGURE 3.4 (a) Single-phase diode bridge rectifier (with parallel *R-C* load) and (b) voltage and current waves.

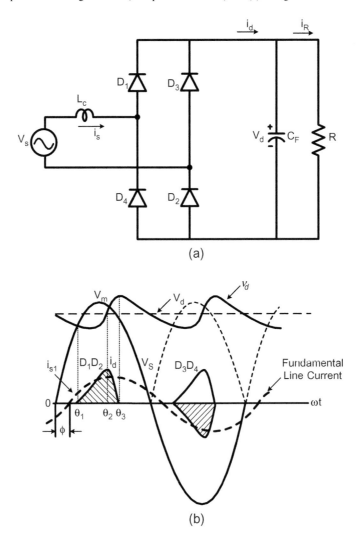

The diode bridge rectifier shown in the figure is extremely important for generation of unregulated dc, which can be used in dc-dc converters to generate regulated dc supply, or in dc-ac converters for ac motor drives. The resistance R can be considered as equivalent load for the output circuit. Without the filter capacitance C_F, the load voltage wave (v_d) is full-wave rectified with the diode pairs D_1D_2 and D_3D_4 conducting alternately for 180°. The C_F filters the v_d wave, tending for it to be pure dc when $C_F \to \infty$. Figure 3.4(b) shows the voltage and current waves at finite values of C_F. The v_d wave is wavy but the mean dc voltage V_d is shown by the dashed horizontal line. The capacitor will charge with a pulse current every half-cycle near the peak supply voltage V_m when $v_d < v_s$, which will be limited by the line leakage inductance L_c (also called commutating inductance). The current i_d will increase during θ_1 and θ_2 angles and then go to zero at θ_3. During the conduction interval, v_d will increase as shown. Then, the capacitor will discharge exponentially with the load time constant $C_F R$. The load current i_R is always proportional to capacitor voltage. The discontinuous pulse current i_d will cause the pulsating line current shown in the figure, and its fundamental component i_{s1} is indicated by the dashed curve.

FIGURE 3.5   Single-phase diode bridge performance characteristics ($C_F \rightarrow \infty$) (not to scale).

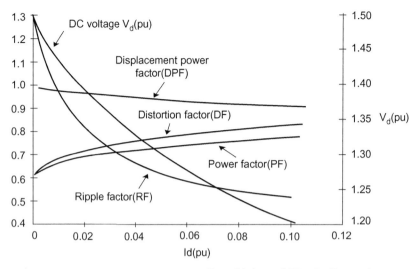

$$\text{RIPPLE FACTOR (RF or THD)} = \frac{Rms\ Value\ of\ Ripple\ Current}{Rms\ Value\ of\ Fund.\ Current}.$$

$$\text{DISTORTION FACTOR (DF)} = \frac{Rms\ Value\ of\ Fund.\ Current}{Rms\ Value\ of\ Total\ Current}.$$

$$\text{DISPLACEMENT POWER FACTOR (DPF)} = \frac{Average\ Power}{Fund.\ Rms\ Voltage \times Fund.\ Rms\ Current}$$

$$\text{POWER FACTOR (PF)} = \frac{Average\ Power}{Fund.\ Rms\ Voltage \times Total\ Rms\ Current}$$
$$= DPF \times DF$$

The figure (upper part) shows the performance characteristics of the diode bridge rectifier at $C_F \rightarrow \infty$, where $I_{d(base)} = V_s/\omega L_c$ (bridge short-circuit rms current) and $V_{d(base)} = V_{do} = (2/\pi)V_m$ (dc voltage with R load). The nonlinear operation of the rectifier causes undesirable pulsating line current, which can be analyzed by Fourier series. The various factors for harmonic-rich line current are shown at the bottom of the figure. The numerical values of these factors can be calculated by simulation of the rectifier with certain source and load conditions. Note that if the line current is sinusoidal at $\varphi = 0$ (see Figure 3.4), $RF = 0$, $DF = 1$, $DPF = 1$, and $PF = 1$. Since $I_R = I_d$ (average value), as the load current goes up, V_d will go down sharply (poor regulation) as shown. At no load ($R = \infty$), C_F will tend to charge at peak voltage, that is, $V_{d(pu)} = \pi/2 = 1.57$. Note that DPF is near unity and goes down slightly with load. At low Id (pu), severe harmonic distortion of the line current occurs, which makes RF or total harmonic distortion (THD) very high. RF improves with higher I_d that increases DF and PF as shown.

FIGURE 3.6 (a) Three-phase diode bridge rectifier (with parallel R-C load) and (b) voltage and current waves ($C_F \to \infty$).

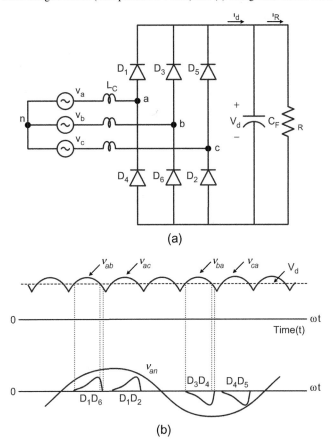

For higher power requirements and where three-phase supply is available, a three-phase bridge rectifier as shown here is used. The performance of the circuit is somewhat similar to that of Figure 3.4. Figure 3.6(b) shows the waveforms for the rectifier with the assumption of infinity filter capacitor (C_F). The six-pulse/cycle voltage wave shown is the ideal load dc voltage (v_d) wave with only R load. With a large filter capacitor, the v_d wave is pure dc (V_d) as indicated. When the line voltage segment v_{ab}, for example, exceeds V_d, the diodes D_1 and D_6 will conduct and contribute a pulse of load current as shown, which will flow in phases a and b. In the same half-cycle of the v_{an} wave, when $v_{ac} > V_d$, another positive current pulse will be contributed with D_1 and D_2 conducting. The load current (i_d) will consist of six identical current pulses/cycle. The two pulses due to conduction of $D_3 D_4$ and $D_4 D_5$ will appear as negative phase current in phase a. All three phase current waves will be identical. The load current wave i_R will always be proportional to v_d wave. If the line voltages are not balanced, the pulses in each half-cycle will be somewhat asymmetrical.

FIGURE 3.7   Performance characteristics of three-phase diode bridge rectifier with parallel *R-C* load ($C_F \rightarrow \infty$) (not to scale).

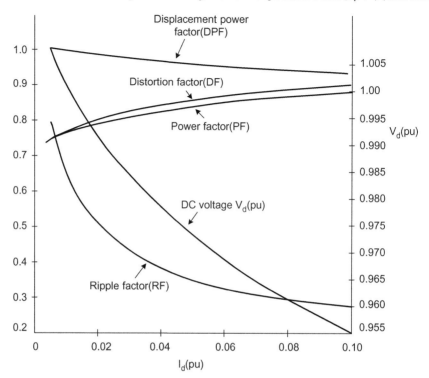

The performance curves in this figure are similar to those in Figure 3.5. However, in this case, $V_{d(base)} = V_{do} = 1.35V_L$ (V_L = rms line voltage) and $I_{d(base)} = V_s/\omega L_c$ (phase short-circuit current) as before. With load current I_d variation, all the parameters can be plotted either by rigorous calculation or by simulation of the converter. Note that the average value of i_d (I_d) will be the same as that of i_R (I_R) at all conditions. As I_d increases, V_d decreases to inject more i_d to the load circuit. When $I_d = 0$, V_d will correspond to the peak line voltage, that is, $V_{d(pu)} = \sqrt{2}/1.35 = 1.05$ as shown in the figure. Because each phase carries two current pulses/half cycle (instead of one), the line current wave will have fewer harmonics. The effect is lower dc voltage drop, smaller *RF*, and better *PF, DF,* and *DPF* with loading compared to a single-phase circuit. (*Note:* The scales in Figures 3.5 and 3.7 are not same.)

FIGURE 3.8   Phase-controlled converter classification.

- AC–to–DC: SINGLE-PHASE
  HALF-WAVE
  FULL-WAVE BRIDGE*
  FULL-WAVE SEMI-BRIDGE
  FULL-WAVE DUAL BRIDGE
  FULL-WAVE BRIDGE WITH CENTER-TAP TRANSFORMER
- AC–to–DC: THREE-PHASE
  HALF-WAVE (3-PHASE AND 6- PHASE)
  SIX-PULSE CENTER-TAP CONVERTER WITH IGR BRIDGE*
  DUAL BRIDGE*
  12-PULSE WITH SERIES BRIDGES*
  12-PULSE WITH PARALLEL BRIDGES
- AC–to–AC: SINGLE-PHASE
  TRANSFORMER TAP CHANGER
  SEMI-CONTROLLER
  FULL-WAVE CONTROLLER*
- AC–to–AC: 3-PHASE
  WYE–CONNECTED LOAD: FULL-WAVE* AND SEMI-CONTROLLER
  DELTA CONNECTED LOAD: FULL-WAVE AND SEMI-CONTROLLER
  NEUTRAL-CONNECTED CONTROLLER

Following Figure 3.3 for diode converters and adding more topolgies, this figure shows a comprehensive list of thyristor phase-controlled converter configurations. These include rectifiers, inverters, and ac-ac controllers (at the same frequency). The phase-controlled cycloconverters are not included in the list. Only a few (indicated by an asterisk) of these topologies will be discussed later in detail. In a phase-controlled converter (PCC), unlike a diode converter, conduction can be delayed by firing angle α. A thyristor converter operating on an ac line uses line commutation for the devices. The phase control is characterized by distortion of the current wave in load and line, and the line DPF is always lagging, which will be explained later. In most cases, a PCC can operate as a rectifier as well as an inverter (called a two-quadrant converter). In a semi-converter, half of the thyristors are replaced by diodes, and it can operate only as a rectifier (one quadrant). A dual-bridge is based on anti-parallel operation of two bridges, and is useful for four-quadrant operation of a dc drive, which will be explained later. Larger pulse numbers in the rectifier reduce the harmonic content in load voltage and line current waves, and are preferred in high-power circuits.

FIGURE 3.9 Phase-controlled converter features.

- LONG HISTORY—CLASSICAL POWER ELECTRONICS
- LOW CONDUCTION LOSS
- NEGLIGIBLE SWITCHING LOSS
- SOFT SWITCHING WITH ZERO CURRENT
- HIGH EFFICIENCY—TYPICALLY 98%
- LINE OR LOAD COMMUTATION
- SIMPLE CONTROL
- GENERATES LINE FREQUENCY-RELATED HARMONICS
  IN LOAD AND SOURCE
- LAGGING LINE DPF PROBLEM
- EMI PROBLEM
- FUSE PROTECTION POSSIBLE
- FAULT CURRENT SUPPRESSION BY GATE CONTROL

Phase-controlled converters have a long history, and have been used since the early 20th century in glass bulb mercury-arc converters. Since the invention of solid-state thyristors in the late 1950s, glass bulbs, steel tanks, and gaseous tubes have been completely replaced by thyristors. Phase control triacs are used in low-power ac power control mainly with *R* load, whereas thyristors are used in all power ranges. Phase-controlled converters are used extensively in utility systems. The converters are very efficient due to low conduction loss and negligible switching loss. The control is very simple. The additional advantages are commutation failure and the fact that load fault can be suppressed by gate control. In addition, high-speed fuses can be used for protection of devices. However, the disadvantages are distortion of load voltage and line current, a low line power factor, and EMI problems. The line harmonics and voltage transients due to commutation cause power quality problems in utility systems.

FIGURE 3.10   (a) Single-phase thyristor bridge with *R-L*-CEMF load, (b) continuous conduction rectification (mode A), and (c) discontinuous conduction rectification (mode B).

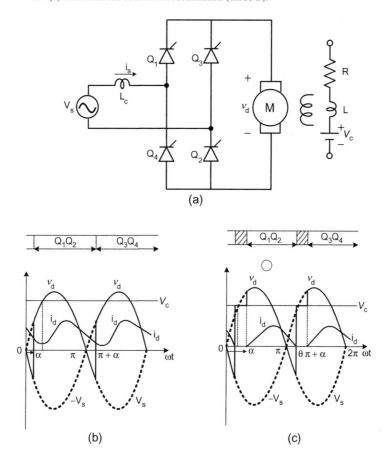

The figure shows a single-phase thyristor bridge converter with a dc motor load and the corresponding voltage and current waves. The motor is assumed to have constant excitation (permanent magnet), and the line leakage inductance (L_c) effect is neglected. Thyristor firing angle α can be controlled to control the dc output voltage v_d that will control the motor speed. The motor is represented by an equivalent circuit with R, L, and counter emf V_c, which is proportional to speed. The thyristor pair Q_1Q_2 can be fired at angle α when their anode voltage is positive. Because of load inductance, current i_d will flow beyond the π angle as shown when their anode voltage is negative. The Q_3Q_4 pair is fired symmetrically at the $\pi + \alpha$ angle when the current is transferred from the Q_1Q_2 to the Q_3Q_4 pair (commutated). In this second half cycle, the line current is reversed but the load current always remains positive. The load (and line) currents will be discontinuous, for example, at high V_c when the speed is high. In these modes of converter operation, the average load voltage (V_d) is positive, which contributes to average positive power with average $+I_d$. Thus, the motor speed can be controlled in motoring mode only (one-quadrant operation).

FIGURE 3.11   (a) Continuous conduction inversion mode (mode C) and (b) discontinuous conduction inversion mode (mode D).

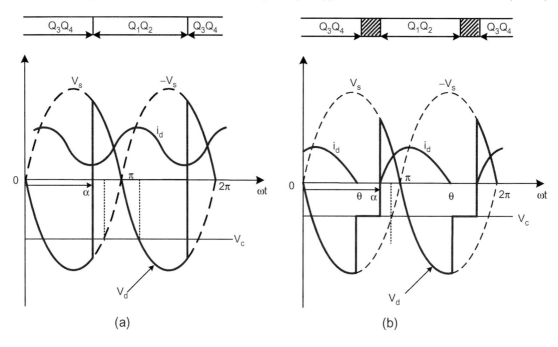

(a)   (b)

The figure explains the inverting mode operation of the single-phase bridge shown in Figure 3.10 for continuous and discontinuous modes of operation. In the continuous conduction rectification mode discussed before, ideally $V_d = 0$ at $\alpha = \pi/2$. Continuous conduction can be maintained at $\alpha > \pi/2$, if the V_c polarity is reversed so that the load and supply voltages will remain balanced. This can be done by reversing the field current when the machine (not for PM machines) is running at a speed. In the initial part of the half cycle, thyristor anode voltage is positive and V_c adds with v_s to build up i_d, but in the latter part v_s opposes the flow of i_d. In the inverting mode, the current i_d can be discontinuous, as shown in Figure 3.11(b), if the V_c or the motor speed is not large enough to maintain continuous conduction. In the inverting mode of operation, $+ I_d$ and $-V_d$ make the average power flow negative. This means that the machine acts as a generator and its stored mechanical energy is converted to electrical energy and fed back to the line. This is defined as regenerative braking mode or second-quadrant operation of the machine.

FIGURE 3.12   Typical torque-speed curves of dc motor with single-phase bridge converter.

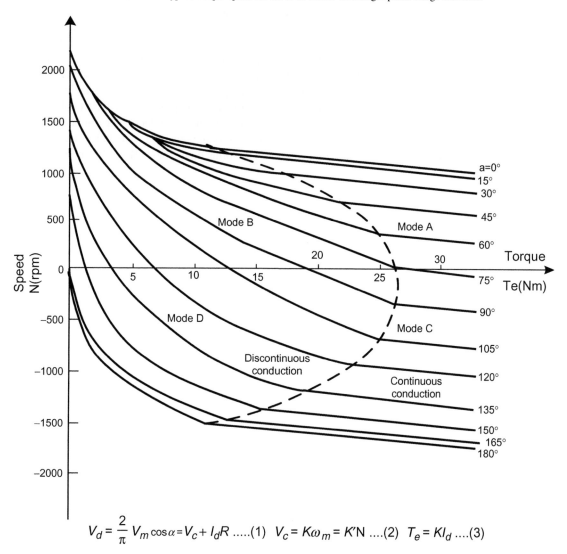

$$V_d = \frac{2}{\pi} V_m \cos\alpha = V_c + I_d R \ .....(1) \quad V_c = K\omega_m = K'N \ ....(2) \quad T_e = KI_d \ ....(3)$$

The performance characteristics of Figures 3.10 and 3.11 can be combined along with the equations shown here to determine the typical torque speed characteristics of the motor as shown in this figure. The figure shows the motoring mode $(+N, +T_e)$ in the first quadrant and regenerative braking mode $(-N, +T_e)$ in the second quadrant, where the boundary between continuous and discontinuous conduction in both the modes is indicated by the dashed curve. The four modes *(A, B, C, and D)* shown in this figure correspond to the modes explained in Figures 3.10 and 3.11. In continuous conduction mode with constant α angle, $V_d = V_c$, if armature resistance R is neglected. Therefore, variation of torque does not vary speed (lines will be horizontal). The straight lines have small negative slope because of finite R value. At T_e or $I_d = 0$, V_c always equals V_d. Therefore, speed will correspond to voltage $V_c = V_m$ for α angle from 0 to 90°, and $V_c = V_m \sin\alpha$ for $\alpha > 90°$ as shown in the figure. Consider, for example, $\alpha = 60°$ in mode A with the torque or I_d gradually being reduced. Below a threshold I_d, the current falls into discontinuous mode or mode B. As I_d decreases further, V_c or speed will increase nonlinearly to satisfy mode B waves. In regenerative mode continuous conduction operation (mode C), as I_d is decreased, the current will fall into discontinuous mode (mode D). Then, as I_d is decreased further, the operation enters into motoring mode (mode B) as shown in the figure.

FIGURE 3.13   Three-phase bridge converter with dc motor load.

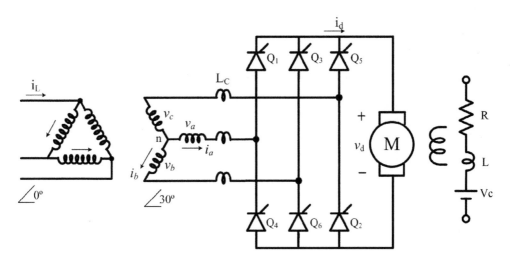

Extending the operating principle of the single-phase bridge shown in Figure 3.10, a three-phase bridge converter can be constructed as shown in this figure. The converter is very widely used, and the applications include dc motor speed control (shown in this figure), general-purpose dc power supplies, uninterruptible power supply (UPS) systems, and current-fed induction and synchronous motor drives. The circuit is also the basic element in phase-controlled cycloconverter drives used for large motors. Basically, the converter consists of six thyristors arranged in the form of three legs with two series thyristors in each leg. The center points of three legs are connected to a three-phase power supply. The transformer is not mandatory, but it provides the advantages of voltage level change, electrical isolation, and phase shift from the primary. The Thevenin leakage inductance L_c constitutes the commutating inductance discussed earlier. In a three-phase bridge, one device in the positive group ($Q_1Q_3Q_5$) and another device from the negative group ($Q_4Q_6Q_2$) must conduct simultaneously to contribute load current i_d. Each thyristor is normally provided with pulse train firing for the desired conduction interval. An unpolarized R-C snubber (not shown) is usually connected across each device to limit di/dt, dv/dt, and undesired transient voltage. The speed of the motor can be controlled by firing angle control of the thyristors.

FIGURE 3.14   Three-phase thyristor bridge waveforms in rectification mode ($\alpha = 40°$) (mode A).

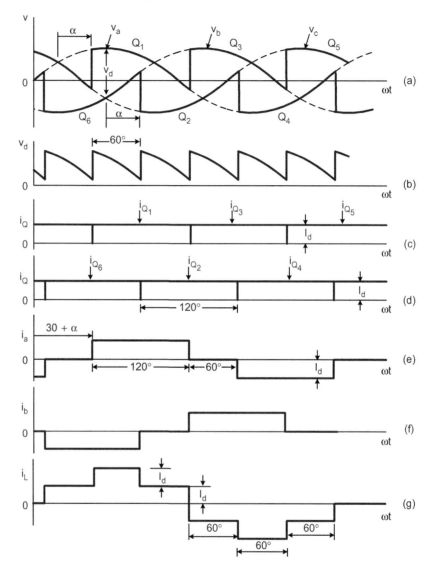

The figure explains the voltage and current waves of three-phase bridge converters assuming load inductance $L \rightarrow \infty$ so that the load current i_d is continuous (continuous conduction mode) and ripple free. The effect of commutating leakage inductance L_c is neglected for simplicity. The thyristors in Figure 3.13 are fired in the sequence Q_1-Q_2-Q_3-Q_4-Q_5-Q_6 at firing delay angle α (in this case 40°) so that each of the positive-side and negative-side devices conducts for a 120° interval with a phase shift of 60° as shown. At $\alpha = 0$, the circuit operates as a diode rectifier. The six-pulse load voltage v_d wave, as shown, is enclosed within the line-to-line voltage envelope, which has an average value $V_d = 1.35V_L \cos \alpha$ (V_L = line voltage). The harmonics of the v_d wave are characterized by $6n$, where $n = 1, 2, 3$, etc. The six-stepped current waves i_a and i_b will flow in the line without a transformer (see Figure 3.13). With a transformer, the line current wave is given by $i_L = N(i_a - i_b)$ which is also a six-stepped wave. The harmonic order of a six-stepped wave is given by $6n \pm 1$. As a controlled rectifier, the average V_d can be controlled by α angle in the range $0 < \alpha < \pi/2$.

FIGURE 3.15   Three-phase thyristor bridge waveforms in inverting mode ($\alpha = 150°$) (mode C).

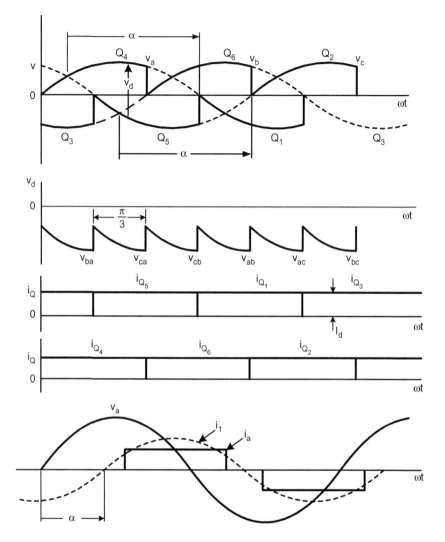

With continuous conduction, if the firing angle $\alpha = 90°$ in Figure 3.14, the v_d wave alternates with average $V_d = 0$. With $\alpha > 90°$ in continuous conduction, V_d is negative. This means that the load counter emf has to be negative. Because i_d is always positive, this must be an inversion mode where the load counter emf supplies power to the source. The two-quadrant operation of a three-phase bridge is similar to that of the single-phase bridge discussed earlier. The load voltage and line current waves remain six stepped as in Figure 3.14. In practical continuous conduction mode, the load current wave may be wavy because of finite load inductance. This will result in wavy current waves in the devices and in the line. Ideally, α can extend up to 180° for the inversion mode, but practically, it should be less than 180° by β angle (advance angle) so that the thyristor can turn off or commutate satisfactorily by a segment of line voltage. A typical phase voltage and line current waves in the rectification mode ($\alpha < 90°$) (without transformer) is shown in the lower part of the figure. The sinusoidal component of the line current shown by the dashed curve indicates that the fundamental current is always lagging and that the displacement power factor (DPF) angle (φ) is the same as the firing angle α.

FIGURE 3.16   Three-phase thyristor bridge waves showing commutation overlap effect.

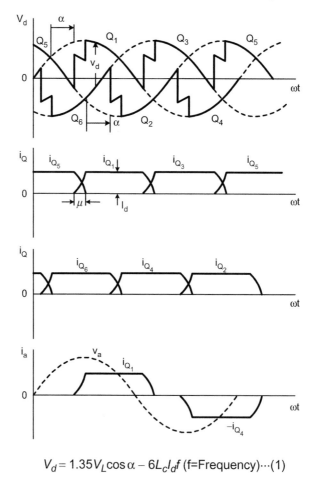

$$V_d = 1.35V_L \cos \alpha - 6L_c I_d f \ (f=\text{Frequency})\cdots(1)$$

So far, the waveforms in Figures 3.14 and 3.15 have been considered without the effect of the commutating inductance L_c in Figure 3.13. With finite L_c, the current cannot commutate instantaneously, for example, from Q_5 to Q_1, when Q_1 is fired. The effect of L_c is shown in this figure in rectification mode. When Q_1 is fired, for example, the negative line voltage v_{ac} will be impressed across Q_5. This will cause gradual current transfer from Q_5 to Q_1 during the overlap angle μ as shown. This will create a notch in the dc voltage wave for every 60° interval. Therefore, average dc voltage V_d will be lower by the notch volt-sec-area for every 60° interval. All the device and line current segments will have gradual rise and fall times, as indicated in the figure. This commutation overlap angle μ is particularly important in the inverting mode because this angle will subtract from the advance angle β for effective commutation of the thyristor (i.e., $\gamma = \beta - \mu$, where γ is the turn-off angle).

FIGURE 3.17 Waveforms of three-phase thyristor bridge converter at discontinuous conduction: (a) rectification mode (mode B) and (b) inversion mode (mode D).

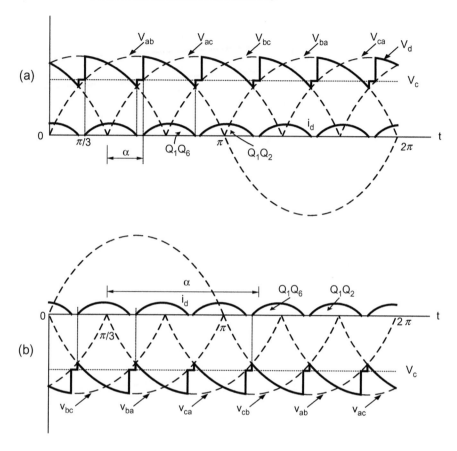

Like a single-phase thyristor bridge, a three-phase thyristor bridge can fall into discontinuous conduction mode with counter emf load as shown in this figure. This is particularly relevant for two-quadrant speed control of a dc motor drive. If armature resistance R is neglected, $V_c = V_d$ in all the modes of operation. Consider, for example, the rectification mode in the top part of Figure 3.17, when V_c is positive and it absorbs power with positive I_d. With discontinuous conduction, a thyristor pair can be fired to conduct if the corresponding line voltage is positive and higher than V_c. In the figure, for example, the Q_1Q_6 pair is fired at angle α when $v_{ab} > V_c$, and then a pulse of i_d will flow in the load. Again, the Q_1Q_2 pair is fired when $v_{ac} > V_c$. A pulse of load current will flow every 60° interval. Each thyristor will carry two consecutive current pulses similar to that shown in Figure 3.6, but a corresponding phase lag to the firing angle will result. In inversion mode, as shown in the lower figure, V_c is negative and the Q_1Q_6 pair can be fired at $v_{ab} > V_c$ so that a positive current pulse flows to pump energy to the source. Similarly, Q_1Q_2 is fired at $v_{ac} > V_c$ for the next pulse as shown.

FIGURE 3.18 Typical torque speed curves of dc motor with three-phase thyristor bridge converter.

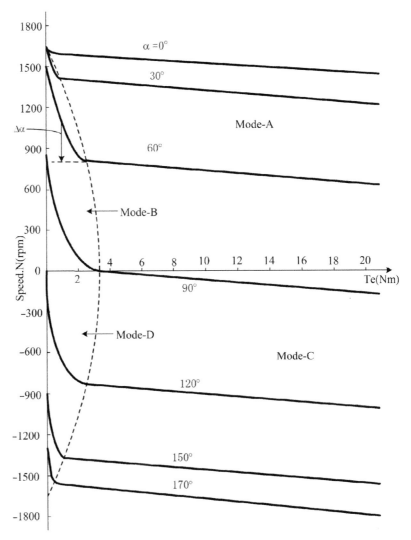

The figure shows the typical two-quadrant torque speed characteristics of a dc motor with rated field excitation. The characteristics are somewhat similar to those of the single-phase bridge shown in Figure 3.12. In quadrant 1, the machine has the usual motoring torque in the forward direction, whereas in quadrant 2, the machine experiences regenerative braking in the reverse direction $(-N, +T_e)$ that will tend to slow down the speed. Each quadrant is characterized by continuous and discontinuous conduction modes, and the dashed curve shows the boundary. Note that the discontinuous conduction zone is somewhat smaller than that shown in Figure 3.12. Consider, for example, $\alpha = 60°$ in quadrant 1. In continuous conduction, if armature resistance R is neglected, $V_d = V_c$, and the lines will be horizontal. A small drop due to R makes the slope slightly negative. The same argument is valid for quadrant 2 also. Now consider ideal no-load operation in quadrant 1 (mode B). No current can flow for $0 < \alpha < 30°$ if $V_c = V_m$ (peak value of line voltage), as is evident from Figure 3.17(a). For $\alpha > 30°$, speed or V_c will decrease at $I_d = 0$ until $V_c = 0$ at $\alpha = 120°$ as shown in the figure. The no-load speed is negative beyond this α angle, which is evident from Figure 3.17(b). As torque increases, the speed will fall in the motoring mode. On the other hand, in the regenerating mode, as explained in Figure 3.17(b), speed increases with torque for constant α angle.

FIGURE 3.19 Three-phase bridge speed control of dc motor showing $\Delta\alpha$ compensation.

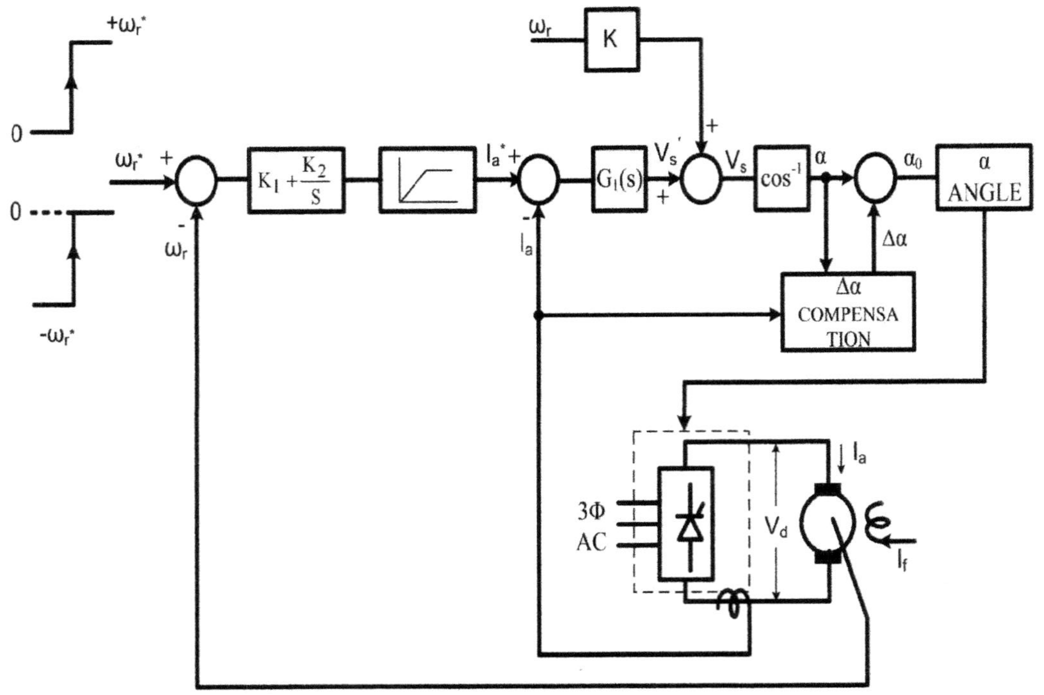

The speed control block diagram using a three-phase bridge converter is shown in this figure. The control can be done independently in quadrant 1 or quadrant 2 of Figure 3.18. If the command speed $+\omega_r^*$ is increased by a step from zero as shown, the operation will be in quadrant 1, whereas if the initial $-\omega_r^*$ is stepped to zero, it will operate in quadrant 2. Consider, for example, the former case only. The feedback speed control loop will generate the current command $+I_a^*$ for the current loop through a proportional-integral (P-I) controller so that steady-state speed error is zero. While accelerating, the rated machine or converter current (i.e., the torque) should be limited as shown in the figure. Within the current loop, a feedforward voltage $K\omega_r$ has been added to increase the loop response. The converter uses a cosine-wave crossing method for α angle control so that V_d is linear with the signal V_s. However, at discontinuous conduction mode, the gain is high and nonlinear. A compensation signal $\Delta\alpha$ has been generated (see Figure 3.18) by a nonlinear look-up table (with I_a and α signals) so that the gain remains the same as in continuous conduction. The system is more stable with this compensation. Quadrant 2 operation is similar. Note that a higher speed range is possible via control (weakening) of the field current I_f but is not discussed here. The field current can also be reversed to reverse the polarity of counter emf and torque ($T_e = KI_fI_a$, $V_c = KI_f\omega_r$). Then speed control becomes possible in both directions in the regenerative mode (four-quadrant operation).

FIGURE 3.20   (a) Three-phase dual converter and (b) output voltage characteristics.

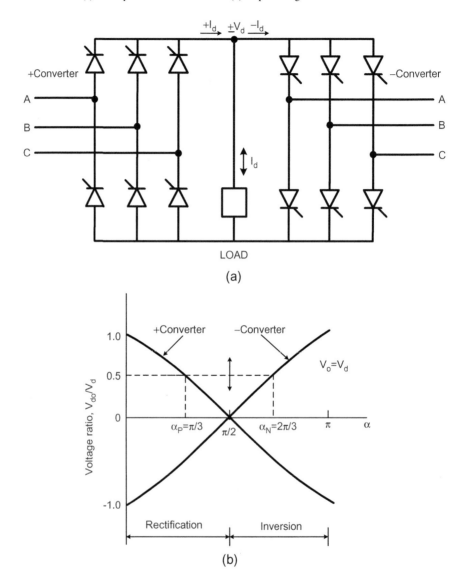

(a)

(b)

Two thyristor bridges can be connected in anti-parallel to constitute a dual-convertor as shown in this figure. The load is connected in the dc link. It is assumed that both the bridges are operating in continuous conduction mode. The + convertor on the left can generate $+I_d$ but $\pm V_d$ giving 2 quadrant operation, as discussed before. The – converter on the right can generate $-I_d$ (sinking current) but $\mp V_d$, where $-V_d$ corresponds to rectification mode and $+V_d$ corresponds to inversion mode. Therefore, the load can have bipolar controllable voltage and current signals, i.e., 4-quadrant operation. When the positive converter is operating, the negative converter should be disabled, and vice versa, to prevent short circuit. The output voltage control characteristics with α angle of the dual converter is shown in (b). Both units are controlled by cosine-wave crossing method to generate identical voltage magnitude with same polarity. The advantage of this control is that it gives linear transfer relation between control and output voltages. The + converter operates in rectification mode within $0 < \alpha < 90°$ and inversion mode within $90° < \alpha < 180°$. The − converter operation is similar but its output is reversed. For example, if $\alpha_p = 60°$ and $\alpha_p = 120°$ (i.e., $\alpha_p + \alpha_n = 180°$), both units will generate $+0.5\ V_d$ as shown, but the former will act as rectifier and the latter will be as an inverter. The load current will be free to flow in any direction.

FIGURE 3.21   Four-quadrant operation of dc drive.

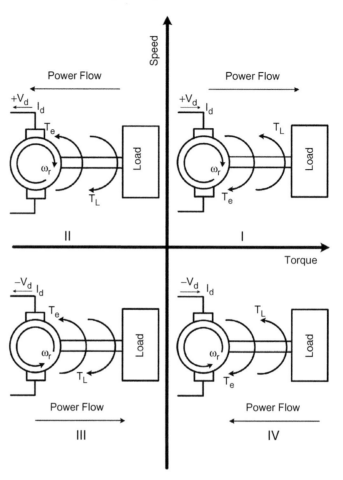

The figure summarizes four-quadrant operation of a dc motor drive by the dual converter discussed in Figure 3.20. Quadrant I is the usual motoring mode when the motor absorbs power, giving speed and developed torque in the same direction (clockwise), but the load torque T_L opposes the developed torque T_e. In quadrant IV (which was defined as quadrant 2 in Figure 3.18), the torque polarity remains the same, but the speed direction is opposite (counterclockwise). This means that the load torque is driving the machine, which is acting as a generator and pumping power back to the source. This requires $-V_d$ at the machine terminal. Operation in quadrants I and IV is made possible by the +Converter only. In quadrant III, both developed torque and speed are negative that make both V_d and I_d negative, i.e., the power flow is positive. In quadrant II, torque is negative and speed is positive (load torque is driving the machine), again resulting in the regenerative mode of operation. Operation in quadrants II and III is made possible by the −Converter only. Thus, the dual converter can cover all the four quadrants of operation. An example of four-quadrant operation is electric vehicle drive, where motoring and regeneration are required in either direction. On the other hand, a subway drive normally operates in quadrants I and II. A simple water pump can operate in quadrant I only.

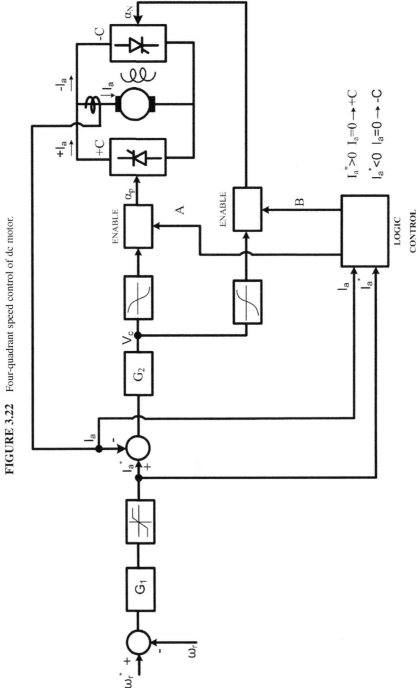

FIGURE 3.22 Four-quadrant speed control of dc motor.

The figure shows a simplified block diagram for four-quadrant speed control of a dc motor using a dual-bridge converter that is operating in blocking mode. The command speed ω_r^* is compared with the actual motor speed ω_r and the loop error sets up the armature current command I_a^* through a P-I controller (G_I) and bipolar armature current limiter. The limiter determines the maximum permissible armature current. The command I_a^* is compared with the actual I_a and the error determines the control voltage V_c. The cosine-wave crossing technique is used, and it is assumed that both the bridges always operate in continuous conduction mode. The α_P and α_N signals are generated so that each converter generates the same voltage (V_d), that is, $\alpha_P + \alpha_N = 180°$, as explained in Figure 3.20. Because there is no intergroup reactor (IGR) between +C and –C, only one bridge is permitted to conduct. The selection of converter is determined by the logic control, and its criteria of operation is shown in the figure. Assume that ω_r^* is a step positive signal. This will enable +C and speed will increase with $+I_a$ limit. If a speed reversal is requested, I_a^* will clamp to negative value. The $+I_a$ will go to zero, the –C will be enabled, $-I_a$ will build up, and the machine speed will decrease in regenerative mode. The control signal V_c is proportional to counter emf, that is, it will always track the actual speed. Then, V_c will be negative and speed will build up in the opposite direction with the –C conducting. It is possible to attain high-speed field-weakening mode speed control, but this is not shown in the figure.

FIGURE 3.23   Twelve-pulse converter: (a) series-connected bridges and (b) parallel-connected bridges.

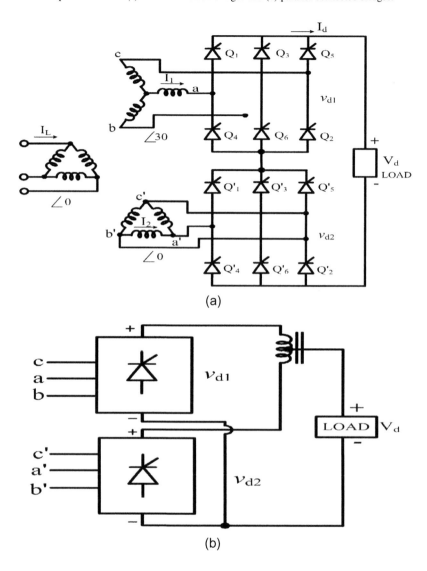

(a)

(b)

A three-phase thyristor bridge is characterized for generation of six-pulse load voltage and line current waves, as discussed earlier. If the load has a higher power need at higher voltage, the thyristors can be connected in series. If, on the other hand, higher power at high current is needed, the devices can be connected in parallel. Series or parallel connection of the devices requires static and dynamic matching of the devices. A more practical approach to get high power at high voltage is to connect two bridges in series through phase-shifting input transformers as shown in Figure 3.23(a). In this case, both the bridges will have identical six-pulse dc voltage output waves (v_{d1} and v_{d2}) except that the v_{d1} wave will be phase-advanced by 30° because of a delta-wye transformer connection. The resulting voltage $v_d = v_{d1} + v_{d2}$ will have a 12-pulse wave, that is, there will be 12 identical pulses per cycle, each with a 30° interval. Although the scheme is expensive because of the transformer, the advantages include lower harmonic content of load voltage and line current, and electrical isolation. With thyristor firing angle (α) control, dc voltage can be smoothly regulated keeping the 12-pulse characteristics. Figure 3.23(b) shows parallel operation of two bridges through a center-tapped reactor to get high power at high current, but $V_d = V_{d1} = V_{d2}$. Because of a phase shift by the input transformer, the v_d wave and the corresponding line current will have a 12-pulse ripple as before. This circuit can be used for high-power dc motor drives, whereas the previous topology is popular for high-voltage dc (HVDC) transmission. A similar phase-shifting technique can be used to construct 24-pulse, 48-pulse, and so on, converters.

FIGURE 3.24   Waveforms for 12-pulse series-connected converters.

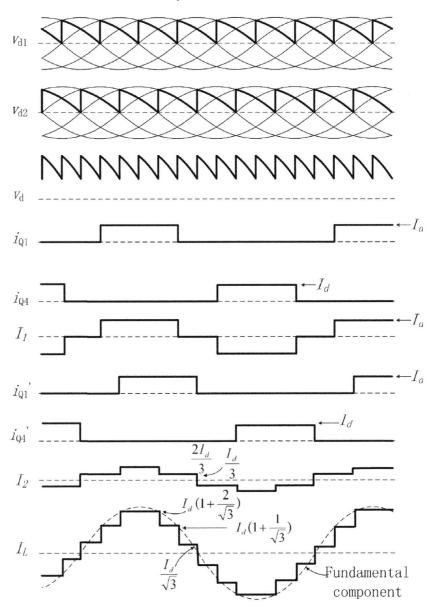

The figure explains the fabrication of voltage and current waves for 12-pulse series-connected bridge converters at a typical firing angle of $\alpha = 60°$. The waveforms v_{d1} and v_{d2} are the usual six-pulse waves of the component converters, where v_{d1} leads the v_{d2} wave by 30° as shown. The resulting $v_d = v_{d1} + v_{d2}$ wave gives 12 pulses per cycle, that is, each pulse is of 30° duration. The harmonic order of such a wave is $12n$, where $n = 1, 2, 3$, etc. The thyristor current waves i_{Q1}, i_{Q4}, $i_{Q1'}$, $i_{Q4'}$ (each pulse for 120° duration with amplitude I_d) from Figure 3.23(a) can be drawn graphically, and correspondingly the phase and line current waves I_1, I_2, and I_L, respectively, can be constructed assuming that the line voltages in primary and secondary of the transformer are equal. Evidently, the line current wave has 12 steps per cycle, which is characterized by the harmonic order $12n \pm 1$, where $n = 1, 2, 3$, etc.

FIGURE 3.25   (a) Single-phase thyristor ac voltage controller with waves and (b) performance characteristics.

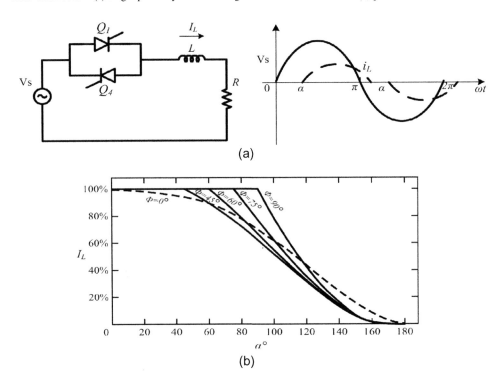

(a)

(b)

Thyristor ac voltage controllers control the rms voltage at the output while keeping the fundamental frequency the same. They can be single-phase or three-phase types. This figure explains the operation of a single-phase controller and gives its performance characteristics. A pair of thyristors is connected in anti-parallel and is in series with R-L load. The thyristors are fired symmetrically at angle α and the corresponding load current (also line current) wave is shown. The load voltage is the same as the supply voltage when the thyristor is on. The natural or line commutation occurs due to the reverse line voltage segment. If we assume that the thyristors are fired by a pulse train that extends from α to π angle, we can visualize that if α is gradually reduced from the π angle, the current pulse will expand until $\alpha = \varphi$, where φ is the load power factor angle. If $\alpha < \varphi$, there is no control on the devices, that is, they always remain closed and sinusoidal current flows in the load. Note that if single-pulse firing is used for the thyristors, at $\alpha < \varphi$ "single phasing" will occur. This means that with a long load current pulse from one device (say, Q_1), which shorts the other device, the Q_2 firing pulse will be missed in the other half cycle. Figure 3.25(b) shows the load rms current (I_L) with variation of α angle for a different φ angle of the load assuming that the load impedance $|Z_L|$ remains constant. With resistive load ($\varphi = 0$), I_L can be varied smoothly in the range $0 < \alpha < \pi$. For a power factor angle, say $\varphi = 60°$, I_L will change smoothly in the range $60° < \alpha < \pi$, but for $\alpha < \varphi$, the sinusoidal I_L will remain clamped to 100% as before because $R = |Z_L|$ The thyristors can be replaced by a triac, particularly for a resistive-type load (see Figure 2.6). The circuit has been used as an on–off switch (circuit breaker) and for heating and lighting control, transformer tap changing, and speed control of universal and single-phase induction motors. For circuit breaker operation, one thyristor can be replaced by a diode (half wave).

FIGURE 3.26   Three-phase ac voltage controllers: (a) full-wave controller, (b) half-wave controller, and
(c) delta-connected controller.

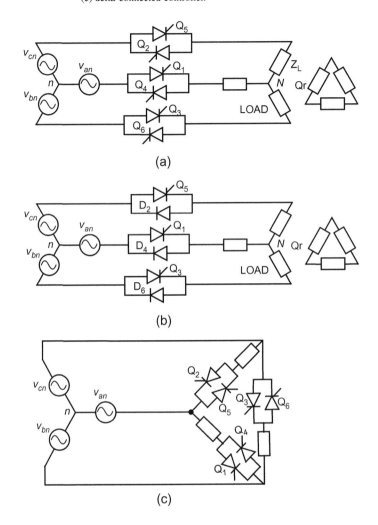

The principle of a single-phase ac voltage controller can be extended to various three-phase
controller configurations as shown in this figure. The upper circuit is defined as a full-
wave controller with anti-parallel thyristors in each line, and the load can be delta or wye
connected. Note that if the load and supply neutrals are connected, the circuit configuration becomes basically a superposition of single-phase controllers through the neutral.
The half-wave controller (middle) replaces a thyristor by a diode in each phase of the
above circuit. It is cheaper but introduces more harmonics in the load and line currents.

Either configuration can be used as a circuit breaker. For circuit breaker operation, even the devices in a phase can be omitted, shorting the supply and load phases. The delta-connected controller with anti-parallel thyristors as shown in the lower figure is somewhat economical because the thyristor current rating is less than that for line connection. For a wye-connected load, the phase loads can be connected in series with the line; only three thyristors can be neutrally connected in the delta configuration (not shown).

FIGURE 3.27   Waveforms for three-phase full-wave controller with *R*-load at $\alpha = 90°$ [see Figure 3.26(a)].

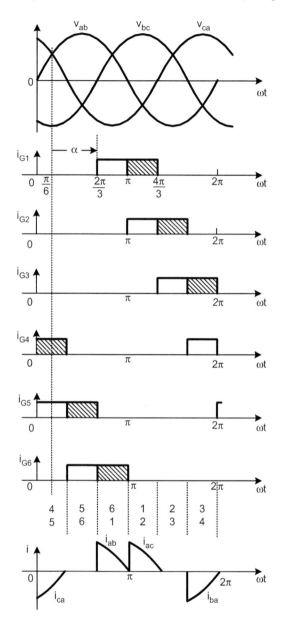

The analysis of three-phase ac controllers is very complex, particularly with inductive load, and therefore, studies can be done conveniently with computer simulation. Since the circuit of Figure 3.26(a) is most practical and frequently used, we will attempt to study it by waveforms with simple resistive load ($\varphi = 0$), as shown in this figure. If all the thyristors are considered as diodes, then all the lines are connected to the load directly. In that case, each diode will conduct sinusoidal current for 180° segment when the respective phase voltage (v_{an}, v_{bn} or v_{cn}) is positive, thus contributing to full-wave sinusoidal load current. In such a case, three devices will conduct at any time with the current starting at positive phase voltage. Thus for thyristor controller, $\alpha = 0$ is defined at zero phase voltage (i.e., $v_{ab} = \pi/6$), as indicated in the figure. The thyristors are fired at firing delay angle α in the sequence Q_1-Q_2-Q_3-Q_4-Q_5-Q_6 for three phase voltages at an interval of 60°, but two thyristors in different phases must conduct at least to contribute to load current. As α angle is increased, the load current or voltage decreases. The figure explains operation at $\alpha = 90°$ when two overlapping conducting devices as shown contribute to load current. It can be seen that α can be controlled in the range $0 < \alpha < 150°$, but the extra shaded region (60°) at the trailing end is needed for this. Otherwise, conduction will cease at $\alpha_{max} = 90°$. The output phase voltage is proportional to phase current I_a.

FIGURE 3.28   Performance characteristics of three-phase full-wave ac controller for Figure 3.26(a).

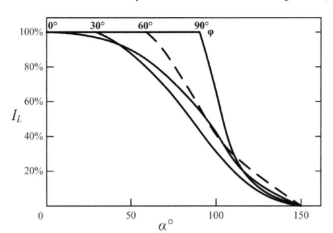

The figure shows the rms load current (I_L) vs. firing angle (α) at different load power factor angle (φ) for a three-phase full-wave controller. These curves can be derived by simulation study because of the complexity in the analysis. The curves are somewhat similar to those of Figure 3.25(b) for a single-phase controller. Again, $|Z_L|$ is considered to be constant at any power factor angle so that 100% load current is sinusoidal and is always of the same magnitude. It is equal to that shown in Figure 3.25(b) if the phase voltage and load impedance are the same. With resistive load, α variation in the range $0 < \alpha < 150°$ gives smooth variation of load current. With a power factor angle of, say, $60°$, I_L is controlled by α in the range $60° < \alpha < 150°$. If $\alpha < \varphi$ (i.e., $60°$), the control is lost and full sinusoidal currents flow in the line at a natural load power factor angle. In such a case, all devices are shorted. The converter is often used as a solid-state starter for cage-type induction motors. In such applications, the machine stator current is controlled by a feedback loop that is created by controlling the α angle. As motor speed develops, eventually, the converter is shorted or bypassed to supply the full voltage. The harmonic currents experienced in line and load are disadvantages of the circuit.

FIGURE 3.29 Phase-controlled cycloconverter classification and features.

CLASSIFICATION
- SINGLE-PHASE TO SINGLE-PHASE
- THREE-PHASE TO SINGLE-PHASE
- THREE-PHASE TO THREE-PHASE
- HALF-WAVE OR BRIDGE CONFIGURATION
- CIRCULATING CURRENT OR BLOCKING MODE FEATURES
- LINE COMMUTATION
- FOUR-QUADRANT OPERATION
- LOW-OUTPUT FREQUENCY RANGE
- LOW LINE DPF AND PF
- COMPLEX LINE AND LOAD HARMONICS
- LINE AND LOAD SUBHARMONIC PROBLEM
- COMPLEX CONTROL

A cycloconverter (CCV) is a direct frequency changer that uses a one-stage power conversion process. This figure summarizes different types of phase-controlled CCVs and their general performance characteristics. PWM-type frequency changers and dc and high-frequency ac-link frequency converters will be discussed later. A thyristor-based CCV basically operates on the phase-controlled converter principle and, therefore, their characteristics, such as line commutation, multiquadrant operation, low line DPF (displacement power factor) and PF (power factor), and line and load harmonics, are retained here. For industrial applications, a CCV generally has an input supply frequency of 60/50 Hz single-phase or three-phase operation. Therefore, CCVs can be classified as 1-φ/1-φ, 3-φ/1-φ, or 3-φ/3-φ configurations, although 3-φ/3-φ configurations are most commonly used. High-power (multi-megawatt) CCVs are used for speed control of induction and synchronous motors, and the applications include rolling mill drives, cement mill drives, and slip power recovery wound-rotor induction motor drives (Scherbius drive). The CCVs are also used in variable-speed constant frequency (VSCF) power supply systems in aircraft, where engine-generator power at several kilohertz is converted to 400-Hz power.

FIGURE 3.30   (a) Three-phase half-wave dual converter and (b) three-phase dual-bridge converter.

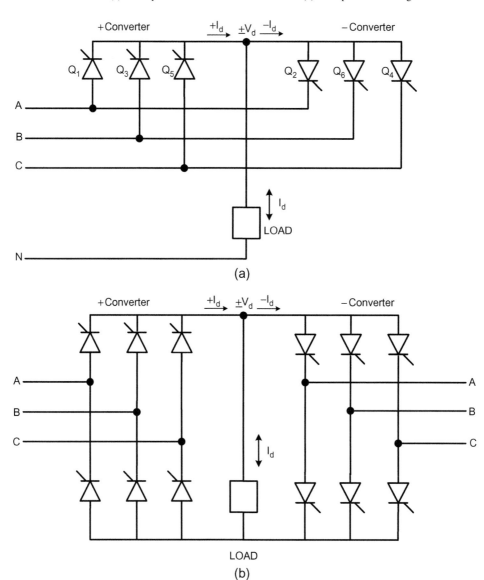

A three-phase dual-converter, as discussed earlier, can be used as a CCV. Figure 3.30(a) shows the configuration of a three-phase half-wave dual converter (assuming the availability of neutral point N), where a 60/50-Hz power supply, for example, can be converted to $+V_d$ or $-V_d$. With the +Converter working only, the three thyristors Q_1, Q_3, and Q_5

conduct in sequence for 120° contributing positive load current I_d and three-pulse load voltage. For firing angle range $0 < \alpha < 90°$, it acts as a rectifier ($+V_d$), whereas for the range $90° < \alpha < 180°$, it acts as an inverter ($-V_d$). The –Converter operation in anti-parallel is similar, but it supplies $-I_d$ in the load as shown. With bipolar power supply voltage and selective polarity of I_d by activating the +Converter or –Converter, the converter has four-quadrant operational capability. The three-phase dual converter shown in Figure 3.30(b) is the same as that shown in Figure 3.20(a). It has a six-pulse load voltage wave and four-quadrant operation capability, which was explained earlier. In both converters, the load can be connected at a center-tapped reactor (often called an intergroup reactor, IGR) to get smooth V_d and simplicity of converter control.

FIGURE 3.31   (a) Thevenin equivalent circuit of dual converter and (b) four-quadrant operation modes.

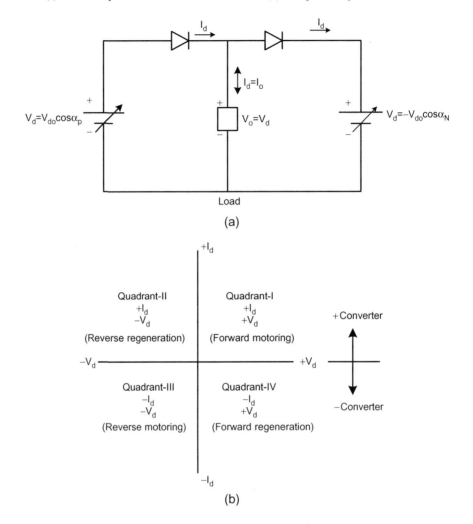

A three-phase dual converter (of either the half-wave or full-bridge type) can be ideally represented by the Thevenin equivalent circuit shown in Figure 3.31(a). It is assumed that the converter operates in continuous conduction mode and that the supply circuit Thevenin impedance and harmonics are neglected. The fictitious diodes indicate the permissible direction of load current I_d. The two converters are controlled by the cosine-wave crossing method ($V_d = V_{do} \cos \alpha$) and their output voltage magnitudes are adjusted such that they are always equal by α angle whether an individual unit is contributing the load current or not. This means that one is acting as a rectifier and the other is acting as an inverter. The load current is free to flow in either converter. Figure 3.31(b) further explains four-quadrant operation, which was discussed in Figures 3.20 and 3.21.

FIGURE 3.32   (a) Dual-converter voltage tracking control and (b) cycloconverter equivalent circuit with IGR.

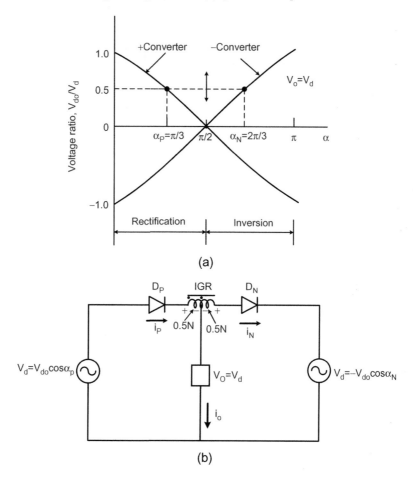

The output dc voltage of a dual converter can be modulated sinusoidally by firing angle control in either polarity to generate sinusoidal output voltage while tracking the output of each unit identically as explained in this figure. For example, if $\alpha_P = 60°$ and $\alpha_N = 120°$ (i.e., $\alpha_P + \alpha_N = 180°$), the +Converter will act as a rectifier and the –Converter will act as an inverter while each generates the identical output $V_d/V_{do} = 0.5$. For $\alpha_P = \alpha_N = 90°$, output voltage is zero, and for $\alpha_P = 120°$ and $\alpha_N = 60°$, $V_d/V_{do} = -0.5$, i.e., the converting units will reverse their roles. Figure 3.32(b) is the equivalent circuit where Thevenin voltages are represented by sinusoidal voltage. An intergroup reactor (IGR) is connected between the +Converter and –Converter that can permit circulating current (in circulating current mode) while absorbing the harmonics. With sinusoidal load voltage, the load current is sinusoidal, and the positive half cycle of current will be taken by the +Converter, whereas the negative half cycle will be taken by the –Converter. This is truly three-phase to single-phase CCV operation of a dual converter.

FIGURE 3.33   (a) Three-phase to three-phase half-wave (18-thyristor) cycloconverter and (b) output voltage fabrication by firing angle modulation.

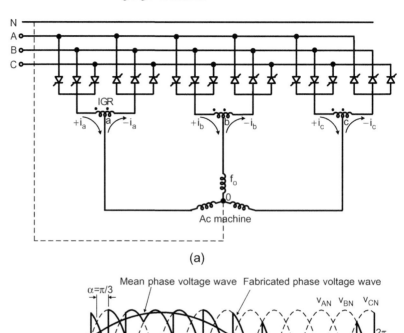

(a)

(b)

Three identical half-wave (or three-pulse) dual converters can be connected as shown in this figure to constitute a three-phase to three-phase CCV. A wye-connected ac machine is shown as the load, and the dashed neutral connection as indicated is not essential. In fact, with neutral connection, each phase group operates independently, creating a large neutral current that borders the machine. The firing angle in each phase group is modulated sinusoidally, and the three phases have identical mean output voltage waves except with a 120° phase shift angle. Figure 3.33(b), for example, shows the firing angle modulation of the +Converter in phase a group to fabricate the raw phase voltage wave (v_{aN}) of which the mean phase voltage wave is shown in the figure. The mean phase voltage of the −Converter is the same but its raw fabricated wave is not the same. The intergroup reactor (IGR) shown permits some circulating current (in circulating current mode) in each phase group and absorbs the harmonic voltage between +Converter and −Converter outputs. The output fundamental frequency and depth of modulation can be varied to generate variable-frequency, variable-voltage-power output. The fabricated output wave contains complex harmonics that are filtered somewhat by the machine inductance.

FIGURE 3.34 Phase voltage and current waves in motoring (upper) and regenerative (lower) modes.

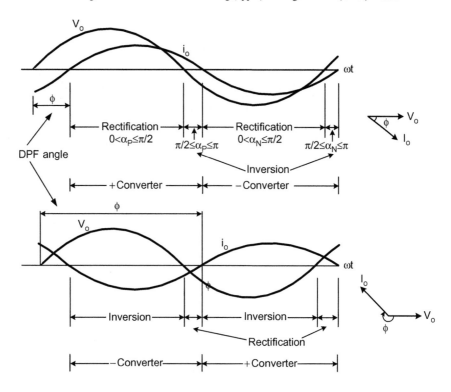

The sinusoidal output voltage (ignoring harmonics) of a CCV creates a sinusoidal load current that can have any arbitrary power factor angle. The upper part of this figure shows the phase voltage and phase current waves in active or motoring mode, where the current lags the voltage wave by DPF angle φ. In this mode, the average power is absorbed by the load. The positive half cycle of current flows in the +Converter, whereas the negative half cycle is taken by the −Converter. Again, with the positive current, the +Converter acts as a rectifier if it generates positive voltage and acts as an inverter if it generates negative voltage (two-quadrant operation). The operation modes of the −Converter are similar. Altogether, therefore, four-quadrant operation is obtained. For the three-phase CCV shown in Figure 3.33, all the phase voltage and current waves are the same except with a 120° mutual phase shift angle. All the phase groups can be controlled simultaneously so as to generate three-phase balanced output at variable voltage and frequency. The regenerative operating mode is shown in the lower part of the figure, where the DPF angle is more than 90° and average power is fed back from the load to the source. Evidently, for the greater segment of the half cycle, a component converter acts in the inverting mode.

FIGURE 3.35 Three-phase to three-phase bridge (36-thyristor) cycloconverter.

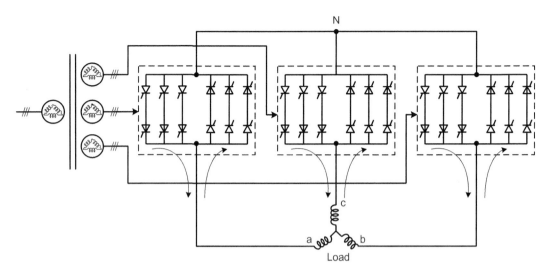

The figure shows a three-phase to three-phase bridge (six-pulse) type CCV, which is commonly used for high-power (multi-megawatt) ac drive applications. The CCV consists of three dual bridges, which are shown without IGRs. The wye-connected machine is supplied from the lower output points of the bridges, whereas the upper output points are shorted to form the neutral point N. The transformer at the input permits electrical isolation and voltage level transformation. No transformer is needed for isolation if the machine has isolated phase windings. The +Converter in each phase group (facing downward) contributes a positive half cycle of current, whereas the negative half cycle is contributed by the –Converter (facing upward). Without an IGR, circulating current is not possible in a phase group. Therefore, the conducting converter component should be enabled only while blocking the other unit (blocking mode) to prevent a short circuit.

FIGURE 3.36 (a) Waveforms explaining circulating current mode of a cycloconverter with IGR and (b) actual voltage and current waves of a bridge cycloconverter in circulating current mode.

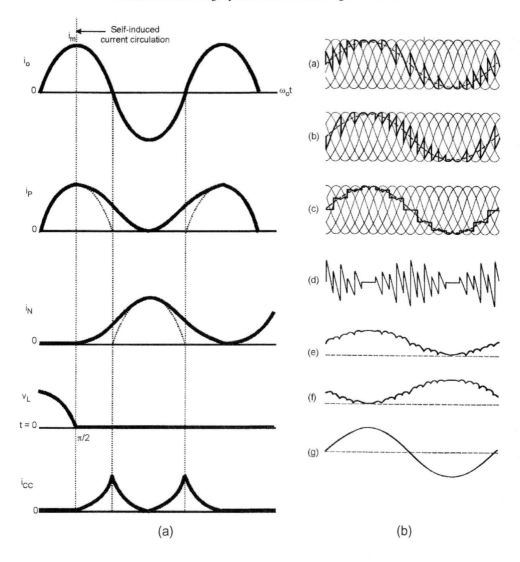

As mentioned before, a CCV with IGR can be operated in circulating current mode, and this figure explains that operation. Consider the CCV equivalent circuit in Figure 3.32(b) with an IGR, and assume that the load circuit is highly inductive when the load current i_0 wave is constrained to be sinusoidal. At $t = 0$, the positive load current i_0 is switched on as shown in part (a) of this figure. The $+ i_0$ will be taken by the +Converter (i.e., $i_P = i_0$) in the first 90° interval. The increasing positive current will create a positive voltage in the

IGR primary and negative voltage in the secondary. The latter will reverse bias the diode inhibiting current flow in the –Converter. However, decreasing the i_0 wave beyond the 90° point will reverse the IGR secondary voltage that will induce circulating current in the –Converter. From that point on, the IGR will be short circuited with both the converters conducting, and its trapped MMF will remain constant due to zero clamping voltage. Figure 3.36(b) shows the practical voltage and current waves of a bridge CCV in circulating current mode. The raw output voltage waves of positive and negative converters and their mean values (i.e., actual v_0 wave) are shown in (a), (b), and (c) of Figure 3.36(b), respectively. The voltage wave across IGR is shown in (d). Then, (e), (f), and (g) show, respectively, the current waves i_P, i_N, and i_0.

FIGURE 3.37   Harmonic families in the output voltage wave in blocking mode.

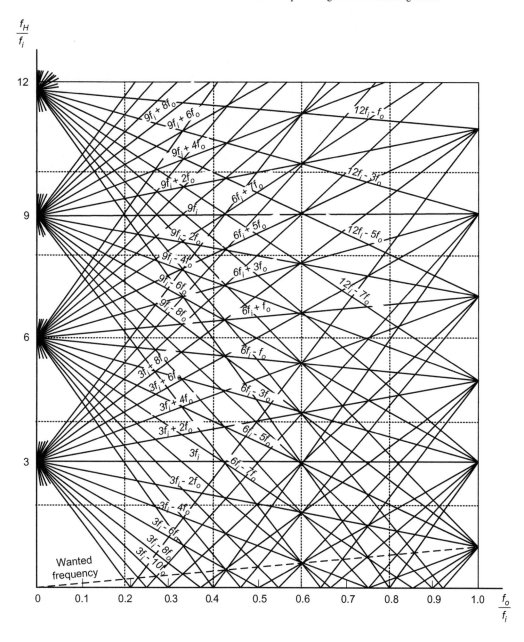

A CCV is basically a phase-controlled converter, and therefore, its output voltage and input current waves are rich in harmonics. In fact, the harmonic patterns are more complex because of sinusoidal modulation of the firing angle. Again, like a phase-controlled converter, the instantaneous power at output and input will balance and will indirectly relate input and output harmonics. The harmonics in CCVs depend on a number of factors, such as circulating or blocking mode of operation, the pulse number, output voltage, output frequency, load DPF, commutating line inductance, and continuous or discontinuous mode of operation. Detailed harmonic analysis by analytical or graphical methods is extremely complex. Computer simulation studies with FFT analysis of the waveform are the only practical way. This figure gives the idealized output voltage harmonic family (but not magnitude) [5] of a CCV operating in blocking mode, where f_H = harmonic frequency, f_0 = output frequency, and f_i = line frequency (60 or 50 Hz). In blocking mode operation without an IGR, the fabricated output voltage wave [see Figure 3.33(b)] appears across the load. The figure shows that if $f_0 = 0$ (i.e., the output is dc), the harmonic order is the same as that of a rectifier. With output frequency, the harmonic family is given by $pnf_i \pm nf_0$, where p = pulse number, n = integer, and $pn \pm n$ = odd integer. The harmonics below the dashed line are subharmonics, which are extremely harmful. The output frequency is usually restricted below $1/3f_i$ to restrict subharmonic output.

FIGURE 3.38   Harmonic families in the output voltage wave in circulating current mode.

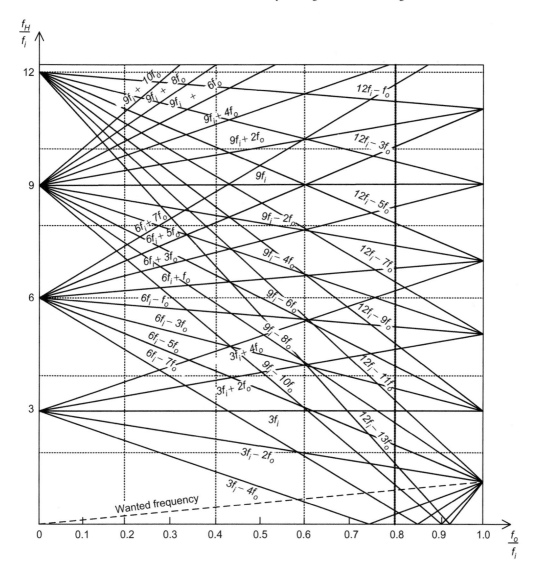

The figure shows the harmonics order of the output voltage wave [5] for a CCV operating in circulating current mode. In circulating current mode with an IGR, the output voltage is smoother (see Figure 3.36), and therefore, the harmonics problem is less severe. Again, for dc output ($f_0 = 0$), the harmonic orders are 3, 6, 9, 12, etc. for $p = 3$ (18 thyristors), and for $p = 6$ (36 thyristors), the corresponding orders are 6, 12, etc. Again, for a three-pulse

CCV, the maximum sideband orders are $3f_i \pm 4f_0$, $6f_i \pm 7f_0$, $9f_i \pm 10f_0$, etc., and for a 36-thyristor CCV, the corresponding orders are $6f_i \pm 7f_0$, $12f_i \pm 13f_0$, $18f_i \pm 19f_0$, etc. As indicated in the figure, the subharmonic problem is less severe in circulating current mode; therefore, output frequency is typically raised to $2/3f_i$ (i.e., 40 Hz with $f_i = 60$ Hz). For machine load, the leakage inductance will provide some filtering for the current wave. However, the harmonics will cause additional copper loss and pulsating torque in the machine.

FIGURE 3.39 Fabrication of output voltage and line current waves for a 36-thyristor cycloconverter in blocking mode.

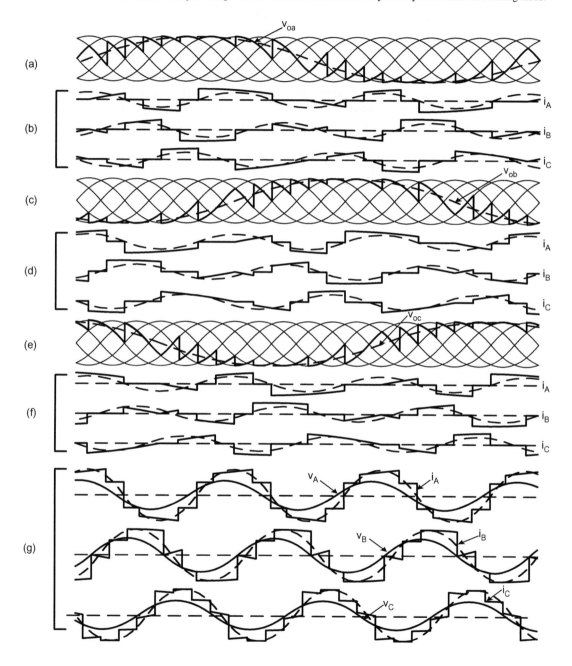

The figure shows the fabrication of line current waves [5] of a 36-thyristor blocking mode CCV that have been constructed graphically from the output voltage and current waves. The assumptions in the figure are $f_0 = 1/3f_i$, modulation factor m_f (ratio of peak and maximum possible peak phase voltage) = 1, and zero load power factor angle ($\varphi = 0$). On the top, phase a group operation is shown only, and its loading contributes to the corresponding line currents i_A, i_B, and i_C, respectively. The second and third group of waves are the respective contributions of phase b and phase c. The component line current waves are symmetrical but are very rich in second harmonics. For three-phase to three-phase CCV operation, these line current components can be added, which are shown as the last group (g) of waves. Note that the even harmonics disappear from these line currents and their fundamental component lags the line voltage although the load power factor is unity. Note that the fundamental in-phase current component only carries the active power to the CCV.

FIGURE 3.40 Harmonic family in the input current of a cycloconverter.

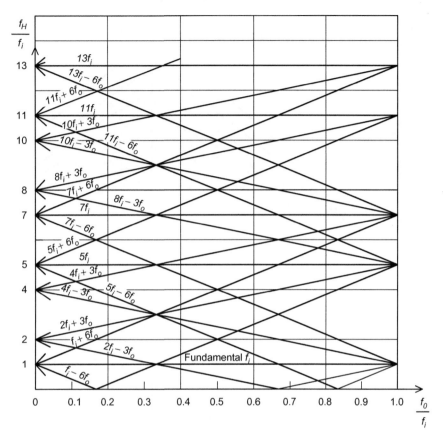

The figure shows the idealized harmonic family (not magnitude) of the input line current [5] of 18- and 36-thyristor CCVs, where f_H = harmonic frequency, f_i = input or line frequency (60 or 50 Hz), and f_0 = output frequency. It can be derived from the current wave in Figure 3.39 by FFT analysis. If the output is dc ($f_0 = 0$), that is, if the CCV acts as a rectifier, the line current harmonic order is given by $np \pm 1$ (n = integer, p = pulse number), which is evident from the figure. The frequency f_0 creates additional sidebands and the general harmonic family is given by the relation $(np \pm 1)f_i \pm mf_0$, where $(np \pm 1) \pm m$ = odd integer. Including the fundamental component f_i, the input frequencies of an 18-thyristor CCV can be summarized as $f_i, f_i \pm 6f_0, f_i \pm 12f_0 \ldots 2f_i \pm 3f_0, 2f_i \pm 6f_0 \ldots$ $4f_i \pm 3f_0$, etc., and for a 36-thyristor CCV, these are $f_i, f_i \pm 6f_0, f_i \pm 12f_0 \ldots 5f_i, 5f_i \pm 6f_0$, etc., which are also indicated in the figure. Note that the fundamental frequency current should always be present to carry the active power. The power associated with harmonic frequencies is alternating for which the average is zero. A small amount of subharmonics is present in the line current, but it is usually neglected.

FIGURE 3.41   Input and output DPF relations with modulation factor m_f.

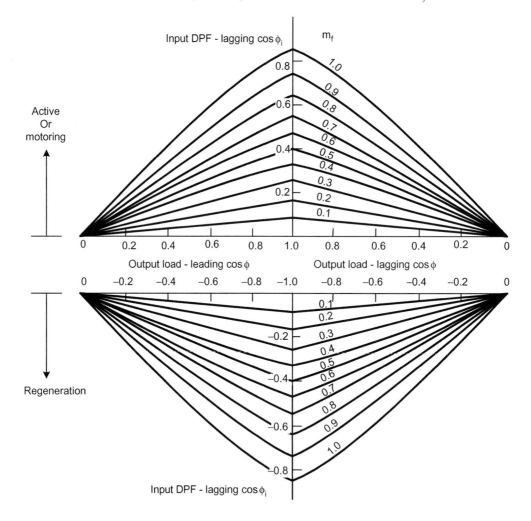

A CCV is basically a phase-controlled line-commutated converter; therefore, it is characterized by lagging DPF at the input. In fact, the modulating firing angles to synthesize the ac output have an additional deteriorating effect on line DPF. This poor DPF, along with complex harmonic patterns of load voltage and line current, is a big disadvantage of CCVs. This figure shows that input DPF is a function of modulation index m_f and load power factor angle φ [5]. With load cos $\varphi = 1$, the input DPF is maximum (0.843) when $m_f = 1$. The DPF deteriorates if cos φ decreases or m_f decreases. In extreme cases, DPF = 0 if $m_f = 0$ or cos $\varphi = 0$. The curves indicate that line DPF decreases even if the load DPF is leading. The curves are symmetrical on both sides of the vertical line. The conditions are also symmetrical in motoring and regenerative modes.

FIGURE 3.42 (a) Firing angle (α) generation of a dual-converter phase group and (b) principle of current control.

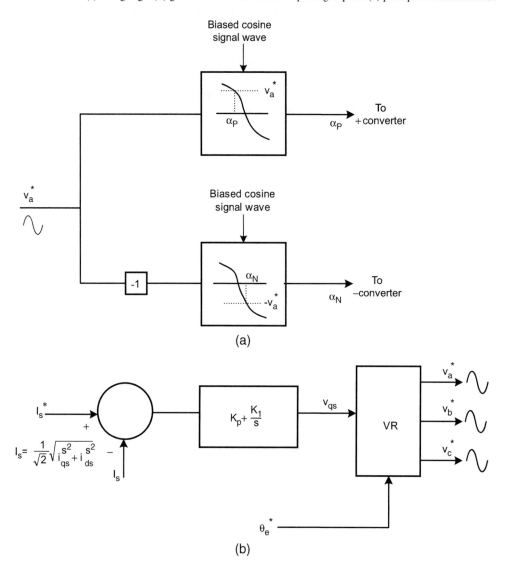

The firing angle control principle of the dual-converter phase a group of a CCV is shown in this figure. It utilizes the modulation principle discussed in Figure 3.32. The sinusoidal phase modulating voltage v_a^* controls the +Converter and −Converter simultaneously by means of the cosine-wave crossing method, which generates the respective firing angles α_P and α_N ($\alpha_P + \alpha_N = 180°$) so that their output voltages are always equal and maintain a linear relationship with the control signal. The biased cosine wave shown for the control is generated from the line voltage waves. Control of phase b and phase c is identical except the voltage waves are at 120° phase difference. Figure 3.42(b) shows the current control loop applied over the voltage control. The command rms phase current I_s^* is compared with the corresponding feedback current (synthesized from the phase current), and the phase voltage commands are generated through the P-I compensator and vector rotator (VR), which will be explained in Chapter 7. The control of CCV drive systems will be discussed further in Chapters 7 and 8.

FIGURE 3.43   (a) A 72-thyristor cycloconverter with asymmetrical α control for DPF improvement and
(b) asymmetrical α control principle.

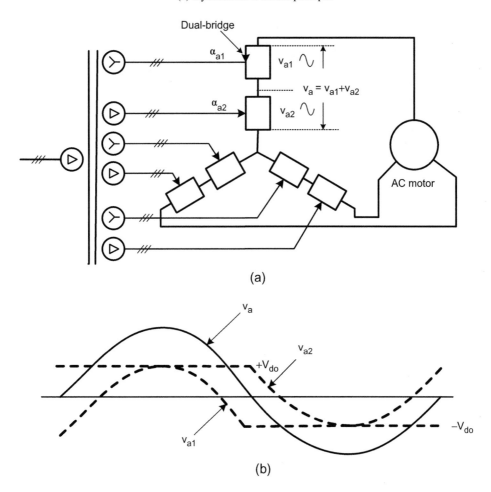

(a)

(b)

The concept of a six-pulse dual-bridge CCV can be extended to a 12-pulse CCV by
connecting two dual bridges in series for each phase as shown in part (a) of this figure.
The component dual bridges in a phase group are supplied by wye- and delta-connected
transformers to get 12-pulse operation that is similar to that of Figure 3.23(a). The 72-
thyristor CCV is wye connected and supplies power to a three-phase machine load for
speed control [7]. The high-voltage, high-power CCV will have considerably fewer harmonics because of the higher pulse number. In the figure, for example, the phase a group
generates the phase voltage $v_a = v_{a1} + v_{a2}$, where v_{a1} and v_{a2} are the component voltages,
and α_{a1} and α_{a2} are the respective firing angles. Instead of symmetrical 12-pulse operation,

it is possible to control the firing angles asymmetrically to get the voltage waves as shown in part (b) of this figure. In the positive half cycle of phase voltage v_a, $\alpha_{a2} = 0$, whereas α_{a1} is controlled to get $v_{a1} = v_a - V_{do}$. The operation is symmetrical in the negative half cycle as shown. The advantage of asymmetrical control is higher input DPF, but the disadvantage is six-pulse CCV operation.

FIGURE 3.44 Input DPF control by circulating current of a 36-thyristor cycloconverter.

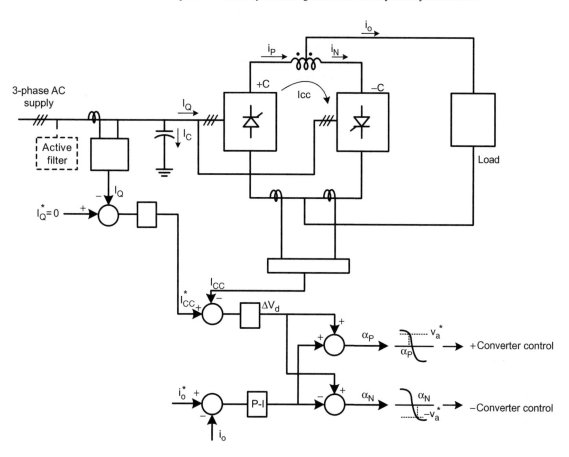

A CCV with circulating current mode (see Figure 3.36) operation has the disadvantages of additional IGR costs and a higher converter rating with the associated losses, but the advantages are simple control and the input DPF control, which is explained here. This figure shows the block diagram [12] for a three-phase to one-phase CCV with the load connected at the midpoint of the IGR. Both the converter components (+C and –C) are always enabled so that circulating current can flow all the time. It is possible to inject additional dc circulating current (I_{CC}) between +C and –C in order to control the total lagging input current I_Q. The lower part of the figure shows the load current (i_0) control being accomplished by means of the cosine-wave crossing technique. The circulating current is increased by injecting symmetrically $+\Delta V_d$ in the +C and $-\Delta V_d$ in the –C. It can be shown that I_{CC} increases the lagging current of both converters equally by ΔI_Q. The I_{CC} can be controlled to control the total CCV lagging current I_Q by a simple closed loop that generates the ΔV_d signal as shown in the figure. A large capacitor is connected at the input that sinks the leading current I_C. The CCV input I_Q is controlled to be equal to I_C so that the line reactive current is zero. The system is shown with an active filter to absorb the line harmonics so that the total PF = 1.0. Active filters will be discussed in the next chapter. The principle of circulating current control to improve line DPF was first demonstrated in a dual-converter (with IGR) controlled dc motor drive [13] (see Figure 3.22).

FIGURE 3.45 High-frequency link power conversion (from [8]).

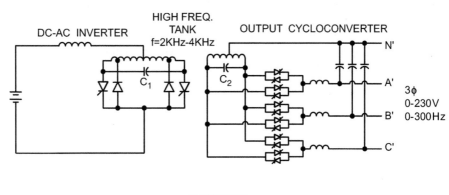

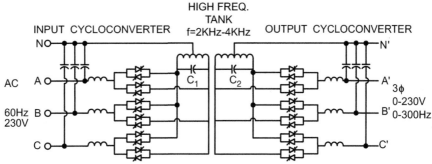

The phase-controlled cycloconverter principle can be extended to a high-frequency link power conversion system, as shown in this figure. Consider first the simpler scheme on the top where the dc voltage from the battery is inverted to high-frequency ac (several kilohertz), which is then converted to variable-voltage, variable-frequency ac for an ac motor drive. The output cycloconverter shown in the figure basically converts single-phase, high-frequency ac to three-phase, low-frequency ac by three single-phase center-tapped independent cycloconverters. Each single-phase unit consists of four thyristors that fabricate low-frequency ac by modulation of firing angles. The commutation is provided by the high-frequency supply. The power to the high-frequency tank is supplied by a current-fed center-tapped inverter (discussed in Chapter 5), where the thyristors are also commutated by the tank voltage. For the line commutation, the inverter frequency is slightly above the tank resonance frequency. Because a phase-controlled CCV can have bidirectional power flow, a second CCV is connected in inverse fashion in the lower figure in order to get 60-Hz ac (high-frequency ac), variable-voltage, variable-frequency ac power conversion for motor drives. The input CCV not only maintains the tank voltage, but its fabricated input voltage (counter emf) magnitude and phase can be varied to control real and reactive power flow at programmable power factor (including unity).

The lightweight, high-frequency transformer permits isolation and voltage level changes. One problem in the system is that the tank frequency drifts and the ac link voltage waveform gets distorted due to nonlinear loading effect. The input CCV only with the tank load can operate as a leading or lagging static VAR compensator (SVC) (similar to no-load operation of a synchronous motor with excitation variation). SVCs will be discussed further in the next chapter.

FIGURE 3.46   Line power quality problems and harmonics standards.

- LARGE GROWTH OF DIODE AND THYRISTOR CONVERTERS ON UTILITY SYSTEM
- LINE VOLTAGE HARMONIC DISTORTION
- POOR LINE POWER FACTOR
- EMI
- LINE AND EQUIPMENT HARMONIC CURRENT LOADING
- COMMUNICATION INTERFERENCE
- METER INACCURACY
- SPURIOUS LINE RESONANCE
- IEEE-519 STANDARD—HARMONIC DISTORTION CONTROL AT COMMON ENTRY POINT
- IEC-1000 STANDARD—CONTROLS HARMONIC DISTORTION OF INDIVIDUAL EQUIPMENT

Utility systems are obligated to supply balanced and sinusoidal line voltage waves of correct magnitude and frequency to the customers. Any deviation from this is defined as a power quality problem. Unfortunately, diode and thyristor converters generate electromagnetic interference (EMI) and large harmonic currents in the line. The harmonic currents flowing through line impedances distort the line voltage wave and create power quality problems for consumers. The additional harmful effects of these harmonics are line and equipment loading and the corresponding increase of power loss, interference with parallel communication lines, and inaccuracy in the measuring instruments. The harmonics may also create spurious resonance problems in the line due to distributed line inductance and capacitance, or extra passive filters installed in the line. A number of harmonic control standards have been developed to limit the maximum line harmonic current injection. The most dominant harmonic standards are IEEE-519 and IEC-1000, which will be discussed later.

FIGURE 3.47   (a) Thyristor converter operation on utility bus and (b) typical line voltage waves of thyristor converter showing commutation spikes.

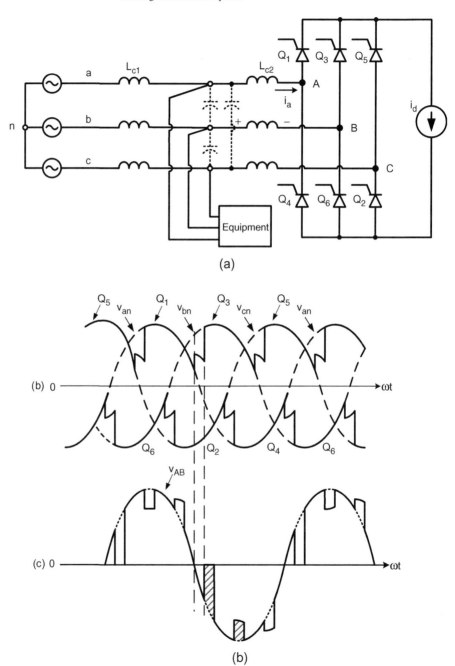

(a)

(b)

The figure explains the transient voltage spike problem in line voltage wave due to thyristor commutation in a bridge rectifier. There are six commutations per cycle of line voltage. During commutation, line current transfers from one phase to another, and creates a transient voltage spike in the line commutating inductance. Thus, there are six commutation voltage spikes in the line voltage v_{AB} as shown in lower part of Figure 3.47(b). A piece of equipment is shown connected on the parallel bus at some intermediate point, where the total commutating (or leakage) inductance L_c is resolved into two segments L_{c1} and L_{c2}. Consider, for example, the commutation from Q_5 to Q_1 in the +Converter when line current $+i_a$ will build up, creating a voltage notch in the line voltage v_{AB}. For commutation from Q_6 to Q_2 after 60° in the –Converter, the notch will be negative, contributing to a positive spike in v_{AB}. Note that twice in the cycle (say, Q_1 to Q_3 and Q_4 to Q_6) the lines will be shorted ($v_{AB} = 0$) during commutation. The connected equipment will have this spiky voltage wave, but the spike will be less severe if the equipment is connected further away from the converter. The sharp dv/dt in the wave creates an EMI problem. The thyristors need a snubber for protection from dv/dt spikes. The parasitic line capacitances shown can cause a spurious ringing effect, which will make the EMI problem worse.

FIGURE 3.48   *dv/dt* and *di/dt* induced EMI coupling.

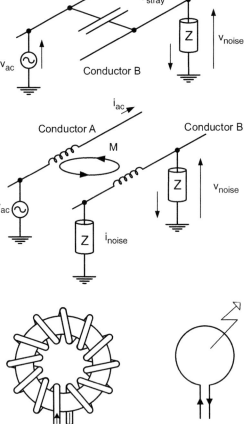

Power electronic apparatuses, in general, are notorious for generating EMI problems. These problems are created by high-frequency voltage and current sources, or *dv/dt* and *di/dt,* due to switching of fast power semiconductor devices (see Figure 2.19). The upper part of this figure shows a circuit with high-frequency voltage source v_{ac}, which may also be *dv/dt* due to switching of the power semiconductor. Conductor A is the main circuit for the voltage source, but conductor B in close vicinity may have capacitive coupling and a noise voltage may appear across the impedance Z as shown. The stray capacitor may be extremely small, but the coupling current due to high *Cdv/dt* may be substantial. The middle figure shows the circuit with high-frequency current or *di/dt*

generated by a high-frequency voltage source. The *di/dt* may also be due to power semi-conductor switching. The current causes inductive EMI coupling to conductor B due to high *Mdi/dt* although the mutual inductance *M* may be very small. While the top figures are illustrations of conducted EMI, the lower figure illustrates radiated EMI due to high-frequency current circulating in the toroidal coil when the toroid is acting like an antenna. Radiated EMI can also occur in power converters with high *dv/dt, di/dt,* or diode recovery current. EMI causes problems in the converter controller and nearby apparatus. (For further discussion, see Chapter 12.)

FIGURE 3.49   EMI problems in power electronics.

- ARISE DUE TO HIGH *dv/dt* AND *di/dt* OF SWITCHING DEVICES
- CONDUCTED AND RADIATED EMI
- DIFFERENTIAL OR COMMON MODE EMI
- MALOPERATION OF CONTROL ELECTRONICS AND OTHER
  SENSITIVE EQUIPMENT

HOW TO SOLVE EMI PROBLEM
    -SNUBBERS
    -SHIELDING
    -GROUNDING
    -FILTERING
    -COMPONENT AND WIRING LAYOUT
    -TWISTED SHIELDED WIRES
    -SOFT SWITCHING

The figure summarizes the types of EMI problems and methods of solving them in power electronics apparatuses. (See also Chapter 12.) EMI problems may be very severe in high-power, high-switching-frequency converters, and low-power control electronics should be well protected from them to prevent maloperation. Conducted EMI can be of the differential or common mode type. Differential mode EMI is a current or voltage between the lines, whereas common mode EMI arises between the lines and ground. The capacitive and inductive couplings, illustrated in Figure 3.48, are examples of common mode EMI. Well-designed power electronics equipment should be free from EMI problems. The methods of solution may be well-designed snubbers, shielding by a metallic enclosure to prevent radiated EMI, effective grounding of the control circuits separately from the power ground with multiple grounds or a ground plane, decoupling of filters, proper component and wiring layout, use of twisted and shielded wires and grounding of the shield to prevent coupling, and soft switching if possible. Soft switching will be discussed in Chapter 4. It is good practice to eliminate the EMI source rather than protecting against it, but the problem may not be simple enough to do that.

FIGURE 3.50   Harmonic current limits by IEEE-519 standard.

AT POINTS OF COMMON COUPLING (PCC)
(VOLTAGE = 2.4 V – 69 kV)

MAXIMUM HARMONIC CURRENT DISTORTION IN % OF FUNDAMENTAL

HARMONIC ORDER n (ODD HARMONICS)

I_{SC}/I_L	<11	11<h<17	17<h<23	23<h<35	35<h	THD
<20	4.0	4.0	1.5	0.6	0.3	5.0
20-50	7.0	3.5	2.5	1.0	0.5	8.0
50-100	10.0	4.5	4.0	1.5	0.7	12.0
100-1000	12.0	5.5	5.0	2.0	1.0	15.0
>1000	15.0	7.0	6.0	2.5	1.4	20.0

I_{SC}, I_L = Short circuit and load fundamental current, respectively, at PCC.
 (1) The table is for the worst case load at time more than 1 hour.
 (2) Short time load (such as solid-state motor starter) permits 50% higher than steady state.
 (3) Even harmonics, if present, are limited to 25% of the add harmonic limits given.

The IEEE-519 standard [9] limits the individual harmonic currents as well as the total harmonic distortion (THD) for nonlinear power electronic loads, as indicated in this figure. The THD is defined as the ratio of total distortion current to fundamental current. The distortion current limit at a point depends on the short circuit current ratio (I_{SC}/I_L) at that point. A bus with lower I_{SC} means higher Thevenin impedance and, therefore, more distortion of bus voltage for a specified load current I_L. This means that for a specified bus voltage distortion, the harmonic current limit should be less at lower I_{SC}. For example, with $I_{SC}/I_L < 20$, IEEE-519 permits THD = 5%, and the individual odd harmonics are given in the top row of the table. Note that this harmonic standard is valid at the point of common coupling (PCC). This means that if a factory consumes both linear and nonlinear loads, the linear load component will compensate the nonlinear load, thus permitting higher harmonics for nonlinear load. The IEEE-519 standard also limits line voltage distortion, commutation voltage notch area, and notch depth [9] (see Figure 3.47), but these are not discussed in this text.

FIGURE 3.51   IEC-1000 harmonics standard.

EQUIPMENT CLASSIFICATION:
   CLASS A – BALANCED THREE-PHASE EQUIPMENT +
      OTHERS EXCEPT B, C, AND D
   CLASS B – PORTABLE TOOLS
   CLASS C – LIGHTING EQUIPMENT (>25 W)
   CLASS D – SPECIAL CURRENT WAVESHAPE +
      75 W < POWER < 600 W

HARMONIC LIMITS FOR CLASS A AND B EQUIPMENT

HARMONIC ORDER n	CLASS A (MAX. PERM. HARM. CURRENT) (A)	CLASS B (MAX. PERM. HARM. CURRENT) (A)
ODD HARMONICS		
3	2.30	3.45
5	1.14	1.71
7	0.77	1.155
9	0.40	0.60
11	0.33	0.495
13	0.21	0.315
$15<=n<=30$	$2.25/n$	$3.375/n$
EVEN HARMONICS		
2	1.08	1.62
4	0.43	0.645
6	0.30	0.45
$8<=n<=40$	$1.84/n$	$2.76/n$

HARMONICS LIMITS FOR CLASS C EQUIPMENT

HARMONICS ORDER n	MAX. % OF INPUT FUND. CURRENT
2	2
3	30.PF
5	10
7	7
9	5
$11 <=n<= 39$	3

HARMONICS LIMITS FOR CLASS D EQUIPMENT

HAMONICS ORDER N	75 W<POWER<600 W (mA/W)	POWER>600 W (A)
3	3.4	2.30
5	1.9	1.14
7	1.0	0.77
9	0.5	0.40
11	0.35	0.22
13	0.296	0.21
$15<=n<=39$	$3.85/n$	$2.25/n$

While IEEE-519 restricts harmonic currents at the common entry point (point of common coupling, PCC) of a building or factory, the IEC-1000 (more specifically 1000-3-2) standards imposed by the International Electrotechnical Commission (IEC) [11] are mainly followed in Europe and are much more rigorous. They restrict steady-state harmonic currents generated by individual equipment, and do not care about cancellation by linear load. These harmonic standards fall under the broad category of EU (European Union) 89/336/EEC on Electromagnetic Compatibility of Industrial Equipment [11]. While applying these standards, equipment is classified into four categories as indicated in this figure. The current consumed by any equipment is restricted below 16 A at 220/380 V, 230/400 V, or 240/415 V, 50/60 Hz. The Class A equipment (which is nonportable) has more rigid harmonics control than Class B, and even harmonics that give asymmetry to waves have more restriction. The Class C equipment (above 25 W) is restricted on percent of fundamental basis. There is no restriction on low-power fluorescent lamps (<25 W). The Class D equipment, which has a current with a special waveshape, is restricted on the basis of power rating unless the power exceeds 600 W. Some harmonic filtering will move them into the Class A category.

Summary

The principles and characteristics of diode and thyristor converters and cycloconverters have been discussed in this chapter. This class of power converters falls into the first generation of power electronics and is often defined as *classical power electronics.* It is an extremely important branch of power electronics since its applications have continuously expanded from the beginning to the modern era.

Diode converters permit only uncontrolled power rectification. Thyristor converters with firing angle control can be operated as rectifiers and inverters, but at zero firing angle their operation reverts to that of a diode rectifier. Generally, single-phase and three-phase converters are used widely in industry, but multiphase (more than three-phase) operation with the help of a transformer is not uncommon in high-power applications. The applications of phase-controlled converters include heating and lighting control, adjustable-speed dc and ac motor drives, electroplating, anodizing, chemical gas production, metal refining, HVDC systems, static VAR control, dc and ac power supplies, utility system-interactive power converters for renewable energy systems, and slip power recovery drives.

A cycloconverter is a direct frequency changer that permits lower output frequency, typically up to one-third (blocking type) to two-thirds (circulating current type) of input frequency. The CCV applications include adjustable-speed ac motor drives, aircraft or shipboard variable-speed, constant-frequency (VSCF) systems, and slip power controlled Scherbius drives. The phase control has the advantages of simplicity of converter control and higher equipment efficiency, but the disadvantages are complex harmonics at input and output, and poor lagging power factor in the line. Because of these disadvantages, recent trends in power electronics are leading to the phase-controlled class of converters being replaced by pulse width modulation (PWM) self-controlled converters that use devices such as power MOSFETs, IGBTs, GTOs, and IGCTs. It is interesting to note that the highest power applications, such as HVDC systems, load-commutated synchronous motor drives, and reactive power compensators use this type of converter.

References

[1] B. K. Bose, *Modern Power Electronics and AC Drives,* Prentice Hall, Upper Saddle River, NJ, 2002.

[2] G. K. Dubey, *Power Semiconductor Controlled Drives,* Prentice Hall, Englewood Cliffs, NJ, 1989.

[3] N. Mohan, T. M. Undeland, and W. P. Robbins, *Power Electronics,* John Wiley, New York, 1995.

[4] M. H. Rashid, *Power Electronics,* Prentice Hall, Upper Saddle River, NJ, 2004.

[5] B. R. Pelly, *Thyristor Phase-Controlled Converters and Cycloconverters,* John Wiley, New York, 1971.

[6] S. B. Dewan and A. Straughen, *Power Semiconductor Circuits,* John Wiley, New York, 1975.

[7] R. Kurowsawa, T. Shimura, H. Uchino, and K. Sugi, "A microcomputer-based high power cycloconverter-fed induction motor drive," *IEEE IAS Annual Meeting Conference Record,* pp. 462–467, 1982.

[8] B. K. Bose and P. Espelage, "High frequency link power conversion," *IEEE Trans. on Ind. Appl.,* vol. 13, pp. 387–393, September/October 1977.

[9] C. K. Duffey and R. P. Stratford, "Update of harmonic standard IEEE-519: IEEE recommended practices and requirements for harmonic control in electric power systems," *IEEE Trans. on Ind. Appl.,* vol. 25, pp. 1025–1034, November/December 1989.

[10] IEEE 519-1981, *IEEE Guide for Harmonic Control and Reactive Compensation of Static Power Converters.*

[11] L. Rossetto, P. Tenti, and A. Zuccato, "Electromagnetic compatibility of industrial equipment," *Proc. of FEPPCON III,* pp. 363–375, 1998.

[12] Hosoda, Tenura, et al., "A new concept high performance large scale ac drive system—cross current type cycloconverter-fed induction motor with high performance digital vector control," *IEEE IAS Annu. Meet. Conf. Rec.,* pp. 229–234, 1986.

[13] J. Rosa and P. Finlayson, "Power factor correction of thyristor dual converter via circulating current control," *IEEE IAS Annu. Meet. Conf. Rec.,* pp. 415–422, 1978.

CHAPTER 4

Voltage-Fed Converters and PWM Techniques

FIGURE 4.1   (a) Buck dc-dc converter for one-quadrant speed control of dc motor and (b) buck-boost converter for two-quadrant speed control.

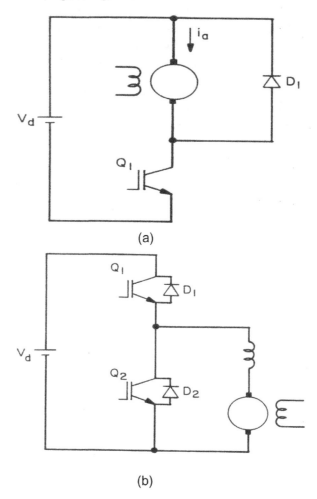

A voltage-fed converter is fed ideally by a zero Thevenin impedance voltage source. Figure 4.1(a) shows a dc-dc buck or step-down converter, where the supply dc voltage (V_d) is converted to variable dc voltage at the motor terminal to control its speed in one quadrant only. The IGBT generally operates on an on–off duty cycle basis at fixed switching frequency. When it is on, the V_d is applied to the load and its current tends to build up. When Q_1 is off, the freewheeling diode D_1 circulates the load inductive current shorting the load (assuming continuous conduction). The average motor voltage is always less than V_d. Figure 4.1(b) shows two-quadrant speed control by a buck-boost hybrid converter.

In the buck mode, as just discussed, the device Q_1 chops the supply voltage and D_2 freewheels. In the boost mode, Q_2 and D_1 are active. If Q_2 is on, the counter emf of the motor tends to builds up the current in the armature inductance, which then flows to the line when Q_2 is off. Thus, the machine operates in the regenerative mode when the motor's mechanical energy is converted to electrical energy and fed back to the source. In this mode, V_d is higher than the load counter emf. Small variable-speed permanent magnet (PM) dc motor drives with dc-dc converters are widely used in industry. Individually, buck and boost converters are also widely used in switching mode power supplies (SMPS). In SMPS, extra filters are normally added to smooth the load harmonics.

FIGURE 4.2 Four-quadrant speed control of dc motor using an H-bridge converter.

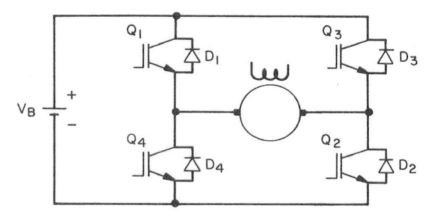

QUADRANT I – BUCK CONVERTER (FORWARD MOTORING)
Q_1 – ON
Q_2 – CHOPPING
D_3, Q_1 – FREEWHEELING

QUADRANT II – BOOST CONVERTER (FORWARD REGENERATION)
Q_4 – CHOPPING
D_2, D_1 – FREEWHEELING

QUADRANT III – BUCK CONVERTER (REVERSE MOTORING)
Q_3 – ON
Q_4 – CHOPPING
D_1, Q_3 – FREEWHEELING

QUADRANT IV – BOOST CONVERTER (REVERSE REGENERATION)
Q_2 – CHOPPING
D_3, D_4 – FREEWHEELING

The buck-boost converter for two-quadrant speed control, as discussed earlier, can be extended to four-quadrant speed control by adding another converter leg to form an H-bridge converter, as shown in this figure. Note that the load voltage can be of either polarity and the current can flow in either direction. All the operation modes of the converter for four-quadrant operation are summarized above. The H-bridge can also be used as inverter, as discussed later.

FIGURE 4.3   (a) Duty cycle modulated buck-boost dc-dc converter and (b) Cu′k dc-dc converter.

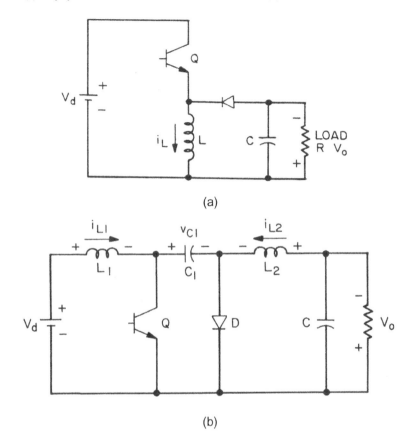

(a)

(b)

Figure 4.3(a) shows a simple topology for a buck-boost converter where buck or boost operation is controlled by duty cycle variation of a single switch (Q). When the switch is on, the source energizes the inductance L, but when it is off, the energy is transferred to the output. No energy feedback is possible to the input. The voltage relationship is given by $V_0/V_d = D/(1 - D)$, where D = duty cycle of the switch. If $D < 0.5$, it acts as buck converter, whereas for $D > 0.5$, it is a boost converter. One disadvantage of the circuit is the polarity reversal at the output. The Cu′k converter shown in part (b) is also a buck-boost converter with output polarity reversal, and it operates on the duality principle of the circuit shown in part (a). The circuit has an extra inductor and a capacitor. The first stage of the circuit is a boost converter because at steady state $V_{C1} = V_d + V_0$ as shown. When switch Q is on, inductor L_1 will be charged by V_d and current i_{L1} will increase. Charged capacitor C_1 will discharge through the load and L_2 thus tends to increase i_{L2} and V_0, and the diode D will remain reverse biased. When Q is off, both i_{L1} and i_{L2} will flow through the diode. Capacitor C_1 will be charged by the source, and i_{L1} and i_{L2} will charge the load. The voltage transfer is given by the same relation mentioned earlier.

FIGURE 4.4 (a) Flyback dc-dc converter and (b) forward dc-dc converter.

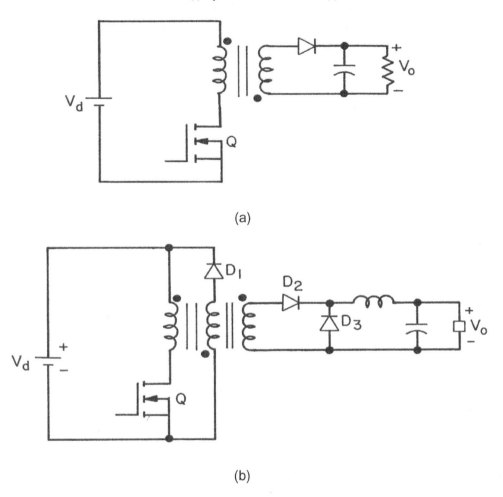

(a)

(b)

The figure shows converters with high-frequency transformer coupling that provides the advantages of electrical isolation as well as voltage level change. In addition, multiple-output power supplies are possible with one-stage power conversion. The isolation and voltage transformation can be provided on the input side with a 50/60-Hz transformer if the primary supply is ac, but the transformer becomes bulky. In the flyback converter shown in part (a), when switch Q is on, the energy is transferred from the source to the self-inductance of the transformer, where it is stored in the form of magnetic energy. Then, when Q is off, the stored energy is released to the output through the diode rectifier. The forward converter shown in part (b) uses a three-winding transformer that has the polarities shown. When the switch is on, the induced output voltage is applied to the load, and at the same time, magnetizing current is built up linearly in the transformer primary self-inductance. When the switch is turned off, the middle winding returns the magnetic energy to the source and, thus, resets the core flux. Diode D_2 remains reverse biased and the load filter current freewheels through D_3. Note that the MOSFET sees double voltage ($2V_d$) because of forward biasing of D_1.

FIGURE 4.5   Single-phase diode rectifier with boost chopper for line power factor control.

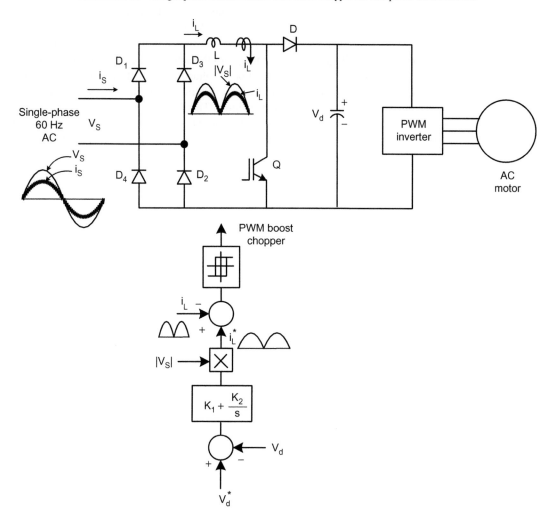

A simple single-phase thyristor bridge rectifier can convert ac to dc and generate regulated dc output by phase control, as discussed in Chapter 3. However, the disadvantages are line harmonics and poor DPF. A diode rectifier with a C or LC filter and a dc-dc converter in cascade can perform the same function, but the line PF still remains poor. The most convenient scheme for ac-to-dc conversion, line PF control, and output voltage regulation is shown in this figure. Basically, it consists of a diode rectifier followed by a boost converter (or chopper) to generate the regulated dc voltage V_d. A PWM inverter with an ac machine is shown as the load. The boost converter has two functions: (1) it regulates the capacitor voltage from the input full-wave rectified dc voltage so that V_d is always greater than the peak line voltage, and (2) it controls the inductor current profile as shown so that the line current is sinusoidal at unity PF. The V_d can be stepped down by another dc-dc converter, if needed. The control principle is shown in the lower part of the figure. The desired output voltage V_d^* is compared with the actual V_d and the error through a P-I controller is multiplied by the absolute supply voltage at the rectifier output to generate the boost converter current command i_L^*. This is compared with actual current i_L within a hysteresis band to generate a gate drive signal. The PWM techniques will be discussed later. Note that in this scheme, power can flow only from the ac to the dc side. The circuit is widely used in spite of additional costs and extra loss in the boost converter.

FIGURE 4.6   Three-phase diode rectifier with boost chopper for line power factor control.

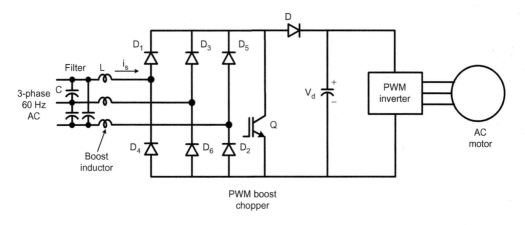

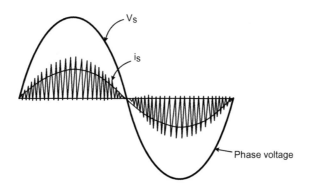

Three single-phase converter units, as discussed in Figure 4.5, can be combined with parallel dc output to interface a three-phase line. However, the circuit shown in this figure is much simpler and more economical. The scheme is somewhat similar to that shown in the previous figure, but the boost converter inductance has been transferred to the line side and distributed equally as phase inductance as shown. The converter switching frequency is constant but its pulse widths are modulated to fabricate the average in-phase sinusoidal current in each phase. When the switch is on, a symmetrical short circuit is created at the rectifier input. Therefore, phase currents build up proportional to the respective voltage magnitude. When the switch is off, the currents are pumped to the capacitor until they fall to zero value, as indicated. The discontinuous phase current pulses at high switching frequency with a sinusoidal locus of the peaks can be filtered with a small LC filter to get approximately sinusoidal line current at unity PF. The inherent leakage inductance in the line is adequate for this filtering. The duty cycle of the converter is modulated to control the line current and to regulate the output voltage V_d. The peaky line current loads the diodes and creates severe EMI problems. Control of the converter is reasonably simple.

FIGURE 4.7   (a) Half-bridge inverter and (b) operation in square-wave mode.

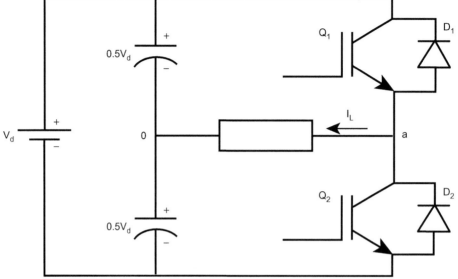

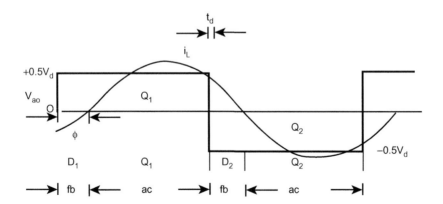

A voltage-fed inverter receives dc voltage at the input and converts to ac at appropriate frequency and voltage. The simplest configuration of inverter is the half-bridge circuit, and its waveforms at square-wave mode operation are shown in this figure. Devices Q_1 and Q_2 switch alternately for 180° to generate square-wave voltage across the load. The load (usually inductive) is connected at the center point of a large split capacitor. There should be a short time gap (t_d) between Q_1 turn-off and Q_2 turn-on to prevent short circuit or "shoot-through" fault. The load current i_L will contain harmonics, but it is assumed to be sinusoidal at a lagging DPF angle of φ. Because of feedback (or freewheeling) diodes, the load current is free to flow in any direction. When load voltage and current have the same polarity, the power flows to the load, but with opposite polarity, power is fed back to the source through the diode. In this case, as shown, average power flows to the load. If DPF angle $\varphi > 90°$, average power will flow from the load to the source, which is defined as the regenerative mode.

FIGURE 4.8   (a) Single-phase H-bridge inverter and (b) load voltage and current waves in square-wave mode.

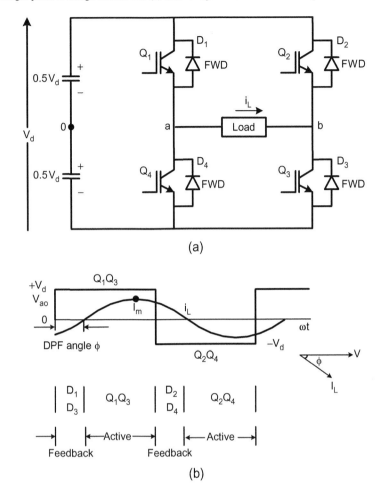

The topology of the single-phase bridge inverter shown in this figure is the same as that for Figure 4.2. Two half-bridges are connected to construct a full or H-bridge inverter. To generate square-wave output voltage, the device pairs Q_1Q_3 and Q_2Q_4 are switched alternately for 180°. Assuming sinusoidal load current (i.e., perfect filtering) at lagging PF angle φ, the load absorbs power when the Q_1Q_3 and Q_2Q_4 pairs are conducting, whereas feedback or regeneration occurs when the diode pairs are conducting, as shown in the figure. In this case, with $\varphi < 90°$, the average power flows to the load. If, on the other hand, $\varphi > 90°$ the average power will flow from the load to source.

FIGURE 4.9 Phase-shift voltage control waves of H-bridge inverter.

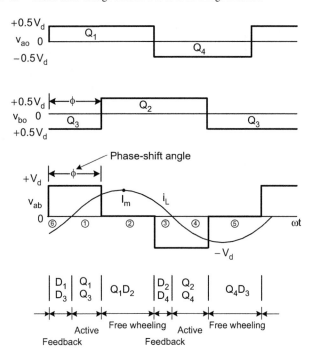

The figure shows a method of controlling the output voltage of Figure 4.8, which is known as *phase-shift control* (also known as *phase-shift PWM*). Both the half-bridges operate in square-wave mode, but the right-side half-bridge operates at phase-shift angle φ as shown in the figure. The output voltage $v_{ab} = v_{a0} - v_{b0}$ can be constructed graphically as indicated. It has a quasi-square-wave shape, where the fundamental voltage can be controlled by the pulse width angle φ. The maximum fundamental voltage is obtained in square-wave mode when $\varphi = 180°$. Note that the phase of the output fundamental voltage also varies by φ angle variation. This phase variation can be eliminated or adjusted to be leading or lagging by distributing the φ angle between the v_{a0} and v_{b0} waves in either direction. The conduction intervals of the devices are summarized in the figure. Besides the active (Q_1Q_3 and Q_2Q_4 conducting) and feedback (D_1D_3 and D_2D_4 conducting) modes, as discussed earlier, there is also a freewheeling mode when the Q_1D_2 and Q_4D_3 pairs are conducting. In this mode, the load is short circuited in the upper side or lower side of the bridge, respectively.

FIGURE 4.10   Spectrum of output voltage with phase-shift control.

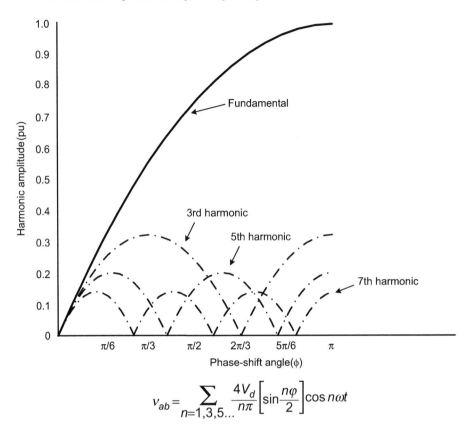

$$v_{ab} = \sum_{n=1,3,5...} \frac{4V_d}{n\pi}\left[\sin\frac{n\varphi}{2}\right]\cos n\omega t$$

The quasi-square wave generated by phase-shift control in Figure 4.9 can be analyzed, and the corresponding Fourier series is shown here, where V_d = dc supply voltage, φ = phase shift angle, and n = odd integer (1, 3, 5, etc.). The fundamental voltage is given by $v_{ab}(f) = 4V_d/\pi \sin \varphi/2$. The frequency spectrum is plotted in this figure. As shown, the fundamental component reaches the maximum value ($4V_d/\pi$) at $\varphi = 180°$, which is defined as 1.0 pu (i.e., the fundamental peak of the square-wave) in this normalized plot. At $\varphi = 60°$, the v_{ab} wave in Figure 4.9 is a traditional six-step wave and its harmonic content can be read from this figure.

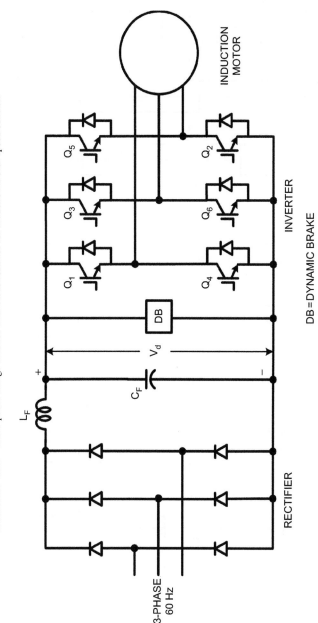

FIGURE 4.11 Three-phase bridge inverter with diode rectifier for induction motor speed control.

DB=DYNAMIC BRAKE

One of the most common converter topologies that is very widely used in industry is shown in this figure. It consists of a three-phase bridge inverter with a three-phase diode rectifier in the front end. The rectifier (which can be single or three-phase) converts ac to uncontrolled dc. The harmonics in the dc link are filtered by an *LC* or *C* filter to generate smooth voltage V_d for the inverter. The inverter consists of three half-bridges or phase legs to generate three-phase ac for industrial motor drives or other applications. For the present, neglect the element DB in the dc link. Instead of generating dc by a rectifier, it can be done with a battery, fuel cell, or photovoltaic dc source. In all such cases, V_d is usually unregulated. The battery-fed inverter drive is commonly used for electric/hybrid vehicle drives. Note that because of the diode rectifier in the front end, the converter system cannot regenerate power. The filter capacitor C_F sinks the harmonics from the rectifier as well as inverter sides.

FIGURE 4.12   Three-phase bridge inverter output voltage waves in square-wave (or six-step) mode.

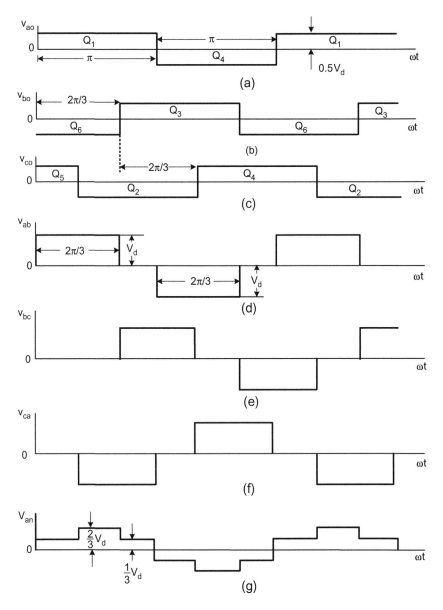

The three-phase inverter of Figure 4.11 can be operated in either the square-wave or PWM mode. The waveforms in square-wave mode are explained in this figure. Three phase legs of the inverter generate square waves at 120° mutual phase-shift angle, where the output phase voltage magnitudes ($\pm0.5V_d$) are shown with respect to the artificial dc link center point. The line voltages v_{ab}, v_{bc}, and v_{ca} are constructed by subtracting the adjacent phase voltages. For an isolated neutral wye-connected load, the phase voltage wave v_{an}, for example, is given by the relation $v_{an}= 2/3\ v_{a0} - 1/3\ v_{b0} - 1/3\ v_{c0}$ due to the absence of triplen (third or multiple of third) harmonics. The line and load phase voltages have characteristic six-step wave shapes with suppression of triplen harmonic voltages. With three-phase balanced load, the line currents are also balanced but may be rich in harmonics. In the square-wave mode, output voltage control is not possible by the inverter and V_d variation reflects to the output.

FIGURE 4.13 Three-phase H-bridge inverter with phase-shift voltage control.

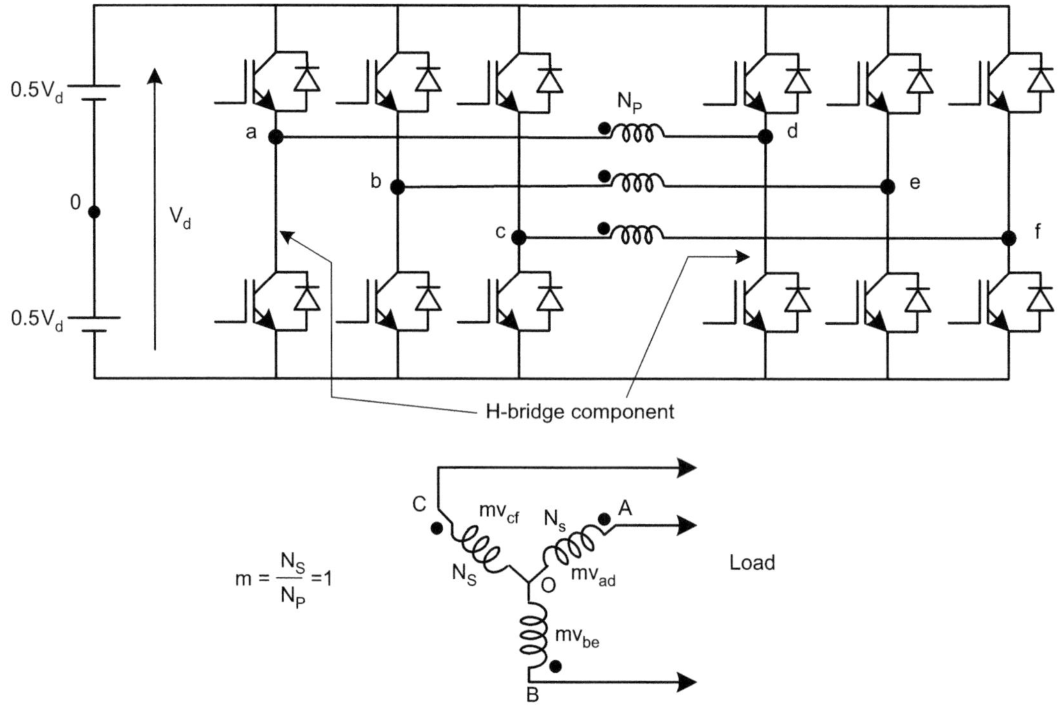

The phase-shift voltage control principle of an H-bridge square-wave inverter was explained in Figure 4.9. The principle can be extended to three-phase inverters as shown in this figure. The scheme uses three component H-bridges for the three phases, but an individual H-bridge operates with phase-shift control as in Figure 4.9. The primary windings of a three-phase transformer are connected as load to each H-bridge. The transformer secondary is connected in wye form as shown, where m is the winding turns ratio.

FIGURE 4.14   Phase-shift control waves of three-phase H-bridge inverter ($\varphi = 150°$).

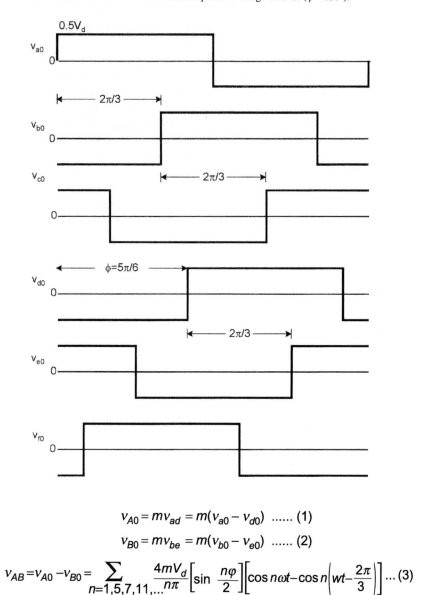

$$v_{A0} = mv_{ad} = m(v_{a0} - v_{d0}) \quad \text{...... (1)}$$

$$v_{B0} = mv_{be} = m(v_{b0} - v_{e0}) \quad \text{...... (2)}$$

$$v_{AB} = v_{A0} - v_{B0} = \sum_{n=1,5,7,11,\dots} \frac{4mV_d}{n\pi}\left[\sin\frac{n\varphi}{2}\right]\left[\cos n\omega t - \cos n\left(wt - \frac{2\pi}{3}\right)\right] \dots (3)$$

The figure explains the waveform synthesis of the three-phase H-bridge inverter shown in Figure 4.13. First, consider the three half bridges on the left. They generate square waves v_{a0}, v_{b0}, and v_{c0}, respectively, with amplitudes $\pm 0.5V_d$ with respect to the dc center point, and the three waves are at 120° phase-shift angle as shown. Next, consider the three half bridges on the right. They also generate similar waves v_{d0}, v_{e0}, and v_{f0} at 120° phase-shift angle, but the right group of waves is phase shifted by φ angle (lagging) with respect to the left group, as shown in the figure. As a result, the transformer primary windings receive the voltages v_{ad}, v_{be}, and v_{cf}, respectively, which are modulated by the φ angle as indicated in Figure 4.9. The corresponding transformer secondary phase voltages, for example, v_{A0}, v_{B0}, and the line voltage v_{AB} can be derived by the equations shown with the figure. The fundamental component of these voltages can be varied by φ angle control. Note that although the transformer primary and secondary phase voltages contain triplen harmonics, the line voltages (v_{AB}, v_{BC}, v_{CA}) have six steps because of the cancellation of triplen harmonics. It is also important to note that the inverter devices switch at line frequency (60/50 Hz); that is, there are altogether six switches per cycle.

FIGURE 4.15 Eighteen-step GTO converter for utility battery peaking service.

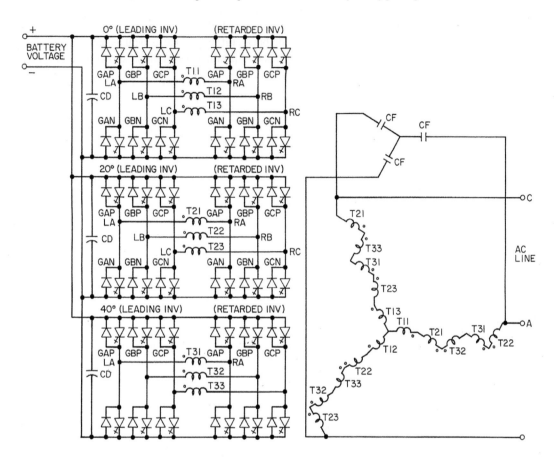

GE installed this 10-MVA GTO converter system [7] for Southern California Edison's electric grid with the object of battery storage of off-peak grid power. At night, the 60-Hz utility system has surplus power, which is converted to dc by the converter system (acting as rectifier) and stored in the lead-acid battery bank. In the day or evening time, when grid power demand increases beyond the generation capacity, the battery energy is converted back to ac (acting as an inverter) to supply power to the grid to meet the demand. The converter system can also act as static VAR compensator or static synchronous compensator (SVC, SVG, or STATCOM) (leading or lagging) that can regulate the system voltage and stabilize the power system. SVC will be described later. The 18-step converter system uses the phase-shift voltage control principle described in Figures 4.13 and 4.14. The system fabricates an 18-step output voltage wave by using three groups of H-bridges, where the second and third groups are at 20° and 40° phase lead, respectively, with respect to the first group. The second and third groups have two secondary transformer windings, and the three-phase transformer secondary connections with a total of 15 segments are as shown in the figure. The harmonics (17th, 19th, etc.) in the 18-step wave are filtered by the capacitor bank C_F. The ac side back emf magnitude as well as phase angle can be controlled by phase shift to control active power flow in either direction and reactive power at the output.

FIGURE 4.16 Features of GTO converter system for battery peaking service.

- 10-MW CAPACITY LEAD-ACID BATTERY STORAGE INSTALLED BY GE FOR SOUTHERN CALIFORNIA EDISON ELECTRIC GRID (1988)
- STORES ENERGY IN OFF-PEAK HOURS AND DELIVERS DURING PEAK DEMAND
- CAN OPERATE AS STATIC VAR COMPENSATOR ON GRID
- CAN CONTROL GRID VOLTAGE AND FREQUENCY
- CAN IMPROVE SYSTEM STABILITY
- THREE-PHASE 60-Hz VOLTAGE MAGNITUDE AND PHASE ANGLE CONTROL BY THE H-BRIDGES
- 60-Hz TRANSFORMER PERMITS COUPLING OF THE PHASE-SHIFTED H-BRIDGES, VOLTAGE BOOST, AND ISOLATION
- GTO SWITCHING FREQUENCY IS LOW AT 60 Hz
- HIGH CONVERTER EFFICIENCY (97%)

The salient features of the GTO converter system of Figure 4.15 are summarized here. The 18-step fabricated line voltage wave is characterized by $18n \pm 1$ (n = integer) order of harmonics at the output (i.e., 17th, 19th, 35th, 37th, etc.), which is not affected by phase-shift voltage and phase angle control. The advantage of phase-shift voltage control over the traditional PWM method (which will be described later) is that the GTO switching frequency is low (60 Hz) and the corresponding switching loss is low. This scheme is particularly important for GTOs that are characterized by high switching losses. In the present scheme, two GTOs are connected in series with an R-C-D snubber (not shown) for each to give GTO failure redundancy. The bulky 60-Hz transformer is definitely a disadvantage of the present system.

FIGURE 4.17   Voltage-fed inverter PWM techniques.

- SINUSOIDAL PWM (SPWM)
- SELECTED HARMONIC ELIMINATION (SHE)
- SPACE VECTOR PWM (SVM OR SVPWM)
- RANDOM PWM (RPWM)
- HYSTERESIS BAND (HB) CURRENT CONTROL
- DELTA MODULATION
- SIGMA DELTA MODULATION

The figure summarizes the principal pulse width modulation (PWM) techniques in voltage-fed inverters. The duty cycle modulated PWM principle was introduced earlier when discussing dc-dc converters. The inverter's fundamental voltage can be controlled and the harmonics can be attenuated (or eliminated) by creating selective notches in the square-wave output. Of course, square-wave inverter voltage control is possible by means of a thyristor rectifier located at the front end or a dc-dc converter at the input; output can be filtered by bulky passive filters. Phase-shift control of square-wave inverters also permits voltage control as well as lower order harmonics cancellation. However, electronic switching of voltage control as well as harmonic elimination is much more efficient and cost effective. However, this increases the device switching frequency, which in turn increases the switching loss. Inverter voltage and frequency control are essential for variable-speed ac drives, whereas voltage control at constant frequency is required for UPS (uninterruptible power supply) systems, which will be described later. In this chapter, we discuss only the important PWM techniques generally used in industry. Comparison of principal PWM techniques is also included.

FIGURE 4.18   Sinusoidal PWM wave fabrication of H-bridge inverter (bipolar switching).

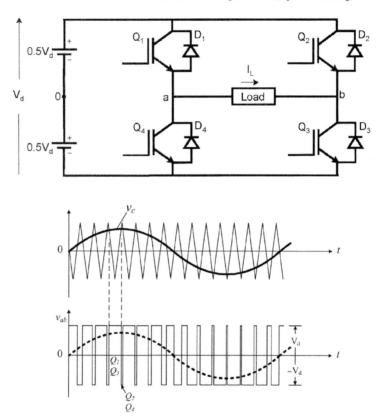

The figure explains the sinusoidal PWM principle for output voltage control and harmonic attenuation of a single-phase H-bridge inverter. Basically, a triangular carrier waveform at the desired switching frequency (f_c) is generated and compared with the sinusoidal command or modulating voltage wave (v_c) as indicated in the upper wave. Device pair Q_1Q_3 is turned on when the modulating voltage exceeds the carrier wave amplitude, whereas pair Q_2Q_4 is on when the carrier amplitude is higher. The resulting PWM output voltage and its fundamental component are shown in the lower part of the figure. A short time gap (t_d) is to be maintained between the switchings to prevent shoot-through fault. The scheme is called *bipolar* because a pair of devices (one from each pole) switches together to generate a bipolar voltage $(\pm V_d)$ wave. Like a square-wave inverter (Figure 4.8), the circuit has two modes of operation: active mode and feedback mode. The PWM wave contains only carrier frequency related harmonics, which are easily filtered by nominal filter inductance. It can be shown easily that the output fundamental frequency is the same as the modulating frequency, and the fundamental amplitude varies linearly with that of the modulating wave until the peaks of the modulating wave and carrier wave are equal.

FIGURE 4.19   Sinusoidal PWM wave fabrication of H-bridge inverter (unipolar switching).

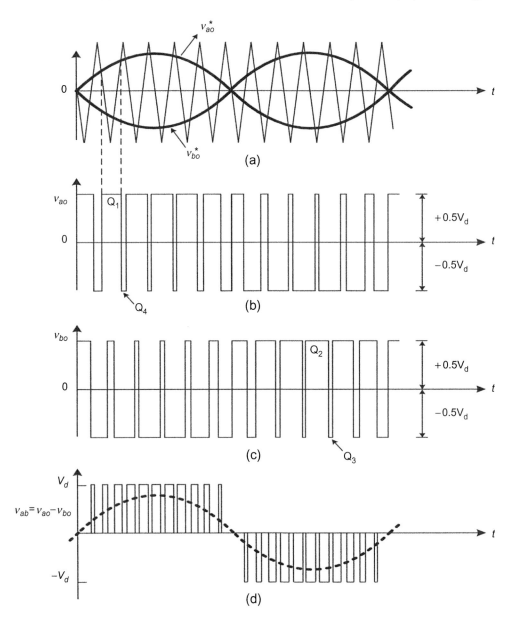

In this method of SPWM, the two legs of the H-bridge inverter are switched separately, as explained in this figure. The modulating signal v_{a0}^* for the left leg is compared with the triangular carrier wave to generate the actual v_{a0} wave shown in part (b). This wave has the same shape as v_{ab} in Figure 4.18, except it has an amplitude variation of $\pm 0.5V_d$ with respect to the dc center point. The modulating signal v_{b0}^* for the right leg is at $180°$ phase angle with v_{a0}^*, and the actual v_{bo} wave is shown in part (c). The output voltage is given by $v_{ab} = v_{a0} - v_{bo}$ as indicated in part (d). The scheme is called *unipolar* because the v_{ab} wave varies between 0 and $+V_d$ in the positive half cycle and 0 and $-V_d$ in the negative half cycle unlike the previous case. Note that when v_{ab} wave has zero gap in the positive half-cycle, Q_1 and Q_2 are closed so that bipolar load current can freewheel in the upper side, whereas for zero gap in the negative half cycle, Q_3 and Q_4 are closed, permitting freewheeling in the lower side. The circuit has three modes of operation (active, freewheeling, and feedback) as in Figure 4.9. The advantages of unipolar switching are lower output harmonics and less switching stress ($\pm 0.5V_d$ jump) on devices.

FIGURE 4.20   Sinusoidal PWM waves of a three-phase bridge inverter.

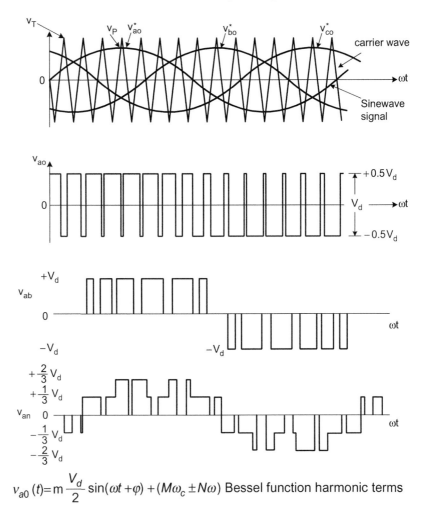

$$v_{a0}(t) = m\frac{V_d}{2}\sin(\omega t + \varphi) + (M\omega_c \pm N\omega) \text{ Bessel function harmonic terms}$$

The principle of unipolar SPWM switching in an H-bridge inverter as discussed before can be extended to a three-phase bridge inverter. The fabrication of the v_{ao} wave from the sinusoidal modulating wave v_{ao}^* is shown in the figure. Similarly, v_{bo} and v_{co} waves can also be fabricated at 120° phase difference from the respective modulating waves v_{bo}^* and v_{co}^* (not shown). The line voltage wave v_{ab}, for example, is given as $v_{ab} = v_{a0} - v_{b0}$, and the corresponding line-to-neutral voltage wave (assuming wye-connected load), for example, is given by $v_{an} = 2/3\, v_{a0} - 1/3\, v_{bo} - 1/3\, v_{co}$, as discussed before. The line voltage has characteristic three levels and phase voltage has characteristic five levels in the full cycle. The Fourier series of the v_{a0} wave is very complex and it contains the fundamental voltage component and carrier frequency related harmonics with modulated wave related sidebands as shown. The modulation index m defined as V_P/V_T, as indicated in the figure. In the undermodulation range $(V_P < V_T)$, the inverter transfer characteristic is linear. This range can be extended by mixing third or triplen harmonics with the signal. Sinusoidal PWM (also called triangulation or the subharmonic or suboscillation method) is most commonly used in industry. In high-power inverters, the switching frequency is low because of large switching losses. Therefore, to avoid subharmonics at low switching frequency, the modulating wave requires synchronization with the modulating wave.

FIGURE 4.21 Sinusoidal PWM overmodulation.

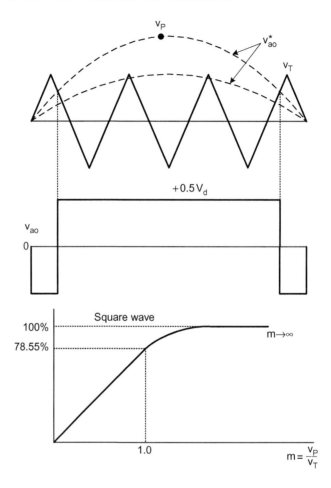

As the modulation index *m* approaches 1.0, the undermodulation region tends to saturate at 78.55% of fundamental voltage (100% corresponds to a square wave). As the modulating wave exceeds the peak of the carrier wave, PWM overmodulation (*m* > 1.0) starts, when some pulses and notches tend to disappear near the middle of the v_{a0} wave as shown in the figure. For any PWM wave, a minimum pulse and notch widths are to be maintained for successful commutation. Therefore, if the minimum pulse and notch widths are reasonably large, a voltage jump will occur and a corresponding current surge will be seen at this transition. This is a special problem in low-switching-frequency, high-power inverters, particularly those with slow switching devices. Modulation index *m* can be increased to a high value until the square-wave mode is reached, as indicated above, but the disadvantages are nonlinear voltage transfer characteristics and fundamental frequency related harmonics at the output, that is difficult to filter.

FIGURE 4.22   PWM inverter dead time effect on output voltage wave.

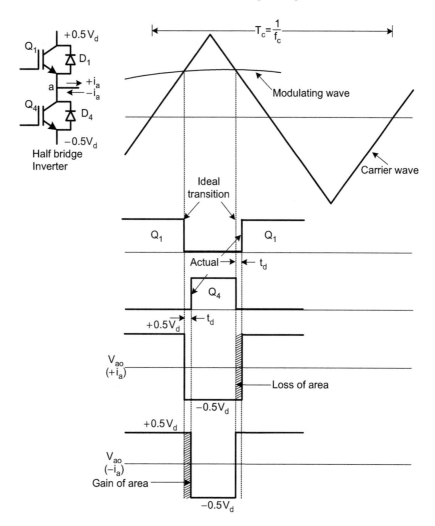

As mentioned before, a dead or lock-out time (t_d) should be provided between the switching of the devices in a leg of the inverter to prevent shoot-through or a short-circuit fault. This figure explains the effect of dead time in a half-bridge PWM inverter. The ideal PWM switching points of Q_1 and Q_4 and their actual switching after the dead time (t_d) are shown in the figure. For positive polarity of load current i_a, there is no effect at the leading edge, but at the trailing edge, delayed turn-on of Q_1 causes load current transfer from D_4 to Q_1 with a delay causing a volt-second area loss as shown. If, on the other hand, the current polarity is negative, D_1 is conducting initially when Q_1 is on. This current continues to flow in D_1 when Q_1 and Q_4 are off, thus causing a gain of volt-second area. On the trailing edge, when Q_4 turns off at the ideal point, the current is transferred to positive bus immediately without causing any effect. For an H-bridge or three-phase inverter, there is a similar dead time effect on all the phase legs.

FIGURE 4.23 (a) Dead time effect on output phase voltage wave and (b) compensation principle.

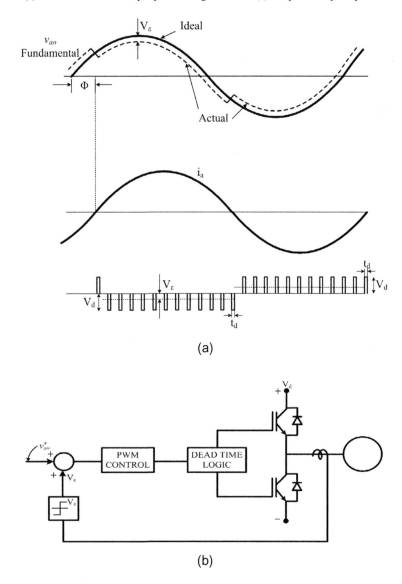

(a)

(b)

The dead time in a PWM inverter has two harmful effects: (1) loss of fundamental voltage and (2) low-frequency harmonic distortion. Consider the fundamental phase voltage wave v_{a0} and the corresponding phase current wave i_a at lagging phase angle φ in the upper part of the figure. In the positive half cycle of current, there is an incremental loss of voltage for every switching cycle, but it does not depend on current magnitude. Similarly, there is an incremental gain of voltage in the negative half cycle of current. The average value V_ε of these increments during the half cycles of current is superposed on the v_{a0} wave, which indicates that the fundamental voltage is slightly decreased. The decrease depends on the switching frequency, delay time, dc voltage, and phase angle. Evidently, there will be distortion, which is related to the harmonics of fundamental frequency. At lower fundamental frequency, this distortion will be more serious. The dead time effect can be compensated easily by current or voltage feedback method [8]. Figure 4.23(b) shows the current control method. The load phase current wave polarity is sensed, and a fixed amount of compensating bias voltage V_ε is added with the modulating voltage wave v_{a0}^* to generate the actual modulating signal. Similar compensation is made in all phases.

FIGURE 4.24 (a) Selected harmonic elimination PWM and (b) notch angle relation with fundamental voltage with α_1, α_2, and α_3 angles only.

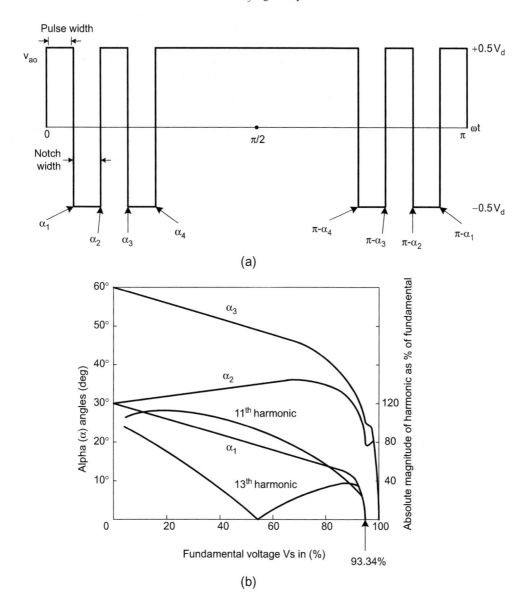

The lower order harmonics from a square wave can be selectively eliminated and fundamental voltage can be controlled as well by creating notch angles as shown in part (a) of this figure. The positive half cycle of voltage wave shown has quarter-cycle symmetry, that is, the negative half cycle has the inverted form of the positive half cycle. It can be shown that the four notch angles α_1, α_2, α_3, and α_4 can be controlled to eliminate three significant lower order harmonics (5th, 7th, and 11th) and control the fundamental voltage of v_{a0} wave. For an isolated neutral load, elimination of triplen harmonics is not necessary. Figure 4.24(b), for example, shows the plot of α angles (α_1, α_2, and α_3) with variable fundamental voltage with elimination of the 5th and 7th harmonics. Note that only 93.34% of the fundamental in the square-wave can be controlled. As a result of this control, the lowest significant harmonics (11th and 13th) are boosted, but being of higher order, their effect may not be that harmful. At low frequency, a large number of α angles can be created to eliminate a large number of lower order harmonics. A disadvantage of the SHE scheme is that a DSP or microcomputer, which should store a large look-up table of angles, is necessary for implementation.

FIGURE 4.25 (a) Summary of three-phase two-level inverter switching states and (b) synthesis of inverter voltage vector $\overline{V}_2$ (110).

state	On devices	V_{an}	V_{bn}	V_{cn}	Space voltage vector
0	$Q_4Q_6Q_2$	0	0	0	$\overline{V}_0(000)$
1	$Q_1Q_6Q_2$	$2V_d/3$	$-V_d/3$	$-V_d/3$	$\overline{V}_1(100)$
2	$Q_1Q_3Q_2$	$V_d/3$	$V_d/3$	$-2V_d/3$	$V_2(110)$
3	$Q_4Q_3Q_2$		.		$\overline{V}_3(010)$
4	$Q_4Q_3Q_5$		.		$\overline{V}_4(011)$
5	$Q_4Q_6Q_5$		.		$\overline{V}_5(001)$
6	$Q_1Q_6Q_5$		.		$\overline{V}_6(101)$
7	$Q_1Q_3Q_5$	0	0	0	$\overline{V}_7(111)$

(a)

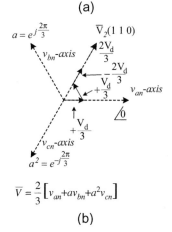

$$\overline{V} = \frac{2}{3}\left[v_{an} + a v_{bn} + a^2 v_{cn}\right]$$

(b)

Space vector PWM (or SVM) is somewhat complex and requires understanding of ac machine space vector or d-q theory, which will be discussed in Chapter 6. This type of PWM gives its best performance for isolated neutral loads. Consider a three-phase bridge inverter (Figure 4.11) supplying power to an isolated neutral induction motor load. The inverter has $2^3 = 8$ switching states, which are summarized in Figure 4.25(a). Consider, for example, state 2 when devices Q_1, Q_3, and Q_2 are closed. In this state, phases a and b are connected to a positive dc bus and c is connected to a negative bus. The corresponding phase voltage magnitudes, shown in the figure (n is the load neutral), can be easily solved. Note that states 0 and 7 correspond to short circuits on the negative and positive buses, respectively. For each inverter state, a space voltage vector can be constructed from the phase voltages using the equation and part (b) of the figure. For example, it illustrates a synthesis of vector $\overline{V}_2$ (110), which has a magnitude of 2/3 V_d (V_d = dc link voltage) and is oriented at a 60° angle.

FIGURE 4.26 (a) Switching state voltage vectors of three-phase bridge (vector symbol omitted) and (b) fabrication of voltage waves in T_s interval.

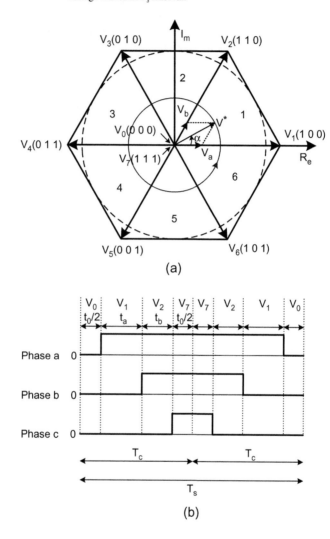

(a)

(b)

The figure explains the SVM principle in the linear or undermodulation region [4] and construction of three-phase PWM waves. In Figure 4.25, all the eight space vectors can be drawn and a hexagon can be constructed with the boundaries, as shown in the figure. The three-phase sinusoidal command voltages for the inverter can be represented by the voltage vector V^* rotating in a counterclockwise direction. For a certain location of V^*, the three nearest inverter vectors can be used on a time-sharing basis so that the average value matches with the reference V^* value, that is, $V^*T_c = V_1 t_a + V_2 t_b + (V_0 \text{ or } V_7)t_0,$

where $T_c = t_a + t_b + t_0$ and $T_s = 2T_c$. The synthesis of symmetrical PWM waves indicates that the inverter states V_0, V_1, V_2, and V_7 have been used for $t_0/2$, t_a, t_b, and $t_0/2$ intervals, respectively, and the sequence has been reversed in the next T_c interval. The under-modulation region is valid until the circle touches the hexagon as shown. At this condition, modulation index m' (note that m' is different from m) with respect to the square wave is 0.907.

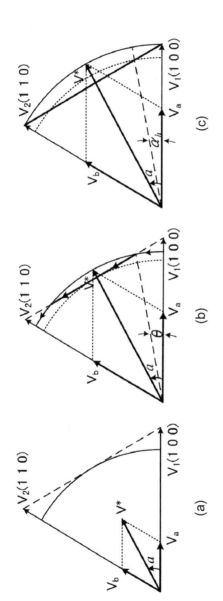

FIGURE 4.27 SVM operation regions: (a) undermodulation mode $(0 < m' < 0.907)$, (b) overmodulation mode -1 $(0.907 < m' < 0.952)$, and (c) overmodulation mode -2 $(0.952 < m' < 1.0)$

The figure explains SVM operation in the overmodulation region [9] and compares it to that in the undermodulation region. Operation in the overmodulation region is important if a higher load voltage beyond the linear or under-modulation mode is required at the cost of lower harmonic quality in the wave. Because the voltage transfer gain is lower in overmodulation (Figure 4.21), a method to linearize the gain is included in the figure. Linear gain is desirable from a system performance viewpoint. Operation in only sector 1 (or A) of the hexagon is explained because operation in other sectors is similar. The command or reference vector V^* in the undermodulation region increases in radius and in the limit describes an inscribed circle in the hexagon, as indicated in part (a). If V^* exceeds this limit, the operation enters into mode 1 of overmodulation, as shown in part (b). Voltage V^* crosses the hexagon boundary at two points and, as a result, there will be a loss of proportional fundamental voltage. To compensate for this loss (i.e., to maintain the linear relationship), the reference voltage is considered with a higher radius that crosses the hexagon at angle θ shown in the figure. The operation is now partly linear on a circular trajectory and partly nonlinear on a hexagonal trajectory. During hexagon tracking, the zero vector vanishes. Mode 1 ends when angle θ is zero and the trajectory is entirely on the hexagon. Then, in mode 2, the command voltage is held at the corner state for angle α_h partly, and partly it tracks the hexagon until square-wave operation is reached at the end. At the square wave, V^* is held for 60° at each hexagon corner (i.e., $\alpha_h = 30°$) to describe the square wave shown in Figure 4.26. The plot of PWM waves is the same as that shown in Figure 4.12, except interval t_0 vanishes.

FIGURE 4.28 SVM operation in overmodulation modes: (a) plot of θ–m' relation in mode 1 and (b) plot of α_h–m' relation in mode 2.

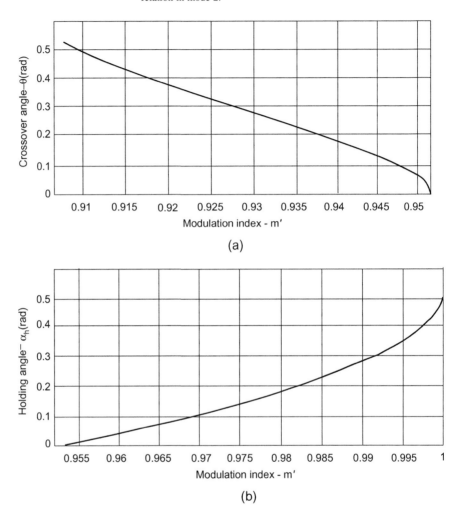

(a)

(b)

The crossover angle θ in overmodulation mode 1, as shown in Figure 4.27, can be calculated as a function of modulation factor m' and plotted in part (a) of this figure. It starts at 30° when $m' = 0.907$ and vanishes at the upper edge when $m' = 0.952$. Similarly, the holding angle α_h profile, which starts at zero and ends at 30° at $m' = 1.0$ (square wave), is shown in part (b). These figures are required for DSP implementation of SVM covering the overmodulation region. The general sequence for SVM implementation is: sense V^* – identify sector – calculate m' – identify mode – calculate time segments of inverter states – synthesize PWM waves.

FIGURE 4.29   (a) Hysteresis band (HB) current control principle of half-bridge inverter and (b) current control block diagram.

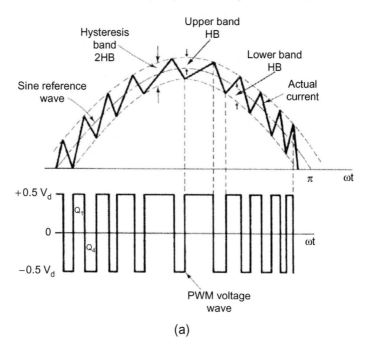

(a)

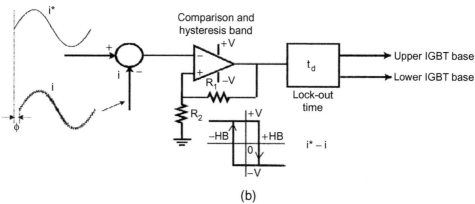

(b)

So far, voltage control PWM methods have been described. Often, systems such as drives require current control. Of course, feedback current control can be provided over the voltage PWM implementation. One of the simplest current control PWM techniques is the hysteresis band (HB) control described in this figure. Basically, it is an instantaneous feedback current control method in which the actual current continuously tracks the command current within a preassigned hysteresis band. As indicated in the figure, if the actual current exceeds the HB, the upper device of the half-bridge is turned off and the lower device is turned on. As the current decays and crosses the lower band, the lower device is turned off and the upper device is turned on. If the HB is reduced, the harmonic quality of the wave will improve, but the switching frequency will increase, which will in turn cause higher switching losses. The PWM implementation is very simple, and the analog control block diagram is shown in part (b). Basically, the current loop error signal generates the PWM voltage wave through a comparator with a hysteresis band. Although the technique is simple, control is very fast, and device current is directly limited, the disadvantages are a harmonically nonoptimum waveform and slight phase lag that increases with frequency. There is, of course, additional distortion in the isolated neutral three-phase load.

FIGURE 4.30 Comparison of SPWM-SVM-SHE-HB PWM techniques.

SPWM

- CARRIER-BASED OPEN LOOP – ASYNCHRONOUS OR SYNCHRONOUS CARRIER
- LOW UNDERMODULATION RANGE (0 < m' < 0.7855) WHERE m' CAN BE INCREASED TO 0.907 BY MIXING TRIPLEN HARMONICS WITH MODULATING WAVE)
- NONLINEAR CHARACTERISTICS IN OVERMODULATION RANGE
- SMOOTH OVERMODULATION UP TO SQUARE-WAVE
- LOWEST HARMONIC RIPPLE FOR 0 < m' < 0.4
- LARGE f_s BASED RIPPLE AT OVERMODULATION
- DC LINK VOLTAGE RIPPLE INTRODUCES ADDITIONAL OUTPUT RIPPLE
- SIMPLE IMPLEMENTATION

SVM

- INDIRECTLY CARRIER-BASED OPEN LOOP – ASYNCHRONOUS OR SYNCHRONOUS CARRIER
- GOOD FOR ISOLATED NEUTRAL 3-PHASE OUTPUT
- LARGE UNDERMODULATION RANGE (0 < m' < 0.907)
- EASY LINEARIZATION IN OVERMODULATION UP TO SQUARE-WAVE
- LOWEST HARMONIC RIPPLE IN UNDERMODULATION RANGE
- LARGE f_s BASED RIPPLE AT OVERMODULATION
- DC LINK VOLTAGE RIPPLE INTRODUCES ADDITIONAL OUTPUT RIPPLE
- COMPLEX COMPUTATION INTENSIVE - NEEDS MICROCOMPUTER/DSP
- CAN NOT BE APPLIED WITH HIGH CARRIER FREQUENCY

SHE

- NO CARRIER BASE - NUMBER OF NOTCHES DETERMINES SWITCHING FREQUENCY
- DIFFICULT TO APPLY AT LOW FREQUENCY
- OUTPUT MAY NOT BE HARMONICALLY OPTIMUM
- MOST USEFUL WHEN SPECIFIC ORDER OF HARMONICS IS HARMFUL
- EASY LINEARIZATION IN WHOLE MODULATION RANGE
- DC LINK VOLTAGE RIPPLE INTRODUCES ADDITIONAL OUTPUT RIPPLE
- MICROCOMPUTER/DSP BASED LOOK-UP TABLE IMPLEMENTATION

HB

- EASY OPERATION FROM ZERO FREQUENCY
- NEEDS CLOSE LOOP CURRENT CONTROL
- DC LINK RIPPLE IS COMPENSATED – PERMITS LOWER C_F IN DC LINK
- FAST TRANSIENT RESPONSE
- SWITCHING FREQUENCY VARIES
- SMOOTH TRANSITION FROM UNDERMODULATION TO OVERMODULATION
- NON-OPTIMUM HARMONIC RIPPLE
- FREQUENCY-SENSITIVE PHASE LAG OF FUNDAMENTAL CURRENT
- VERY SIMPLE TO IMPLEMENT

FIGURE 4.31   MAGLEV linear synchronous motor drive system.

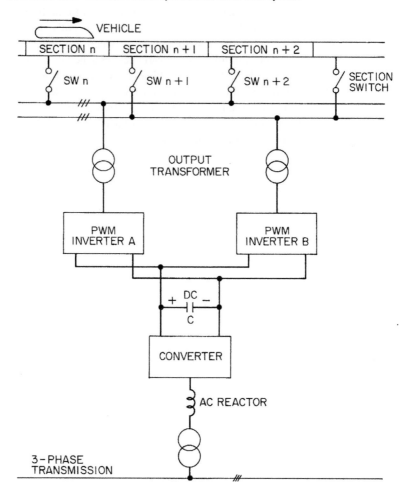

Japan has recently developed and is now testing a magnetically levitated (MAGLEV) high-speed (361-mph) railway transportation system [10, 11] to be installed for 320 miles between Tokyo and Osaka. The block diagram for this system is shown in this figure. In a MAGLEV system, the moving vehicle is suspended in the air by magnetic force to eliminate track friction. The vehicle has a linear synchronous motor (LSM) drive in which the three-phase stator winding armature coils are placed on the track in a section-alized manner, and liquid helium–cooled superconducting magnets for levitation and machine excitation are mounted on the vehicle. The power from a three-phase trans-mission line (154 kV) is stepped down in voltage (66 kV) and then converted to dc (9 kV) by two units of 24-pulse thyristor converters (69 MW) that incorporate phase-shifting step-down transformers. A higher harmonics filter and reactive power compensator are installed on the transmission line, but not shown in the figure. The output dc is filtered by a capacitor and fed to two PWM inverters (recently, a third inverter and feeder have been added to supplement the system), which are described in the next figure. The inverters supply variable-voltage variable-frequency power to the track winding through distribution feeders. When the vehicle moves in a section, that particular section is kept energized through the switch, as shown in the figure, keeping the other sections deenergized. This loads the inverters alternately. In fact, the sections on both sides of the track are laid out with a half section overlap so that even if one of the inverter units fails, the other two continue to operate the vehicle.

FIGURE 4.32 Forty-MVA GTO inverter unit for MAGLEV-LSM drive system.

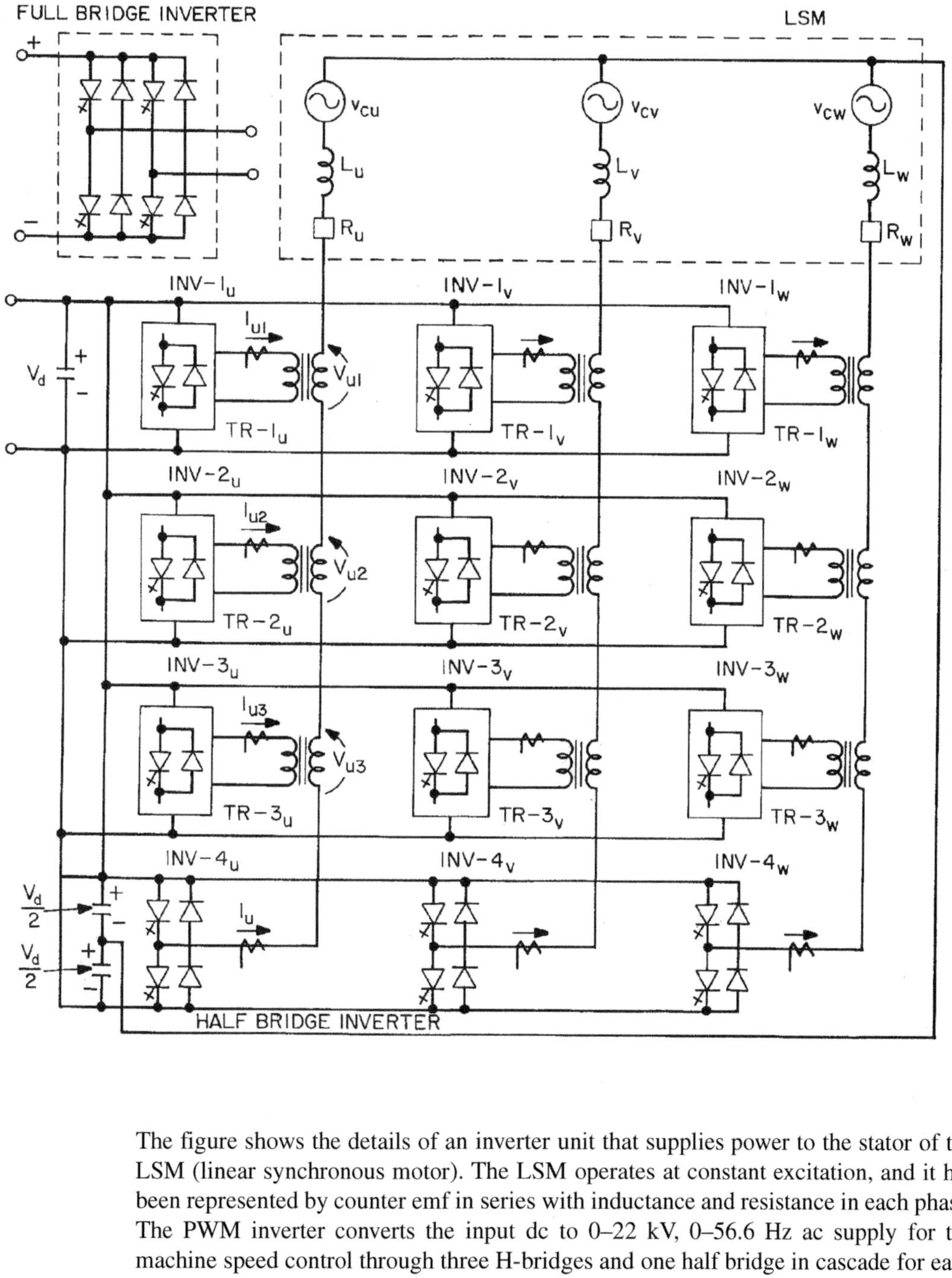

The figure shows the details of an inverter unit that supplies power to the stator of the LSM (linear synchronous motor). The LSM operates at constant excitation, and it has been represented by counter emf in series with inductance and resistance in each phase. The PWM inverter converts the input dc to 0–22 kV, 0–56.6 Hz ac supply for the machine speed control through three H-bridges and one half bridge in cascade for each phase, as shown in the figure. The H-bridge outputs are connected in series by transformers to boost the phase voltage. Since the machine supply includes very low frequency (including dc), which is not possible to handle through transformer, a half-bridge converter output is directly connected to the phase to satisfy this requirement. The transformer dc saturation is avoided by the inverter output voltage. Each inverter bridge operates with sinusoidal PWM at 500-Hz carrier frequency in the unipolar mode, for which the carrier wave of each leg in the H-bridge is phase shifted by 180°. The triangular carrier waves for the four bridges are again mutually phase shifted by 45° so that the equivalent PWM frequency for output current is 4 kHz. When the LSM operates in braking mode, the regenerative energy is not recovered in the source. It can be absorbed in a parallel machine operating in the system, or dissipated in a chopper-controlled resistance (not shown) to control the dc bus voltage.

FIGURE 4.33 (a) Three-level diode-clamped inverter and (b) typical phase voltage wave.

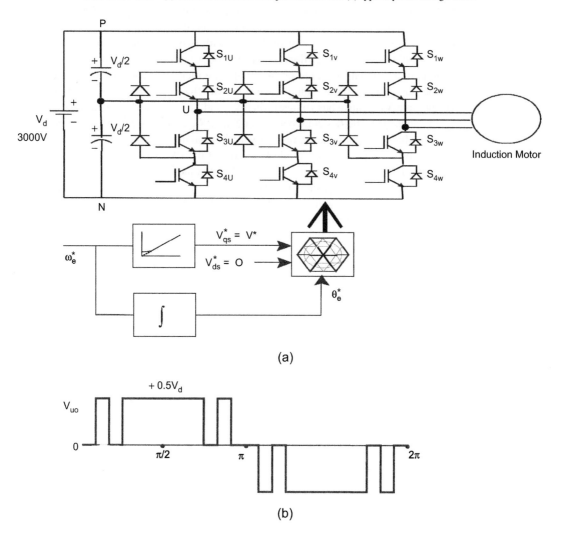

(a)

(b)

So far, we have discussed two-level inverters where the phase voltage has only two levels ($\pm 0.5 V_d$). For high power levels (multi-megawatts), the devices can be connected in series and/or parallel, but the problem of matching arises. Of course, the multistepped inverter discussed previously can be a solution, but it requires a bulky transformer. The problems may be solved by means of multilevel (more than two levels) inverters. Generally, there are three types of multilevel inverters: the diode-clamped type, flying capacitor type, and cascaded H-bridge type. This figure shows the most commonly used three-level diode-clamped inverter, which is also known as a neutral-point clamped (NPC) inverter. The other types will be discussed later. Although, IGBT is used here, GTO or IGCT also can be used. In the figure, the dc link capacitor has been split to create the neutral point O. A pair of devices with bypass diodes is connected in series with an additional clamping diode connected between the neutral point and the center of the pair, as shown. All three phase groups are identical. The phase U, for example, gets the state P (positive bus voltage) when the switches S_{1U} and S_{2U} are closed, whereas it gets the state N (negative bus voltage) when S_{3U} and S_{4U} are closed. At neutral-point clamping, the phase gets the O state when either S_{2U} or S_{3U} conducts depending on positive or negative phase current polarity, respectively. For balancing of neutral-point voltage, the average current injected at O should be zero. All the PWM techniques discussed so far can be used, but we will emphasize the SVM technique in the following pages. A typical PWM phase voltage wave, shown in part (b), indicates the three voltage levels. It can be shown that each device withstands $0.5V_d$ voltage in the clamping mode. Otherwise, it shares statically the full voltage V_d. But the voltage step across the device never exceeds $0.5V_d$. Multilevel inverters have the advantages of easy voltage sharing of devices, lower dv/dt, and improved PWM quality, but the disadvantage is the difficulty of neutral-point voltage balancing and more numbers of devices.

FIGURE 4.34 (a)Three-level inverter switching states: (b) space vector diagram showing the switching states and (c) sector A space vectors indicating switching times.

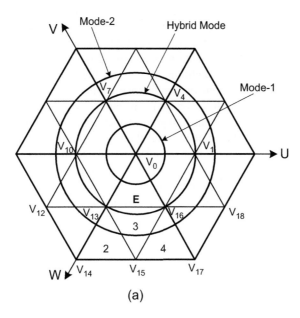

(a)

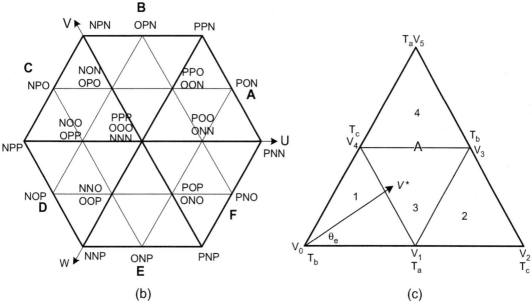

(b)

(c)

A three-level inverter has $3^3 = 27$ switching states, because a leg can have the P, N, or O state. These states and the corresponding space vectors generated by each state are summarized in part (a) of this figure [12]. The large hexagon corner states (*PNN, PPN, NPN, NPP, NNP,* and *PNP*) in part (b) correspond to those of a two-level inverter, where each vector magnitude is $2/3V_d$. There are three zero states (*PPP, OOO,* and *NNN*), and each space vector of the inner hexagon has two possible switching states, as indicated in the figure. There are, in addition, six vectors that correspond to the middle of the large hexagon sides. The sinusoidal reference voltage $V*$ describes a circle that should be limited within the large hexagon for the linear or undermodulation mode of operation. There are six sectors (*A–F*) in the hexagon of which sector A is shown in part (c) separately. For the location of $V*$ shown, the nearest three state vectors should be selected for computing the time segments so that the average voltage matches with that of $V*$. These states can be *POO/ONN, PON,* and *PPO/OON*. Note that neutral current can flow in any state that includes O except the zero states. The positive current (outgoing) lowers the neutral voltage, whereas the negative current (incoming) raises it. The manipulation of redundant inner states helps neutral-point voltage balancing. As Figure 4.34(c) indicates, the switching vectors V_1, V_3, and V_4 should be selected for intervals T_a, T_b, and T_c, respectively, to synthesize $V*$ voltage.

FIGURE 4.35 Waveforms showing the switching sequence in undermodulation (sector A): region 1, (b) region 2, (c) region 3, and (d) region 4.

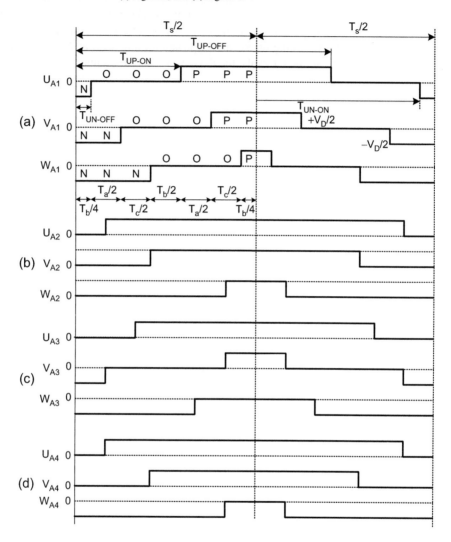

The figure shows the PWM wave synthesis within the sampling period (T_s) in all four regions of sector A shown in Figure 4.34(c). The waveforms are symmetrical about the center line, and the device switching frequency is the same as the sampling frequency. The region 1 waves for the three phases are shown in detail for clarity. For vector V^* location in region 1, according to Figure 4.34, the states *POO/ONN* should be for T_a, *PPP/OOO/ NNN* should be for T_b, and *PPO/OON* should be for T_c, respectively. All the states have been utilized and appropriately distributed within the first $T_s/2$ interval considering the neutral-point voltage balancing, and the sequence is repeated in inverse order for the next $T_s/2$ interval for the waveform symmetry. The waveform for phase U (U_{A1}), for example, is fabricated by turning on device S_{1U} at time $T_{UP\text{-}ON}$ and turning it off at $T_{UP\text{-}OFF}$, and similarly, turning on S_{4U} at $T_{UN\text{-}ON}$ and turning it off at $T_{UN\text{-}OFF}$, as shown in the figure. During all the intervals, the devices S_{2U} and S_{3U} remain on.

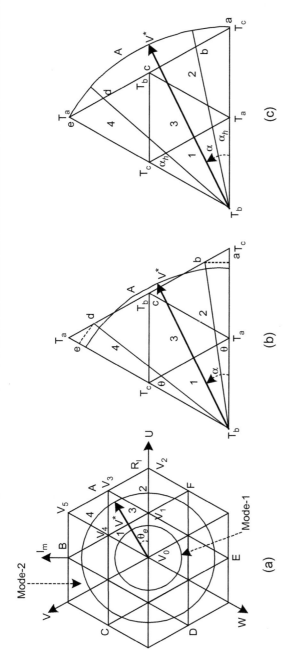

FIGURE 4.36 Three-level inverter operation in overmodulation mode: (a) undermodulation, (b) overmodulation mode 1, and (c) overmodulation mode 2.

If operation of a drive is desired in the high-speed field-weakening mode, the inverter should be operated in the overmodulation region [13] to increase the output voltage at the cost of nonlinearity and lower order harmonics. The undermodulation region, shown in part (a) and described in Figure 4.34, ends when the command vector V^* trajectory describes an inscribed circle of the large hexagon. At this point, the modulation factor m' can be calculated as 0.907 as in a two-level inverter. The overmodulation starts when the trajectory exceeds the hexagon boundary. The SVM overmodulation strategy, described here, is somewhat similar to that of a two-level inverter. In overmodulation mode 1, shown in part (b) for sector A only, V^* crosses the hexagon side at two points. To compensate for the loss of fundamental voltage, a modified reference trajectory is selected at larger radius that remains partly on the hexagon and partly on a circle. The circular parts ab and de cross the hexagon at angle θ. Mode 1 ends when angle θ is zero, that is, when the modified trajectory is entirely on the hexagon. At this point, the modulation factor $m' = 0.952$. In mode 2, $\alpha_h = 30°$ gives square-wave output. In practical operation, m' is slightly less than 1.0 to maintain three-level operation. The θ and α_h curves with m', shown in Figure 4.28, are also valid here.

FIGURE 4.37   (a) Time segment derivation in hexagon tracking, (b) PWM waves for region 2, and (c) PWM waves for region 4.

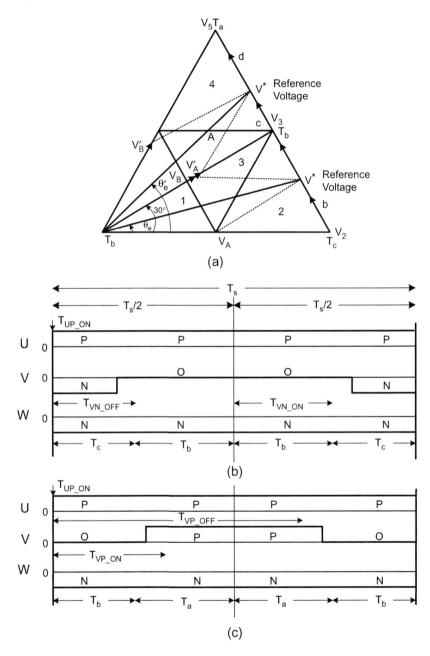

In overmodulation modes 1 and 2, a significant part of operation occurs on hexagon side *bd,* as discussed in Figure 4.36. Figure 4.37(a) explains the derivation of inverter vector time segments during the hexagon tracking for sector A only. The segment *bd* can be resolved into *bc* and *cd* segments. During tracking of *bc*, which is located in region 2, V^* is resolved into V_A and V_A' components. Therefore, vectors V_2 and V_3 are selected, respectively, for time segments T_c and T_b, where $T_c + T_b = T_s/2$. Similarly, during tracking of the *cd* segment located in region 4, vectors V_3 and V_5 are selected for time segments T_b and T_a, respectively, where $T_b + T_a = T_s/2$. The PWM waveforms for all the phases are shown for regions 2 and 4 in the lower part of the figure. Note that in both the regions, the U wave is saturated at the P state (S_{1U} and S_{2U} are on) and W is saturated in the N state (S_{3W} and S_{4W} are on).

FIGURE 4.38   Flow diagram for SVM implementation of three-level inverter (covering the UM and OM regions).

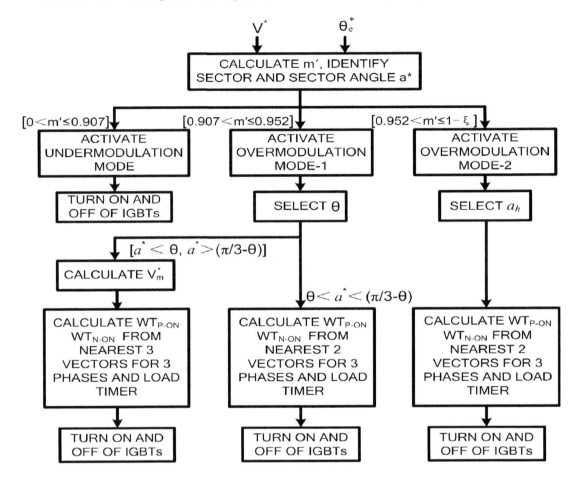

The figure shows the simplified flow diagram for implementation of the SVM algorithm with the help of DSP, and covers both undermodulation and overmodulation regions. In the beginning, the command voltage vector magnitude V^* and its angular orientation with respect to horizontal axis θ_e^* are detected. These permit calculation of modulation factor m', sector angle α^*, and identification of the sector. Based on these parameters, one of the three paths as shown is selected. In undermodulation mode, the digital words for turn-on and turn-off times of the switches can be calculated and activated with the help of timers. To simplify calculation, look-up tables can be created for these parameters. In the overmodulation mode1, the path for the linear or hexagon tracking segment is selected on the basis of angle α^*. The curves in Figure 4.28 are required for mode 1 and mode 2 implementation. Because of the computational complexity of the SVM algorithm, the typical limit of PWM sampling frequency is 10 kHz. This is particularly convenient for high-power inverters where the sampling frequency is lower.

FIGURE 4.39 Motor voltage and current waves for three-level inverter: (a) undermodulation (40 Hz), (b) overmodulation mode 1 (56 Hz), and (c) overmodulation mode 2 (59 Hz).

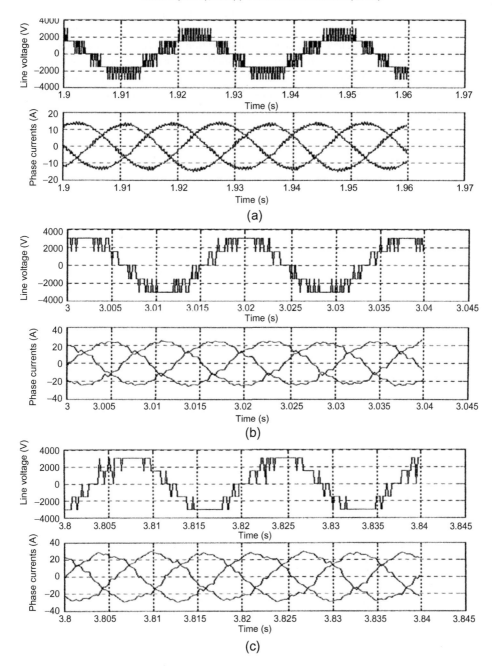

A three-level inverter (see Figure 4.33) is operated in an open-loop volts per hertz control mode with SVM to generate a variable-frequency, variable-voltage supply for induction motor load. This figure shows the motor line voltage and phase current waves at under-modulation and at overmodulation modes 1 and 2. The SVM carrier frequency is 1.0 kHz, and the base frequency for square-wave operation is 60 Hz. The line voltage wave has five steps ($\pm V_d$, $\pm 0.5 V_d$, 0), where $V_d = 3000$ V, as indicated. In the undermodulation mode, the current waves are smooth with carrier frequency related harmonics, whereas in overmodulation, some amount of lower frequency harmonics (more in mode 2) is evident.

FIGURE 4.40 Five-level diode-clamped inverter.

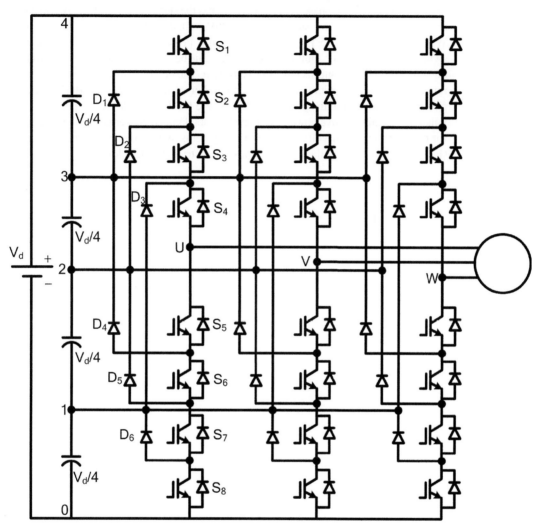

Extending the principle of the three-level diode-clamped inverter shown in Figure 4.33, a general $N(N > 2)$-level inverter [14] can be constructed. The higher number of levels provides the advantages of a higher power rating and lower output harmonics. This figure shows a five-level inverter that uses IGBT devices. In general, an N-level inverter requires $2(N-1)$ switching devices and $2(N-2)$ clamping diodes for each phase leg, and $(N-1)$ dc-side capacitors, where each switching device shares $V_d/(N-1)$ voltage. It produces N-level phase voltage and $2N-1$ level line voltage waves. With the dc negative bus as the reference, the switching table is given in Figure 4.42. This means that with respect to dc center point 2, the phase voltage wave v_{U2} has the steps of 0, $\pm 0.25 V_d$, $\pm 5 V_d$ (five steps), and the line voltage wave steps are 0, $\pm 0.25 V_d$, $\pm 0.5 V_d$, $\pm 0.75 V_d$, and $\pm V_d$ (nine steps). Within each step of 0.25 V_d, PWM operation is possible (see Figure 4.39) at a certain switching frequency, or multi-stepped waves can be generated at a low switching frequency (50/60 Hz) to interface the utility system. The dc-side capacitor voltage balancing remains a serious problem with a larger number of levels if the inverter handles real power (as in drive), but the problem is less severe for the static VAR compensator, where it handles mainly reactive power. Multilevel diode-clamped converters show great promise for multiterminal high-voltage dc (HVDC) systems and future ac transmission systems (FACTS), but currently their applications are restricted to five levels.

FIGURE 4.41   Multi (N)-level diode-clamped converter features.

- HIGH-VOLTAGE, HIGH-POWER CONVERTER
  - VARIABLE-FREQUENCY DRIVE
  - STATIC VAR GENERATOR (STAT COM)
  - MULTITERMINAL HVDC SYSTEM
- N LEVELS OF PHASE VOLTAGE AND ($2N-1$) LEVELS OF LINE VOLTAGE (ΔV STEP SIZE = $V_d/(N-1)$)
- VOLTAGE SHARING OF SWITCHING DEVICE $=\Delta V$
- PWM OR MULTISTEP OPERATION POSSIBLE
- IMPROVED PWM QUALITY
- MULTISTEP 60/50-Hz OPERATION WITH UTILITY INTERFACE (WITH OR WITHOUT TRANSFORMER)
- LESS dv/dt AND COMMON MODE CURRENT
- CAPACITOR VOLTAGE BALANCING PROBLEM WITH REAL POWER FLOW
- NEEDS ADDITIONAL CAPACITORS AND CLAMPING DIODES

Multilevel diode-clamped converters are extremely important for high-voltage, high-power applications, such as ac drives, static VAR generators, and multiterminal HVDC systems, although currently, their applications (three or five levels) are limited to the former two areas only. All the features discussed earlier are summarized here. The converter can be operated in stepped-wave or PWM mode in any application. For drive application, PWM mode is preferred, where the switching loss is higher but the machine harmonic loss is considerably reduced. Again, space vector PWM is definitely preferred. Because of the small ΔV step size, the EMI is reduced, machine insulation stress is lower, and the common mode bearing current will be low. These will be discussed later. For utility interface applications, the stepped wave is preferred because of its low switching frequency (60/50 Hz), which gives smaller device switching losses. Capacitor voltage balancing is a big problem with a larger number of levels when the converter handles real power (P), but with reactive power (Q) flow (as in SVG), it is not a serious problem. Therefore, in comparison with a high-voltage, high-power converter with simple series connection of devices, the diode-clamped converter has a lot of advantages. The application of a multilevel converter in STATCOM and unified power flow control (UPFC) utility systems will be described in Chapter 12.

TABLE 4.1   Switching States of a Phase Group for a Five-Level Inverter

STATE	V_{U0}	S_1	S_2	S_3	S_4	S_5	S_6	S_7	S_8
4	V_d	ON	ON	ON	ON	OFF	OFF	OFF	OFF
3	$0.75V_d$	OFF	ON	ON	ON	ON	OFF	OFF	OFF
2	$0.5V_d$	OFF	OFF	ON	ON	ON	ON	OFF	OFF
1	$0.25V_d$	OFF	OFF	OFF	ON	ON	ON	ON	OFF
0	0	OFF	OFF	OFF	OFF	ON	ON	ON	ON

FIGURE 4.42   Space vectors of five-level diode-clamped inverter.

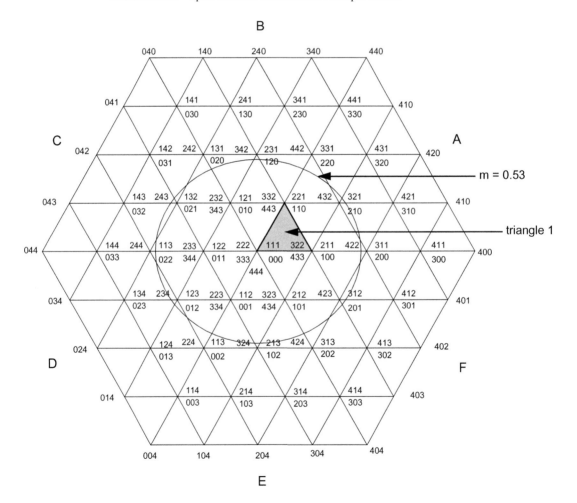

Table 4.1 shows the device switching table for a phase group and the corresponding phase voltage level or state. Each state corresponds to the dc-side capacitor tap as indicated. For example, state 3 corresponds to output voltage $0.75V_d$ and the switch group is closed such that positive current can flow through S_2, S_3, and S_4, and negative current can return through S_5, thus clamping the phase voltage at state 3. The inverter space vector diagram with its large number of states is shown in Figure 4.42 [15], which is important for SVM algorithm development. In general, an N-level inverter has N^3 states, $N^3 - (N-1)^3$ voltage space vectors, and $6(N-1)^2$ number of triangles. Therefore, for the five-level inverter, there are 125 states, 61 space vectors, and 96 triangles, as shown. For example, state 301 means that the U phase is in state 3, the V phase is in state 0, and the W phase is in state 1. The inverter has five zero states (222/111/333/000/444) at the center and a multiplicity of hexagon states as shown in the figure. A typical command voltage vector trajectory at modulation factor $m' = 0.53$ is shown in the figure.

FIGURE 4.43   Five-level flying capacitor inverter.

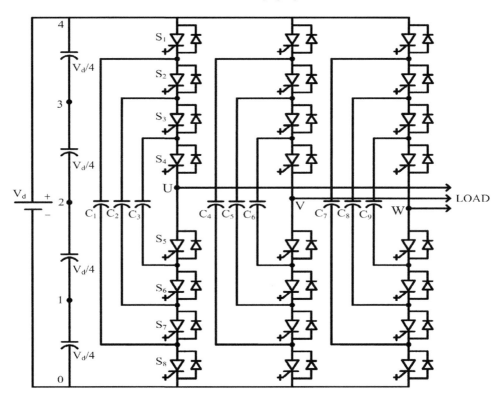

TABLE 4.2   One Possible Switching Table for a Flying Capacitor Inverter

STATE	V_{U0}	S_1	S_2	S_3	S_4	S_5	S_6	S_7	S_8
4	V_d	ON	ON	ON	ON	OFF	OFF	OFF	OFF
3	$0.75V_d$	OFF	ON	ON	ON	OFF	OFF	OFF	ON
2	$0.5V_d$	OFF	OFF	ON	ON	OFF	OFF	ON	ON
1	$0.25V_d$	OFF	OFF	OFF	ON	OFF	ON	ON	ON
0	0	OFF	OFF	OFF	OFF	ON	ON	ON	ON

Figure 4.43 shows the topology of a flying capacitor-type multilevel inverter [14], for which the number of levels is five. The circuit is somewhat similar to a diode-clamped type, except charged capacitors are used for voltage clamping. Capacitors C_1, C_2, and C_3 are charged to voltages $0.75V_d$, $0.5V_d$, and $0.25V_d$, respectively. Again, the voltage levels in the flying capacitor inverter are similar to those of a diode-clamped inverter. However, voltage level fabrication has much more flexibility than that of the diode-clamped type. Table 4.2 shows one possible device switching table. While states 4 and 0 are unique, the intermediate states (3, 2, and 1) can be fabricated by a number of switching combinations. For real power transfer, clamping voltage balancing is a difficult problem. The large amount of capacitance storage provides some ride-through capability in power outage, and switch combination redundancy permits voltage balancing capability, but the converter control is very complex. The circuit can be used for reactive power compensation.

FIGURE 4.44   Five-level cascaded H-bridge inverter.

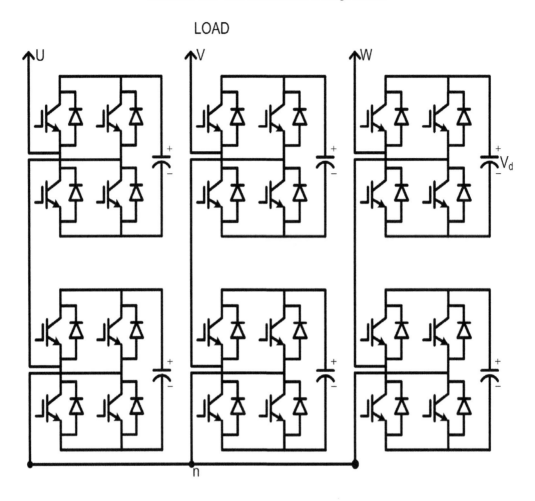

The multilevel cascaded H-bridge inverter [14] shown in this figure has a somewhat simple topology compared to the diode-clamped and flying capacitor types. In this case, a number of single-phase H-bridge inverters are connected in series to fabricate the phase voltage. Assuming the dc-side capacitor voltage to be V_d, each H-bridge can fabricate three voltage levels: 0, $+V_d$, and $-V_d$. Therefore, with two bridges in a cascade, as shown, there are five voltage levels per phase: 0, $\pm V_d$, and $\pm 2V_d$. The general expression is $N = 2B + 1$, where N = number of levels in phase voltage and B = number of bridges. The dc-side precharged capacitors are satisfactory ideally for reactive power transfer, but for real power transfer, they should be dc sources, such as batteries or photovoltaic or fuel cells. The output can be connected in delta or wye as shown. Modular cells with a low number of components make the scheme somewhat attractive.

FIGURE 4.45   Uninterruptible power supply (UPS) system—why is one needed?

- RELIABLE POWER SUPPLY FOR CRITICAL
  LOADS:
     COMPUTERS
     COMMUNICATION EQUIPMENT
     HOSPITALS
     CRITICAL PROCESS CONTROL DRIVES,
     ETC.
- LINE POWER QUALITY PROBLEMS:
  OUTAGE (BLACKOUT)
  UNDERVOLTAGE (BROWNOUT)
  OVERVOLTAGE
  VOLTAGE SPIKES
  DISTORTED VOLTAGE WAVE
  EMI

A reliable and high-quality dc or ac power supply is of vital importance for many critical loads, such as computers, communication equipment, hospital power, and important process control drives. Ideally, the utility system should maintain sine wave, single-or three-phase, and balanced supply voltage with regulated magnitude and frequency. In practice, problems occur such as outages (blackouts), undervoltage (voltage sag or brownouts), overvoltage, voltage transients with EMI, and distorted voltage waves. For example, the outage may be due to a system fault, undervoltage may be due to a sudden switch-in of load or starting a large induction motor, overvoltage may be due to sudden rejection of a load, voltage spikes and EMI may be due to sudden switching in or rejection of a load, and waveform distortion may be due to parallel operation with a large phase-controlled converter. Except for the power outage, any power quality problem can be solved by processing the line power through power electronics before supplying it to the load. For power outages, a UPS system is needed with a battery or alternate energy source. A UPS system can have dc or ac output.

FIGURE 4.46   Three-phase UPS system with 60-Hz line backup.

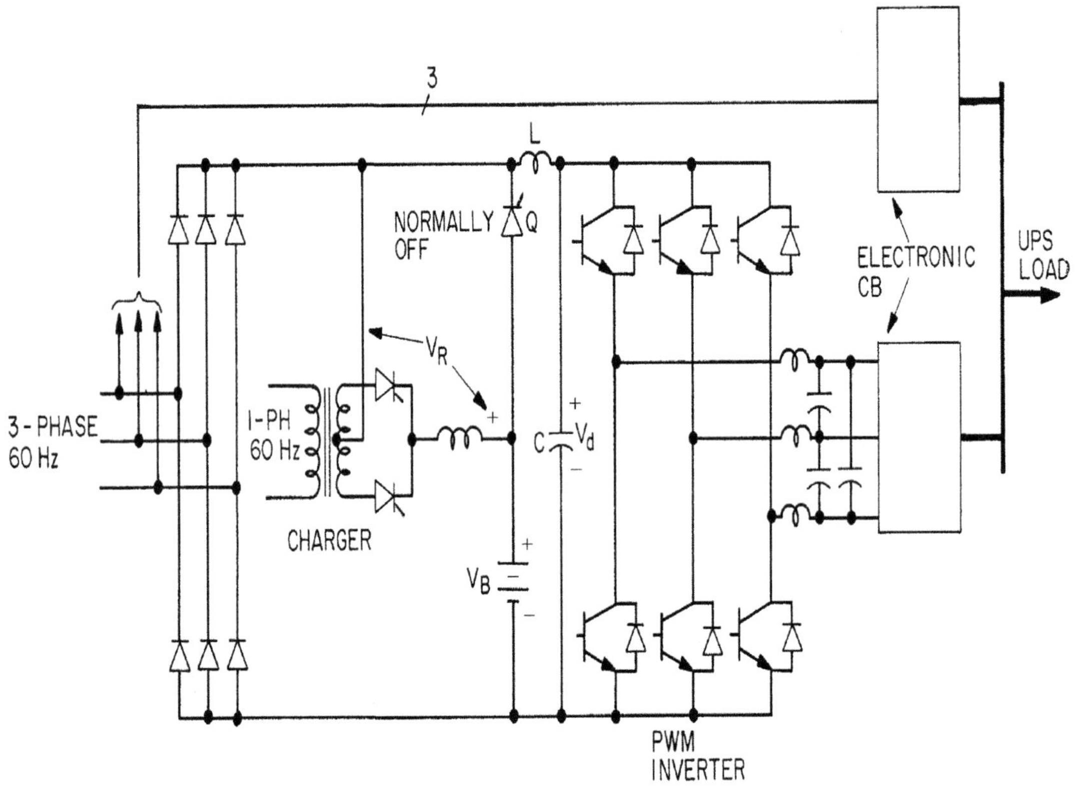

A typical three-phase UPS system is shown in this figure, where the load can be supplied either from the utility line or a battery-backed inverter. The electronic circuit breakers consisting of anti-parallel thyristors (see Figure 3.26) can help in the fast transfer of the source. In the system, the inverter normally supplies the load with line CB remaining open. This system consists of a diode rectifier (three or single phase), battery charger, storage battery, LC filter in the dc link, PWM inverter, and output LC filter. The inverter generates sinusoidal regulated voltage at 60 Hz at the output with unregulated and inferior ac supply at the diode rectifier input, and the thyristor Q normally remains off. At this condition, the charger generates trickle charging current to the battery. At utility power interruption, Q is turned on and the battery takes over the supply to the inverter. When the line power is restored to supply the inverter, the battery is disconnected by turning off Q. If there is a problem in the inverter system, the load is transferred to the ac line. The transfer causes power interruption for a subcycle period when the load storage elements maintain the continuity. Alternately, the line power can be normally used, and the inverter system can be used as a standby. If the load is single phase, an H-bridge inverter can be used. If the load is dc, the dc link voltage V_d can be used directly or it can be converted through a dc-dc converter.

FIGURE 4.47 Versatile UPS system with backup of battery and engine-generator.

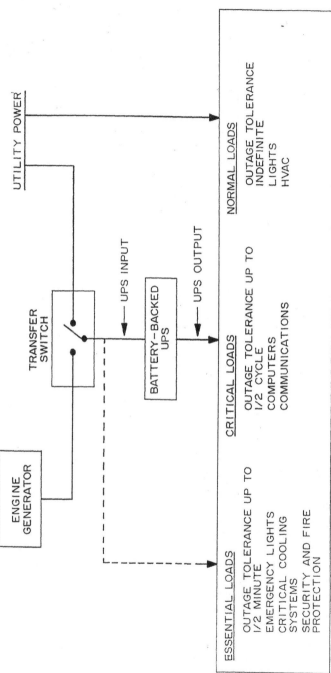

A more versatile UPS system with an additional engine-generator backup is shown in the figure. The normal loads, such as lighting and the HVAC system, that can withstand interruption of utility supply are supplied directly from the utility bus without any backup. But the loads that are essential and critical are supplied through UPS systems. At interruption of utility power, the critical loads that have only subcycle outage tolerance are switched to battery-backed UPS systems. If the outage continues and battery storage runs out, the standby engine-generator is started and the load is transferred to it. The essential loads, such as emergency lights and cooling systems that have an outage tolerance of up to 0.5 minute and are not backed up by battery, are also transferred to the engine-generator system. The start-up time of the engine-generator is assumed to be less than 0.5 minute. A fuel cell can also be used as a backup generator.

FIGURE 4.48   Progression of voltage-fed converter systems for ac drives.

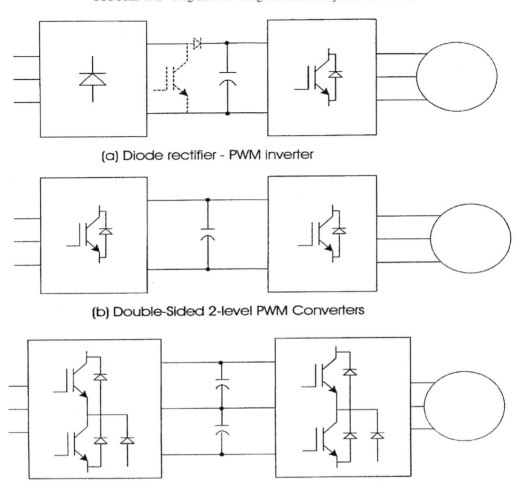

(a) Diode rectifier - PWM inverter

(b) Double-Sided 2-level PWM Converters

(c) Double-Sided 3-level PWM Converters

The figure summarizes the progression of three classes of voltage-fed converter systems for industrial ac motor drives. Although IGBTs are shown as switching devices, for low-power ranges, power MOSFETs can be used; for high-power ranges, IGCT or GTO devices can be used. The simple diode-rectifier PWM inverter topology was discussed in Figure 4.11. For reasonably low power, single-phase supply with an H-bridge rectifier is satisfactory, but for high power, a three-phase supply with a three-phase bridge rectifier is essential. The drive is nonregenerative, but a dynamic brake (DB), as shown in Figure 4.11, can be used in the dc link. A boost chopper in the dc link, as shown in the figure, can also be used for line power factor correction (see Figure 4.5 or 4.6). Instead of a boost chopper, a large inductor filter can be used in the dc link to produce stepped line current wave. Two three-phase diode bridges with series or parallel connection and phase-shifting transformers can also be used for 12-step line current wave. This is the common converter topology for nonregenerative ac drives. Figures 4.48(b) and (c) are regenerative drives that are used for moderate and high powers, respectively. These two topologies will be described in detail in the following pages.

FIGURE 4.49   Electric locomotive drive with parallel induction motors.

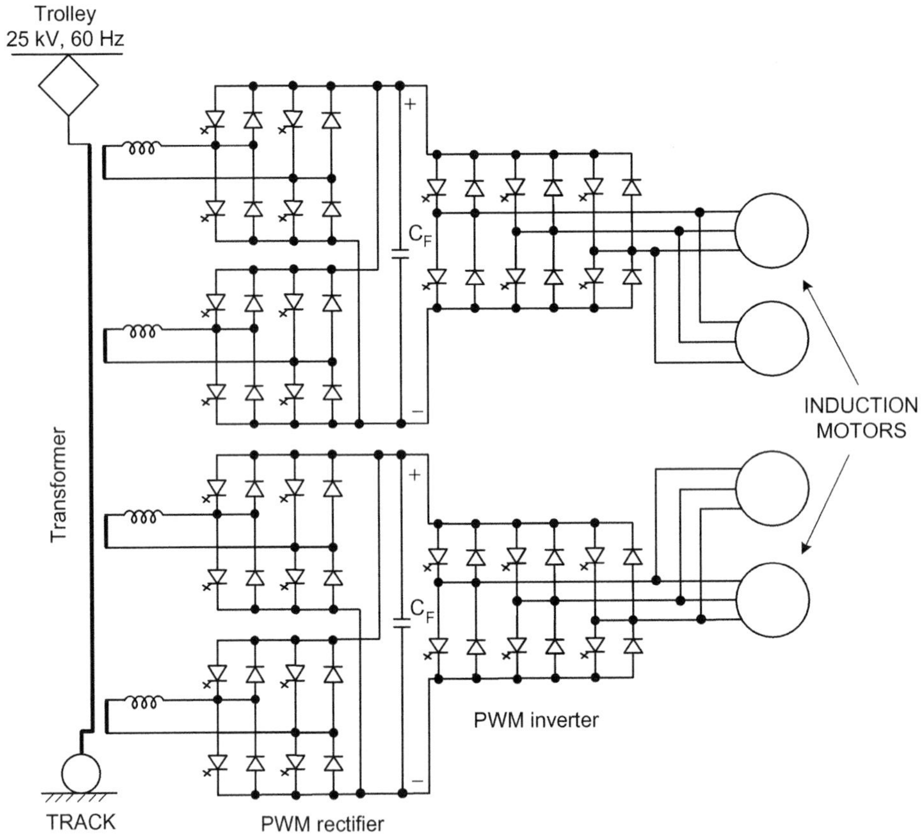

The figure shows the typical electric locomotive drive with induction motors from the single-phase catenary supply at 25 kV and 60 Hz. If the locomotive is diesel-electric, an engine-coupled synchronous generator generates constant-voltage, constant-frequency (CVCF) ac, which is converted to variable-voltage, variable-frequency (VVVF) supply for the motor through a diode rectifier and PWM inverter. In a subway drive, the primary dc supply is converted to ac (VVVF) for the motor. This figure shows a four-quadrant regenerative drive. The front end has two H-bridge PWM rectifiers (inverter operating in boost rectifier mode) supplied by step-down isolated winding transformers, one three-phase PWM inverter, and two parallel induction motors in each group. The four motors drive the respective axles in a buggy. Evidently, parallel operation requires close matching of the machine characteristics and identical wheel diameters (for equalizing motor speed with the same linear wheel velocity). Although GTOs are shown, IGBTs can also be used. The line current is sinusoidal at unity power factor because of PWM control of the H-bridges. However, for the power balance, there will be a 120-Hz ripple in the dc link that requires a large filter capacitor. In some drives, only one converter group with four machines is used in parallel. The Japanese Shinkansen system (bullet train), for example, uses this type of drive with four parallel machines using three-level IGBT inverter and three-level IGBT rectifier (see Fig. 12.14).

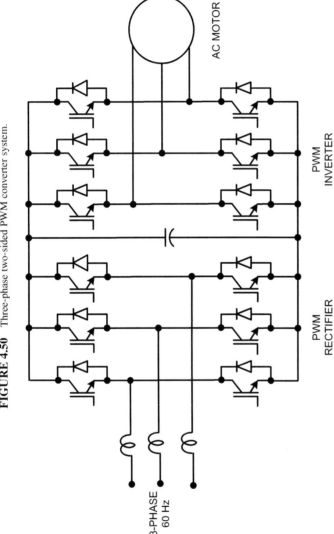

FIGURE 4.50 Three-phase two-sided PWM converter system.

A PWM inverter can have active power flow in either direction, as discussed earlier. Therefore, a similar unit can be connected on the line side in inverse manner to act as a rectifier. A two-sided converter system is shown in this figure for a four-quadrant ac motor drive. In motoring mode, the machine demands active power; therefore, the line-side converter acts as a rectifier, whereas the machine-side converter acts as an inverter. In regenerative mode operation, the stored mechanical energy is converted to electrical energy, rectified to the dc link, and the line-side converter (acting as inverter) converts the energy to ac and pumps back to the line. The application of such a converter system is very popular for rolling mill drive-type applications, where frequent stopping and speed reversal are needed. The line-side converter is controlled to regulate the dc-link voltage to be constant. Note that the dc voltage level has to be higher than the peak values of line and load voltages, because each converter has to operate in buck mode for dc-to-ac conversion. For this reason, the line-side converter is often called a *boost rectifier*. A small line-side inductance converts the PWM wave into sinusoidal voltage, the magnitude and phase of which can be controlled. Therefore, the additional advantages of the system are sinusoidal line current and programmable (leading or lagging) line power factor. If the line voltage fluctuates, or becomes low (for instance, during voltage sag or a brownout), the rectifier can regulate the dc voltage, hence maintaining reliability of the drive. The scheme can also be used in a UPS system with an energy storage device (battery, ultracapacitor, or fuel cell) coupled in the dc link. An important control feature of the system is that the machine-side converter can be operated in either the undermodulation or overmodulation mode, but the line-side converter must operate in the undermodulation mode. For single-phase line and load, three-phase converters can be replaced by H-bridge units.

FIGURE 4.51 Two-sided two-level converter system features.

- REAL POWER FLOW IN EITHER DIRECTION
- CONVERTER SUPPLIES MACHINE EXCITATION
- SINUSOIDAL LINE CURRENT AT UNITY POWER FACTOR
- LINE POWER FACTOR CAN BE PROGRAMMED TO BE LEADING OR LAGGING (VAR COMPENSATOR IN EXTREME CASE)
- LINE-SIDE RECTIFIER OPERATES IN BOOST MODE
- LINE VOLTAGE SAG COMPENSATION TO IMPROVE SYSTEM RELIABILITY
- DC LINK VOLTAGE IS HIGHER THAN LINE AND MACHINE PEAK COUNTER EMFs
- INVERTER CAN OPERATE IN OVERMODULATION MODE, BUT RECTIFIER MUST OPERATE IN UNDERMODULATION MODE
- DC LINK CAPACITOR REQUIRES PRECHARGING
- SOMEWHAT MORE EXPENSIVE THAN WITH FRONT-END DIODE RECTIFIER – BUT VERY POPULAR FOR FOUR-QUADRANT MOTOR DRIVE

The figure summarizes the features of the two-sided PWM converter system discussed previously, where each unit is a two-level converter (inverter or rectifier). Although the scheme is somewhat expensive compared to the traditional system with a diode rectifier in the front end, it is very popular because of its many advantages. The inverter can supply leading or lagging excitation current to the ac machine. With the same principle, the line current can be leading, lagging, or at unity power factor. Therefore, for light machine loads (when the rectifier real power loading is light), it can carry reactive current to compensate the ac system power factor. The dc link capacitor can be charged from the line through the diode rectifier with a series resistor and a bypass contactor, but can be overcharged by regenerative operation of the drive. Instead of two-level converters, multilevel converters can be used in high-power systems. Also, IGBTs can be replaced by GTOs or IGCTs.

FIGURE 4.52 Back-to-back utility system intertie with 300-MW two-sided GTO converter system.

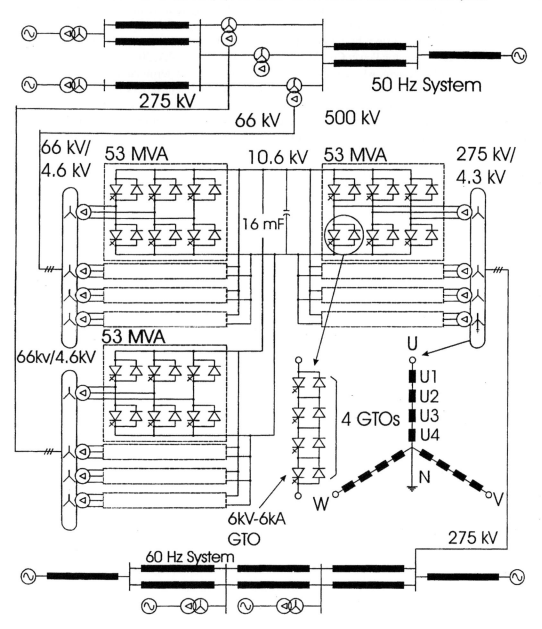

Traditionally, high-voltage dc (HVDC) systems use current-fed thyristor phase-controlled converters at each end. The disadvantages of this system are line current harmonics, lagging DPF, and only two-terminal capability. This figure shows a voltage-fed PWM GTO converter system based on a three-terminal HVDC intertie, which was developed by Tokyo Electric Power Co. (TEPCO) [18]. It links two 66-kV, 50-Hz systems with a 275-kV, 60-Hz system, where active power can flow from any terminal to any terminal. Unlike the thyristor-based system, the voltage-fed PWM converters easily permit multiple ac/dc terminals, line DPF control, and near sinusoidal line current. For each terminal, four bridge converters are paralleled on each dc side with a 53-MVA (37.5-MW and 37.5-MVAR) rating. Each arm of the converter unit has four GTOs (6 kV, 6000 A) in series. Each converter unit is fed by a wye-delta step-down transformer, where the primary wye windings are connected in series to interface the high-voltage line. The GTO switching frequency is around 500 Hz (nine pulses per cycle), and the switching losses are reduced by using regenerative snubbers. The turn-off time of series GTOs is adjusted to improve transient voltage sharing of the devices. Instead of GTOs, IGCTs can be used, and multilevel converters can be used to avoid series connection of devices.

FIGURE 4.53 Features of GTO-based utility intertie system.

- THREE-TERMINAL HVDC SYSTEM BACK-TO-BACK INTERTIE
- LINKS TWO 66-kV, 50-Hz TERMINALS WITH ONE 275-kV, 60-Hz TERMINAL
- NINE-PULSE SINUSOIDAL SYNCHRONIZED PWM FOR EACH CONVERTER
- NEAR SINUSOIDAL LINE CURRENT WITH UNITY, LEADING, OR LAGGING POWER FACTOR FOR SYSTEM VAR CONTROL
- FOUR SERIES-CONNECTED GTOs (6 kV, 6000 A) WITH REGENERATIVE SNUBBER TO IMPROVE CONVERTER EFFICIENCY
- GTOs CAN BE REPLACED BY IGCTs
- MULTILEVEL PWM OR STEPPED WAVE CONVERTERS CAN AVOID SERIES CONNECTION OF DEVICES

The figure summarizes all the features discussed in Figure 4.52. A similar intertie has also been developed by ABB that uses IGCTs. Instead of PWM converters, multistepped converters, as discussed previously, could also have been used.

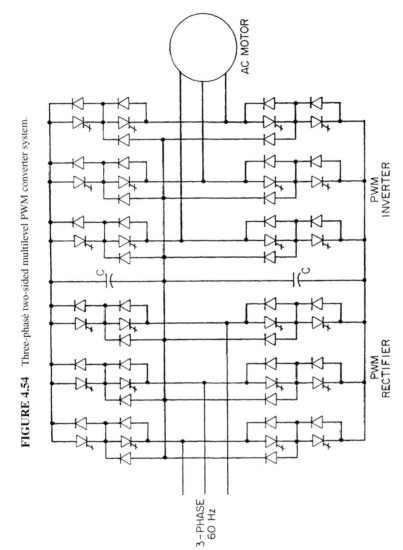

FIGURE 4.54 Three-phase two-sided multilevel PWM converter system.

For high-voltage, high-power applications, the multilevel ($N = 3$) two-sided converter system shown in this figure can replace the two-level converter system discussed before. All basic features of the latter system are retained here. Also, the additional advantage is that the problem of series connection of devices is avoided. However, the disadvantage is the difficulty of balancing the capacitor voltages. Although GTOs are shown in the figure, IGBTs or IGCTs can also be used, which will permit higher PWM switching frequencies. The scheme is recently finding favor as a replacement for multi-megawatt thyristor cycloconverter drives. With single-phase ac supply, either an H-bridge or half-bridge (one line connected to capacitor center-tap) converter can be used.

FIGURE 4.55   Features of three-phase two-sided multilevel converter system.

- HIGH-VOLTAGE, HIGH-POWER CONVERSION SYSTEM
  (DC LINK CYCLOCONVERTER)
- AVOIDS DIFFICULTY OF SERIES OPERATION OF DEVICES, BUT NEEDS
  EXTRA DIODES AND CAPACITORS
- INSTEAD OF GTO, IGBT OR IGCT CAN BE USED WITH
  HIGHER SWITCHING FREQUENCY
- IMPROVED HARMONIC QUALITY, PARTICULARLY WITH SPACE
  VECTOR PWM
- FOUR-QUADRANT DRIVE OPERATION
- SINUSOIDAL LINE CURRENT
- LINE DPF CAN BE CONTROLLED TO BE UNITY, LEADING, OR LAGGING
- DIFFICULTY OF CAPACITOR VOLTAGE BALANCING

This figure summarizes all the features discussed in Figure 4.54. Five-level converters, as discussed earlier, can also be used for higher power and higher voltage systems. For GTO PWM converters, regenerative snubbers are usually used to improve efficiency.

FIGURE 4.56   (a) Three-phase to three-phase matrix converter, (b) line voltage waves, and (c) PWM fabrication of output line voltage wave.

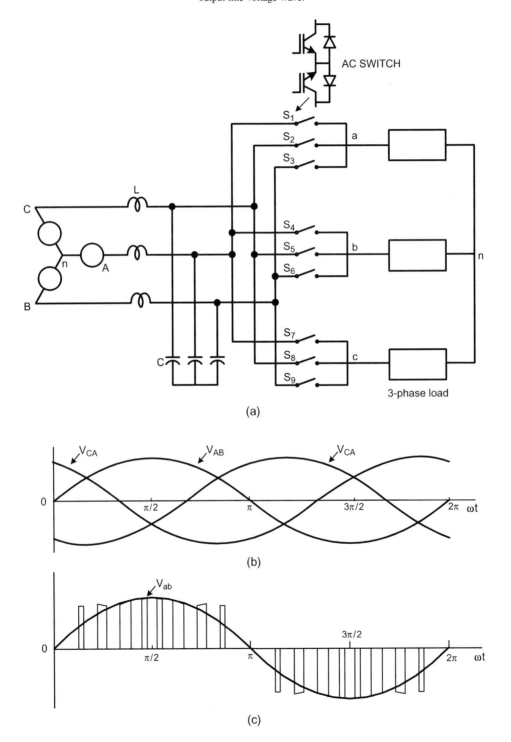

(a)

(b)

(c)

The name *matrix converter* arises because the converter consists of a matrix of self-controlled switches without any reactive elements within it. Basically, it is a frequency changer or cycloconverter under PWM control of ac switches. Therefore, the inherent disadvantages of a phase-controlled cycloconverter are avoided. The converter, as shown above, is a matrix of nine ac switches, where a typical ac switch connection is shown in the figure. The switches are controlled by PWM to fabricate the output fundamental voltage, which can vary in magnitude and frequency to control the speed of an ac motor. The principle of output line voltage (v_{ab}) synthesis from the input line voltages is shown in the figure. Phases *a* and *b* can be connected between the positive and negative envelope of input line voltages or shorted to fabricate the v_{ab} wave. For example, if S_3 and S_5 switches are closed, the v_{BC} segment appears in the v_{ab} segment. If S_3 and S_6 are closed, v_{ab} is shorted. There are $3^3 = 27$ switching states to fabricate the output PWM waves. Up to 95% of line voltage and a several hundred hertz fundamental frequency are obtainable. An ac capacitor bank is required at the input for commutation of switches and filtering of PWM harmonics on the line side. The converter is regenerative and line PF can be controlled to be unity, leading, or lagging. Besides the input ac capacitor bank, the converter needs 36 devices compared to the 24 devices in a two-sided PWM converter system. The converter has not been successful commercially in spite of the vast literature available on it. Further discussion on a matrix converter is given in Chapter 12.

FIGURE 4.57   Static VAR compensator (SVC) and active harmonic filter (AHF).

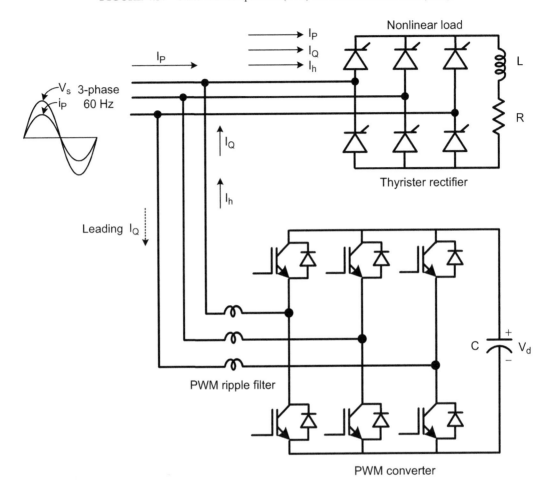

It was mentioned earlier that a PWM rectifier can draw controllable active (I_P) and reactive current (I_Q) at the input. If active power is zero (i.e. $I_P = 0$), then it draws I_Q only. In this mode, it is defined as a static VAR generator (SVG), static VAR compensator (SVC), or static synchronous compensator (STATCOM) [27]. Its operation is similar to ideal no-load operation of a synchronous machine with variable excitation. In this mode, the converter acts as a three-phase variable capacitor or inductor load. Note that, traditionally, an SVG is realized by means of a thyristor-controlled reactor (TCR) in parallel with a capacitor. The PWM-type SVG, shown in this figure, acts as variable capacitor load and compensates the phase-controlled converter lagging reactive current (I_Q) so that the line DPF is unity. Since the line current of SVG is ideally reactive, no active power flows to the dc side, and dc voltage V_d does not change. In practice, a small ac active current supplies the loss of the converter, and only harmonic current flows in the capacitor. The thyristor converter I_Q component can be sensed and the SVG I_Q can be controlled to be opposite in phase within the V_d control loop. The same SVG can also sink the harmonic current of the nonlinear load if the PWM converter is switched at reasonably high frequency. In this function, it is defined as an active harmonic filter (AHF) [28]. Basically, AHF line current I_h is synthesized to be opposite in phase with the rectifier I_h. The operation principle is similar to that of an SVG. In fact, combined operation of the SVG and AHF (universal power line conditioner or UPC) is possible so that the line current is sinusoidal at unity PF. The SVC can be used as a voltage regulator, dynamic system stabilizer, and for compensation of fluctuating reactive and negative sequence currents. SVG and AHF operation is also possible by means of a PWM current-fed converter, which will be discussed in Chapter 5.

FIGURE 4.58   Forty-eight-MVA static VAR generator for electric railways.

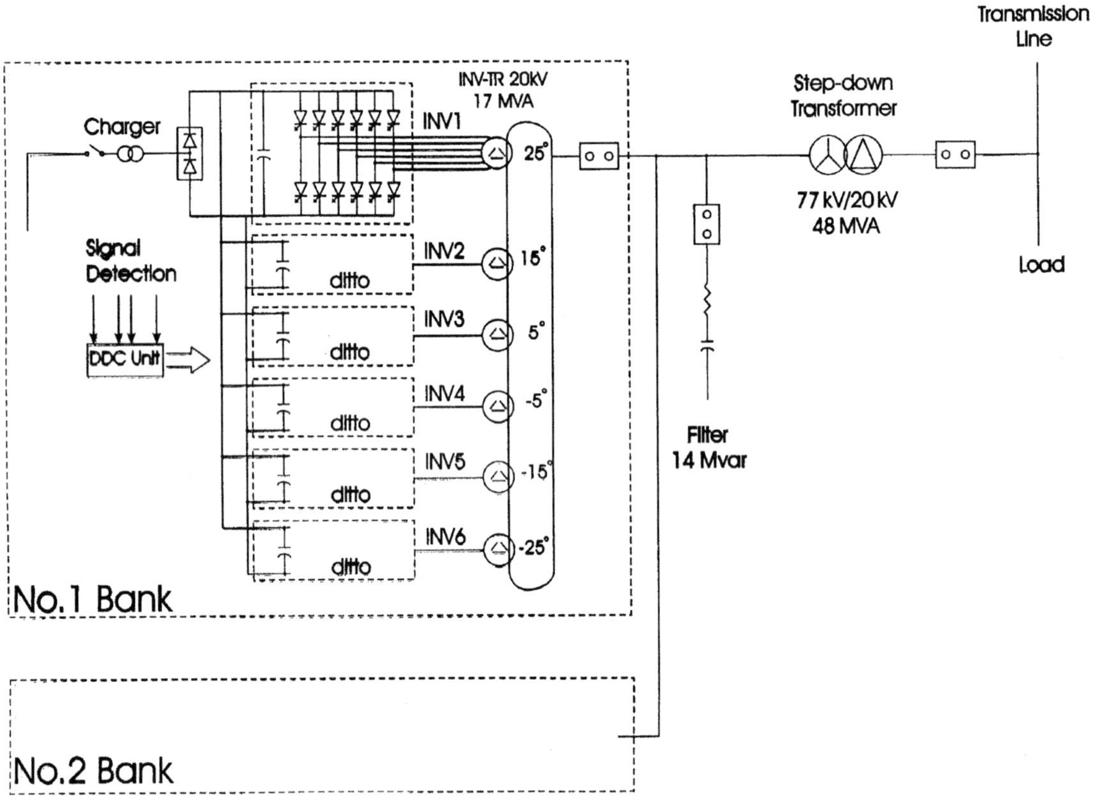

Fuji Electric Co. of Japan recently installed a 48-MVA SVC on the Tokaido Shinkansen railway system [19] (Japanese bullet train), the schematic of which is shown in this figure. Its function is to regulate the line voltage and compensate line voltage imbalances due to fluctuating single-phase train loads. The system configuration is somewhat similar to that of Figure 4.15, except that ideally it does not carry any active power. Basically, the converter system acts as a three-phase variable capacitor or inductor bank on the 20-kV bus, which is stepped down from a 77-kV line. The reactive MVA supplied by the SVC compensates the ac bus voltage that fluctuates with the reactive loading. Each three-phase H-bridge converter converts the dc voltage to three-phase six-step voltage waves that can be altered in magnitude and phase, and six converter banks, each with 10° phase shift, synthesize a 36-pulse stepped wave through a multiwinding delta-wye transformer. The high-frequency harmonics are removed by a passive *R-C* filter. The bus voltage imbalance is compensated by generating a three-phase negative sequence current at the output. The capacitor bank is precharged from the ac line through the transformer and diode rectifier, but its dc voltage is regulated from the converter side. There are similar installations in the U.S.

FIGURE 4.59   Forty-eight-MVA static VAR compensator features.

- VOLTAGE-FED PHASE-SHIFTED MULTISTEP WAVE SVC ON JAPANESE SHINKANSEN RAILWAY SYSTEM – INSTALLED BY FUJI
- REGULATES AC BUS VOLTAGE (WITHIN ±2%) AND COMPENSATES LINE VOLTAGE IMBALANCES DUE TO SINGLE-PHASE LOAD
- 20-MVA LAGGING VAR TO 48-MVA LEADING VAR CAPABILITY
- 36-PULSE STEPPED WAVE OUTPUT WITH MAGNITUDE AND PHASE CONTROL
- SINGLE REVERSE CONDUCTION GTO (4.5 kV, 3000 A) IN EACH H-BRIDGE
- TRANSFORMER WITH DIODE CHARGER PRECHARGES THE CAPACITOR
- ±10% DC VOLTAGE REGULATION BY CONVERTER
- 14-MVA CAPACITIVE HARMONIC LINE FILTER
- HIGH EFFICIENCY (97%)

The salient features of the 48-MVA SVC are summarized in this figure. The two inverter banks (shown within the dotted enclosure in Figure 4.58), each with 17-MVA capacity, and the capacitive filter of 14 MVA constitute a 48-MVA total capacity, which permits the controllable range of 20 MVA lagging to 48 MVA leading. The negative sequence current generation induces second-harmonic voltage ripple in the dc link capacitor but the worst case ripple is limited to ±10%. Because the GTOs are reverse conducting with built-in diodes, no diode is shown separately. The GTO switching frequency is low (50/60 Hz) and, therefore, the efficiency is very high. The inverter voltage is sufficiently low in the dc link such that no series connection of devices is needed. Altogether 144 GTOs are used in the system. The dc capacitor could be charged from the line, but because of higher voltage, it is charged separately through a transformer and diode rectifier. Because only reactive current is supplied by the converter, capacitor voltage does not change. The bulky low-frequency transformer is a great disadvantage of the scheme. It is possible to replace the system with multi-level inverters contributing large numbers of levels (see Chapter 12).

FIGURE 4.60   Hard switching effects of converter.

- HIGH DEVICE SWITCHING LOSS OR SNUBBER LOSS
- BURDEN ON THE CONVERTER COOLING SYSTEM
- DEVICE STRESS
- EMI PROBLEMS
- EFFECT ON MACHINE INSULATION
- MACHINE BEARING CURRENT PROBLEM
- MACHINE TERMINAL OVERVOLTAGE WITH LONG CABLE

So far in this chapter, we have discussed converters using only hard switching. Hard switching without any snubber was shown in Figure 2.19. Because of the overlapping of voltage and current waves, for every on–off cycle of switching, there will be some amount of power loss. As the PWM switching frequency increases, this loss goes up and decreases the converter efficiency. With a normal *R-C-D* snubber, the turn-off switching loss in the device is lower, but the stored energy in the capacitor is dissipated in the resistance at turn-on. The turn-on switching loss tends to be lower due to series inductance (turn-on snubber), but its stored energy is transferred to the capacitor, which is lost finally. The overall effect of the snubber may be increased switching loss [4]. Because the switching locus of the device is in the active region of *V-I* plane, the device is stressed and its life is shortened. The EMI problem due to high *dv/dt* and *di/dt* is high. The high *dv/dt* has a deteriorating effect on the machine insulation because of *Cdv/dt* displacement current, and the insulation life becomes shorter. The high *dv/dt* also causes common mode current that flows through machine bearing, thus shortening its life (explained later). If the PWM inverter is connected to the machine with a long cable, the high *dv/dt* also boosts the machine terminal voltage by the reflection of a high-frequency traveling wave. High-frequency ringing may also occur at the machine terminal due to stray inductances and capacitances. This overvoltage can threaten the machine insulation.

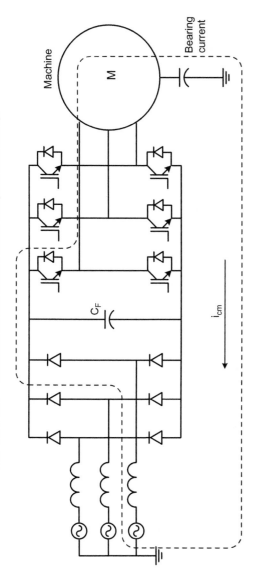

FIGURE 4.61 Common mode *dv/dt* induced current through machine bearing.

The figure shows the *dv/dt* induced common mode current flow through the machine bearing. As the switching speed of modern power semiconductor devices (such as IGBTs) increases, the problems due to high *dv/dt* are worsening. PWM inverters can be represented by common mode equivalent circuits as shown with high *dv/dt* sources. In the figure, assume that one of the IGBTs is creating a high *dv/dt* source that will cause *Cdv/dt* displacement current through the grounded source voltage, supply cable leakage inductance, diode rectifier, machine cable leakage inductance, coupling to the rotor, and creation of a circulating current flow to the ground through machine shaft and stray capacitance of the insulated bearing as indicated. This current will tend to shorten the bearing life. The circulating current can be contributed similarly by the other switching devices. Even if the source is not grounded, the stray capacitances to ground through converter heat sinks and cables can cause the path for the circulating current. Of course, the bearing current can be prevented by short circuiting to the frame with a brush, breaking the path by insulation, or it can be diverted through a low-pass filter at the machine terminal.

FIGURE 4.62   Machine stator winding stray capacitor coupling to stator and rotor by *dv/dt* effect.

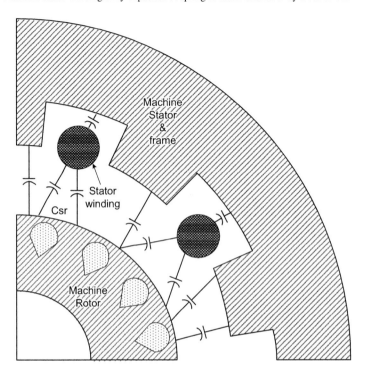

The figure shows the coupling of the stator winding of an induction motor with the stator and rotor irons through the stray capacitances. The stator winding is located in the stator slots. The winding insulation provides stray capacitance to the stator iron and the frame as shown. There is also stray capacitance between the stator winding and the rotor iron due to the air gap, and directly between stator and rotor irons, as indicated. The *Cdv/dt* induced current to the stator iron and frame shunts to the ground directly through the machine earthing. Similar current induced in the rotor iron flows through the shaft, machine bearings, and then to ground causing reduction of bearing life. It is possible to analyze the bearing current by modeling the system and simulating it in the computer, or it can be measured in the actual installation.

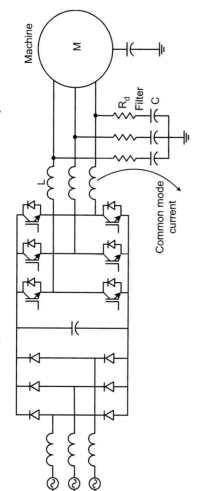

FIGURE 4.63 Low-pass filter at machine terminal to solve *dv/dt*-induced problems.

It was discussed before that hard switching of high-speed power devices (such as IGBTs) creates high *dv/dt* and thus causes *dv/dt*-induced problems, such as machine winding insulation deterioration, bearing current, and machine terminal overvoltage. One of the ways to solve the problem is to reduce the *dv/dt* by soft switching of the devices, which will be discussed later. The problem can also be solved by installing a low-pass *L-C* filter at the machine terminal shown in this figure. Additional inductance is connected in series with each machine phase as shown. There is a wye-connected capacitor bank (*C*) at the machine terminal with the neutral point connected to the ground. This *L-C* filter softens *dv/dt* at the machine terminal and shunts away the common mode current through the bearing. The filter also absorbs the inverter-generated harmonics, thus making the machine current nearly sinusoidal. As a result, machine efficiency is improved and torque pulsation and acoustic noise in the iron are attenuated. These will be discussed in later chapters. However, such filters tend to be bulky in high-power drives although the scheme has been adopted by some drive manufacturers (see Figure 7.23). Another problem is the resonance effect in the *L-C* filter induced by inverter harmonics. The damping resistance (*R*$_d$) prevents the resonance effect, but causes additional loss in the circuit.

FIGURE 4.64   Hard and soft switching of power semiconductor devices.

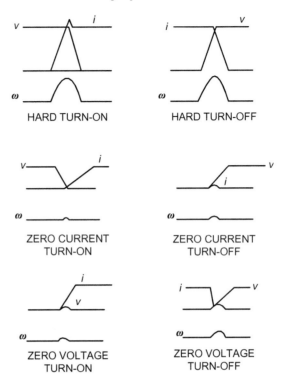

The disadvantages of hard switching effects, as discussed before, can be practically eliminated by soft switching of power devices, as explained in this figure. The hard turn-on
and turn-off, shown at the top, are basically repeated from Figure 2.19. The large overlap
between the voltage and current waves during switching creates a large switching loss,
thus reducing the efficiency of PWM converters. The main idea in soft switching is to
prevent or minimize the switching overlap so that the switching loss is minimal. There
are two principal types of soft switching: zero current switching (ZCS) and zero voltage
switching (ZVS). Consider the zero current turn-on and turn-off shown in the figure. The
turn-on current in a device can be slowed down by a series inductance (series snubber)
as indicated. Most devices have limited di/dt ratings. With such a delayed turn-on, the
switching loss is very low. Similarly, the current can be reduced to zero first before turning
off the device. A small recovery current overlapping with the voltage wave gives a small
turn-off loss. A good example of ZCS is the thyristor commutation in the phase-controlled
converter. That is why thyristor converter efficiency is so high. ZVS is explained at the
bottom of the figure. A device can be turned on, for example, when its bypass diode is
conducting and then it takes the forward current. The small voltage across the device

will cause small turn-on loss. Similarly, for ZVS turn-off, the bypass diode may also conduct when the device is turned off. With a large snubber capacitor, if the device is turned off, voltage across the device builds up slowly, causing a small tail current, as shown in the figure. The stored energy in snubber capacitor in such ZVS turn-off is absorbed by the converter (not dissipated in the snubber resistor as in an *R-C-D* snubber). In ZVS and ZCS, *dv/dt* and *di/dt* are usually low, as shown in the figure.

FIGURE 4.65 Features of soft-switched converters.

- REDUCES SWITCHING LOSSES
- SOFTENS EMI PROBLEM
- REDUCES SNUBBER SIZE OR SNUBBER ENERGY RECOVERY
- REDUCES *dv/dt* EFFECT ON MACHINE INSULATION
- ELIMINATES MACHINE BEARING CURRENT
- MINIMIZES ACCOUSTIC NOISE
- ELIMINATES MACHINE TERMINAL OVERVOLTAGE WITH LONG CABLE
- ADDITIONAL CONVERTER COMPLEXITY
- ADDITIONAL CONTROL COMPLEXITY
- ADDITIONAL LOSSES

The figure summarizes the advantages and disadvantages of soft-switched converters, i.e., the converters that use soft switching of power devices. These points are essentially valid for motor drive applications. All the advantages tend to solve the disadvantages of hard-switched converters indicated in Figure 4.60. Soft-switched converters often may not use a snubber at all, or a minimal size snubber can be used for device protection. A simple capacitive snubber can be used, but its stored energy is absorbed in the converter (often called a *resonant snubber* or *energy recovery snubber*). In the class of converters that uses regenerative snubbers, the capacitor-stored energy is recovered through active or passive devices as indicated in Figure 2.9. The soft-switched converters, however, have some disadvantages. Often, the converter topology becomes more complex with additional control complexity. As a result, there are additional losses and the converter system reliability is impaired. Although soft-switched converters for motor drives were proposed a long time ago, and there is a vast amount of literature in this area, hardly any industrial drive so far uses soft-switched converters. Soft-switched converter technology will be described briefly for completeness of the subject. However, soft-switched converters have several popular applications in nondrive areas, which will be described later.

FIGURE 4.66   Soft-switching converter classification for motor drives.

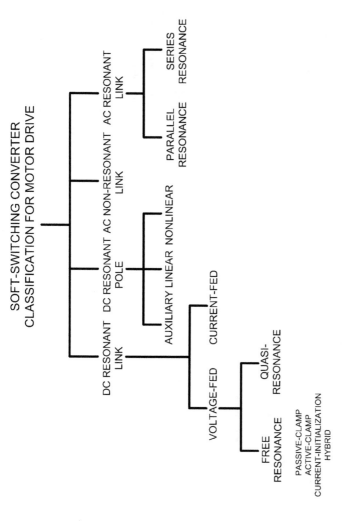

The modern era of soft-switched converters for ac motor drives was initiated by the introduction of dc resonant link inverters in 1986 [20]. Since then, many different soft-switched converter topologies have been proposed [21]. This figure shows the classification of the principal types of soft-switched converters. In general, they may be dc link or ac link types. The dc link types can be classified as resonant link dc (voltage fed and current fed) and resonant pole types. These are further classified as shown in the figure. The ac link types can be classified as nonresonant link and resonant link, where the latter again is classified into parallel and series resonance types. Detailed discussion of all the soft-switched converters is beyond the scope of the book. Only a few important types will be discussed briefly in the following pages.

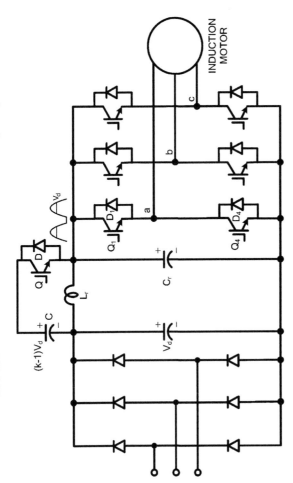

FIGURE 4.67 Soft-switched resonant link dc converter (RLDC).

A voltage-fed inverter operating on free-running dc link resonance (tens of kilohertz) is shown in this figure [20]. The input dc voltage V_d, obtained from a battery or ac line through a rectifier, is converted to unidirectional sinusoidal voltage pulses (v_d) with zero voltage gap on the inverter dc bus through a L_r-C_r resonance circuit, as shown in the figure. The gap permits zero voltage soft switching of the inverter devices. The variable-voltage, variable-frequency supply at the machine terminal can be fabricated by the delta or sigma-delta PWM modulation principle using the integral voltage pulses. To establish the resonant bus voltage pulses, an initial current in the resonant inductor L_r is needed for compensation of the resonant circuit loss and the reflected inverter input current. The initial current is established during the zero voltage gap by shorting the inverter legs. The bus voltage ($v_d > V_d$) can be controlled either by passive or active clamping. Using an active clamping method with IGBT Q, as shown in the figure, the peak value of v_d is typically limited to $1.5V_d$ with the help of precharged capacitor C. One disadvantage of the scheme is that the slow discharge of L_r current in the clamped capacitor causes fluctuation of resonant pulse widths, which reduces PWM resolution, thus causing harmonic penalty in the machine. The inverter can be made regenerative by replacing the diode rectifier with a similar converter unit. The RLDC converter is essentially the forerunner of all modern soft-switched converters for motor drives. Although initially the converter looked very attractive, the additional power circuit components, control complexity, and additional losses in the resonant circuit did not favor it for drive application.

FIGURE 4.68   RLDC inverter with current initialization control.

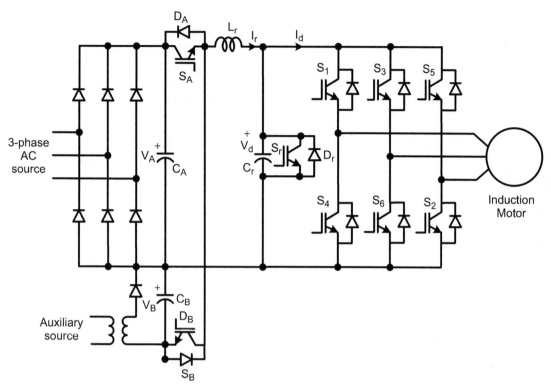

Current initialization circuit

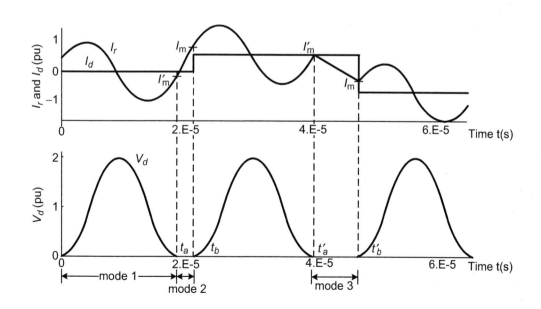

The programmable initial current control in the resonant inductor L_r proposed in this figure [22] adds some complexity but improves the PWM resolution, which was a disadvantage in the previous scheme. The fluctuating inverter load current at the input can be predicted by load current measurements and, accordingly, the initial current of L_r can be programmed so that the bus voltage is always limited to $2V_d$. This corresponds to the no-load operation of the inverter. The explanatory waveforms are shown in the lower part of the figure. Mode 1 indicates that the inverter is at no-load condition, and thus requires small positive initial current in L_r. This current is established by turning on devices S_A and S_r and then turning S_r off at the beginning of the resonant cycle. In mode 2, the input current is higher and, therefore, higher initial current is established. The input current may decrease or be negative. The decrease of initial current in mode 3 is possible by turning off S_A and turning on S_B so that current is established in L_r from the auxiliary dc voltage across the capacitor C_B. In all conditions, the peak value of resonant voltage is the same ($2V_d$). If active clamping is now applied to the inverter, the pulse widths will be uniform.

FIGURE 4.69 Soft-switched auxiliary resonant commutated pole (ARCP) converter.

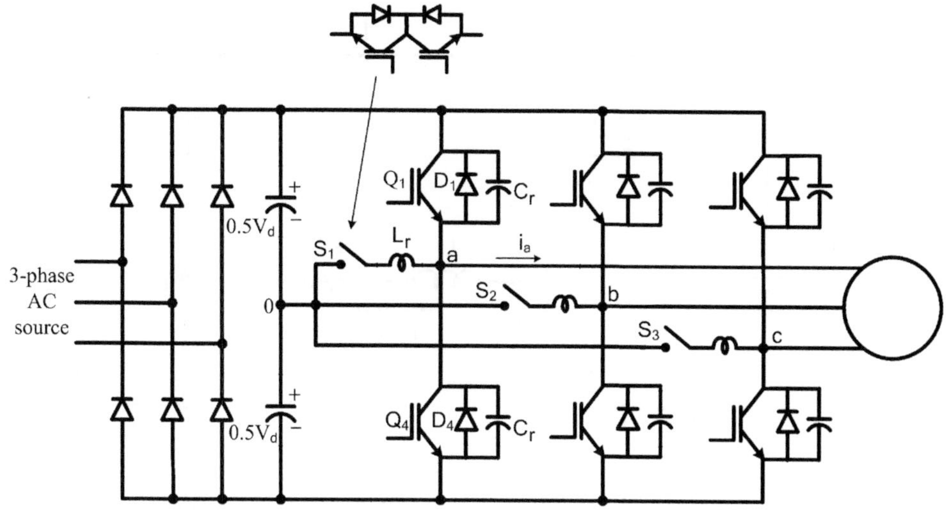

Instead of free resonance in the dc link, as discussed in the previous circuit, the resonance can be activated at the instant of device switching (quasi-resonance) with the help of an auxiliary circuit that establishes zero voltage, as shown in this figure. Each pole or half bridge of the inverter has a common resonant circuit consisting of inductor L_r and shunt capacitors C_r which are connected to the dc link center point through an auxiliary switch S. The capacitor is of the energy recovery type. The inverter has three modes of zero voltage soft-switched commutation: OTOF, RTOF, and DCOM. In OTOF mode, for example in phase a, the large current i_a is initially flowing in Q_1. It is turned off to commutate i_a to the diode D_4. The gradual charging of bypass capacitor C_r provides soft turn-off of the device. If i_a is initially low, resonance circuit-assisted commutation is performed by closing the switch S_1 to boost the Q_1 current. If the phase current is initially flowing through bypass diode D_4, then the resonant circuit is activated to force the current through D_1 so that Q_1 can be turned on at zero voltage. The inverter does not have any voltage or current penalty on the devices, but the disadvantages are extra circuit components, a split-capacitor power supply, and large switching losses in S due to diode recovery current. A regenerative converter unit can easily be added at the input. Again, initially the circuit looked very promising, but it is hardly used because of the circuit complexity and extra losses.

FIGURE 4.70   Soft-switched resonant inverter and explanatory waves.

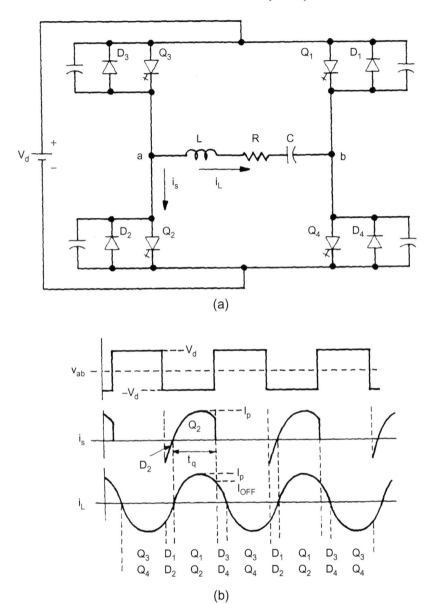

The figure shows a voltage-fed H-bridge square-wave inverter with series resonance circuit load. This is often called a *resonant inverter*. The inverter is soft switched if the inverter frequency is higher than the resonance frequency of the circuit. Although GTOs are used in this circuit, for medium- and low-power applications, other self-controlled devices, such as IGBTs and power MOSFETs, respectively, can be used. The circuit can be used in high-frequency induction heating, resonant link dc-dc converter, etc. The equivalent resistance R constitutes the load, and i_L is the load current. The explanatory waveforms are given in the lower part of the figure. Consider, in the beginning, devices Q_3–Q_4 are conducting and the output voltage v_{ab} is positive. Because $f > f_R$, the load is inductive and the load current is lagging with the voltage v_{ab} wave. For simplicity, assume that the load current is sinusoidal due to filtering effects. At the end of the positive half cycle of v_{ab}, Q_3–Q_4 are turned off, and with a short time delay, Q_1–Q_2 are turned on. At turn-off of Q_3–Q_4, the respective snubber capacitors charge impressing low dv/dt across each device. When charging is complete, v_{ab} becomes negative and inductive load current is fed back to the source through diodes D_1–D_2. Devices Q_1–Q_2 are switched on when their bypass diodes are conducting, thus eliminating the turn-on switching loss. Thus, the inverter uses the ZVS switching principle and the snubber capacitors are lossless. Load current i_L can be regulated either by varying the inverter frequency or supply voltage V_d.

FIGURE 4.71   High-frequency induction heating using SITs.

FIGURE 4.71   High-frequency induction heating using SITs.

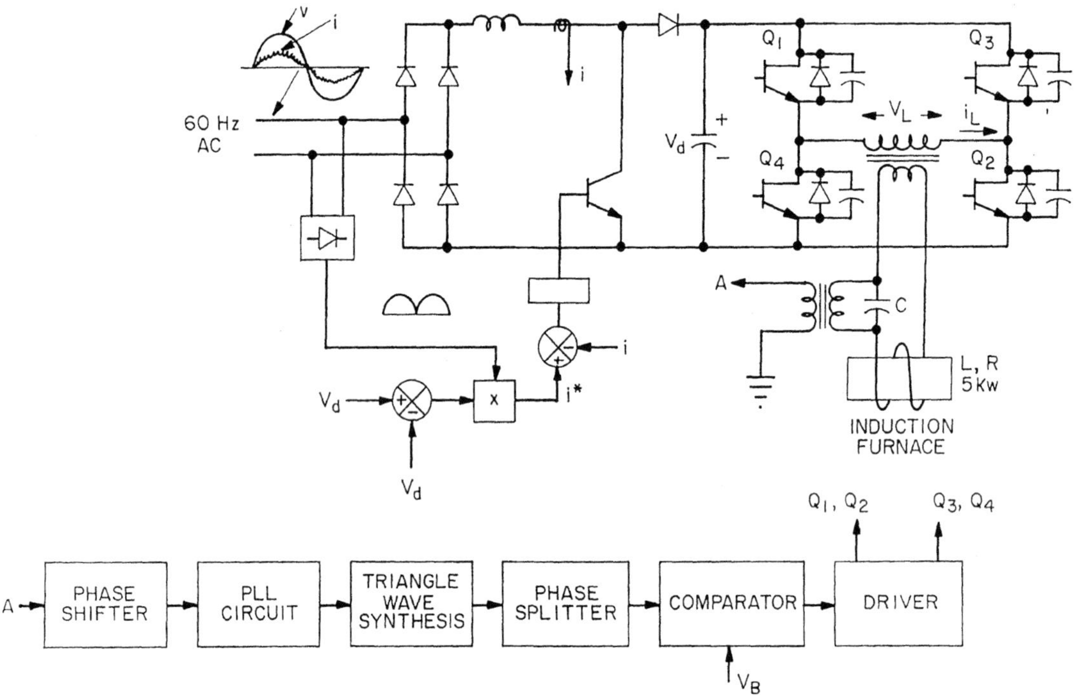

The resonant inverter described in Figure 4.70 is used here in a practical induction heating system (100 kHz, 5 kW) using static induction transistor (SIT) devices. An SIT is a solid-state version of a vacuum triode and it is a normally ON device. The dc supply voltage (V_d) is rectified from a single-phase ac supply through a power factor corrected (PFC) rectifier, described in Figure 4.5. The output of the H-bridge inverter is stepped down by a transformer and applied to the work coil of an induction heating furnace through a series tuning capacitor C. The work coil parameters L and R, the transformer leakage inductance, and the capacitor C constitute the resonance circuit. The inverter is controlled such that its frequency f tracks the resonance frequency f_R ($f > f_R$) that varies with load variation. Thus, the lagging load PF permits ZV soft switching, as explained before. As the work-piece gets heated by the iron loss and its temperature rises, the inductance L will decrease, causing a rise of both f and f_R. This will raise the power output of the inverter. However, V_d can also be regulated to control the power flow. The basic control elements of the inverter are shown in the figure. The capacitor voltage (V_A) that lags the coil current by 90° and the load voltage (V_L) by an additional lag angle is phase shifted to control the inverter. The signal is converted to high frequency by the phase-locked loop (PLL) circuit and then used to synthesize a triangular wave by an up–down counter. This wave is phase splitted and compared with a bias voltage V_B to generate the logic gate drive signals.

FIGURE 4.72 Resonant link dc-dc converter.

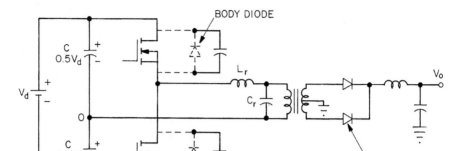

The PWM-type hard-switched dc-dc converters have been discussed in the earlier part of the chapter. If the switching frequency in PWM converter is raised too high for reduction of size of the passive components, the switching loss will be excessive, which will not only adversely affect the converter efficiency, but the additional heatsink size will increase the overall size and weight. The class of high frequency resonant-link dc-dc converters fulfill the need of high efficiency and reduced size, thus promoting their applications in computers, telecommunications, etc. There are wide varieties of this class of converters, but only a typical but popular type, as shown in this figure, will be described. In the circuit, the primary dc source may be a battery, or a rectified dc supply. The supply is split with the help of capacitors and then inverted to high frequency (several MHz) ac by a power MOSFET half-bridge resonant inverter. The resonant capacitor voltage is transformer-coupled, rectified by high frequency Schottky diodes, and then filtered to get the output dc voltage. The inverter is soft-switched because $f > f_R$. Therefore, the efficiency is improved as well as the size of the transformer and the filter parameters are reduced. The transformer provides isolation and buck/boost function. The output voltage is usually controlled by the inverter frequency. For higher power output, an H-bridge inverter can be used. Again, instead of parallel loading as shown, the resonant circuit can be series-loaded, i.e., the transformer with the output circuit can be placed in series with the tuned circuit.

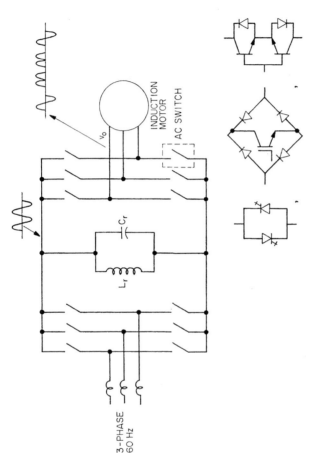

FIGURE 4.73 High-frequency resonant link (ac-HFAC-ac) power conversion.

The concept of resonant ac link power conversion using thyristor phase-controlled cycloconverters was discussed in Chapter 3 (Figure 3.45). The main difference in the present scheme is that thyristors are replaced by self-controlled ac switches. The ac switches (similar to those of a matrix converter) can carry current in either direction and have ac voltage blocking capability. Three different topologies for ac switches are shown of which the inverse-series connection (on the right) is commonly preferred. In the scheme shown, 60-Hz power is first converted to high-frequency ac (typically 20 kHz), which is then converted back to variable-voltage, variable-frequency ac for ac motor drives. Both the input and output converters are soft switched near zero voltage to synthesize the low-frequency sinusoidal voltage wave by the integral half-cycle PWM method illustrated for the output converter. The machine and line currents are nearly sinusoidal by nominal filter inductance, and power can flow in either direction giving the drive a four-quadrant speed control capability. The line-side converter regulates the resonant tank voltage and also controls the input power factor that can be unity, leading, or lagging. Note the performance analogies with the dc link system discussed earlier. Since the resonant tank energy capacity is small, a small mismatch between the input and output instantaneous power will tend to modulate the link voltage, thus tending to cause system instability during fast transients. The link frequency will drift some and waveform deterioration will be seen due to a nonlinear loading effect on the tank, which will cause harmonic deterioration of input and output currents.

FIGURE 4.74 Features of high-frequency resonant link (ac-HFAC-ac) converters.

- HIGH-FREQUENCY RESONANT AC LINK (INSTEAD OF DC LINK)
- ZERO VOLTAGE SOFT SWITCHING
- NEEDS 12 AC SWITCHES OR 6 AC PHASE LEGS (24 IGBTs + 24 DIODES)
- LINK FREQUENCY IS TYPICALLY 20 kHz, BUT CAN BE HIGHER
- FOUR-QUADRANT SPEED CONTROL
- LINE VOLTAGE BROWNOUT COMPENSATION CAPABILITY
- INPUT DPF CAN BE UNITY, LEADING, OR LAGGING
- CAN BE USED AS STATIC VAR GENERATOR (SVG)
- DIFFICULTY OF SWITCHING ZV AT HIGH LINK FREQUENCY
  (QUASI-SOFT SWITCHING)
- PROBLEM OF LINK FREQUENCY DRIFT AND DISTORTION OF WAVEFORM
- STABILITY PROBLEM DURING FAST TRANSIENTS

All essential features of a high-frequency resonant link conversion system are summarized in the figure. Note the performance analogy of the system with a dc link two-sided PWM converter system. In the dc link system, a large electrolytic capacitor stores energy and provides decoupling between the input and output converters. Because of the dc link, the switches are unidirectional but hard switched. The ac link scheme requires a large number of switches, but soft switching may provide some efficiency advantage. However, quasi-soft switching and resonant tank loss tend to offset the efficiency. The scheme was considered for NASA space station application, but today hardly any drives use this scheme.

FIGURE 4.75 Soft-switched nonresonant link (dc-HFAC-ac) converter with integral pulse modulation (IPM).

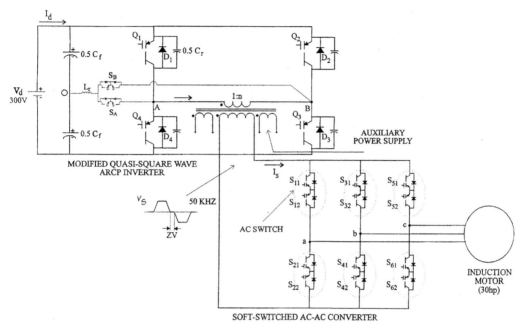

The figure shows a dc-HFAC-ac nonresonant link soft-switched power conversion system [23] in which the input dc voltage is converted to variable-voltage, variable-frequency ac for motor drives through a high-frequency link. The scheme, although somewhat complex, is particularly attractive for electric vehicle drives. The circuit uses a P-type MCT, but IGBTs can also be used. An MCT has lower conduction drop and lower SOA, which are not disadvantages for soft switching. The input dc is first inverted to high-frequency (typically 20 kHz), nonresonant, single-phase ac by a voltage-fed H-bridge inverter that uses a somewhat modified version of an ARCP inverter. The two-phase legs generate a trapezoidal voltage wave with variable zero voltage gap by phase-shifting control. A matrix ac-ac converter operating on the integral half-cycle sigma-delta modulation principle generates the machine voltage waves. The zero voltage gap in the v_s wave provides soft switching of the ac-ac converter. A high-frequency, lightweight, ultra-low-leakage inductance power transformer provides the advantages of galvanic isolation, fewer EMI problems, elimination of the path for common mode machine bearing current, and multiple windings for auxiliary power supplies. The variation of the high-frequency pulse width supplements motor voltage control with a harmonic advantage. Of course, for soft or quasi-soft switching, extremely low Thevenin leakage inductance is desirable. The sluggish rise and fall time of MCT softens the *Ldi/dt* voltage during commutation.

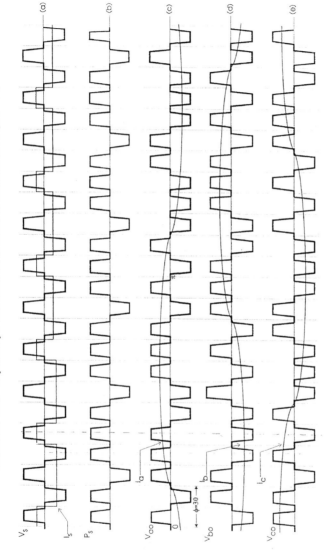

FIGURE 4.76 Explanatory PWM waves for dc-HFAC-ac converter (see Figure 4.75).

The figure explains the fabrication of PWM voltage waves at the machine terminal by the ac-ac converter using the integral half-cycle PWM principle. Each phase group operates independently to fabricate the phase voltages v_{a0}, v_{b0}, and v_{c0} with respect to the transformer center tap using the V_s wave. The fundamental phase currents are shown as perfectly filtered waves at lag angle 30°. The commutation of phase current between the two ac switches of a phase leg occurs at zero voltage with a short overlap. The converter has altogether $3^3 = 27$ switching states. One, two, or three ac switches can commutate at any instant typically at the middle of zero voltage gap, as shown in the figure, and the corresponding waveform of the reflected current I_s (not in correct magnitude) in the transformer is shown. The trapped energy in the transformer leakage inductance (extremely small) during commutation is partially absorbed in the devices and partially in the single R-C snubber.

FIGURE 4.77 Sigma-delta PWM waveform synthesis of dc-HFAC-ac IPM converter (see Figure 4.75).

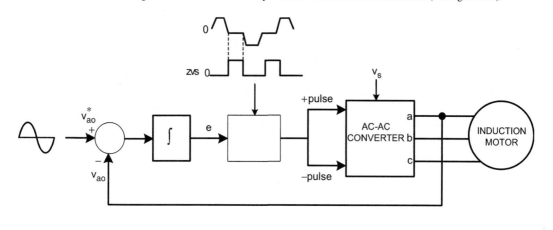

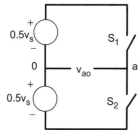

$$e = \int v_{ao}^* \, dt - \int v_{ao} dt$$

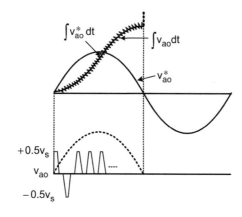

The sigma-delta modulation principle for waveform fabrication in Figure 4.76 is explained in this figure. The command sinusoidal voltage wave is compared with the actual phase voltage wave to generate the integrated loop error signal $e = \int v_{a0}^* \, dt - \int v_{a0} \, dt$ as shown. When the magnitude of e exceeds a threshold value, the positive or negative voltage pulse is generated at the output for the corresponding polarity of the e signal. The positive pulse can be generated by closing switch S_1 in the positive half cycle or by closing S_2 in the negative half cycle, and vice versa. The switches are constrained to operate during the zero voltage gap as indicated. Instead of voltage control by sigma-delta modulation, the fundamental current can also be controlled by delta modulation where the feedback current is compared with the command current and the error signal activates the switches at zero voltage gap.

FIGURE 4.78 Features of dc-HFAC-ac IPM converter (see Figure 4.75).

- ALL THE SWITCHES ARE SOFT SWITCHED (INVERTER – ZVS, HFAC-AC CONVERTER – ZVS)
- NONRESONANT HFAC LINK
- NO OVERVOLTAGE OR OVERCURRENT PENALTY
- MODIFIED BUT SIMPLE ARCP-TYPE INVERTER
- NEEDS TWO DC PHASE LEGS AND THREE AC PHASE LEGS
- ADVANTAGES OF HIGH-FREQUENCY LIGHTWEIGHT TRANSFORMER: ISOLATION, VOLTAGE BOOST, AUXILIARY AC SUPPLY
- FOUR-QUADRANT SPEED CONTROL
- SIGMA-DELTA OR SPACE VECTOR PWM POSSIBLE
- PROGRAMMABLE ZV GAP OF INVERTER PERMITS OUTPUT VOLTAGE CONTROL WITH HARMONIC OPTIMIZATION

All the salient points of a nonresonant high-frequency link IPM converter system are summarized in this figure. Although the system is expensive, there are some special advantages for ac drives, particularly for EV applications.

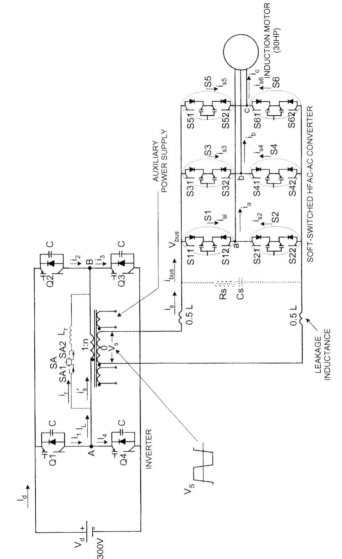

FIGURE 4.79 Soft-switched nonresonant link dc-HFAC-ac phase-controlled converter.

This is the topology of another soft-switched dc-HFAC-ac nonresonant link conversion system [24] for ac motor drives with similar virtues as in Figure 4.75. In this case, a square (or trapezoidal) wave (typically at 20-kHz frequency) is generated by the modified ARCP single-phase inverter. The ac-ac converter generates the variable-voltage, variable-frequency output waves by the conventional phase control principle. The phase control permits zero current soft switching of the devices, where the transformer leakage inductance is gainfully utilized. In addition, phase control generates lagging input current (i_s), which helps zero voltage soft switching of the inverter. At light load, the inverter requires resonance-assisted commutation (RTOF), as explained previously. Since the ac link frequency is high, self-controlled ac switches are used instead of conventional anti-parallel thyristors, which are too slow. Again, P-type MCTs are shown, which can also be replaced by IGBTs. The high-frequency link transformer provides the usual advantages discussed before.

FIGURE 4.80 Explanatory waves of half-bridge dc-HFAC-ac phase-controlled converter (see Figure 4.79).

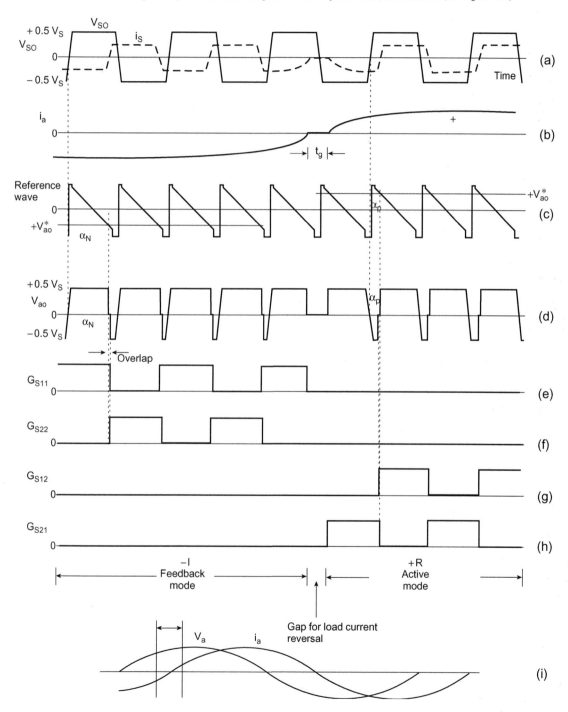

The figure shows explanatory voltage and current waves for a half-bridge ac-ac converter (phase a) with respect to the transformer center tap for a high-frequency, nonresonant link phase-controlled converter system. The waveforms are shown only when the phase current changes from $-i_a$ to $+i_a$ polarity. The operation of the other phase groups is similar. A sawtooth reference wave with advance and retard limits is compared with the phase voltage command v_{a0}^* to generate the firing angle (α) control signals, and the device gate signal waves are shown in parts (e)–(h) of the figure. The fundamental voltage is always positive when the converter transitions from negative converter inversion mode to positive converter rectification mode. In the inversion mode, $\alpha_n > 90°$, whereas in the rectification mode, $\alpha_p < 90°$. Note that in part (c), v_{a0}^* polarity is reversed when i_a is negative. During commutation overlap, the bus is shorted, creating a zero voltage gap in the v_{a0} wave. When i_a changes polarity, a time gap (t_g) is provided to prevent a short circuit. Evidently, the reflected current in the high-frequency link will lag, as shown in the figure.

FIGURE 4.81 Features of dc-HFAC-ac phase-controlled converter (see Figure 4.79).

- ALL DEVICES ARE SOFT SWITCHED (INVERTER – ZVS, HFAC-AC CONVERTER – ZCS)
- NONRESONANT HFAC LINK
- NO PENALTY FOR OVERVOLTAGE OR OVERCURRENT
- LAGGING INVERTER LOAD CURRENT PERMITS EASY COMMUTATION
- SIMPLE RESONANCE-ASSISTED COMMUTATION AT LOW CURRENT
- NEEDS TWO DC PHASE LEGS AND THREE AC PHASE LEGS
- HIGH-FREQUENCY, LIGHTWEIGHT TRANSFORMER – ISOLATION, VOLTAGE BOOST, AUXILIARY SUPPLIES
- SOME KVA PENALTY OF TRANSFORMER DUE TO PHASE CONTROL
- FOUR-QUADRANT SPEED CONTROL
- IMPROVED EFFICIENCY, POWER DENSITY, AND RELIABILITY

The figure summarizes all of the features of a dc-HFAC-ac nonresonant link system using the phase control principle in the ac-ac converter. All essential features of Figure 4.75 are also valid here. Note that self-controlled, high-frequency ac switches are used for phase control instead of the traditional thyristors. The only disadvantage is that the phase control generates lagging VA that loads the transformer. Therefore, the transformer rating is somewhat higher.

Summary

This chapter deals very comprehensively with the voltage-fed-type converters which are most commonly used in power electronics. The power devices in voltage-fed converters are only forward blocking with a bypass diode (fast recovery) so that current can freely flow in the reverse direction. Dc-dc converters constitute a very important subject in power electronics, and they are extensively used in switching mode power supplies (SMPS), chopper drives, etc. However, this coverage in the chapter is very brief. First, dc-dc converters are discussed with PWM control. With high switching frequency, the size of the passive components tends to be small, but efficiency tends to decrease because of high switching losses. Besides, the size of the heat sink tends to be bigger. These problems are overcome to some extent in resonant link dc-dc converters, which will be discussed later. Power-factor-corrected nonregenerative rectifiers are discussed that solve the poor form factor problem of line current wave at the cost of converter complexity and extra losses. Voltage-fed inverters can be either square-wave type or PWM type. Control of a square-wave inverter is simple and switching loss is low, but the penalty is the higher output harmonics demanding bulky filters. A number of phase-shifted bridge inverters can synthesize a multistepped wave that has low harmonic content and the switching loss is low because the devices switch at fundamental frequency.

However, there is the penalty of a heavy, low-frequency transformer. Because GTO switching loss is high, multistepped GTO inverters of high capacity have found popular applications in utility systems. PWM inverters are extremely important for motor drives where the nominal motor leakage inductance filters the harmonics. They are also used in UPS systems that have an extra filter, but the filter size is small. Among the different PWM techniques, sinusoidal and space vector PWM are very common. The latter is used for an isolated neutral load. Sigma-delta and delta modulation methods are used in soft-switched converters. A standard bridge inverter is a two-level inverter. By definition, a multilevel converter has more than two levels. This class of inverters can have, in principle, any number of levels. It is a hot topic of current research in power electronics. Currently, two-sided three-level diode-clamped PWM inverters are finding increasing applications in large power ac drives tending to replace the cycloconverter drives. However, the five-level inverters on the horizon are looking attractive for higher power levels. In utility systems, multi-level converters (as rectifiers) show large promise for applications as static VAR generators, which are the core elements of flexible ac transmission systems [27]. A multistepped, multilevel converter system or utility system has the advantages of low converter switching losses and small harmonic output. All the PWM techniques are also applicable to multilevel converters. PWM inverters can be operated in undermodulation as well as in overmodulation regions to extract the maximum fundamental voltage, but with the penalty of lower order harmonics in overmodulation. The chapter discusses three-level inverter SVM control in undermodulation and overmodulation extensively. The same principle can be extended to inverters with a larger number of levels. Two-sided PWM converter systems, although expensive, have the advantages of four-quadrant motor drives, line harmonics and power factor improvements, and line voltage sag compensation. Although GTO devices have been used in high-power converters, currently IGBTs and IGCTs are tending to replace GTOs because of higher switching frequency (lower switching loss). The same PWM rectifier topology can be used for static VAR generators and active harmonic filters. However, the latter requires higher device switching frequencies. Active filters can again be combined with passive filters (hybrid filters). A unified power line conditioner with the PWM rectifier topology can perform multiple functions, such as harmonic filtering, damping instability, load balancing, reactive power control, and voltage flicker compensation. Matrix converters with high–frequency, self-controlled ac switches have essentially same features as two-sided PWM converters, but the former use more devices and expensive ac capacitors (instead of dc capacitors) at the input. In spite of repeated attempts to use them commercially over a long period, it is doubtful whether they will ever fly successfully. A converter with self-controlled switches can be hard switched or soft switched. The traditional converters are based on hard switching in spite of the disadvantages of high switching losses, device stress, EMI problems, motor insulation deterioration, and voltage boost at the machine terminal with long cables. In spite of great promise over a long period of R&D, soft-switched converters are hardly used in motor drives today. However, their applications are very common in resonant link dc-dc converter, high frequency induction heating, etc. High-frequency resonant link power converters have the usual advantages of soft switching, but the additional advantages are transformer isolation, auxiliary power supply, and voltage level change. However, these converters are expensive, and have control complexity and difficulty maintaining stable and high-quality link voltage waves. Both resonant and nonresonant type conversion systems with their advantages and disadvantages were discussed in the chapter. A number of interesting but practical converter applications in the high-power range have been discussed including battery storage, locomotive propulsion drive, static VAR generator, high-voltage dc-link utility interties, induction heating, and MAGLEV transportation systems. Advances and trends of converters will be discussed at the end of the next chapter.

References

[1] B. K. Bose, "Power electronics—a technology review," *Proc. IEEE*, vol. 80, pp. 1303–1334, August 1992.

[2] H. Akagi, "The state-of-the-art of power electronics in Japan," *IEEE Trans. Power Electronics*, vol. 13, pp. 345–356, March 1998.

[3] B. K. Bose, "Recent advances and trends in power electronics and drives," *Proc. NORPIE*, Helsinki, pp. 170–181, August 1998.

[4] B. K. Bose, *Modern Power Electronics and AC Drives*, Prentice Hall, Upper Saddle River, NJ, 2002.

[5] N. Mohan, T. M. Undeland, and W.P. Robbins, *Power Electronics*, John Wiley, New York, 1995.

[6] B. K. Bose, *Modern Power Electronics*, IEEE Press, New York, 1992.

[7] L. H. Walker, "10-MW GTO converter for battery peaking service," *IEEE Trans. Ind. Appl.*, vol. 26, pp. 63–72, January/February 1990.

[8] T. Sukegawa, K. Kamiyama, K. Mizuno, T. Matsui, and Okuyama, "Fully digital, vector-controlled PWM VSI-fed ac drives with an inverter dead-time compensation strategy," *IEEE Trans. Ind. Appl.*, vol. 27, pp. 552–559, May/June 1991.

[9] J. Pinto, B. K. Bose, L. daSilva, and M. Kazmierkowski, "A neural network based space vector PWM controller for voltage-fed inverter induction motor drive," *IEEE Trans. Ind. Applicat.*, vol. 36, pp. 1628–1636, November/December 2000.

[10] S. Tadakuma, S. Tanaka, and H. Tnoguchi, "Consideration on large capacity PWM inverter for LSM drives," *Proc. Intl. Power Electronics Conference,* Tokyo, pp. 413–420, 1990.

[11] H. Ikeda, S. Kaga, Y. Osada, J. Kitano, K. Ito, Y. Mugiya, and K. Tutumi, "Development of power supply system for Yamanashi MAGLEV test line," *Proc. PCC-Nagaoka*, pp. 37–41, 1997.

[12] S. K. Mondal, B. K. Bose, V. Oleschuk, and J. Pinto, "Space vector pulse width modulation of three-level inverter extending operation into overmodulation region," *IEEE Trans. Power Elec.*, vol. 18, pp. 604–611, March 2003.

[13] C. Wang, B. K. Bose, V. Oleschuk, S. Mondal, and J. Pinto, "Neural network based SVM of a 3-level inverter covering overmodulation region and performance evaluation on induction motor drives," *IEEE IECON Conference Record,* pp. 1–6, 2003

[14] J. S. Lai and F. Z. Peng, "Multilevel converters—a new breed of power converters," *IEEE Trans. Ind. Appl.*, vol. 32, pp. 509–517, May/June 1996.

[15] N. P. Filho, J. Pinto, B. K. Bose, and L. da Silva, "A neural-network-based space vector PWM of a five-level voltage-fed inverter," *IEEE IAS Annual Meeting Conference Record,* pp. 2181–2187, 2004.

[16] D. C. Griffith, *Uninterruptible Power Supplies*, Marcel Dekker, New York, 1989.

[17] M. M. Bakran and H. G. Eckel, "Evolution of IGBT converters for mass transit applications," *IEEE IAS Conference Record,* pp. 1930–1935, 2000.

[18] T. Nakajima, "Development and testing of prototype models of a 300 MW GTO converter for power system interconnection," *IEEE IECON Conference Record,* pp. 123–129, 1997.

[19] M. Hirakawa, N. Eguchi, M. Yamamoto, S. Konishi and Y. Makino, "Self-commutated SVC for electric railways," *Intl. Conference Record Power Electronics and Drive Systems (PEDS)*, pp. 732–737, 1995.

[20] D. M. Divan, "The resonant DC link converter—a new concept in static power Conversion," *IEEE IAS Annual Meeting Conference Record,* pp. 648–656, 1986.

[21] T. S. Wu, M. D. Beller, A. Tehamdjou, J. Mahadavi, and M. Ehsani, "A review of soft-switched dc-ac converters". *IEEE Trans. IAS*, vol. 34, pp. 847–860, July/August 1998.

[22] B. K. Bose, "Soft-switched power conversion for motor drives," *Intl. Aegean Conference Elec. Machines and Power Elec. (ACEMP) Record,* Kusadasi, Turkey, pp. 437–446, June 2001.

[23] L. Hui, B. Ozpineci, and B. K. Bose, "A soft-switched high frequency non-resonant link integral pulse modulated dc-dc converter for ac motor drive," *IEEE IECON Conference Record,* pp. 726–732, 1998.

[24] B. Ozpeneci and B. K. Bose, "A soft-switched performance enhanced high frequency non-resonant link phase-controlled converter for ac motor drive," *IEEE IECON Conference Record,* pp. 733–739, 1998.

[25] H. Ogiwara et al., "Development of SIT high frequency resonant inverter for metal melting uses," *Proc. PCI,* pp. 140–155, 1987.

[26] T. Ohmae and K. Nakamura, "Hitachi's role in the area of power electronics for transportation," *IEEE-IECON Conference Record,* pp. 714–718, 1993.

[27] S. Bhattacharya et al., "Convertible static compensator: voltage source converter based FACTS application in the New York 345 kV transmission line," *Intl. Power Electronics Conference Record,* pp. 2286–2294, Niigata, Japan, April 2005.

[28] H. Akagi, "Active harmonic filters," *Proceedings of the IEEE*, 2006 (to be published).

CHAPTER 5

Current-Fed Converters

FIGURE 5.1 General features of current-fed converters.

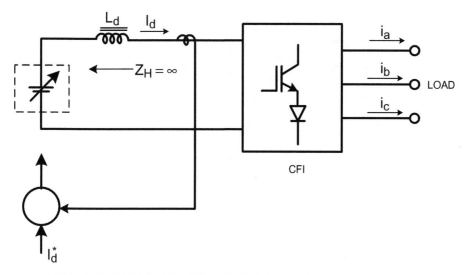

- NEEDS BULKY IRON-CORED DC LINK INDUCTANCE
- PROBLEMS OF COST, SIZE, WEIGHT AND LOSS
- NEEDS DEVICES WITH SYMMETRIC BLOCKING CAPABILITY
- LOAD IMPEDANCE DOES NOT AFFECT THE LOAD CURRENT WAVE
- FOR PWM OPERATION, ONE SIDE SHOULD BE CAPACITIVE AND OTHER SIDE INDUCTIVE. FOR PHASE CONTROL, BOTH SIDES CAN BE INDUCTIVE
- SLOW DYNAMIC RESPONSE
- CONVERTER SYSTEM IS INHERENTLY REGENERATIVE WITH 4-QUADRANT DRIVE CAPABILITY
- INABILITY TO OPERATE AT NO-LOAD CONDITION
- PARALLEL CONVERTER OPERTION IS DIFFICULT
- MORE RUGGED AND RELIABLE OPERATION—NO SHOOT-THROUGH PROBLEM
- SIMPLER FAULT PROTECTION

The PWM current-fed inverter requires a dc current source as indicated in the figure. The current source can be generated from a variable dc voltage source (via a dc-dc converter from dc voltage or phase-controlled rectifier from ac voltage) by feedback control of dc current. The salient features of current-fed converters (inverters or rectifiers) are summarized in this figure. The phase-controlled converters discussed in Chapter 3 also fall in this class. The power devices that are reverse blocking, such as thyristors, GTOs, IGCTs, or IGBTs, with a series blocking diode can be used. The other features will be evident later.

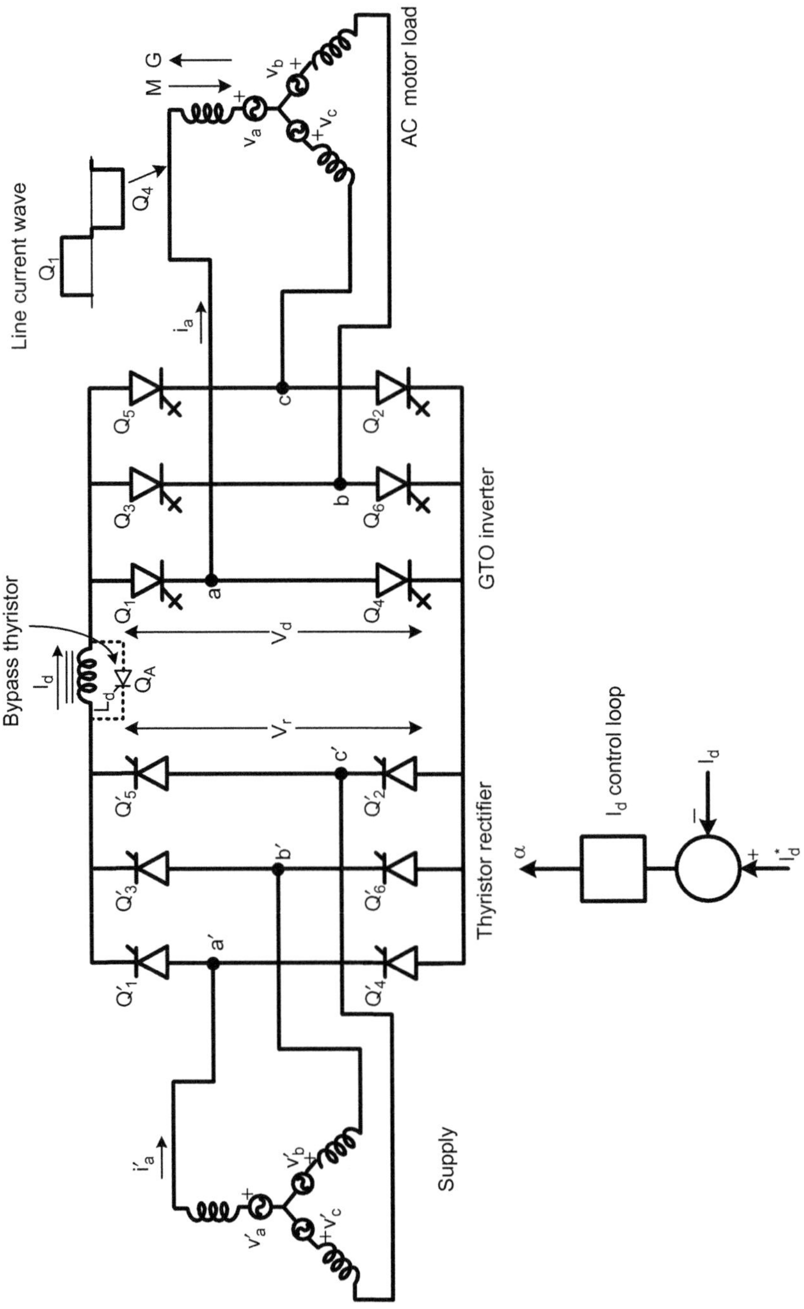

FIGURE 5.2   General power circuit of current-fed converter systems.

The figure shows a general power circuit for a three-phase current-fed inverter that uses reverse or symmetric blocking GTO devices. The inverter operates in square-wave mode to generate six-step (or six-pulse) current wave to the load. The load is shown as an equivalent circuit of induction or synchronous motor, which is represented by counter emf in series with leakage inductance in each phase. The devices can also be thyristors, reverse-blocking IGCTs, or IGBTs with series blocking diodes. The inverter dc current I_d is supplied by a thyristor rectifier through a series inductor L_d as indicated. The function of the bypass thyristor Q_A will be explained later. The thyristors are phase controlled to generate a variable dc voltage at the output, and a feedback current control loop generates the variable dc current for the inverter, as needed. For ideal dc current without any harmonics, inductor L_d should be infinity. If L_d is to be of a practical size, some ripple will have to be introduced in I_d, which can be tolerated. The rectifier thyristors are commutated by the ac line voltages (line commutation). The dc current I_d is steered through the inverter by selectively switching on its devices so as to construct symmetrical six-step current in each phase at 120° phase difference. Basically, the inverter has the same topology as the rectifier, and their operational modes are somewhat similar. The GTOs can be commutated by the load, or they can be self-commutated by the gate. Only two devices (one on the positive side and another on the negative side) of the inverter conduct at the same time with a conduction period of 120°, like that of the rectifier. For a variable-speed motor drive, the inverter can generate variable-frequency, variable-current power supply. At steady state, the current is continuous and $V_r = V_d$ if the inductor resistance is neglected.

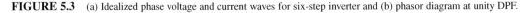

FIGURE 5.3   (a) Idealized phase voltage and current waves for six-step inverter and (b) phasor diagram at unity DPF.

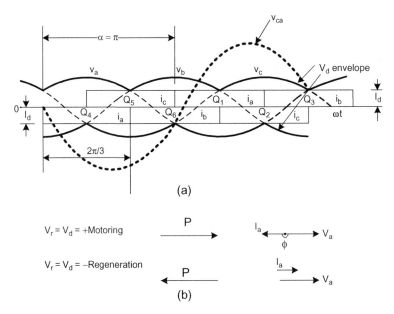

The figure shows the idealized six-step current waves in relation to the respective phase counter emf waves at firing angle $\alpha = 180°$ so that the fundamental current wave is in phase opposition with the voltage wave, i.e., the inverting mode of operation. For example, when Q_1 is fired, the current is transferred to it from Q_5, and then $+i_a$ flows through Q_1 and $-i_b$ flows through Q_6. The power is absorbed by the phase counter emfs because the respective phase voltage polarities are opposite. At this condition, the line-side converter is acting as a rectifier supplying power to the inverter. Since I_d is always positive, both V_d and V_r are positive at this condition. For this motoring mode operation of the inverter, the fundamental frequency phasor diagram is shown in the figure. If firing angle α of the inverter is reduced to zero, the current waves are advanced by 180° so that the inverter acts as a rectifier pumping power to the dc link from the machine, which acts as a generator. Under this condition, the line-side converter acts as an inverter (making both V_d and V_r negative), and regenerated power is fed to the line. The phasor diagram of the machine in regenerative mode is shown at the bottom of the figure. Let us consider the commutation of line-side converter (rectifier) in Figure 5.2 again. This converter is always line commutated irrespective of rectification or inversion mode. The line commutation implies lagging VAR demand at the input of the converter. From the same consideration, we can infer that the load-side converter can be commutated by the load counter emfs (load commutation) if it can supply lagging VAR to the converter, i.e., the load power factor is leading. With this condition, an outgoing device turns off by a negative voltage segment of the counter emf (see Figure 5.17). The conclusion is that an overexcited synchronous motor load at a leading power factor can provide load commutation, whereas an induction motor load at lagging power factor requires some type of forced or self-commutation.

FIGURE 5.4   Idealized phase a voltage and current waves showing different modes of operation ($\alpha=\phi$).

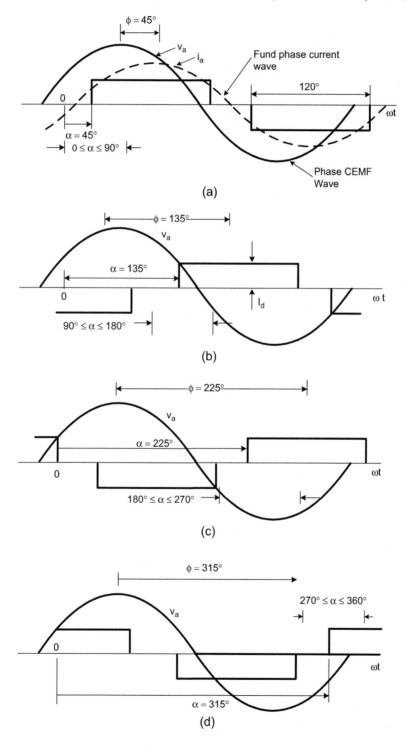

The figure explains the different modes of converter operation when the firing angle α is varied in the range of $0°$ to $360°$ with respect to the phase counter emf waves. If the firing angle is between $0°$ and $90°$, as shown in (a), the machine supplies real power to the inverter, but it absorbs leading reactive power (i.e., supplies lagging VAR). This is the load-commutated rectifier mode of operation, which means that an overexcited synchronous machine is operating in regenerative mode. At this condition, the line-side converter operates as an inverter and pumps power to the line. If the firing angle range is between $90°$ to $180°$, as shown in (b), the machine receives active power (motoring mode), and the load power factor is leading. This is the load-commutated inverter mode of operation with overexcited synchronous motor load. The load power factor is lagging in the waveforms of (c) and (d) and, therefore, the converter requires forced or self-commutation. The former case can be considered an induction machine load operating in the motoring mode, whereas the latter case is an induction machine operating in the regenerating mode.

FIGURE 5.5 Summary of ac machine operation modes.

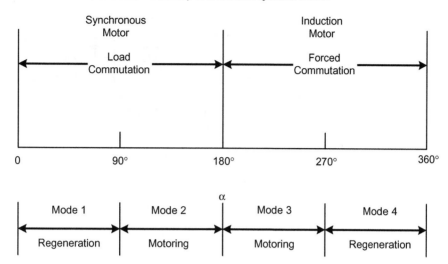

Mode 1: LOAD-COMMUTATED RECTIFIER
(SYNCHRONOUS MACHINE REGENERATION)

Mode 2: LOAD-COMMUTATED INVERTER
(SYNCHRONOUS MACHINE MOTORING)

Mode 3: SELF OR FORCE-COMMUTATED INVERTER
(INDUCTION MACHINE MOTORING)

Mode 4: SELF OR FORCE-COMMUTATED RECTIFIER
(INDUCTION MACHINE REGENERATION)

All the converter operation modes discussed previously have been summarized in this figure. It is assumed that a synchronous machine always operates in an overexcitation mode to give a leading power factor. Although machine operation is illustrated in each mode, the load can be simple passive with inductance or capacitance. Since load-commutated inverter operation is equivalent to line-commutated rectifier operation, simple thyristors can be used in the inverter. Thyristors can also be used in modes 3 and 4, but more complex forced commutation is needed. Self-commutation with the help of a GTO, IGCT, or IGBT (with series diode) is more preferable in modes 3 and 4. Different types of current-fed converters with forced, load, and self-commutation will be discussed in the remaining part of the chapter.

FIGURE 5.6   Progression of current-fed converters.

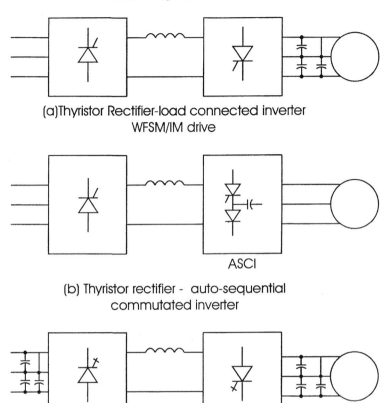

(a)Thyristor Rectifier-load connected inverter
WFSM/IM drive

ASCI

(b) Thyristor rectifier - auto-sequential
commutated inverter

(c) Double-Sided PWM Converters

Different classes of current-fed converters are summarized in this figure. A simple thyristor phase-controlled rectifier generates a variable dc current source in (a) and (b). With overexcited synchronous machine load in (a), a simple phase-controlled inverter can be used. Similar load commutation with leading power factor load is possible with induction motor load if a suitable capacitor bank, as shown, is connected at the machine terminal. In (b), the load is an induction motor and, therefore, forced commutation of inverter thyristors is needed. The type of forced commutation used is called ASCI (autosequential commutated inverter), which will be described later. Figure 5.6(c) shows a PWM current-fed converter system, which requires self-commutated devices. PWM operation requires a capacitor bank on the ac side. All the topologies can give active power flow in either direction. The current-fed converters are normally used when larger power ranges are required. Current-fed converter drives, except for the multimegawatt load-commutated synchronous motor drives, have recently fallen into disfavor.

FIGURE 5.7 Autosequential current-fed inverter (ASCI) with CEMF equivalent circuit of induction motor.

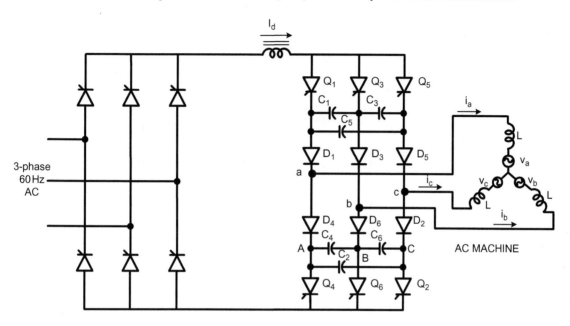

A current-fed inverter for controlling the speed of induction machines requires either forced or self-commutation of the devices. This figure shows the once-popular ASCI inverter, which uses thyristor devices with forced commutation. The circuit is used for large power drives. Although a large number of ASCI inverter drives still exist in the industry, they have recently become obsolete because of the disadvantages of force commutation, however, we discuss them briefly for completeness of the subject. The dc current source I_d is generated with the help of phase-controlled rectifier, as usual. The inverter uses six thyristors $(Q_1 - Q_6)$ that conduct in sequence for 120° to establish the six-stepped load current wave as indicated in Figure 5.3. The series diodes and the capacitor banks (all with equal values), which are connected to each of the upper and lower groups of thyristors, are for forced commutation. The capacitors with stored charge of correct polarity and the isolating diodes help forced commutation. Every 120° interval has a commutation to transfer current from the outgoing device to the incoming device in upper and lower banks, i.e., there are six commutations per cycle. If commutation effect is neglected, the phase voltage and current waves are ideally the same as in Figure 5.3.

FIGURE 5.8   ASCI inverter equivalent circuit with induction motor load during commutation from Q_2 to Q_4.

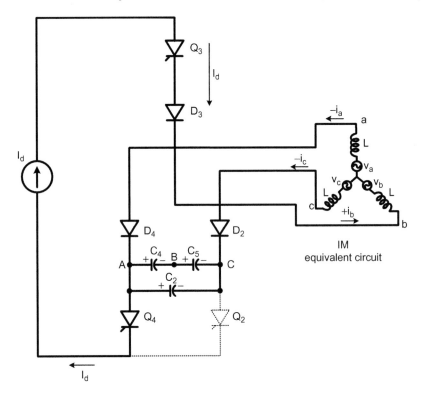

The figure explains ASCI inverter forced commutation for current transfer from thyristor Q_2 to Q_4 in the lower group of the bridge with the correct polarity of the capacitor bank voltage shown in the figure. The other five commutations in the cycle are similar. When incoming thyristor Q_4 is fired, Q_2 is turned off instantaneously by the negative capacitor voltage. The dc link current I_d then flows through Q_3 and D_3 in the upper group, phases b and c of the machine, device D_2, the delta capacitor bank, and Q_4 to the negative polarity. The capacitor bank then charges linearly with the dc current I_d. During the constant current charging mode of the capacitor bank, there is no drop in machine inductance L and diode D_4 remains reverse biased by the dominating machine line counter emf v_{ca}. The linear charging period ends when the capacitor bank voltage reverses and equals the machine line voltage (negative polarity), and then diode D_4 begins to conduct. Current I_d then resonantly transfers to D_4, which completes the commutation process. During fast resonant current transfer, a large voltage spike (Ldi/dt) is induced and adds in series with the machine CEMF. This is a serious problem for ASCI inverters. For this reason, the thyristor voltage should be overrated, or the machine should be designed with low leakage inductance. The spike voltage can also be absorbed externally by a clamping circuit.

FIGURE 5.9   Features of ASCI inverters.

- POWER FLOW IN EITHER DIRECTION (FOUR-QUADRANT DRIVE OPERATION)

- SIX-STEP LINE CURRENT AT LAGGING DPF

- SIX-STEP LOAD CURRENT AT VARIABLE FREQUENCY

- PROBLEM OF MACHINE HEATING AND TORQUE PULSATION

- COMPLEXITY OF POWER CIRCUIT

- BULKY DC LINK INDUCTOR

- MACHINE PARAMETER INTERACTION WITH COMMUTATION CIRCUIT

- PROBLEM OF OVERVOLTAGE SPIKE AT MACHINE TERMINAL AND STABILITY

- SIMPLE CONTROL

- SOMEWHAT OBSOLETE IN NEW INSTALLATION

The salient features of the force commutated ASCI inverter shown in Figure 5.7 are summarized here. The circuit is used for a lagging power factor load, such as an induction motor. As usual, the power flow is bidirectional, giving the drive system four-quadrant characteristics. For regenerative braking, the machine acts as an induction generator, the ASCI inverter acts as a rectifier, and the line-side rectifier acts as an inverter. The control of the inverter is simple. The line and load currents are ideally six-stepped waves with large harmonic content. Since machine frequency is variable, the low-frequency current at low speed gives severe torque pulsation and harmonic heating problems. The limited PWM mode of operation of the inverter can alleviate these problems. The line power factor is low due to phase control. Besides the complexity of the power circuit, the dc link inductance is bulky for reasonable smoothness of dc current I_d. ASCI inverter drives have been widely used in medium to large capacity equipment in industry. The advent of self-controlled devices with reverse blocking gradually made ASCI inverters obsolete.

FIGURE 5.10 Single-phase parallel resonant inverter with load commutation.

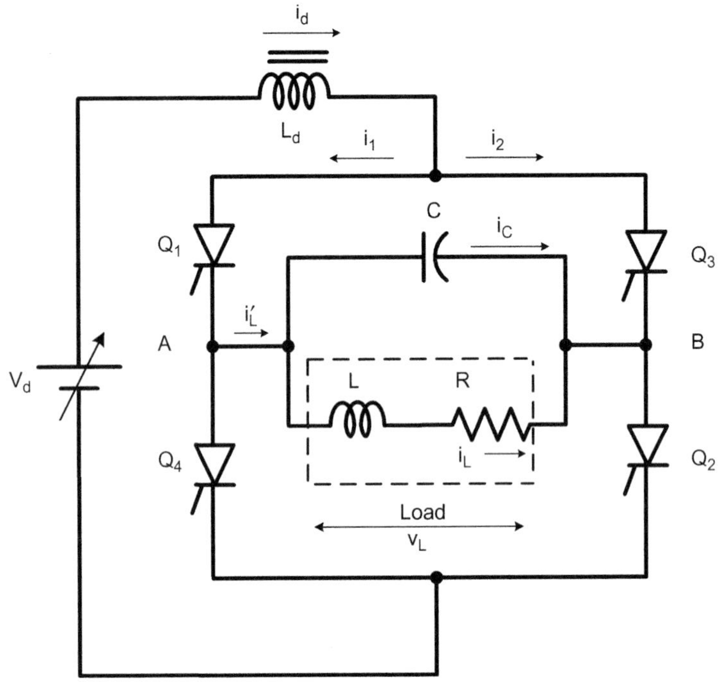

A single-phase H-bridge current-fed inverter that feeds a parallel resonant circuit load is shown in this figure. The variable dc current source I_d is generated by a phase-controlled rectifier (single or three phase) through a dc link inductor L_d, as indicated in the figure. The circuit has been widely used for high-frequency induction heating applications, where it is a competitor to a voltage-fed inverter with series resonant load (see Figure 4.70). Basically, the load in this case is a series R-L circuit. A capacitor C of sufficiently high value is connected across the load so that at fundamental frequency, the effective load has a leading power factor. The inverter operates in square-wave mode, feeding a square-wave current to the resonance circuit. If the inverter frequency is higher than the load resonance frequency, the load power factor will be leading, permitting load commutation of the thyristors. Normally, the load parameters are variable, which will cause variations in the resonance frequency. With a fixed value of C, the inverter frequency can be varied, tracking the variation of resonance frequency so that a fixed load power factor can be maintained. The dc current I_d can be varied to control the output power. Again, with load commutation, thyristors can be replaced by GTOs, IGCTs, or IGBTs with a series diode.

FIGURE 5.11 (a) Load voltage and current waves and (b) phasor diagram.

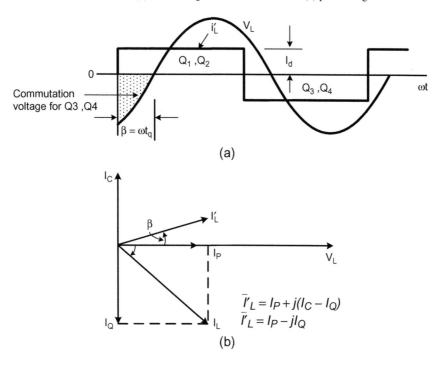

(a)

(b)

$$\bar{I'}_L = I_P + j(I_C - I_Q)$$
$$\bar{I'}_L = I_P - jI_Q$$

The figure shows the load voltage and current waves, and the corresponding phasor diagram for Figure 5.10. The square current wave of ideally constant magnitude I_d is established across the load by alternate switching of the Q_1Q_2 and Q_3Q_4 pairs. There will be some harmonic distortion of the load voltage wave v_L across AB, but it is assumed to be sinusoidal. All the odd harmonics of the square current wave (3rd, 5th, 7th, etc.), particularly the higher harmonics, will essentially short circuit through C causing little distortion on the voltage wave. The load DPF is leading by cos β, where β is the leading angle of the fundamental of the wave with respect to the wave, as shown in the figure. When thyristor pair Q_1Q_2 is switched on , outgoing pair Q_3Q_4 is impressed with a negative voltage segment for duration $\beta°$, causing load commutation. Since, the time t_q should be sufficient for turning off the outgoing thyristors. Note that a load-commutated current-fed inverter is equivalent to a line-commutated phase-controlled rectifier, where the device switching loss is zero by zero current soft switching (ZCS).

FIGURE 5.12   One thousand-kilowatt, 50-kHz parallel resonant inverter system for induction heating.

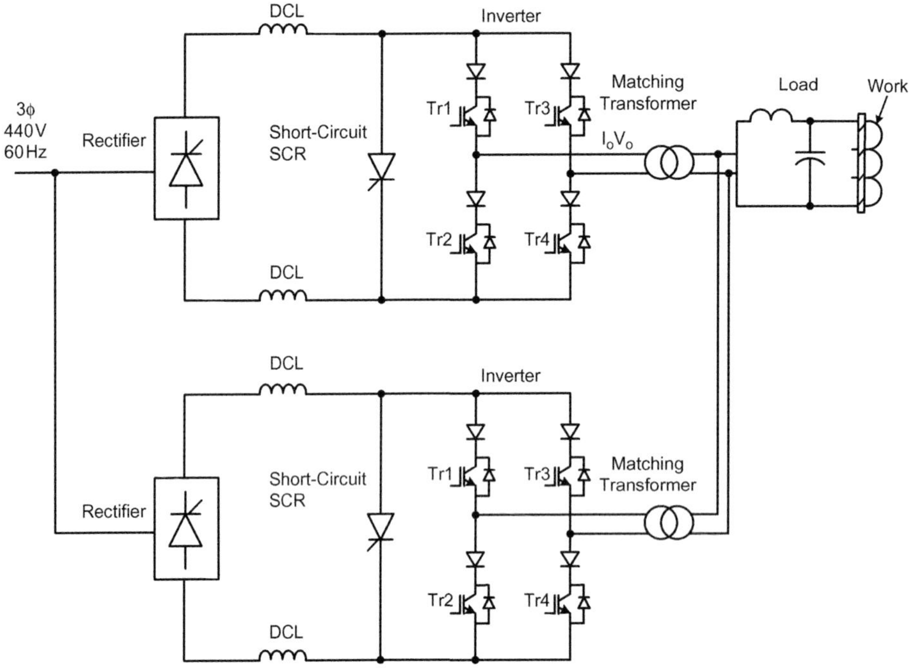

The figure shows a schematic for a 1000-kW, 50-kHz induction heating system [4] that was developed for metal melting, surface heat treatments, and tube welding applications. It uses current-fed H-bridge inverters with parallel resonance circuit load, as discussed earlier. The metal to be heated is placed in the work coil, where the high-frequency current causes hysteresis and eddy current losses (iron loss) that generate heat in the metal. In fact, two inverter circuits, each of 500-kW capacity, are paralleled through a matching transformer (1:2) to couple to the load. The resonant circuit is created by equivalent L and R of the work coil, the capacitor, and the equivalent leakage inductance of the transformer. Since the frequency is very high, thyristors and GTOs are not suitable. IGBTs with series blocking diodes are used in parallel in each arm of the H-bridge. The dc current source for each inverter is supplied by a six-pulse thyristor rectifier, as shown in the figure. Although a current-fed inverter uses bulky dc link inductance, one advantage is that the short circuit protection is easy. In case of load short, a short-circuit silicon-controlled rectifier (SCR) in each circuit is fired to bypass the current I_d from the inverter; inverter firing pulses are inhibited and then rectifier firing pulses are suppressed. The control and protection of the system are microcomputer based. The resonance circuit voltage and current signals are sensed and, correspondingly, the turn-off time angle (see Figure 5.10) is detected. There will be a small overlap angle μ due to inverter leakage inductance, which is considered ($\gamma = \beta - \mu$) for calculation of time t_q. The command $t_q{}^*$ is compared with actual t_q, and an error signal generates the inverter firing signals through a phase-locked loop (PLL). The rectifier is controlled to increase dc link currents when higher power is demanded in the load.

FIGURE 5.13   Features of 1000-kW induction heating inverter systems.

- APPLICATIONS:
  MELTING
  SURFACE HEAT TREATMENT
  TUBE WELDING, ETC.

- 1000-kW LOAD AT 50-ktHz FREQUENCY

- MATCHING LIGHTWEIGHT TRANSFORMERS COUPLE TWO 500-kW UNITS

- PARALLEL RESONANT CURRENT-FED INVERTERS AT LEADING
  PF LOAD – PROVIDE LOAD COMMUTATION (NO SWITCHING LOSS)

- MULTIPLE IGBTs (1200 V, 500 A) IN PARALLEL WITH SERIES
  DIODE – HIGHER FREQUENCY THAN THYRISTOR OR GTO

- SIX-PULSE THYRISTOR RECTIFIER CONTROLS DC CURRENT

- IMPROVED RELIABILITY

- EASIER TO PROTECT THAN VOLTAGE-FED INVERTER

- CONSTANT TURN-OFF TIME CONTROL BY PHASE-LOCKED LOOP

- HIGH INVERTER EFFICIENCY (97%)

All the features of a 1000-kW, 50-kHz induction heating system are summarized in this figure. The system could have been designed using a voltage-fed inverter and series resonance circuit also, but reliability of operation with an easy protection feature dictates this type of system. The power circuit efficiency is claimed to be 97% (excluding transformer losses) in spite of the losses in dc link inductor and additional losses in series diodes of the IGBTs. The load power factor is near unity because turn-off time is very low for IGBTs in comparison with thyristors or GTOs. An IGCT is possibly the better choice, but it did not exist when the system was developed (1995). Note that closer to load unity DPF has been possible by controlling turn-off time t_q instead of turn-off angle γ.

FIGURE 5.14   Three-phase thyristor inverter with load commutation for passive *R-L* load.

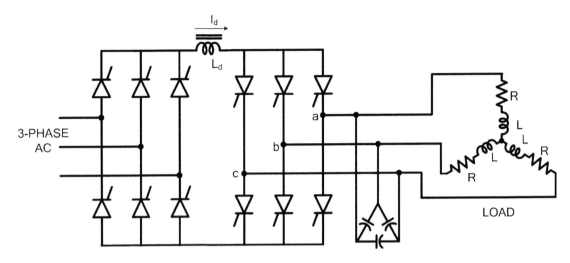

The single-phase parallel resonant current-fed inverter concept discussed earlier can be extended to a three-phase system. This figure shows a three-phase thyristor inverter with passive *R-L* load that can be varied. For such a lagging power factor load, the thyristor inverter requires forced commutation, such as an ASCI inverter. However, with a parallel capacitor bank as shown, resonance is created with the load circuit and load commutation is possible. The inverter frequency should be slightly above the resonance frequency so that the effective load power factor is leading. With fixed capacitance bank, if the load parameters vary, the resonance frequency will vary. Therefore, inverter frequency should track with the load resonance frequency $(f > f_r)$ for satisfactory load commutation.

FIGURE 5.15 Thyristor inverter load commutation with induction motor load, assisted by voltage-fed SVG.

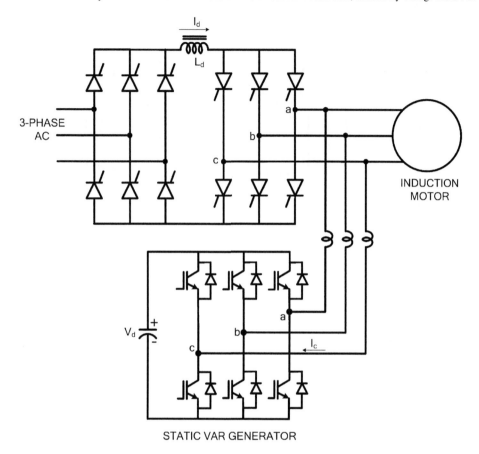

If the load of a current-fed inverter is an induction motor that requires speed control, it is necessary for the inverter to supply a variable-frequency, variable-magnitude current supply. This is made possible by employing a force commutated ASCI inverter, as discussed earlier. If a fixed capacitor bank is connected at the machine terminal as in Figure 5.14, the capacitor current increases with frequency by the relation $I_C = V\omega C = K\omega^2 C$, where $V \propto \omega$ because of the constant flux condition of the machine. Since ω is dictated by the machine speed, the parameter C has to be varied for the correct leading power factor condition at the inverter output terminal. Instead of a variable tuning capacitor, a voltage-fed static VAR generator (see Figure 4.57) can be connected at the machine terminal, as shown in this figure. The SVG can be controlled to act as a variable capacitor load to control the inverter terminal leading power factor to the correct value at all operating conditions of the machine. Although such a scheme is expensive, practical drives that use this configuration have been applied in industry. One advantage of the scheme is that a universal current-fed converter system can be designed for both induction and synchronous motor drives of comparable capacity.

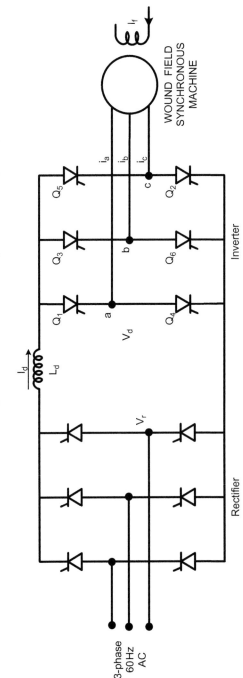

FIGURE 5.16 Load-commutated thyristor inverter with wound-field synchronous motor (WFSM) drive.

The load-commutated thyristor inverter multi-megawatt synchronous motor drives are extremely popular in industry for applications such as pump and compressor drives, ship propulsion, rolling mill drives, and variable-frequency starters for synchronous motors. For high-power applications, instead of 6-pulse converters as shown, 12-pulse converters are used, as discussed later. It is easy to maintain the desired leading power factor angle at the wound-field machine terminal by adjusting the field excitation current. The load commutation with thyristors at high power makes the inverter simple, reliable, cost effective, and more efficient. Slow speed phase-controlled type thyristors can be used in both converters. Of course, the disadvantages of 6-pulse (or 12-pulse) operation, particularly at low frequency, are retained. Note that at low speed, the inverter needs force commutation [5] because the counter emf may not be enough.

FIGURE 5.17 Phase voltage and current waves at synchronous motor terminal for load commutation.

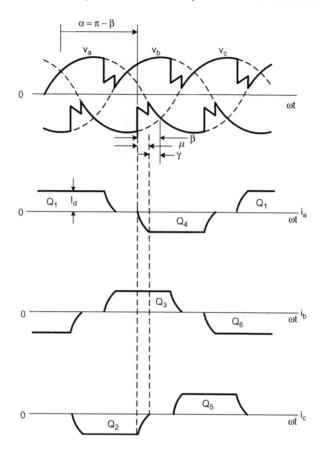

The figure shows the machine terminal voltage and current waves explaining the load commutation effect. These waves are essentially identical with those of the standard phase-controlled inverter that was discussed in Chapter 3. In fact, Figure 5.17 is identical to Figure 3.15, except the latter shows instantaneous current transfer without considering the commutation overlap effect. In this figure, the commutation from Q_2 to Q_4 is shown in detail. When incoming device Q_4 is fired at firing angle $\alpha = 180° - \beta$ (β = advance angle as indicated in the figure), the dc link current I_d begins transferring from Q_2 to Q_4, and at μ angle (defined as overlap angle), this transfer is complete. During the angle γ (defined as turn-off angle), outgoing thyristor Q_2 turns off by a segment of the negative line voltage v_{ac}. The machine phase voltage and current waves indicate that the leading power factor angle φ is slightly less than β ($\gamma < \varphi < \beta$). The machine will go into the regenerative mode of operation if the α angle is reduced to zero, at which point the inverter will ideally act as a diode rectifier maintaining the load commutation. Note that for load commutation to be effective, the machine should have sufficient counter emf, i.e., speed (typically 5%). Therefore, for startup operation, the inverter requires some type of forced commutation, such as discontinuous current pulsing [5].

FIGURE 5.18   Twelve-pulse rectifier-inverter system for load-commutated inverter (LCI) synchronous motor drive.

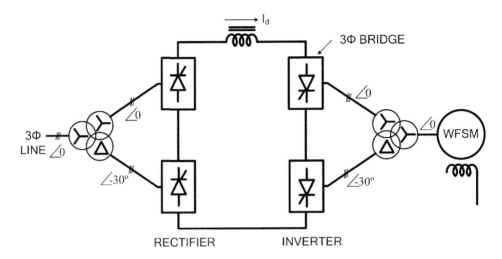

A simplified block diagram of a 12-pulse converter system for high-power synchronous motor drive is shown here. Increasing the number of pulses in a converter is an obvious way to improve the current wave to be near sinusoidal and solve harmonics and machine torque pulsation problems. A 12-pulse, phase-controlled thyristor converter using two 6-pulse bridges in series and phase-shifting transformers was discussed in Figure 3.23. The same circuit has been used as a rectifier and inverter in this figure. The machine with counter emfs is equivalent to three-phase ac line, where load commutation has been substituted for line commutation. A similar converter system is traditionally used in HVDC intertie or transmission systems. In an intertie, the line side and load side have dissimilar frequencies, such as 60 and 50 Hz, respectively. The power flow in the converter system is bidirectional, thus permitting four-quadrant drive capability of the machine. The current harmonics of both the line and load sides are 11th, 13th, 23rd, 25th, etc. The higher harmonics with lower amplitudes reduce harmonic heating and torque pulsation problems in the machine. However, at low speed, the fundamental frequency is small; therefore, the torque pulsation effect can be considerable. With multipulsing, the dc link inductor has to absorb less harmonic voltage and, therefore, its size becomes small. It is evident that to justify the complexity of the converter system, the drive system power rating should be quite high.

FIGURE 5.19 Twelve-pulse load-commutated inverter drive with asymmetric six-phase synchronous machine.

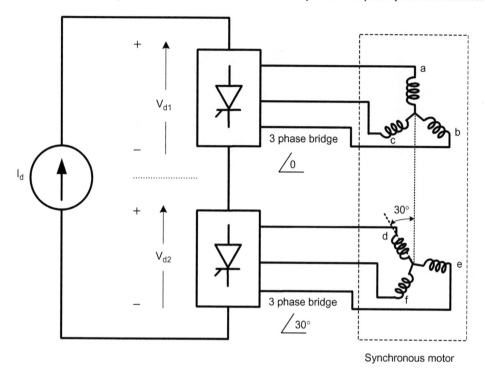

Synchronous motor

Twelve-pulse inverters with phase-shifting transformers, as discussed previously, become very expensive, although the advantage of a standard three-phase machine can be retained. The transformer is usually eliminated by using an asymmetric six-phase machine, as shown in this figure. In this machine, the three-phase winding group *def* is phase advanced by 30° with respect to the *abc* winding group. In a conventional six-phase machine, the three-phase winding groups are displaced by 30°. The advantage of asymmetric connection is that if the *abc* and *def* groups are supplied by the respective three-phase bridge inverters at 30° phase shift angle, as shown in the figure, the resultant magnetic field has 12-pulse harmonics. Therefore, as far as harmonic torque pulsation is concerned, the machine has 12-pulse operation. Of course, the windings carry the usual 6-pulse current wave. The machine counter emfs for each 3-phase group are at 30° phase-shift angles, thus giving 12-pulse operation of the inverter. The same power circuit configuration (12-pulse rectifier, 12-pulse inverter and asymmetric six-phase machine) can be used for high-power induction motor drives except the load-commutated inverter is replaced by a force-commutated or self-commutated inverter drive.

FIGURE 5.20   Six-step self-commutated IGBT inverter with induction motor load and phasor diagram.

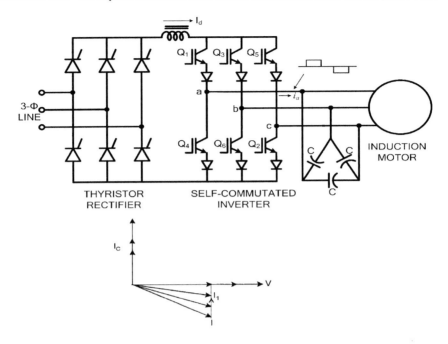

Our study so far indicates that the speed of an induction motor can be controlled by a six-step thyristor inverter if the thyristors are force commutated (such as ASCI inverter) or if load commutation is used with the help of an expensive SVG at the machine terminal. Let us now consider a six-step, current-fed inverter using self-controlled devices like GTOs or IGBTs with series diode, as shown in this figure. The IGBTs can switch to generate a six-step current wave for lagging power factor induction motor load at any arbitrary phase position. However, a capacitor bank is needed at the machine terminal that has the following functions: (1) It provides current transfer from the outgoing phase to the incoming phase when the IGBT is switched without causing undue voltage overshoot. (2) It acts as a low-pass filter so that the machine voltage and current waves are nearly sinusoidal. The distinct advantages of this type of inverter are that the machine harmonic copper loss is substantially low (thus giving better efficiency), there is no pulsating torque problem at any speed, and acoustic noise due to harmonics practically disappears. However, if a harmonic current component falls near the resonant frequency (due to machine leakage inductance and capacitor bank), the corresponding voltage component will be boosted. The inverter can also be operated in PWM mode, as described later. A phasor diagram at the inverter terminal is drawn at increasing frequency (or speed), assuming constant active and reactive currents in the machine. The machine voltage increases proportionally with frequency. At low speeds, both frequency and voltage are low, which make the capacitor current I_C low. Therefore, the inverter terminal power factor is low. At higher speeds, the power factor improves, giving less loading of the inverter. At maximum speed, the power factor is unity with minimum loading of the inverter. With higher capacitance, it may be possible to operate the inverter at load commutation.

FIGURE 5.21 Equivalent circuit during commutation from Q_1 to Q_3.

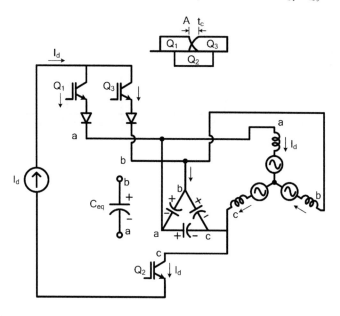

To explain the self-commutation capability of an inverter with the help of an external capacitor bank, consider this figure, which gives, for example, the equivalent circuit for commutation from Q_1 to Q_3 in the upper group. Initially, devices Q_1 in the upper group and Q_2 in the lower group are conducting and the dc link current I_d is flowing through Q_1, phase a, phase c, and Q_2, as indicated. The equivalent capacitance C_{eq} between lines a and b and the polarity of line voltage v_{ba} across C_{eq} are indicated. When the incoming IGBT is turned on, the current will not automatically transfer from Q_1 to Q_3 because of the indicated capacitor voltage polarity. But turning off Q_1 will quickly transfer current to Q_3, which will flow through C_{eq}. This capacitor charges, quickly overcoming the machine counter emf voltage, and the current gradually transfers to phase b with total commutation time t_c, as shown in the figure. Once the commutation is complete, current can be transferred back to Q_1 again, if desired. This back-and-forth commutation can create a PWM current wave in the load, as discussed later.

FIGURE 5.22   Inverter frequency variation to excite machine resonance at harmonics.

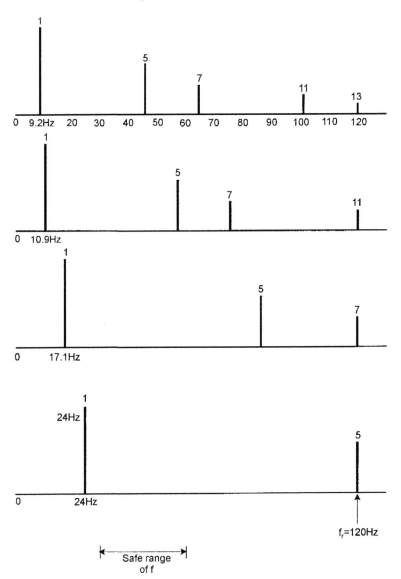

One problem with the six-step self-commutated inverter of Figure 5.20, as mentioned before, is that a harmonic component of the current wave can excite destructive voltage and current in the machine due to the resonance effect. The machine can be represented approximately by equivalent inductance, which forms a resonant circuit with the commutating capacitance bank. The resonance effect is also accompanied by excessive copper loss, core loss, and pulsating torque. Consider, for example, a typical resonance frequency of 120 Hz. The six-step current wave is characterized by a convergent Fourier series with harmonic orders of 5th, 7th, 11th, 13th, etc., the spectrum of which is shown in this figure. This means that the 13th harmonic of 9.2 Hz (fundamental frequency), 11th harmonic of 10.9 Hz, 7th harmonic of 17.1 Hz, and 5th harmonic of 24 Hz coincide with the resonance frequency, as indicated in the figure. As the machine starts from zero speed, the order of the harmonic that coincides with the resonant frequency will gradually decrease until it is above the 24-Hz fundamental frequency, and there will be no resonance effect. If the drive is designed for maximum frequency of 60 Hz, then for operation typically in the range of 30–60 Hz, there is no possibility of resonance. During start-up, it is possible to ride through the resonance effect by fast acceleration of the drive.

FIGURE 5.23   Selected harmonic elimination PWM current waves showing five pulses per half cycle ($M = 5$).

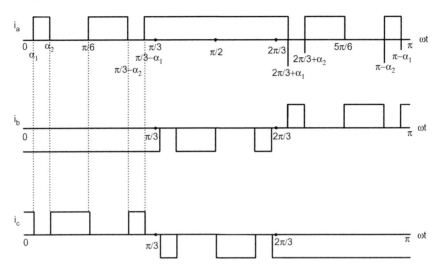

FIGURE 5.24   Selected harmonic elimination PWM current waves for $M = 3$ and $M = 7$.

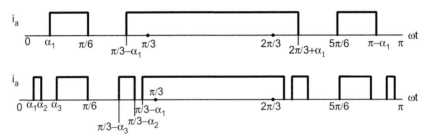

The selected harmonic elimination (SHE) PWM method [6, 7] has the advantages that it can not only improve the harmonic content of the current wave, but it can also eliminate the harmonic component that will tend to cause resonance problems. SHE-PWM for a current-fed inverter is somewhat different from that of the voltage-fed inverter discussed in Chapter 4. Consider, for example, the PWM current waves shown in Figure 5.23 with five pulses per half cycle. In addition to the symmetry of the waveform, the first 30° and last 30° in a half cycle are inverse images of each other. In the first 30° interval, the positive phase currents i_a and i_c are switched back and forth while $-i_b$ remains constant. There are two notch angle variables (α_1 and α_2) in the wave by which the two most significant harmonics (5th and 7th) can be eliminated. The analytical solutions for the corresponding angles are $\alpha_1 = 7.93°$ and $\alpha_2 = 13.73°$. Figure 5.24 shows two phase a current waves with $M = 3$ and 7, respectively. For $M = 3$, there is only one variable and, therefore, only one significant harmonic (5th) can be eliminated, and the corresponding angle $\alpha_1 = 18°$. For $M = 7$, i.e., seven-pulse wave, three significant harmonics (5th, 7th, and 11th) can be eliminated and the corresponding α values are $\alpha_1 = 2.24°$, $\alpha_2 = 5.6°$, and $\alpha_3 = 21.26°$, respectively. The value of M can be increased at low fundamental frequency to eliminate more significant harmonics.

FIGURE 5.25 A 7200-hp, 4160-V GTO current-fed inverter drive for an induction motor with selected harmonic elimination.

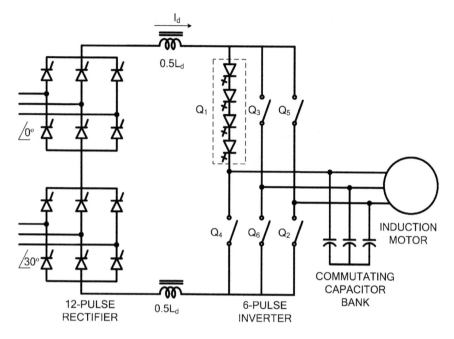

The figure shows a variable-speed industrial drive that uses a self-commutated current-fed GTO inverter for retrofitting the constant-speed applications [9]. Traditionally, constant-speed induction motor drives are used in boiler ID and FD fans with variable throttle flow control. Variable-frequency drives provide substantial energy savings in such applications. The drives are manufactured in the 350- to 7000-hp range with a corresponding voltage range of 2300 to 4160 V. At higher power, a 12-pulse phase-controlled thyristor rectifier is used with phase-shifting transformers (not shown). The dc link inductor is used in two sections as shown to limit common mode stray current. The GTO inverter is six pulse with four devices in series in each arm. The static and dynamic voltages of the devices are shared by series resistors and *R-C-D* snubbers, respectively, which are not shown in the figure. The inverter generates PWM current waves with selected harmonic elimination so that the terminal capacitor bank does not excite resonance with the machine leakage inductances at any harmonic frequency. The capacitor bank is large so that the machine current is nearly sinusoidal to reduce harmonic losses. At high speeds, the terminal power factor is leading, which provides the flexibility of load commutation. The efficiency of the inverter can be improved by using IGCTs (symmetric blocking) instead of GTOs.

FIGURE 5.26   Features of a 7200-hp, 4160-V CFI-IM drive.

- RETROFIT FIXED-SPEED MOTOR FOR VARIABLE-SPEED FAN DRIVE TO SAVE ENERGY

- 0–66 Hz, 350–7200 HP, 2300–4160 V INDUCTION MOTOR

- 12-PULSE THYRISTOR RECTIFIER

- 6-PULSE GTO INVERTER WITH FOUR (4500 V, 2500 A) SYMMETRIC BLOCKING GTOs IN SERIES

- GTO STATIC VOLTAGE SHARING BY RESISTOR AND DYNAMIC VOLTAGE SHARING BY R-C-D SNUBBER

- SELF- OR LOAD COMMUTATION AT HIGH SPEED

- RESONANCE FREQUENCY – 167 Hz

- CAPACITOR BANK – 1.25 PU

- SELECTED HARMONIC ELIMINATION PWM:
    f = 0–30 Hz. M = 5th, 11th, 13th
       7th, 11th
       5th, 7th
    f = 30–40 Hz, M = 3rd, 7th
       5th
    f = 40–66 Hz SIX-STEP OPERATION

- VECTOR-CONTROLLED DRIVE

- DRIVE EFFICIENCY – 96.7%

The figure summarizes the salient features of the GTO-CFI drive shown in Figure 5.25. It can be applied either to fan drives (torque $\propto$ speed^2) or constant torque drives starting from zero speed. For fan drives in the 10- to 66-Hz frequency range, no speed sensor is required. The drive has vector or field-oriented control, which will be described in Chapter 7. The resonance frequency of the machine with the terminal capacitor bank is 167 Hz. As the inverter frequency varies for variable-speed control, selected harmonics from the square current wave are eliminated to prevent a resonance effect. A small hysteresis band is provided in the harmonics elimination control. The terminal capacitor bank is very large because of the machine current filtering requirement to improve efficiency. Self-commutation is used in the whole frequency range, although load commutation may be inherent at higher speeds because of the leading power factor.

FIGURE 5.27 (a) Trapezoidal PWM principle and (b) harmonic relation with modulation index ($M = 21$).

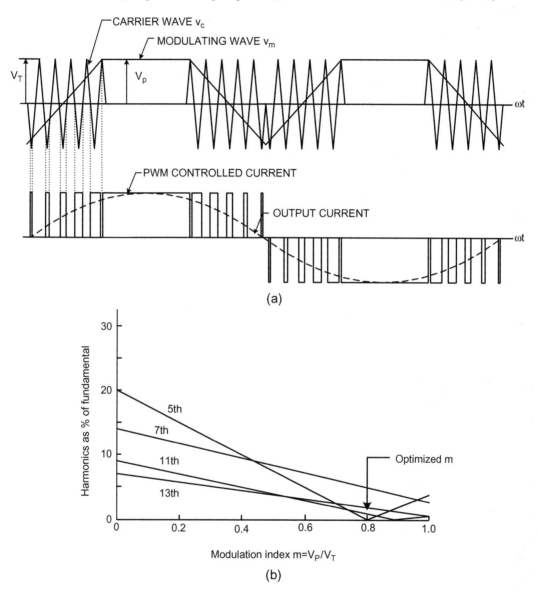

Although SHE-PWM is popular in high-power current-fed drives from the viewpoint of resonant suppression, the triangular PWM technique [10], as shown in this figure, is generally applicable for harmonically optimized waveform generation. A trapezoidal modulating wave of maximum amplitude V_p and at load fundamental frequency is compared with a triangular carrier wave of peak amplitude V_T, and the points of intersection generate the desired PWM pattern of a phase current, as shown on the upper figure. The pattern has quarter-cycle symmetry, and the middle 60° segment does not have any modulation. There are two variables for the PWM pattern generation: One is the modulation index $m = V_p/V_T$, and the other is pulse number M in the half-cycle (for example, 11 in this figure). Harmonics in the PWM pattern can be varied by changing these parameters. The lower figure shows the variation of magnitudes of lower order harmonics in the PWM pattern with variation of modulation index for a particular value of $M = 21$. The harmonic magnitudes decrease with higher modulation index, and the value $m = 0.82$ is considered optimum, where the 5th harmonic is zero, and the 7th, 11th, and 13th harmonics have magnitudes of 4%, 1%, and 2%, respectively. The fundamental component of the pattern is relatively insensitive to the variation of m.

FIGURE 5.28 (a) Three-phase PWM current waves and (b) GTO switching frequency vs. inverter output frequency.

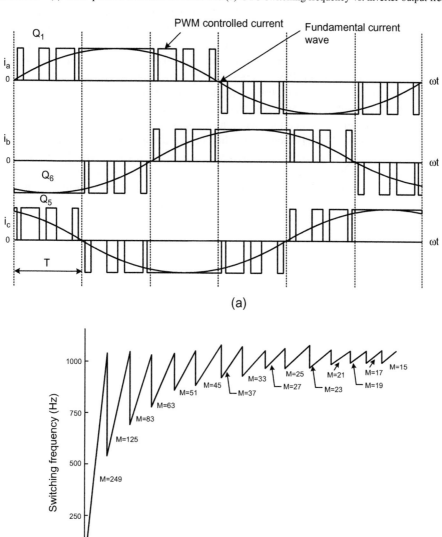

(a)

(b)

The figure shows the fabrication of three-phase current waves [10] using the PWM pattern discussed in Figure 5.27. The fundamental components are also indicated in the figure. With the harmonically optimum wave pattern remaining constant, the magnitude of fundamental current is controlled by the dc link current I_d. Since only two devices conduct (one in the upper group and another in the lower group) at any instant in a current-fed inverter, PWM is possible by switching the current back and forth between two phases. The middle 60° segment of a wave does not have any modulation, but the 60° segments in leading and trailing edges are modulated with an adjacent device in the same group. Since the switching frequency is to be regulated to be nearly constant (in this case 1.0 kHz) for a GTO-like device irrespective of the variation of fundamental frequency, a synchronous PWM technique is used, where the parameter M (pulse number in half-cycle) is varied widely, as indicated in the lower figure. A look-up table of optimized PWM patterns for different values of M can be stored in a microcomputer for implementation. In a practical multi-megawatt GTO inverter, the switching frequency hardly exceeds several hundred hertz.

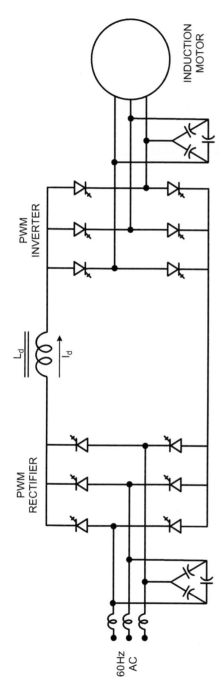

FIGURE 5.29 Two-sided IGCT (reverse-blocking) PWM current-fed converter system for induction motor drives.

Since a PWM current-fed inverter can operate either in rectification or inversion mode, the line-side phase-controlled thyristor converter can be replaced by a PWM converter, as shown in this figure. Such a converter system has four-quadrant control capability. The two-sided PWM current-fed converter system [11] is dual to the PWM voltage-fed converter system shown in Figure 4.50. The rectifier controls the dc link current I_d by modulating the dc link voltage, and the phase position of the input PWM fabricated current wave is controlled such that the line current is not only near sinusoidal like that of machine current, but its DPF is also controllable to unity, leading or lagging. Since the dc link inductor sees reduced voltage ripple compared with that in Figure 5.16, its value is substantially reduced (typically 10%). Again, the converter system can use GTOs or IGBTs with series diode (for reduced power rating). A two-sided IGBT-based induction motor drive system has been used in elevator speed control. For high-power wound-field synchronous motor drives, thyristor converters with load commutation are invariably used.

FIGURE 5.30   Features of two-sided PWM current-fed inverter systems.

- POWER FLOW IN EITHER DIRECTION – FOUR-QUADRANT DRIVE CAPABILITY

- SINUSOIDAL LINE CURRENT AT UNITY POWER FACTOR

- PROGRAMMABLE LINE PF – LEADING OR LAGGING (SVC IN EXTREME CASE)

- NEAR SINUSOIDAL MACHINE CURRENT – HIGH EFFICIENCY

- NO MAGNETIC NOISE IN MACHINE

- SMALL DC LINK INDUCTOR

- SLUGGISH TRANSIENT RESPONSE

- POSSIBILITY OF HARMONIC RESONANCE IN LOAD

- AC CAPACITORS WITH DC LINK INDUCTOR MAKE IT MORE EXPENSIVE THAN VFI SYSTEM

The salient features of the two-sided current-fed converter system described in Figure 5.29 are summarized here. Either the trapezoidal PWM or SHE-PWM technique can be used in the converters with the merits and demerits discussed before. The four-quadrant drive system can easily operate from zero speed, but in order to have near-sinusoidal machine current, the capacitor size is reasonably large with a finite number of pulses (M) in the half cycle. With large capacitor size, the commutation is sluggish, which limits the M value. Near-sinusoidal machine current has the obvious advantages of improved machine efficiency, less burden on the cooling system, and minimal torque pulsation and magnetic noise. The current-fed system has the disadvantage of sluggish transient response of the drive because of the delay in I_d control, but the system is more reliable and the converter short circuit can be easily recovered. Overall, the system is more expensive than the equivalent voltage-fed system, and the latter is more preferred in modern drives.

FIGURE 5.31 Current-fed PWM rectifier for dc motor speed control.

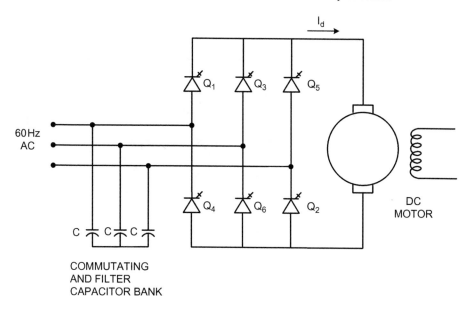

In the two-sided PWM converter system shown in Figure 5.29, the inverter side can be disconnected and replaced by a dc motor load, as shown in this figure. Again, although GTOs have been used in the circuit, IGCTs or IGBTs with series diode can also be used. The PWM rectifier generates controllable dc current for the armature circuit to generate the motor torque as needed. Since PWM rectifiers can also operate as inverters with the same I_d polarity, the motor regenerative braking power can be fed back to the line. In fact, the drive can operate in four-quadrant mode with the reversal of field current. Or else, another converter unit can be connected in inverse parallel. The advantages of such a drive system in comparison with the traditional phase-controlled thyristor converter are that the line current is nearly sinusoidal and line DPF can be near unity (also leading or lagging). Either the SHE-PWM or trapezoidal PWM technique can be used. However, SHE-PWM is preferable because the line-side harmonic resonance can be avoided. The SHE-PWM is simpler to implement because the line frequency is constant. One difficulty with a GTO circuit is that the minimum pulse and notch widths are significant, which makes PWM operation difficult near zero voltage, i.e., near zero speed. This problem is avoided by using phase control at low voltage. The problem is less significant with IGBTs because of the higher switching frequency. The scheme has been used successfully in elevator speed control systems. An alternate scheme, in which the lower side GTOs are replaced by thyristors, has also been used [13].

FIGURE 5.32   PWM rectifier application in static VAR generator (SVG), active harmonic filter (AHF), and superconducting magnet energy storage (SMES).

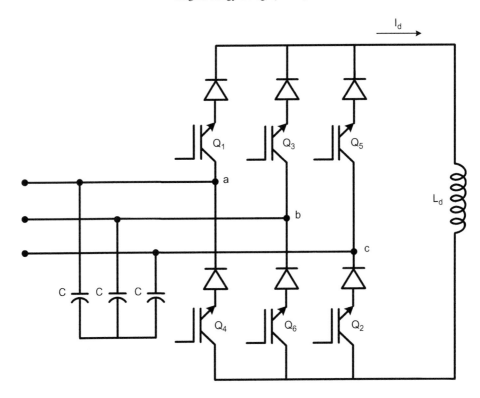

The two-sided PWM current-fed converter system shown in Figure 5.29, the inverter section with the machine can be removed and the dc link output can be shorted as shown in this figure. Such a converter configuration can be used as a static VAR generator or active harmonic filter. The scheme can be considered as the dual to the voltage-fed scheme shown in Figure 4.57. The dc current I_d can be regulated to be constant by the PWM rectifier, and the fabricated PWM current wave at the input side is held in 90° phase position with the voltage so that no active power is drawn to the converter (except the losses). Either a voltage-fed or current-fed rectifier can be designed for a single-phase line, but the size of the capacitor or inductor, respectively, will be larger because of a large second harmonic component. The current-fed scheme is rarely used because of high losses, extra costs, and the need for high-frequency operation (particularly for AHFs). The converter scheme is also applicable for superconductive magnet energy storage. The load inductor can be cooled cryogenically so that it becomes lossless with zero resistance. The stored energy in L_d can be varied by controlling the dc current I_d, and the energy can be pumped back to the line by inverter mode operation of the circuit. The line current can be maintained sinusoidal at unity DPF during the operation. Such an SMES scheme has been considered for off-peak utility energy storage.

FIGURE 5.33   Current-fed series resonance converter system for induction motor drives.

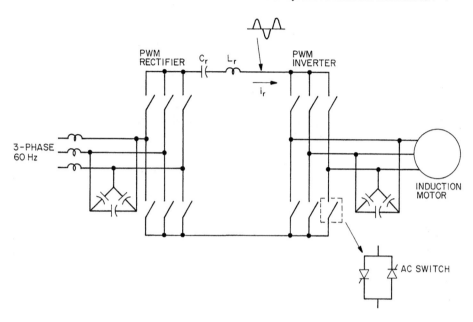

The series resonant ac current link converter system shown in this figure [14] can be considered as the dual of the voltage-fed scheme given in Figure 4.73. Normally, anti-parallel thyristors have been used as ac switches, as shown in the figure, and these are commutated at zero current (ZCS) of discrete resonant current pulses. The power flow is controlled by appropriate gating of the devices, which sets up high-frequency resonant current pulses. At the end of the current pulse, the outgoing thyristors become reverse biased and turn off. The magnitude, polarity, and duration of the resonant current pulse depend on the tank circuit elements L_r and C_r and on the line-to-line dc voltage segment (positive or negative) selected on the input and output side converters. The PWM current wave consisting of a train of discontinuous resonant current pulses is filtered by the capacitor banks to constitute near sinusoidal current waves at both input and output. Again, the active power flow can be in either direction, and the line-side power factor can be controlled to be near unity, leading or lagging. Application of the scheme seems to be rare.

FIGURE 5.34   Performance comparison of ASCI-CFI, PWM CFI, two-sided PWM-CFI, and two-sided PWM-VFI.

	ASCI-CFI (Figure 5.7)	PWM-CFI (Figure 5.20)	TWO-SIDED PWM-CFI (Figure 5.29)	TWO-SIDED PWM-VFI (Figure 4.50)
1. Power devices:	Symmetric thyristors in rectifier and inverter	Symmetric thyristors in rectifier. Self-controlled devices in inverter	Symmetric self-control devices in both units	Asymmetric devices with fast recovery diodes
2. Converter rating:	Hundreds of kW	Hundreds of kW–MWs	Hundreds of kW–MWs	Up to hundreds of kWs–MWs
3. Power flow:	Bidirectional	Bidirectional	Bidirectional	Bidirectional
4. Load harmonics:	High; possible torque pulsation at low freq.	Low. Possible load resonance problem	Low. Possible load resonance problem	Low with high switching frequency
5. Line harmonics:	High with low-order harmonics	High with low-order harmonics	Low. Possible line resonance problem	Low with high switching frequency
6. Line PF:	Low lagging	Low lagging	Near unity; may be lagging or leading	Near unity; may be lagging or leading
7. M/C acoustic noise:	Moderate to low with multisteps	Low	Low	Somewhat high; low at high switching frequency
8. Efficiency:	High, but additional inductor loss	High, but extra switching and inductor loss	Moderately high; high switching and inductor loss	Moderately high; high switching loss
9. Cost	Somewhat high	Somewhat high	High, but small inductor	Somewhat high
10. Control comp.	Simple	Somewhat complex	Complex	Complex
11. Comments	Currently obsolete although many are in field	Popular in multi-MW drive especially in retrofit applications	Very attractive in high power drive but expensive	Very popular in high-power regenerative drive

The figure gives a general comparison of the key types of current-fed converter schemes. The two-sided voltage-fed converter system is included at the end to show how it compares with current-fed converter systems. In overall comparison, the voltage-fed system is much better than the current-fed system. The comparisons shown are only approximate, because exact comparison depends on many factors including power rating, exact configuration, devices used, switching frequency, etc. By using 12-pulse converters on the line side, the harmonics can be significantly improved for ASCI-CFI and PWM-CFI, but will not help the low-power factor. High-power VFI invariably use a multilevel topology where the PWM harmonic quality is much better. Note that load-commutated CFI is extremely popular in high-power WFSM drives.

FIGURE 5.35   Advances and trends in converters.

- POWER QUALITY AND LAGGING PF PROBLEMS ARE MAKING PHASE-CONTROLLED CONVERTERS OBSOLETE –PROMOTING PWM-TYPE CONVERTERS ON THE LINE SIDE

- VOLTAGE-FED CONVERTERS ARE SUPERIOR TO CURRENT-FED CONVERTERS IN OVERALL FIGURE-OF-MERIT CONSIDERATIONS

- TWO-SIDED VOLTAGE-FED GTO/IGBT/IGCT THREE-LEVEL PWM CONVERTERS ARE REPLACING HIGH-POWER PHASE-CONTROLLED CYCLOCONVERTERS

- MULTILEVEL CONVERTERS HAVE HIGH PROMISE FOR HIGH POWER DRIVES AND UTILITY SYSTEMS

- SPACE VECTOR PWM HAS THE BEST PERFORMANCE

- SOFT-SWITCHED CONVERTERS FOR MOTOR DRIVES DO NOT SHOW ANY PROMISE

- CONVERTER TECHNOLOGY HAS NEARLY REACHED SATURATION

- FUTURE EMPHASIS WILL BE ON INTEGRATED PACKAGING AND DESIGN AUTOMATION

The advances and trends of converter technology are given in this figure considering the various types of converters described in Chapters 3, 4, and 5. Historically, phase-control type converters evolved in the first generation of power electronics. Although, they are simple, have high efficiency, and are extensively used, the line DPF is poor and harmonic currents cause PQ problems. It is expected that they will be progressively replaced by PWM type wave shaping or PWM regenerative rectifiers, as needed. Temporarily, for existing phase-controlled installations, SVGs/AHFs will tend to alleviate these problems. Large multi-megawatt cycloconverters are already being replaced by two-sided voltage-fed PWM converters. The voltage-fed and current-fed topologies are somewhat dual to each other, but generally voltage-fed schemes have better features and they are used more popularly. Multilevel converter technology is in the process of evolution, and such converters are expected to have wide applications in multi-megawatt drives and other high-power applications, particularly in high-voltage utility systems. Currently, SPWM and SVM techniques are widely used. However, for isolated neutral loads, such as machine drives, SVM has established its superiority. Soft-switched PWM converters have a long history of evolution for ac drive applications, but due to additional cost, loss, and control complexity, the scheme has not shown much promise for the future. Although, currently, much developmental activity is going on for multilevel converters, it appears that converter technology is nearing saturation. Future emphasis will be mainly on integration and automated design and manufacturing, which are similar to the trend of VLSIs.

Summary

The chapter deals with the principles and applications of a class of converters that are fed by a current source instead of voltage source. The phase-controlled line-commutated thyristor converters that were discussed in Chapter 3 are popular examples in this class. In many respects, a current-fed converter is similar to a voltage-fed converter, i.e., the voltage source is replaced by a current source, the parallel dc-link capacitor is replaced by a series inductor, and forward blocking devices are replaced by symmetric blocking devices. Generally, they are preferred in medium- to high-power applications (hundreds of kilowatts to megawatts). Examples of high-power applications are HVDC systems and ship propulsion. The devices in current-fed converters may be self-commutating (but reverse-blocking) GTOs, IGCTs, forward blocking devices with series diode, or line or load commutating thyristors depending on the power rating and switching frequency. Current-fed converters have the virtues of more reliability and easy recoverability from short-circuit faults, but the disadvantages are higher cost and less efficiency, mainly due to the bulky dc-link inductor and series diodes (if used). The line harmonic problems in a converter can be alleviated by multipulsing (12 pulses instead of 6 pulses) or PWM control, whereas the low-power factor can be improved by PWM only. One of the problems of using self-commutated converter drives for induction motors is the harmonic resonance problem, which has been solved by SHE-PWM. In an overall comparison for general applications, a voltage-fed converter system is always preferred. Both load-commutated and self-commutated circuits are widely used in industry. However, load-commutated synchronous motor drives are the highest rated power drives in industry today. Recently, cycloconverters and load-commutated inverters for multi-megawatt synchronous motor drives are being replaced by two-sided multilevel (three-level) PWM voltage-fed converters. Performance comparisons for several converter types along with a voltage-fed converter system have been given for clarity of understanding. A number of example applications in industry have been included in the chapter. Finally, the general advances and trends of converters are included. Further discussions relating to current-fed induction and synchronous motor drives will be given in Chapters 7 and 8, respectively.

References

[1] B. K. Bose, "Power electronics—a technology review," *Proc. IEEE,* vol. 80, pp. 1303–1334, August 1992.

[2] B. K. Bose, *Modern Power Electronics and AC Drives,* Prentice Hall, Upper Saddle River, NJ, 2002.

[3] B. K. Bose, *Power Electronics and AC Drives,* Prentice Hall, Englewood Cliffs, NJ, 1986.

[4] T. Ueno, H. Akagi, N. Uchida, and H. Nanba, "A 1000 kW 50 kHz current source inverter using IGBTs for induction heating applications," *Proc. IPEC,* Tokyo, pp. 947–950, 1995.

[5] B. Mueller, T. Spinanger, and D. Wallstein, "Static variable frequency starting and drive system for large synchronous motors," *IEEE/IAS Annual Meeting Conference Record,* pp. 429–438, 1979.

[6] C. Namuduri and P. C. Sen, "Optimal pulsewidth modulation for current source inverters," *IEEE Trans. Ind. Appl.,* vol. 22, pp. 1052–1072, November/December 1986.

[7] H. Karshenas, H. Kojori, and S. B. Dewan, "Generalized techniques of selective harmonic elimination and current control in current source inverters/converters," *IEEE Trans. Power Elec.,* vol. 10, pp. 566–573, September 1995.

[8] Stemmler, H., "High power industrial drives," Chapter 7 in *Power Electronics and Variable Frequency Drives,* edited by B. K. Bose, IEEE Press, 1997.

[9] P. M. Espalage, J. M. Novak, and L. H. Walker, "Symmetrical GTO current source inverter," *IEEE/IAS Annual Meeting Conference Record,* pp. 302–307, 1988.

[10] M. Hombu, S. Ueda, A. Ueda, and Y. Matsuda, "A new current source GTO inverter with sinusoidal output voltage and current," *IEEE Trans. Ind. Appl.,* vol. 21, pp. 1192–1198, September/October 1985.

[11] M. Hombu, S. Ueda, and A. Ueda, "A current source GTO inverter with sinusoidal inputs and outputs," *IEEE Trans. Ind. Appl.,* vol. 23, pp. 247–255, March/April 1987.

[12] N. R. Zargari, Y. Xiao, and B. Wu, "Near unity displacement factor for current source PWM drives," *IEEE Ind. Appl. Magazine,* pp. 19–25, July/August 1999.

[13] H. Inaba, S. Shima, A. Ueda, T. Ando, T. Kurosawa, and Y. Sakai, "A new speed control system for dc motors using GTO converter and its application to elevators," *IEEE Trans. Ind. Appl.,* vol. 21, pp. 391–397, March/April 1985.

[14] F. C. Schwartz, "A double-sided cycloconverter," *IEEE Trans. Ind. Elec.,* vol. 28, pp. 282–291, 1981.

CHAPTER 6

Electrical Machines for Variable-Speed Drives

FIGURE 6.1   Classification of machines for drives.

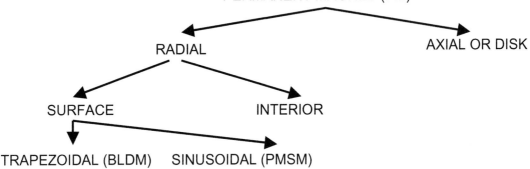

1. DC MACHINES
      SEPARATELY EXCITED
      SHUNT
      SERIES
      COMPOUND

2. AC MACHINES
      A. INDUCTION MACHINES: (ROTATING OR LINEAR)
            CAGE
            WOUND ROTOR (WRIM) OR DOUBLY FED

      B. SYNCHRONOUS MACHINES: (ROTATING OR LINEAR)
            WOUND FIELD (WFSM)
            RELUCTANCE MACHINE (SyRM)
            PERMANENT MAGNET (PM)

          RADIAL   AXIAL OR DISK

      SURFACE   INTERIOR

TRAPEZOIDAL (BLDM)   SINUSOIDAL (PMSM)

      C. VARIABLE RELUCTANCE (VRM) (ROTATING OR LINEAR)
            SWITCHED RELUCTANCE (SRM)
            STEPPER

An electrical machine is the workhorse in a variable-speed drive system, and its function is to convert electrical energy into mechanical energy in various industrial applications. A machine can also operate as a generator to convert mechanical energy into electrical energy. In industrial drives, generator operation permits use of the regenerative braking mode, by which the machine speed slows down while its mechanical energy is recovered. Generally, machines can be classified as dc and ac machines. Traditionally, ac machines, particularly induction motors, have been used in constant-speed applications, whereas dc machines were used in variable-speed applications. Chapter 3 discussed phase-controlled dc drives, whereas Chapter 4 included brief descriptions of dc drives fed by dc-dc converters. AC machines, particularly induction and synchronous machines, will be discussed comprehensively in this chapter, and variable-speed drives with these machines will be treated in Chapters 7 and 8, respectively. Electrical machines constitute a vast and complex subject. But an engineer designing a high-performance drive must have thorough knowledge of machines. The primitive machines, since their invention more than a century ago, have gone through a process of continuous evolution to achieve the goals of cost, size, and performance, and this evolution will continue into the future.

FIGURE 6.2 Features of a dc machine.

- HIGH COST, VOLUME, AND WEIGHT
- HIGH INERTIA
- LOWER EFFICIENCY
- RELIABILITY AND MAINTAINABILITY PROBLEM
- EMI PROBLEM
- LIMITED HIGH SPEED
- CANNOT OPERATE AT HIGH ALTITUDE
- CANNOT OPERATE IN DIRTY AND EXPLOSIVE ENVIRONMENT
- INHERENTLY FAST TORQUE RESPONSE
- SIMPLE CONVERTER AND CONTROL
- HAVE BEEN POPULAR IN TRADITIONAL VARIABLE-SPEED DRIVES
- TEND TO BE OBSOLETE IN FUTURE

Our discussion of machines would remain incomplete without a brief review of the features of dc machines, because dc drives are still widely used in industry. This figure summarizes the general features of a dc machine and its drives. Generally, separately excited machines are selected for the drives. DC machines have a large number of disadvantages compared to ac machines, particularly when compared with induction machines. The main disadvantages are that the machines are bulky, expensive, and have lower efficiency; they also have problems with commutators and brushes. However, the drive configuration is simple with a simple converter, and torque control is very fast because of the inherent decoupling of field flux and armature MMF. Below the base or rated speed, the dc armature current is controlled at constant field flux to control the torque for speed regulation (constant-torque region), whereas above the base speed, the field current is weakened at the rated armature current (field-weakening or constant-power region) to control the speed at reduced torque. DC drives have been used extensively in various industrial applications. However, in recent years, they have begun to become obsolete so that in new installations and replacement drives for old installations, ac drives are invariably used.

FIGURE 6.3   (a) Idealized three-phase induction motor and (b) rotating magnetic field with sinusoidal MMF distribution.

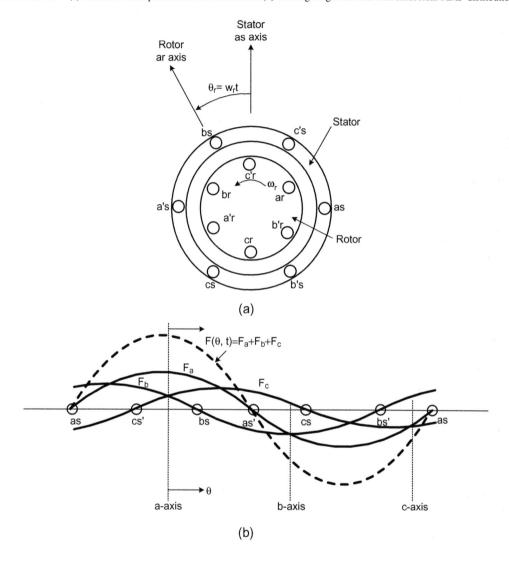

(a)

(b)

Induction machines, particularly the cage type, are by far most commonly used in constant- and variable-speed drives among all types of ac machines. The machines are simple in construction, economical, rugged, reliable, and are available in a wide power range, including FHP (fractional horse power) to multi-megawatt capacities. Single-phase FHP motors are used, for example, in residential applications. However, three-phase machines are commonly used in variable-speed drives. Figure 6.3(a) shows an idealized three-phase, two-pole machine, in which each stator and rotor phase winding is shown by a concentrated coil, although in practice they are distributed sinusoidally. In wound rotor or doubly fed machines, the rotor winding is similar to stator winding, but in a cage motor, the rotor has slot-embedded bars that are shorted by end rings. The machine is basically a transformer with short-circuited secondary windings. When three-phase sinusoidal voltages are supplied to the stator, the resulting currents create a synchronously rotating magnetic field or MMF as explained in (b) and as shown by the equation below the figure. Each phase creates a sinusoidal MMF wave, as shown, which is pulsating about its own axis, but the resultant MMF wave $F(\theta,t)$ with peak value NI_m (N = turns/ phase, I_m = peak current) moves at synchronous speed (rpm) $n = 120\,f\,I\,P$ (f = supply frequency, P = number of poles). The rotating MMF induces rotor current (or MMF) that drags the rotor along by the developed torque.

FIGURE 6.4 (a) Stator-referred equivalent circuit and (b) phasor diagram.

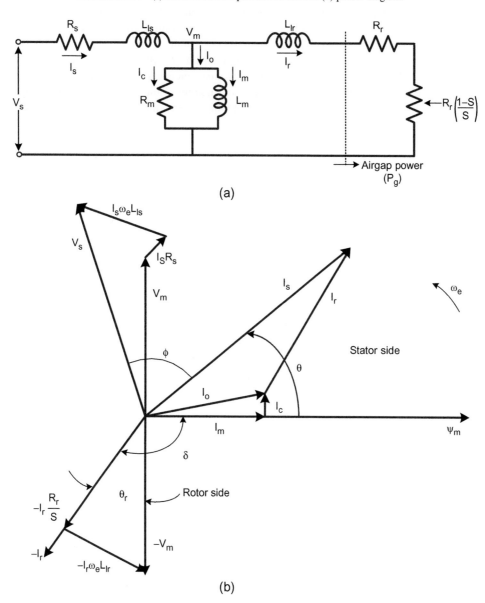

(a)

(b)

The steady state per phase equivalent circuit shown in this figure is extremely important for calculating the performance of the machine. It is a transformer-like equivalent
circuit in which all machine parameters are transferred to the stator side. The equivalent
circuit shows a load resistance $R_r(1 - S)/S$, where S is the per unit (pu) slip $(0 < S < 1.0)$ of

the machine. The power absorbed in this resistance is the output power (per phase). The machine has a magnetizing current component I_m, which flows in the magnetizing inductance L_m, and no-load iron loss current component I_c, which flows through the equivalent resistance R_m. The total stator current (I_s), which is the sum of the no-load exciting current (I_0) and reflected rotor current (I_r), flows through the stator impedance, as shown. At standstill ($S = 1$), the stator current is very high, whereas at no load ($S = 0$), the stator current is small. The phasor diagram of the equivalent circuit with rms variables is shown in the lower figure. It indicates that the terminal DPF (cos φ) is lagging.

FIGURE 6.5 Power and torque expressions from per-phase equivalent circuit.

- *INPUT POWER: $P_{in} = 3V_s I_s \cos \varphi$*

- *STATOR COPPER LOSS: $P_{ls} = 3I_s^2 R_s$*

- *CORE LOSS: $P_c = 3V_m^2 / R_m$*

- *POWER ACROSS AIRGAP: $P_g = 3I_r^2 R_s / S$*

- *ROTOR COPPER LOSS: $P_{lr} = 3I_r^2 R_r$*

- *OUTPUT POWER: $P_0 = P_g - P_{lr} = 3I_r^2(1-S)/S$*

- *SHAFT POWER: $P_{sh} = P_0 - $ friction & windage loss*

- *EFFICIENCY $\eta = P_0 / P_{in}$*

- *DEVELOPED TORQUE:*

$$T_e = 3\left(\frac{P}{2}\right)\left(\frac{R_r}{S\omega_e}\right) \cdot \frac{V_s^2}{(R_s + R_r/S)^2 + \omega_e^2 (L_{ls} + L_{lr})^2}$$

$$= \frac{3}{2}\left(\frac{P}{2}\right) \hat{\psi}_m \hat{I}_r \sin\delta$$

$$= 3\left(\frac{P}{2}\right)\frac{1}{R_r} \psi_m^2 \omega_{sl}$$

An induction motor can be analyzed and its performances can be calculated if the parameters of the equivalent circuit are known. The traditional blocked rotor and no-load tests can determine these parameters approximately, or some automated tests can be performed for this. The value of slip S can be determined from operating speed. The equivalent circuit can be solved to determine the various power expressions given above. The friction and windage or mechanical loss at a particular speed can be subtracted from output power P_0 to calculate the shaft power P_{sh} and the corresponding machine efficiency η. Typical machine efficiency may be near 90%. The developed torque can be calculated by $T_e = P_0/\omega_m$, where ω_m = mechanical speed (rad/s). It can also be determined from the phasor diagram or expressed as function of slip frequency (ω_{sl}) and airgap flux rms (ψ_m) or peak value ($\hat{\psi}_m$) as indicated.

FIGURE 6.6   (a) IEEE-recommended equivalent circuit and (b) simplified equivalent circuit.

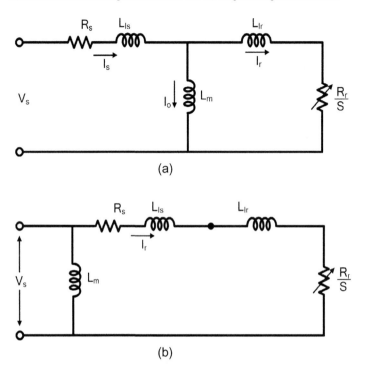

If the motor is operated at constant supply voltage and frequency for constant speed application, the equivalent circuit in Figure 6.4 can be somewhat simplified. In such a case, as
shown in (a) of this figure, the parameter R_m can be omitted, and the constant core loss P_c
can be lumped together with the friction and windage (F&W) loss. This equivalent circuit
is known as an IEEE-recommended equivalent circuit, which is widely used for analysis
in constant-speed applications. In a further simplification, the magnetizing inductance L_m
can be shifted at the input, as shown in (b) of this figure. Since the parameters R_s and L_{ls}
correspond to the stator winding resistance and leakage inductance, respectively, and
they are very small, this shift may be justified. However, note that induction motor magnetizing current is high (25–30% of full load current) because of the airgap; therefore,
the drop in stator impedance may be significant, causing a large error. Often, a Thevenin
equivalent circuit is drawn for Figure 6.4 to simplify calculations.

FIGURE 6.7 Torque-speed curve for an induction motor.

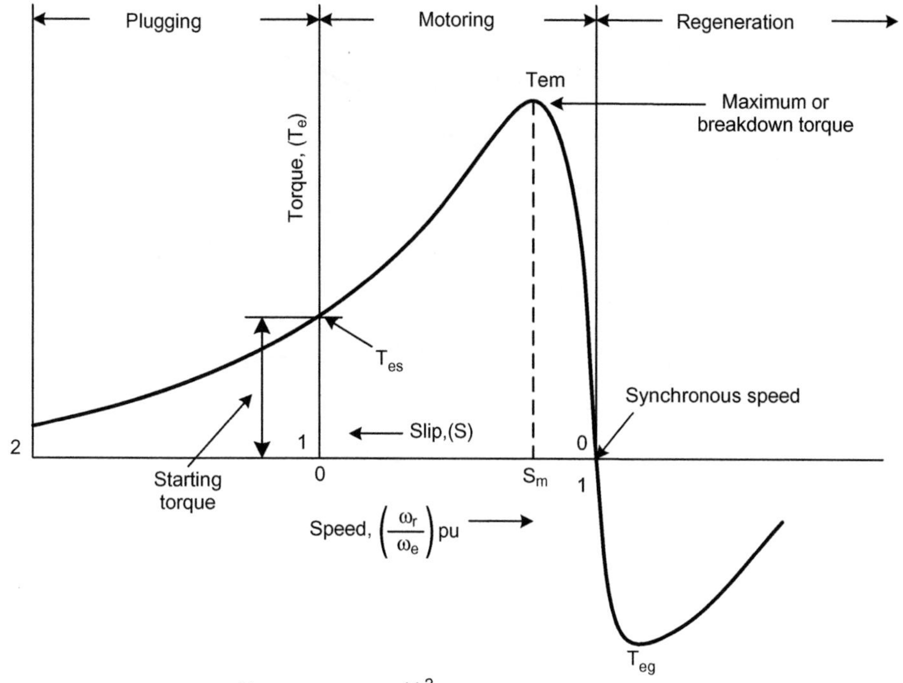

$$T_e = 3\left(\frac{p}{2}\right)\frac{R_r}{S\omega_e} \times \frac{V_s^2}{(R_s + R_r/S)^2 + \omega_e^2(L_{ls} + L_{lr})^2} \quad \text{........(1)}$$

$$T_e = \frac{3}{4}\left(\frac{p}{\omega_e}\right)\frac{V_s^2}{\sqrt{R_s^2 + \omega_e^2(L_{ls} + L_{lr})^2} - R_s} \quad \text{........(2)}$$

$$S_m = \pm \frac{R_r}{\sqrt{R_s^2 + \omega_e^2(L_{ls} + L_{lr})^2}} \quad \text{........(3)}$$

The developed torque expression T_e derived from the equivalent circuit has been plotted in this figure as a function of slip S or pu speed ω_r/ω_e. There are three regions of the curve: motoring, regeneration, and plugging. In motoring mode, the motor starts with initial torque T_{es}, reaches the maximum value T_{em} at slip S_m, and then reaches the synchronous speed ($S = 0$) at zero torque. At starting ($S = 1$), the current is high (typically 500%), and at $S = 0$, the machine takes only the excitation current. In the normal operating region, S is very small and the airgap flux (ψ_m) is nearly constant. The machine goes into regenerative mode at supersynchronous speed, when the slip is negative ($S < 0$). The peak torque in regeneration mode T_{eg} is somewhat small ($T_{eg} < T_{em}$), and normal operation occurs at a low value of negative slip. For sustained induction generator operation, the slip remains at a negative value. In plugging mode, the motor rotates in a direction opposite that of the rotating field ($S > 1$). The shape of the curve depends on the machine's parameters.

FIGURE 6.8   Variable-voltage, constant frequency speed control of cage motor.

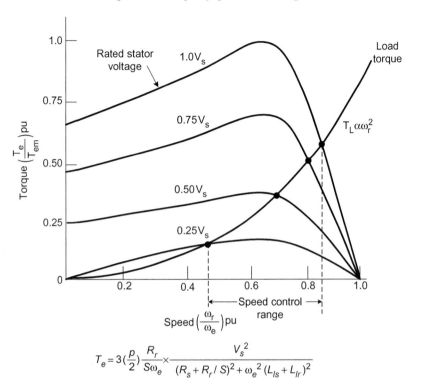

$$T_e = 3\left(\frac{p}{2}\right)\frac{R_r}{S\omega_e} \times \frac{V_s^2}{(R_s + R_r/S)^2 + \omega_e^2\,(L_{ls} + L_{lr})^2}$$

The developed torque T_e of a motor is a function of voltage (V_s) and slip (S) at constant frequency, as shown in the above equation. Therefore, for a fixed slip, voltage can be varied to vary the torque as shown in this figure. The load torque characteristics for a fan or pump load are parabolic ($T_L \propto \omega_r^2$), and are superimposed on the figure. The points of intersection indicate that speed can be controlled by variation of the supply voltage at constant frequency. The three-phase ac controller shown in Figure 3.26(a) can be used for such simple speed control of a cage machine. Note that a large range of speed control are possible if the machine has high starting torque, which is possible if the cage resistance (R_r) is high. Such a machine can be classified as a NEMA (National Electrical Manufacturers Association) Class D motor. The disadvantages of this type of drive are higher machine copper loss (i.e., poor efficiency) and flow of large harmonic currents in line with poor PF. Single-phase, low-power appliance-type drives, where cost is a dominant consideration, often use this type of drive. The converter is often used as a solid-state "soft-starter" for constant-speed three-phase cage motors.

FIGURE 6.9   Complex mechanical load for a motor with equations for a resilient shaft.

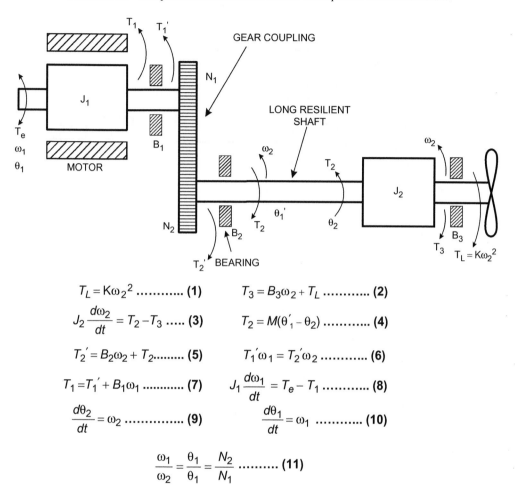

$$T_L = K\omega_2{}^2 \dots\dots (1) \qquad T_3 = B_3\omega_2 + T_L \dots\dots (2)$$

$$J_2\frac{d\omega_2}{dt} = T_2 - T_3 \dots (3) \qquad T_2 = M(\theta'_1 - \theta_2) \dots\dots (4)$$

$$T_2' = B_2\omega_2 + T_2 \dots\dots (5) \qquad T_1'\omega_1 = T_2'\omega_2 \dots\dots (6)$$

$$T_1 = T_1' + B_1\omega_1 \dots\dots (7) \qquad J_1\frac{d\omega_1}{dt} = T_e - T_1 \dots\dots (8)$$

$$\frac{d\theta_2}{dt} = \omega_2 \dots\dots (9) \qquad \frac{d\theta_1}{dt} = \omega_1 \dots\dots (10)$$

$$\frac{\omega_1}{\omega_2} = \frac{\theta_1}{\theta_1} = \frac{N_2}{N_1} \dots\dots (11)$$

The motor is connected to a mechanical load, which may be simple or complex. The figure shows an example of complex mechanical load. The machine rotor with inertia J_1 develops torque T_e while running at mechanical speed ω_1 and instantaneous position θ_1. The motor speed is reduced by gear coupling, where the primary and secondary gears have numbers of teeth N_1 and N_2 ($N_2 > N_1$), respectively. The secondary gear is connected to a long resilient shaft which acts like a spring of constant M. The torque transmitted through the shaft goes through another inertia load J_2 and is then connected to a fan load, which has parabolic characteristics with speed ($T_L \propto \omega_2{}^2$). The system has three bearings, which have friction coefficients B_1, B_2, and B_3, respectively, as shown in the figure. Each bearing has a torque loss that is proportional to the respective speed. The gears are assumed to have no loss. The step-by-step dynamic equations of the mechanical system are included in the figure.

FIGURE 6.10   Motor mechanical load dynamic equations block diagram.

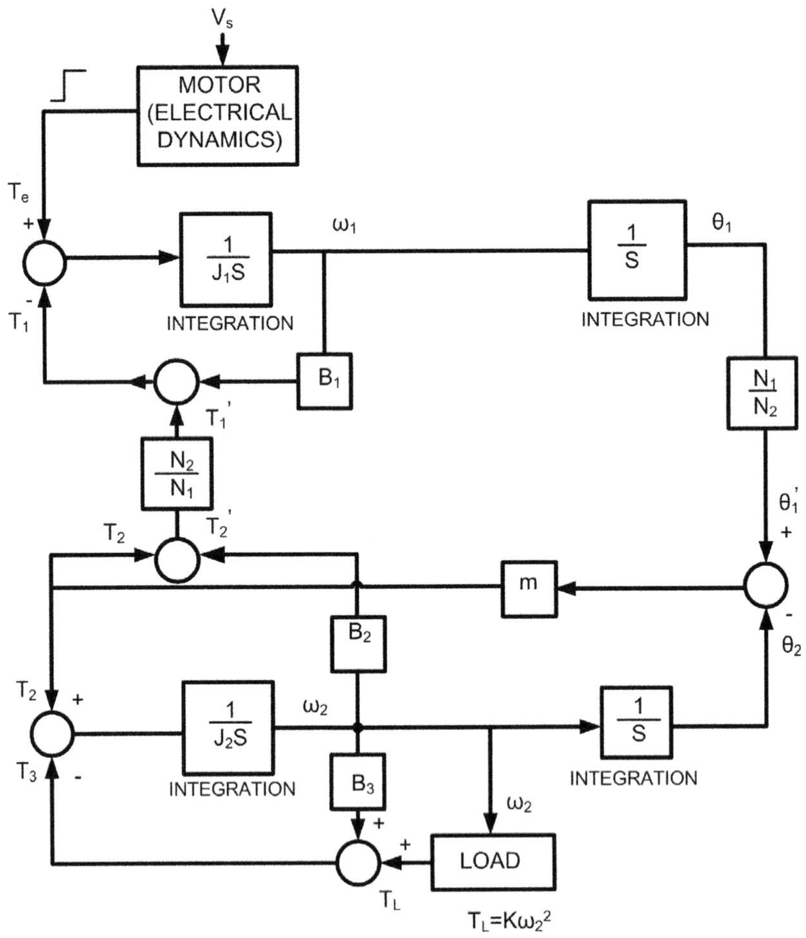

The equations that describe the dynamic load in Figure 6.9 have been used to draw the block diagram of this figure. The machine generates the developed torque T_e with its own dynamics (electrical dynamics), which will be described later. The electrical dynamics are combined with the mechanical dynamics in this figure to constitute the total system. Such a block diagram helps with the creation of a computer simulation study, which will be described later. The equations in the previous figure could also have been combined and expressed in state variable form. There are four integrators (i.e., four state variables) in the system indicating that it is a dynamic system of fourth order. The system is nonlinear because load torque is nonlinearly related with speed. The block diagram indicates that the system can have mechanical resonance (torsional oscillation) without sufficient damping. At resonance, the position, speed, and torque signals will be magnified.

FIGURE 6.11 Induction motor torque-speed curves with variable-frequency, variable-voltage power supply (covers field-weakening region).

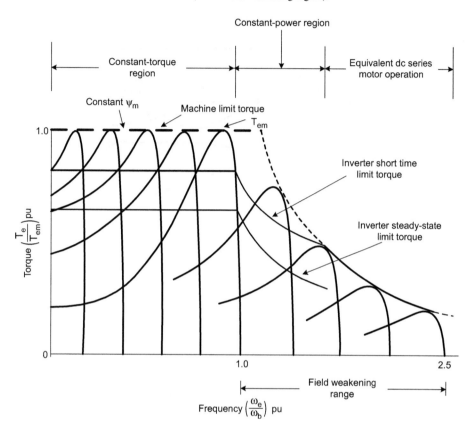

The figure shows how the torque curve varies with variations in the supply voltage and frequency. With the rated voltage, if the frequency is increased above the base value (1.0), the peak torque (T_{em}) decreases because of the decrease of airgap flux ($\psi_m = V_s/\omega_e$). This is basically dc series motor-like operation ($T_e\, \omega_e^2 = K$). If, on the other hand, frequency is decreased, the voltage has to decrease proportionately so that flux remains constant. In this region (called the constant torque region), peak or pull-out torque remains constant, and the curve shifts to the left, as shown in the figure. Evidently, volts/Hz control is the usual method for speed control, and the drives operate mostly in this constant torque region. This means that, as the frequency is varied to control the speed, the voltage has to vary proportionately in the constant torque region. With some margin for machine parameter variation, the inverter short-time-limit torque is somewhat less than machine limit torque. For steady-state operation, the inverter permits a lower limit, which is defined as the rated torque-speed curve. In the field or flux weakening region, operation is usually limited by the constant power curve, as shown in the figure. Note that the motor can be accelerated from zero speed with full rated torque within the limit of rated current.

FIGURE 6.12   (a) Induction motor characteristics with variable-frequency, variable-voltage power supply and
(b) four-quadrant operation of motor.

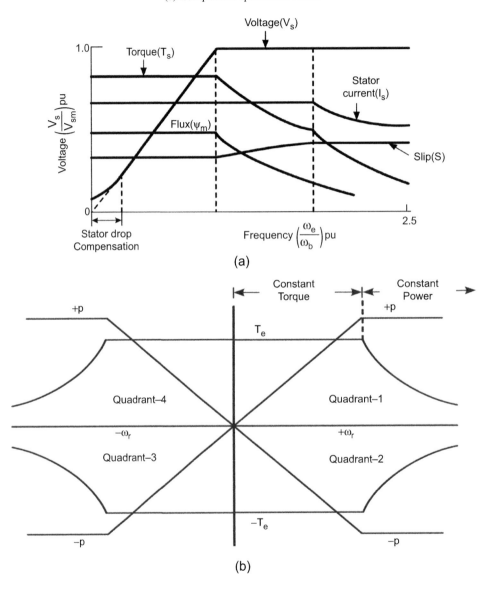

(a)

(b)

For the different operating regions of Figure 6.11, the upper figure shows the plot of different variables as a function of frequency. With constant flux in the constant torque region, the machine current and slip are limited. Of course, the operating point can be anywhere within the envelope. For example, the machine might accelerate at the rated torque, but at steady state ($T_e = T_L$), the developed torque may be low. At very low frequency, the airgap flux may be low due to a large stator drop. Therefore, additional boost voltage is needed as shown to overcome the stator drop. The four-quadrant operation characteristics of a drive (explained in Chapter 3) are shown in the lower figure. At constant torque, the output power linearly increases with speed, but is constant in the field-weakening region. Examples of four-quadrant drive operation are rolling mills, electric vehicles, and elevators.

FIGURE 6.13 (a) Equivalent circuit with current source and (b) torque-speed curves with variable current.

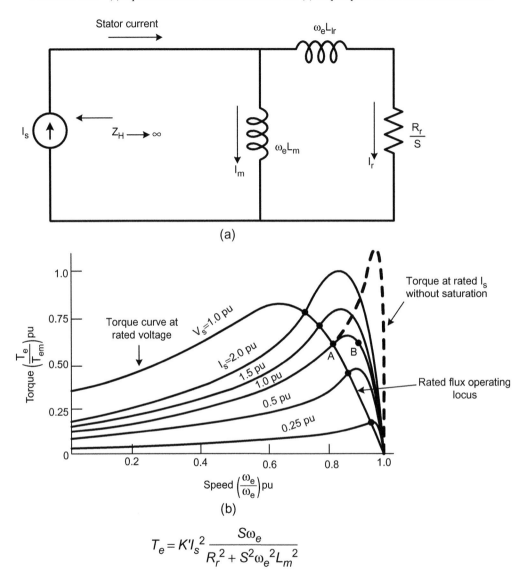

(a)

(b)

$$T_e = K'I_s^2 \frac{S\omega_e}{R_r^2 + S^2\omega_e^2 L_m^2}$$

If a machine is fed by an ideal current source I_s, the Thevenin impedance is infinity, and the equivalent circuit can be represented by the figure here. In fact, at the operating condition, slip S is very small ($\omega_e L_{lr} \ll R_r/S$) and, therefore, leakage inductance L_{lr} can be omitted from the circuit. The stator current I_s will be divided into magnetizing current I_m and rotor current I_r in the parallel paths depending on the value of S. Since developed torque is related to the product of I_m and I_r, an expression of developed torque can be derived, as given below the figure. The torque equation is plotted as a function of S for various values of current I_s. Near $S = 1$, most of the current flows in the rotor circuit, contributing very small starting torque. Similarly, as S approaches 0, torque approaches zero from a very high value. Practical torque curves considering magnetic saturation are shown by solid curves. The equivalent torque curve at rated voltage is superimposed on the figure. Because practical machines operate at constant rated flux, the points of intersection are the actual operating points. For a 1.0-pu torque curve, the developed torque is the same at points A and B, but point B is not preferred because of magnetic saturation. This means that as the current I_s is increased to increase the torque, slip S has to increase, as shown in the figure. Note that unlike voltage-fed drives, a current-fed drive operates in the positive slope unstable region.

FIGURE 6.14   Features of a variable-frequency inverter-fed machine.

- HIGH-EFFICIENCY MACHINE CAN BE USED FOR ENERGY
  SAVING
- SMOOTH STARTING
             NO IN-RUSH CURRENT
             CAN START AT RATED TORQUE
             NO STRESS ON ROTOR BARS – LONGER LIFE
- SHIELDED FROM LINE VOLTAGE TRANSIENTS
             LESS INSULATION STRESS – LONGER LIFE
- DESIGN WITH HIGHER LEAKAGE INDUCTANCE FOR VFI
  BUT LOWER INDUCTANCE FOR CFI
- FORCED COOLING MAY BE NEEDED AT LOW SPEED
- VARIATION OF EQUIVALENT CIRCUIT PARAMETERS
  MAY CREATE CONTROL COMPLEXITY AND STABILITY PROBLEMS
- HIGH *dv/dt* STRESS ON MACHINE INSULATION
- POSSIBILITY OF VOLTAGE BOOST AT MACHINE TERMINAL
  WITH LONG CABLE

Traditionally, for fixed-speed operation, an induction motor is fed from a 50/60-Hz line power supply. There are some differences when the machine is supplied from a converter, particularly for variable-speed drives. Because there is no in-rush starting current, a high-efficiency low-slip machine can be used for energy savings. The machine can be started smoothly within the limit of the stator's rated current. The problem of high rotor bar stress due to large in-rush current does not exist. The machine is shielded from high line voltage transients. Unlike a standard fixed-speed motor, low-speed sustained operation at high torque (such as EV) requires forced cooling. However, for pump or fan load, load torque is low at low speed and, therefore, no forced cooling may be needed. However, the disadvantages of converter-fed machines are additional harmonic loss and *dv/dt* and voltage boost problems with the PWM inverter, as discussed in Chapter 4. If the machine is designed with higher leakage inductance, the current filtering is better for a voltage-fed converter (with some loss of torque), but the same machine with a current-fed inverter will give longer commutation times and voltage transients during commutation. Note that machine equivalent circuit parameters are not constant. Both stator and rotor resistances vary with temperature. Besides, there is a skin effect problem in rotor bars with harmonic currents that increases the resistance but decreases the leakage inductance. The magnetizing and leakage inductances can saturate at higher currents. This parameter variation is a problem in closed-loop control, particularly in high-performance drives.

FIGURE 6.15 Harmonic equivalent circuit and losses.

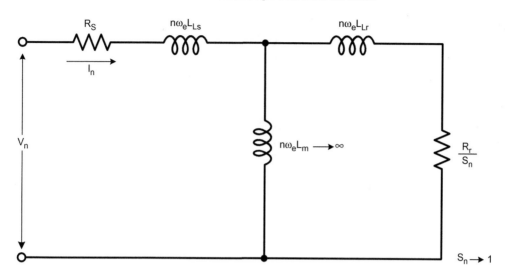

$$v_{as} = V_{1m} \sin \omega_e t + V_{5m} \sin \omega_e t + ...(1)$$

$$v_{bs} = V_{1m} \sin(\omega_e t - 120°) + V_{5m} \sin(5\omega_e t + 120°) + V_{7m} \sin(7\omega_e t - 120°) + ...(2)$$

$$v_{cs} = V_{1m} \sin(\omega_e t + 120°) + V_{5m} \sin(5\omega_e t - 120°) + V_{7m} \sin(7\omega_e t + 120°) + ...(3)$$

$$S_n = \frac{n\omega_e + \omega_r}{n\omega_e} ....(4)$$

$$Copper\ Loss : P_{lc} = 3(I_{sl}^2 + I_h^2)R_s + 3(I_{rl}^2 + I_h^2)R_r .....(5)$$

$$Copper\ Loss : P_c = \left[K_h\left(\frac{1+S}{\omega_e}\right) + K_e(1 + S^2)\right]V_m^2 + \left(\frac{K_h}{n\omega_e} + K_e\right)V_m^2 ......(6)$$

The converter-fed harmonics flowing in the machine create two problems: additional heating and pulsating torque generation. For a square-wave inverter, the phase voltages can be expressed by a Fourier series with odd harmonics, as shown above. The triplen harmonics are of no significance in an isolated neutral machine. The harmonic currents corresponding to harmonic voltages can be solved by the equivalent circuit shown in the figure, and the resulting rms currents can be calculated. The total copper and core loss expressions are included in the figure, where K_h = coefficient for hysteresis loss and K_e = coefficient for eddy current loss. Note that the fundamental and 7th harmonic will create a rotating magnetic field in a counterclockwise direction, whereas the 5th harmonic rotating field has a clockwise direction. The slip at harmonic field (S_n) is near unity.

FIGURE 6.16   (a) Rotating phasors for generation of pulsating 6th harmonic torque and (b) converted phasor diagram with fundamentals stationary.

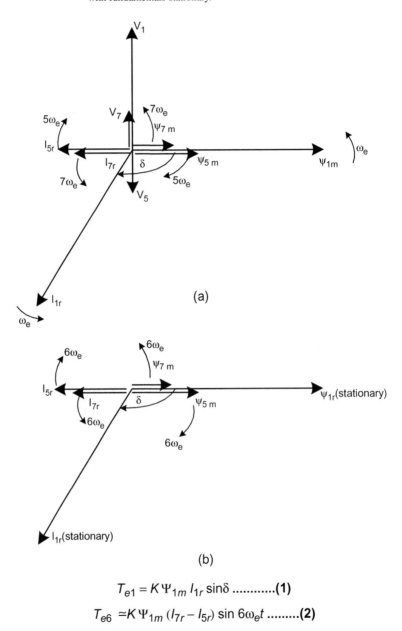

$$T_{e1} = K \Psi_{1m} I_{1r} \sin\delta \dots\dots\dots(1)$$

$$T_{e6} \simeq K \Psi_{1m} (I_{7r} - I_{5r}) \sin 6\omega_e t \dots\dots(2)$$

A machine develops pulsating torque if the airgap flux at one frequency interacts with rotor MMF at a different frequency. Consider the fundamental airgap flux (ψ_{Im}) and rotor current (I_{Ir}) of (a), which are at phase angle δ (see Figure 6.4) and rotating counterclockwise at fundamental frequency. Their interaction will result in unidirectional torque (T_{el}) given by Eq. (1). The effects of 5th and 7th harmonic voltages are superimposed on the same figure assuming that the respective fluxes are co-phasal at this instant. Neglecting the resistance at these harmonics, the current, flux, and voltage phasors are shown in the figure. The rotational speeds and direction of rotation are also indicated in the figure. In (b), the rotational speed ω_e has been subtracted to make the fundamental phasors stationary. The diagram indicates that 6th harmonic torque is contributed by the interaction of the fundamental flux with 5th and 7th harmonic currents, and the fundamental current with the 5th and 7th harmonic fluxes. Because the harmonic fluxes are small, an approximate expression of 6th harmonic torque is given in Eq. (2). The pulsating torque on the machine has the harmful effect of speed jitter, and may induce mechanical resonance with complex load (see Figure 6.9). However, the high-frequency torque components are filtered by the shaft inertia.

FIGURE 6.17   Two-phase induction motor showing concentrated stator and rotor windings.

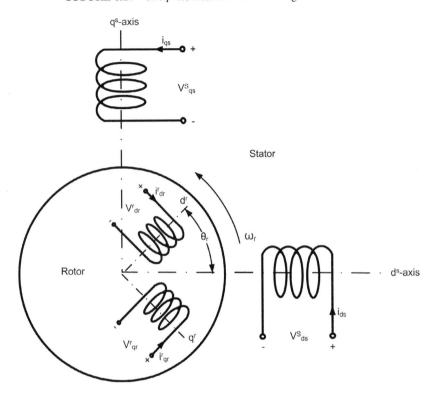

The induction motor per-phase equivalent circuit discussed thus far is only valid for steady-state operation. The dynamic model of the machine is important for transient analysis. When the machine is placed in a feedback control loop for controlling its speed, the dynamics of the machine model dictate the stability of the system. The machine's dynamics are complex because the rotor windings move with respect to stator windings, creating a transformer with time-changing coupling coefficient. A three-phase motor with three stator windings and three rotor windings (see Figure 6.3) can be substituted by an equivalent two-phase machine with two stator (in d^s-q^s axes) and two rotor windings (in d^r-q^r axes), as shown in this figure. If a two-phase power supply (with 90° phase difference) is fed to the stator, a rotating magnetic field will be created as in a three-phase machine. The operating characteristics of the machine are identical to those of a three-phase machine. The time-changing coupling problem was first solved by Park [4] for a synchronous machine when he formulated the transformation of all stator variables to a synchronously rotating frame (d^e-q^e) that moves with the rotor. Later, Stanley [5] showed that a similar solution is possible in induction motors if the rotor circuit variables are referred to the stationary reference frame (d^s-q^s). Then, Kron [6] proposed transformation of both stator and rotor variables to the d^e-q^e frame that moves with the rotating magnetic field. We will develop the dynamic d^e-q^e and d^s-q^s models of the machine and use them extensively.

FIGURE 6.18 Stationary frame (*as-bs-cs* to *d*ˢ-*q*ˢ) transformation.

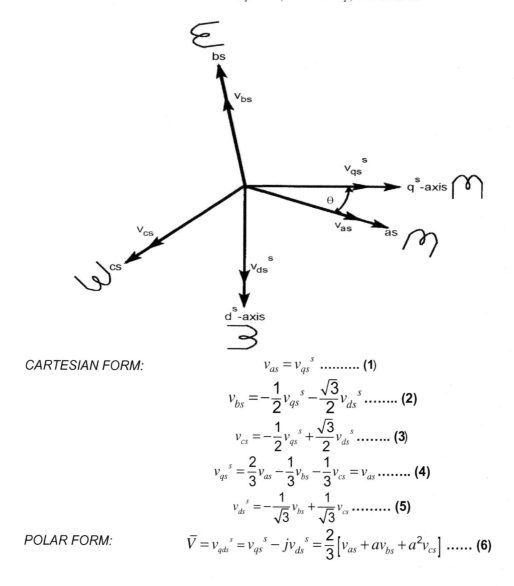

CARTESIAN FORM:

$$v_{as} = v_{qs}^{s} \quad \cdots \cdots \cdots \text{ (1)}$$

$$v_{bs} = -\frac{1}{2}v_{qs}^{s} - \frac{\sqrt{3}}{2}v_{ds}^{s} \quad \cdots \cdots \text{ (2)}$$

$$v_{cs} = -\frac{1}{2}v_{qs}^{s} + \frac{\sqrt{3}}{2}v_{ds}^{s} \quad \cdots \cdots \text{ (3)}$$

$$v_{qs}^{s} = \frac{2}{3}v_{as} - \frac{1}{3}v_{bs} - \frac{1}{3}v_{cs} = v_{as} \quad \cdots \cdots \text{ (4)}$$

$$v_{ds}^{s} = -\frac{1}{\sqrt{3}}v_{bs} + \frac{1}{\sqrt{3}}v_{cs} \quad \cdots \cdots \text{ (5)}$$

POLAR FORM:

$$\overline{V} = v_{qds}^{s} = v_{qs}^{s} - jv_{ds}^{s} = \frac{2}{3}\left[v_{as} + av_{bs} + a^{2}v_{cs}\right] \quad \cdots \cdots \text{ (6)}$$

Since a three-phase machine is equivalent to a two-phase machine, the variables of a three-phase machine can be converted into those of a two-phase machine, and vice versa. Consider a symmetrical three-phase machine with the *as* axis aligned at lagging angle θ with respect to the horizontal line. The equivalent two-phase machine stator axes *d*ˢ and *q*ˢ (also defined as α, β axes) at 90° phase difference are shown in the figure, where the q^s axis is aligned horizontally with the d^s axis lagging. If three phase voltages v_{as}, v_{bs}, and v_{cs} are applied in the respective stator phases, the corresponding two-phase machine stator phase voltages $v_{ds}{}^s$ and $v_{qs}{}^s$ in Cartesian form can be derived by resolving the three phase voltages into the respective axes components. For convenience, it can be assumed that the q^s and as axes are aligned ($\theta = 0$). In that case, the v_{as}, v_{bs}, and v_{cs} voltages can be expressed in terms of $v_{ds}{}^s$ and $v_{qs}{}^s$ voltages, as given in Eqs. (1)–(3). From these equations, $v_{ds}{}^s$ and $v_{qs}{}^s$ expressions can be solved in terms of v_{as}, v_{bs}, and v_{cs}, as shown in Eqs. (4) and (5). The q and d axis components can be combined into the complex polar form shown in Eq. (6), where $a = e^{j2\pi/3}$. Similar expressions are also valid in three-phase (ar, br, cr) to two-phase (dr^r, qr^r) rotor phase voltages transformations, and vice versa.

FIGURE 6.19 Stationary (d^s-q^s) frame to synchronously rotating (d^e-q^e) frame transformation.

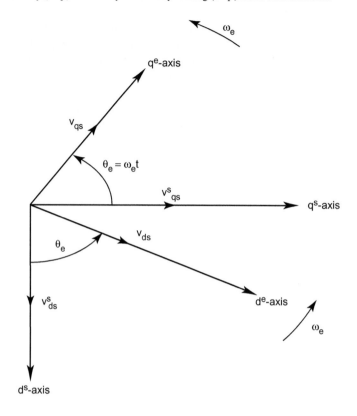

CARTESIAN FORM:
$$v_{ds} = v_{qs}{}^s \sin\theta_e + v_{ds}{}^s \cos\theta_e \quad\text{...... (1)}$$
$$v_{qs} = v_{qs}{}^s \cos\theta_e - v_{ds}{}^s \sin\theta_e \quad\text{..... (2)}$$
$$v_{ds}{}^s = -v_{qs}\sin\theta_e + v_{ds}\cos\theta_e \quad\text{..... (3)}$$
$$v_{qs}{}^s = v_{qs}\cos\theta_e + v_{ds}\sin\theta_e \quad\text{...... (4)}$$
POLAR FORM: $\quad v_{qds} = v_{qs} - jv_{ds} = (v_{qs}{}^s - jv_{ds}{}^s)e^{-j\theta_e} = \bar{V}e^{-j\theta_e} \quad\text{.... (5)}$
$$\bar{V} = v_{qds}e^{j\theta_e} \quad\text{...... (6)}$$

According to Park's theory, the stationary frame d^s-q^s axes windings can be transformed into fictitious windings mounted on synchronously rotating d^e-q^e axes at speed ω_e, as shown in the figure. This transformation eliminates time-changing coupling coefficients between the two sets of windings. The variables in the d^s-q^s frame can be transformed into

d^e-q^e axes variables, and vice versa, by resolving the variables into the respective axes components. Equations (1) and (2) show transformation of the v_{ds}^s, v_{qs}^s variables to v_{ds}^e-v_{qs}^e variables and the reverse equations are solved and shown in Eqs. (3) and (4), where θ_e = $\omega_e t$. For convenience, the superscript "e" is dropped from the synchronous frame variables. These transformations are often known as *Park's transformation*. The equations can also be expressed in polar form as shown in Eq. (5). It can be easily shown that if the d^s-q^s variables are sinusoidal variables, which is true for a practical machine, then d^e-q^e variables are dc values. This transformation is of great advantage because transient analysis with dc variables is very convenient.

FIGURE 6.20 Synchronous frame (d^e-q^e) dynamic model equivalent circuits.

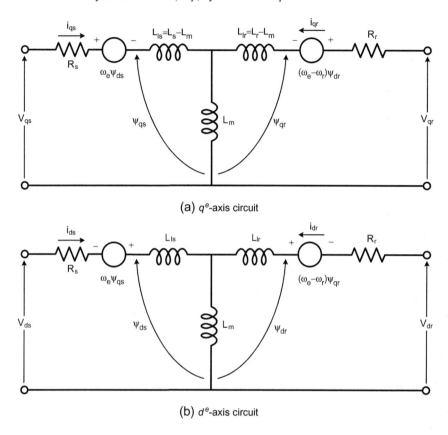

(a) q^e-axis circuit

(b) d^e-axis circuit

The figure shows the transient equivalent circuits of an induction motor in a synchronously rotating d^e-q^e frame, where the frame is moving at speed ω_e with respect to the stator. In the two-phase machine shown in Figure 6.17, first, the stator winding voltage ($v_{ds}{}^s$, $v_{qs}{}^s$) equations (in terms of resistance and time rate of flux linkage drops) are transformed into synchronous frame (d^e-q^e) equations. Then, the rotor voltage ($v_{dr}{}^r$, $v_{qr}{}^r$) equations in the d^r-q^r frame are transformed into the d^e-q^e frame and merged into the respective equivalent circuits shown above. Since the stator windings are stationary ($\omega_e = 0$) and rotor windings move at speed ω_r, the respective speed emf terms $[-\omega_e \psi_{qs}, \omega_e \psi_{ds}$ and $(\omega_e - \omega_r) \psi_{qr}, -(\omega_e - \omega_r) \psi_{dr}]$ shown in the figure appear after transformation into the d^e-q^e frame. Generally, the machine is considered doubly fed with rotor supply voltages v_{dr} and v_{qr}. For cage machines, these variables are zero. The equivalent circuit stator and rotor flux linkages are indicated in the figures. Note that d^e and q^e circuits have mutual coupling. In steady state, all the variables are dc, and there is no voltage drop across inductances.

FIGURE 6.21 Dynamic model (d^e-q^e) equations for an induction motor with voltages and currents.

$$\begin{bmatrix} v_{qs} \\ v_{ds} \\ v_{qr} \\ v_{dr} \end{bmatrix} = \begin{bmatrix} R_s + L_s & \omega_e L_s & SL_m & \omega_e L_m \\ -\omega_e L_s & R_s + SL_s & -\omega_e L_m & SL_m \\ SL_m & (\omega_e - \omega_r)L_m & R_s + SL_m & (\omega_e - \omega_r)L_r \\ -(\omega_e - \omega_r)L_m & SL_m & -(\omega_e - \omega_r)L_r & R_s + SL_m \end{bmatrix} \begin{bmatrix} i_{qs} \\ i_{ds} \\ i_{qr} \\ i_{dr} \end{bmatrix} \quad \cdots\cdots(1)$$

$$T_e = \frac{3}{2}\left|\frac{p}{2}\right|(\psi_{ds}i_{qs} - \psi_{qs}i_{ds}) \quad \cdots\cdots \textbf{(2)}$$

$$T_L + \frac{2}{p}\left|\frac{p}{2}\right|JS\omega_r = T_e \quad \cdots\cdots \textbf{(3)}$$

The dynamic equivalent circuits in the previous figure have four loops, and the corresponding loop equations can be written in the matrix form shown in this figure. In this equation, S = Laplace operator and $\omega_r = (p/2)\omega_m$, where ω_m = mechanical speed and ω_r = electrical speed. For a singly fed machine (such as a cage motor), $v_{qr} = v_{dr} = 0$. If the speed is considered constant (moment of inertia $J \to \infty$), the electrical dynamics of the machine are given by a 4th-order linear system. The expression of developed torque and mechanical load equation with constant load torque are included in the figure. Altogether, the machine model is nonlinear and is given by a 5th-order system. The order will be higher with complex load dynamics. By knowing the applied variables v_{qs}, v_{ds}, ω_e, and T_L for a singly fed machine, the currents i_{qs}, i_{ds}, i_{qr}, i_{dr}, speed ω_m, and the corresponding flux linkage expressions can be solved. Note that for an ideal current-fed cage machine, the variables v_{qs}, v_{ds}, i_{qr}, i_{dr}, and ω_m can be solved from the equations for applied i_{qs}, i_{ds}, ω_e, and T_L variables. Note that Eq. (1) is not in state space form and therefore is not convenient to use for computer simulations.

FIGURE 6.22   Complex synchronously rotating frame (*dqs*) equivalent circuit model.

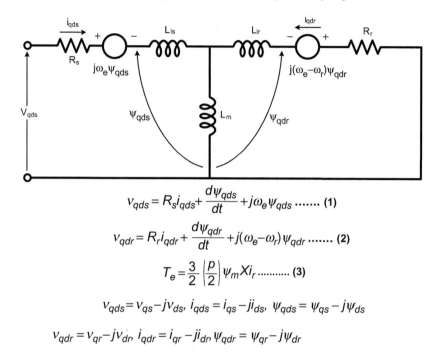

$$v_{qds} = R_s i_{qds} + \frac{d\psi_{qds}}{dt} + j\omega_e \psi_{qds} \dots (1)$$

$$v_{qdr} = R_r i_{qdr} + \frac{d\psi_{qdr}}{dt} + j(\omega_e - \omega_r)\psi_{qdr} \dots (2)$$

$$T_e = \frac{3}{2}\left|\frac{p}{2}\right|\psi_m X i_r \dots (3)$$

$$v_{qds} = v_{qs} - jv_{ds}, \ i_{qds} = i_{qs} - ji_{ds}, \ \psi_{qds} = \psi_{qs} - j\psi_{ds}$$

$$v_{qdr} = v_{qr} - jv_{dr}, \ i_{qdr} = i_{qr} - ji_{dr}, \ \psi_{qdr} = \psi_{qr} - j\psi_{dr}$$

Instead of handling *d* and *q* equations and the corresponding equivalent circuits separately, it is convenient to combine them and represent them with a single complex vector equation and the corresponding single complex equivalent circuit shown in this figure. Consider a cage machine, in which the rotor supply voltage is shorted ($v_{qdr} = 0$). All the complex vector variables are expressed in the form $x_{qds} = x_{qs} - jx_{ds}$, but the sign of vector is omitted for simplicity. The developed torque is defined as the cross-product of airgap flux (or flux linkage) $\psi_m = \psi_{dm} - j\psi_{qm}$ and rotor current $i_r = i_{qr} - ji_{dr}$. The steady-state, per-phase equivalent circuit relations can be derived from this circuit, where inductive voltage drops are zero.

FIGURE 6.23   Synchronous frame state space equations in terms of flux linkages ($\psi_{qs}, \psi_{ds}, \psi_{qr}, \psi_{dr}$).

$$\frac{dF_{qs}}{dt} = \omega_b \left[v_{qs} - \frac{\omega_e}{\omega_e} F_{ds} - \frac{R_s}{X_{ls}} \left(F_{qs} - F_{qm} \right) \right] \quad \text{........(1)}$$

$$\frac{dF_{ds}}{dt} = \omega_b \left[v_{ds} + \frac{\omega_e}{\omega_b} F_{qs} - \frac{R_s}{X_{ls}} \left(F_{ds} - F_{dm} \right) \right] \quad \text{........(2)}$$

$$\frac{dF_{qr}}{dt} = -\omega_b \left[-\frac{(\omega_e - \omega_r)}{\omega_b} F_{dr} + \frac{R_r}{X_{lr}} \left(F_{qr} - F_{qm} \right) \right] \quad \text{........(3)}$$

$$\frac{dF_{dr}}{dt} = -\omega_b \left[-\frac{(\omega_e - \omega_r)}{\omega_b} F_{qr} + \frac{R_r}{X_{lr}} \left(F_{dr} - F_{dm} \right) \right] \quad \text{........(4)}$$

$$F_{qm} = \frac{X_{m1}}{X_{ls}} F_{qs} + \frac{X_{m1}}{X_{lr}} F_{qr}$$

$$F_{dm} = \frac{X_{m1}}{X_{ls}} F_{ds} + \frac{X_{m1}}{X_{lr}} F_{dr}$$

$$X_{m1} = \left(\frac{1}{X_m} + \frac{1}{X_{ls}} + \frac{1}{X_{lr}} \right)^{-1}$$

$$T_e = \frac{3}{2} \left| \frac{p}{2} \right| \frac{1}{\omega_b} \left(F_{ds} i_{qs} - F_{qs} i_{ds} \right) = T_L + \frac{2}{p} J \frac{d\omega_r}{dt} \quad \text{........(5)}$$

The induction motor model given before in terms of voltages and currents is not conven-ient for computer simulation studies. This figure gives machine dynamic model equations in terms of flux linkages in state variable form. However, all inductances and fluxes are multiplied by base frequency ω_b (376.8 r/s at 60 Hz) to convert them into reactances (X) and voltage signals (F), respectively, because of easy manipulation of small signals. Once the flux signals are solved in the above equations, the current components can be determined easily. Then, developed torque and speed can be calculated by Eq. (5). Evidently, the system is a nonlinear 5th-order system including the mechanical dynamics. The flux equations are widely used for dynamic simulation of induction machines.

FIGURE 6.24   Stationary frame (d^s-q^s) dynamic equivalent circuit for an induction motor.

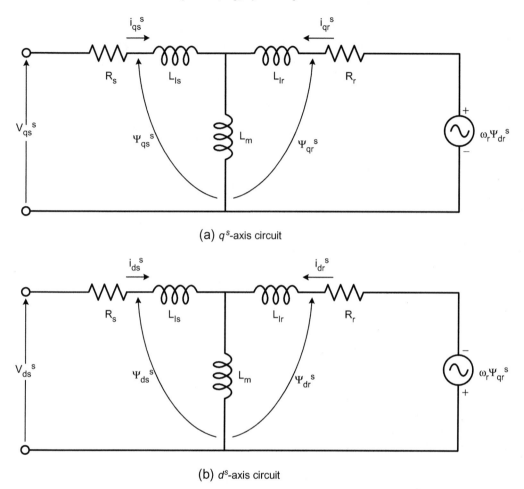

(a) q^s-axis circuit

(b) d^s-axis circuit

The synchronous frame model equations given in Figure 6.21 are valid when the frame axes (d^e-q^e) are moving at synchronous speed ω_e. Therefore, the same model can be converted to the stationary frame (or with respect to the stator) by substituting $\omega_e = 0$. The corresponding equivalent circuits for a singly fed machine ($v_{qr}^s = v_{dr}^s = 0$) is shown in this figure. It is also evident by comparing this with Figure 6.20. In the stationary frame, all variables are sinusoidal in the steady state with sinusoidal supply. Note the cross-coupling effect, where the d^s axis rotor flux rotating at speed ω_r creates sinusoidal counter emf in the q^s axis. Similar counter emf also appears in the d^s axis as shown. Chapter 7 will show that the stationary frame model is extremely important for solving the circuit variables in real time by means of a microprocessor or DSP (digital signal processor) and use for adaptive (model-following) control of speed.

FIGURE 6.25 Stationary frame (d^s-q^s) dynamic model for an induction motor.

$$\frac{d}{dt}\begin{bmatrix} i_{qs}^s \\ i_{ds}^s \\ i_{qr}^s \\ i_{dr}^s \end{bmatrix} = \frac{1}{\sigma L_s L_r}\left\{\begin{bmatrix} -R_s L_r & -\omega_r L_m^2 & R_r L_m & -\omega_r L_r L_m \\ \omega_r L_m^2 & -R_s L_r & \omega_r L_r L_m & R_r L_m \\ R_s L_m & \omega_r L_s L_m & -R_r L_s & \omega_r L_s L_r \\ -\omega_r L_s L_m & R_s L_m & -\omega_r L_s L_m & -R_r L_s \end{bmatrix}\begin{bmatrix} i_{qs}^s \\ i_{ds}^s \\ i_{qr}^s \\ i_{dr}^s \end{bmatrix}\right.$$

$$\left. +\begin{bmatrix} L_r & 0 & -L_m & 0 \\ 0 & L_r & 0 & -L_m \\ -L_m & 0 & L_s & 0 \\ 0 & -L_m & 0 & L_s \end{bmatrix}\begin{bmatrix} v_{qs}^s \\ v_{ds}^s \\ v_{qr}^s \\ v_{dr}^s \end{bmatrix}\right\}\cdots\cdots \textbf{(1)}$$

$$\frac{2}{p}J\frac{d\omega_r}{dt}=T_e-T_L=\frac{3}{2}(\frac{p}{2})L_m\,(i_{qs}^s\,i_{dr}^s-i_{ds}^s\,i_{qr}^s)-T_L\cdots\cdots \textbf{(2)}$$

$$\sigma=1-\frac{L_m^2}{L_s L_r}$$

The four loop equations described from the equivalent circuits in Figure 6.24 in terms of the variables i_{qs}^s, i_{ds}^s, i_{qr}^s, and i_{dr}^s describe the mathematical model in the stationary frame. The synchronous frame equations in Figure 6.21 can also be directly converted to the stationary frame by substituting $\omega_e = 0$. The stationary frame model can be used for computer simulation studies or for real-time solutions by DSP for adaptive control of drives, as mentioned earlier. The state variable form for the stationary frame model, shown here, is very convenient for simulation. Note that v_{qs}^s, v_{ds}^s, v_{qr}^s, and v_{dr}^s are the applied voltages. For a singly fed cage motor, $v_{qr}^s = v_{dr}^s = 0$. The developed torque expression given in Figure 6.21 is also valid with stationary frame variables, i.e.,

$T_e = \frac{3}{2}\left(\frac{p}{2}\right)(\psi_{ds}^s\,i_{qs}^s - \psi_{qs}^s\,i_{ds}^s)$. The stator fluxes in this expression can be written in terms of stator currents as shown in Eq. (2). The total electromechanical dynamics of the system are 5th order, as mentioned earlier.

FIGURE 6.26 Stationary frame dynamic model with stator currents and rotor fluxes ($i_{qs}{}^s$, $i_{ds}{}^s$, $\psi_{qr}{}^s$, $\psi_{dr}{}^s$).

$$\frac{d}{dt}(i_{qs}{}^s)=-\frac{(L_m{}^2R_r+L_r{}^2R_s)}{\sigma L_s L_r{}^2}\,i_{qs}{}^s-\frac{L_m\omega_r}{\sigma L_s L_r}\,\psi_{dr}{}^s+\frac{L_m R_r}{\sigma L_s L_r{}^2}\psi_{qr}{}^s+\frac{1}{\sigma L_s}v_{qs}{}^s \cdots (1)$$

$$\frac{d}{dt}(i_{qs}{}^s)=-\frac{(L_m{}^2R_r+L_r{}^2R_s)}{\sigma L_s L_r{}^2}\,i_{qs}{}^s+\frac{L_m R_r}{\sigma L_s L_r{}^2}\,\psi_{dr}{}^s+\frac{L_m\omega_r}{\sigma L_s L_r}\psi_{qr}{}^s+\frac{1}{\sigma L_s}v_{ds}{}^s \cdots (2)$$

$$\frac{d}{dt}(\psi_{qr}{}^s)=-\frac{R_r}{L_r}\psi_{qr}{}^s+\omega_r\psi_{dr}{}^s+\frac{L_m R_r}{L_r}i_{qs}{}^s \cdots (3)$$

$$\frac{d}{dt}(\psi_{dr}{}^s)=-\frac{R_r}{L_r}\psi_{qr}{}^s-\omega_r\psi_{qr}{}^s+\frac{L_m R_r}{L_r}i_{ds}{}^s \cdots (4)$$

$$J(\frac{2}{p})\frac{d\omega_r}{dt}=T_e-T_L=\frac{3}{2}(\frac{p}{2})\frac{L_m}{L_r}(\psi_{dr}{}^s i_{qs}{}^s-\psi_{qr}{}^s i_{ds}{}^s)-T_L \cdots (5)$$

$$\sigma=1-\frac{L_m{}^2}{L_s L_r}$$

For the d^s-q^s equivalent circuits shown in Figure 6.24, the mathematical model equations can also be described in terms of fluxes, or by combining fluxes and currents. The fluxes can be easily converted to currents, and vice versa. Several combinations of variables are ($i_{qs}{}^s$, $i_{ds}{}^s$, $\psi_{qr}{}^s$, $\psi_{dr}{}^s$), ($i_{qs}{}^s$, $i_{ds}{}^s$, $\psi_{qs}{}^s$, $\psi_{ds}{}^s$), ($i_{qr}{}^s$, $i_{dr}{}^s$, $\psi_{qs}{}^s$, $\psi_{ds}{}^s$), ($i_{qr}{}^s$, $i_{dr}{}^s$, $\psi_{qr}{}^s$, $\psi_{dr}{}^s$), or ($\psi_{qs}{}^s$, $\psi_{ds}{}^s$, $\psi_{qr}{}^s$, $\psi_{dr}{}^s$). The model shown in this figure is with the stator currents and rotor fluxes for a stator-fed machine ($v_{qr}{}^s=v_{dr}{}^s=0$). The torque expression included in Eq. (5) has been transformed to relate with the same variables. It will be shown later that this model is of particular importance for induction motor drives. Generally, note that synchronous frame model equations (see Figure 6.21) can also be described with similar sets of state variables. The synchronous frame model with stator and rotor fluxes in state variable form was given in Figure 6.23. In fact, for computer simulations or real-time solutions, any one of these stationary or synchronous frame models can be used.

FIGURE 6.27   (a) Stationary frame complex equivalent circuit and (b) simplified per-phase equivalent circuit.

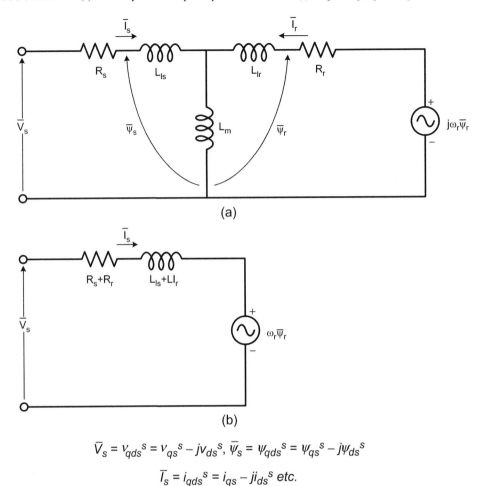

$$\overline{V}_s = v_{qds}{}^s = v_{qs}{}^s - jv_{ds}{}^s,\ \overline{\psi}_s = \psi_{qds}{}^s = \psi_{qs}{}^s - j\psi_{ds}{}^s$$

$$\overline{I}_s = i_{qds}{}^s = i_{qs}{}^s - ji_{ds}{}^s \ etc.$$

The complex stationary frame equivalent circuit shown in (a) of this figure can be drawn for convenience by combining the d^s and q^s equivalent circuits in Figure 6.24. The circuit and the corresponding equations shown in Figure 6.22 can be directly converted to complex form by substituting $\omega_e = 0$. All the circuit variables are in vector form and are denoted by simpler symbols with "bar" notation. The "bar" is often omitted for simplicity. A simplified Thevenin equivalent circuit is shown in (b), assuming magnetizing inductance L_m is too large. This simplified stationary frame transient equivalent circuit has often been used for converter analysis in the literature.

FIGURE 6.28   Classification of single-phase induction motors and applications.

CLASSIFICATION:

- SPLIT-PHASE MOTOR
- CAPACITOR-START MOTOR
- PERMANENT-SPLIT-CAPACITOR MOTOR
- CAPACITOR-START, CAPACITOR-RUN MOTOR
- SHADED-POLE MOTOR

SOME APPLICATIONS:

* PUMP	* TOY
* COMPRESSOR	* AIR CONDITIONER
* FAN	* FURNACE BLOWER
* HAIR DRYER	* VACUUM CLEANER
* WASHING MACHINE	* BLENDER
* DRYER	* ELECTRIC LAWN MOWER
* REFRIGERATOR	* HAND TOOL, ETC.

So far, we have discussed polyphase (three-phase) induction motors with symmetrical three-phase sinusoidally distributed stator and rotor windings. Of course, in cage motors, the rotor windings are replaced by a cage. All the discussions are valid also for a two-phase symmetrical machine, where the supply voltages (or currents) are phase-shifted by 90°. A three-phase machine can be represented by an equivalent two-phase machine, as shown in Figure 6.17. We now discuss single-phase machines [1] for completeness of the subject in this chapter. Single-phase motors are generally of miniature FHP (fractional horse power) range, and are extensively used in residential and commercial applications. Some applications are given in this figure. The applications may be at constant speed or variable speed. The general classification of single-phase induction motors is given on the top of this figure. We will show later that single-phase motors are basically two-phase motors but with unsymmetrical stator windings. The theory of single-phase motors is much more complex than that of polyphase machines, and we will discuss only the basic principles. The performance of single-phase motors, particularly the efficiency, is inferior. However, the low cost is given more importance than is performance for extensive applications.

FIGURE 6.29   (a) Schematic view of single-phase induction motor and (b) torque-speed characteristics.

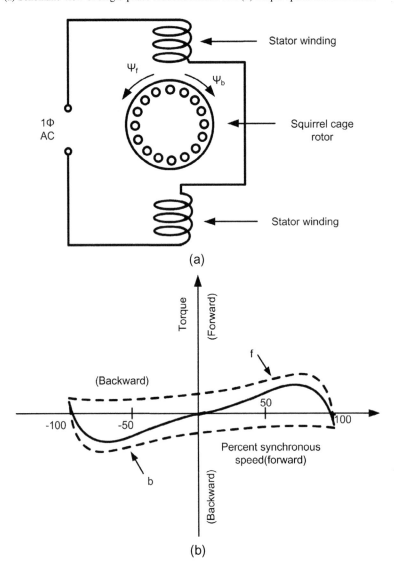

(a)

(b)

The schematic view of the single-phase induction motor shown in (a) indicates that it has one stator winding, and the rotor has a cage structure. The stator winding is sinusoidally distributed with a vertical axis and produces a pulsating sinusoidal MMF wave similar to phase a in Figure 6.3. The pulsating stator flux can be resolved into forward and backward rotating flux components (ψ_f and ψ_b), as shown in the figure. The torque is developed independently by these component fluxes, and the resultant motor torque is given by the superposition principle. The lower part of the figure shows the component torque curves (*f* and *b*) and the corresponding resultant torque. This means that a single-phase motor does not have starting torque, but once it is started (by hand or other means), the motor will continue to rotate with load torque in either the forward or backward direction. Basically, the method of starting by an auxiliary winding classifies the motor type. Note that in starting and running conditions, the motor will experience large pulsating torque at double the stator frequency (120 Hz with 60-Hz supply).

FIGURE 6.30   (a) Split-phase motor connections with phasor diagram and (b) torque-speed characteristics.

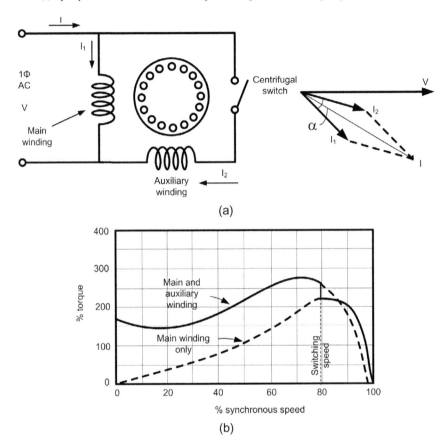

(a)

(b)

The split-phase motor, shown at the top of this figure, has a main (or run) winding and an auxiliary (or start) winding, both with sinusoidal distribution, and the axes are at 90° phase difference. The auxiliary winding has thinner wire, i.e., its winding resistance is high. The phasor diagram on the right indicates that the current (I_2) in the auxiliary winding will have a small lagging angle with supply voltage compared to that (I_1) of the main winding. The resulting phase difference between I_1 and I_2 is α angle. Basically, it is an unsymmetrical two-phase motor with unbalanced power supply. The developed torque curve shown in the lower figure indicates that the motor has large starting torque by the combined effect of the two windings. The motor starts satisfactorily, and typically at 78% of the synchronous speed, a centrifugal switch isolates the auxiliary winding. Then the developed torque falls to that due to the main winding only. With load torque T_L, the motor operates near synchronous speed on the negative slope of the torque curve. The motors are very low cost and are used in the power range of 50–500 W for applications, such as fans, blowers, centrifugal pumps, and office equipment. The vibration of the motor at 120-Hz pulsating torque is not appreciable if inertia is high.

FIGURE 6.31 (a) Capacitor-start motor and (b) torque-speed characteristics.

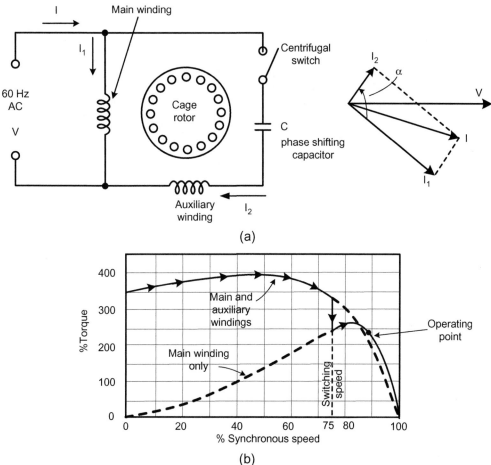

(a)

(b)

The capacitor-start motor shown here has two symmetrical stator windings. The auxiliary winding is connected with a series capacitor (C) and a centrifugal switch. The phasor diagram shows that the main winding current I_1 lags the voltage V, but the auxiliary winding current I_2 leads the voltage. With proper design of capacitor value, I_2 can be near 90° leading with respect to I_1, thus developing very high starting torque as shown. Of course, to get such high starting torque, the rotor resistance (R_r) of the machine should be high. At about 75% of synchronous speed, the centrifugal switch isolates the auxiliary winding and the motor settles down on the negative slope of the torque curve at steady state. These motors are commonly used for pumps, compressors, and refrigeration and air-conditioning equipment where high starting torque is needed.

FIGURE 6.32 (a) Permanent-split-capacitor motor and torque-speed characteristics and (b) capacitor-start, capacitor-run motor and torque-speed characteristics.

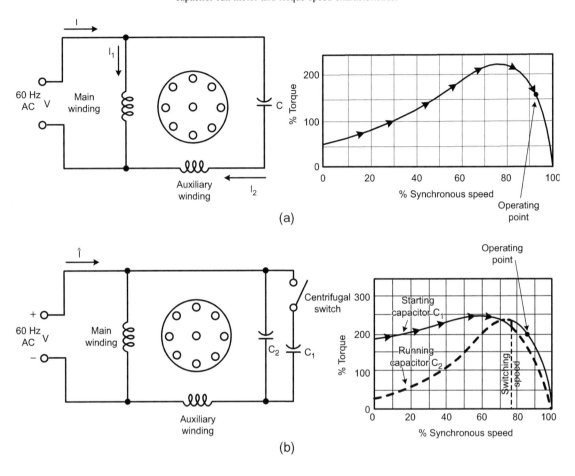

(a)

(b)

Instead of using the capacitor in auxiliary winding mode for starting only, it can be kept permanently in a permanent-split capacitor motor as shown in (a). This will save the centrifugal switch and eliminate the transient from starting to running mode. Besides, the power factor, efficiency, and torque pulsation will improve. Reduced pulsating torque makes the motor operation quieter. This is basically a two-phase motor operation. The capacitor is designed to compromise starting and running conditions, as shown by the torque curve on the right. It is possible to use two capacitors, as shown in the lower figure, where the capacitor C_1 in series with the switch is used for starting, and capacitor C_2 remains permanently as the running capacitor. Although the scheme is more expensive than other single-phase motors, optimum starting and running conditions are possible. The starting capacitor for intermittent duty can be of a special ac electrolytic type, but the running capacitor should be ac paper, foil, or oil type.

FIGURE 6.33   (a) Shaded-pole motor and (b) torque-speed characteristics.

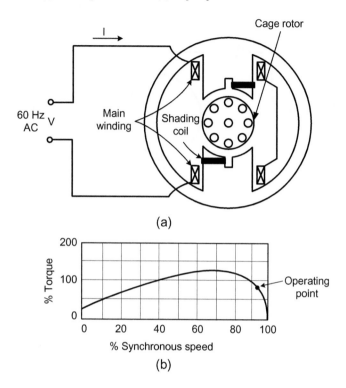

The shaded pole induction motor has salient stator poles, where the stator winding in each pole is connected in series, as shown in the figure. One portion of each pole has a short-circuited turn of copper called the shading coil. The induced current in the shading coil causes the flux under the shading coil to lag behind the flux in the unshaded portion. As a result, a rotating magnetic field is created that moves from the unshaded to the shaded portion. A typical torque-speed curve is shown in (b). Shaded pole motors are least expensive, have very low efficiency, and are typically available at below 50-W ratings. They are widely used in industry.

FIGURE 6.34   (a) Universal motor and torque-speed characteristics and (b) speed control of single-phase motor.

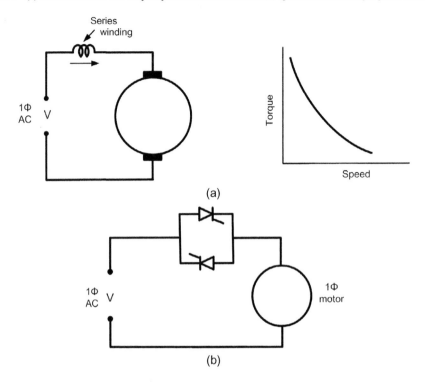

Figure 6.34(a) shows the schematic diagram of a single-phase ac series motor (not an induction motor). Basically, it is the same as a dc series motor. Because the motor operates with dc as well as ac, it is called a *universal* motor. With ac supply, the direction of both the armature flux and series field flux changes as the polarity of current changes, and thus the developed torque is unidirectional. The torque-speed characteristics of a universal motor, shown on the right, are similar to those of a dc series motor ($T_e\omega_m^2 = K$). To reduce core loss, the poles as well as the rotor are made of laminated steel. The disadvantages of ac operation of a universal motor are lower available torque, torque pulsation, lower efficiency, and poor power factor. These motors are commonly used in vacuum cleaners, kitchen appliances, and portable tools. A universal motor provides the highest power per dollar in the low-power range, at the expense of noise, relatively short life due to commutators and brushes, and high speed at no load. High-power traction applications have also used this type of motor with a single-phase trolly power supply. So far, we have discussed single-phase motor principles without speed control. A simple and inexpensive method of speed control is to use anti-parallel thyristors in line as shown in (b), where the firing angle is controlled to control the supply voltage. Lower voltage will develop lower torque (see Figure 6.8), and thus permit control of speed. The thyristor pair can also be replaced by a triac in many cases. However, the disadvantages of phase control are lower line power factor and higher losses in the machine, as mentioned earlier.

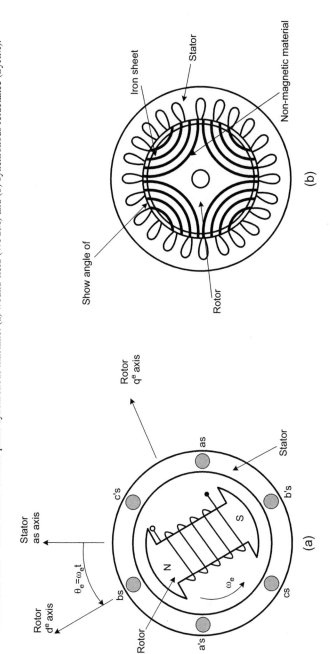

FIGURE 6.35 Cross section of three-phase synchronous machine: (a) wound-field (WFSM) and (b) synchronous reluctance (SyRM).

The three-phase wound-field synchronous machine shown in (a) has the same stator winding as that of an induction motor and produces a similar rotating magnetic field. The rotor has a field winding that carries dc current and produces an airgap flux, which is locked with the stator-induced rotating flux. The rotor always moves at synchronous speed (i.e., slip $S = 0$), and the synchronously rotating d^e-q^e axes as shown are locked on the rotor. Since there is no stator-induced excitation on the rotor, machine power factor can be arbitrary (leading, lagging, or unity) by rotor excitation control. A low-speed hydroelectric generator is normally a salient pole machine, as shown in the figure, but high-speed steam power station generators are normally nonsalient pole-like induction machines. There can be a damper or amortisseur winding also on the rotor. Although such machines are more expensive than the induction type, their efficiency is higher. A section of an SyRM is shown in (b). The three-phase stator winding is the same, but there is no winding (or permanent magnet) on the rotor. The iron lamination shown gives saliency to the rotor and, due to reluctance torque, the stator poles drag the rotor poles at synchronous speed. The machine is simple and robust, and the input power factor is lagging as in an induction motor (and may be somewhat poorer). The machine has been used widely in low-power applications.

FIGURE 6.36   Features of WFSM and SyRM.

WFSM

- SALIENT OR NONSALIENT POLE
- HIGH-POWER, MULTI-MEGAWATT SIZE
- TERMINAL DPF CAN BE CONTROLLED TO BE UNITY, LEADING, OR LAGGING
- DAMPER OR AMORTISSEUR WINDING TO PREVENT HUNTING OR REDUCE COMMUTATING INDUCTANCE
- HIGH EFFICIENCY
- CENTRAL POWER STATION GENERATORS INVARIABLY USE WFSM
- USED IN HIGHEST POWER DRIVES (ROLLING MILLS, CEMENT MILLS, MINING DRIVES, PUMPS, COMPRESSORS, ETC.)

SyRM

- BASICALLY SYNCHRONOUS MACHINE – BUT RELUCTANCE TORQUE ONLY DUE TO ROTOR SALIENCY
- GENERALLY SMALL MACHINE WITH SIMPLE LOW-COST CONSTRUCTION
- LAGGING DPF (SLIGHTLY LOWER THAN IM)
- ROBUST MACHINE FOR HIGH-SPEED OPERATION
- COMPETITOR OF SWITCHED RELUCTANCE MACHINE (SRM)

General features of WFSM and SyRM are summarized in the figure. Although both are basically three-phase synchronous machines, there are many differences. WFSMs are the largest of all machines with high performance. But SyRMs are generally of miniature size with poorer performance. In many respects, SyRM machine is comparable with SRM. In many applications, permanent magnet synchronous machines (will be discussed later) have found favor over SyRMs.

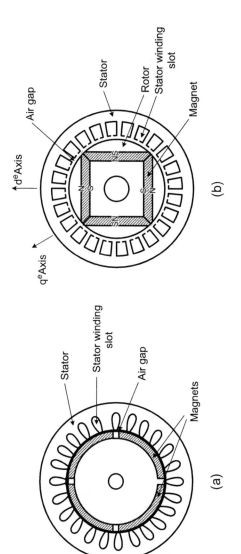

FIGURE 6.37 Cross section of sinusoidal permanent magnet machine: (a) surface PM (SPM) and (b) interior PM (IPM).

In a three-phase sinusoidal permanent magnet synchronous machine (PMSM), the stator winding is the same as that of WFSM, but the rotor winding is replaced by a permanent magnet (generally ferrite, neodymium-iron-boron or cobalt-samarium). The advantage is the elimination of rotor loss, but the flexibility of field control is lost. With a high-energy magnet (such as NdFeB), the machine size is smaller with lower inertia, which can be an advantage in many drive applications. PMSMs are more expensive than induction motors, but have the advantage of higher efficiency. However, recently, the prices of NdFeB are falling, promoting more applications of PM machines. Typical power ranges are below 100 kW, but the size is increasing recently. In a line-start PMSM, there is a cage winding on the rotor for starting it as an induction motor. In the SPM shown in (a), the magnets are glued to the surface of a solid or laminated rotor. The machine effective airgap is large (magnet $\mu_r \approx 1$) and uniform, giving the machine nonsalient pole characteristics. In the IPM shown in (b), the magnets are buried inside the rotor, giving it salient pole characteristics, a lower effective airgap, and robust high-speed operation. Currently, SPM machines are more commonly used in industry. Further features of these machines and comparison of drives with these machines will be discussed in Chapter 8.

FIGURE 6.38 Per-phase equivalent circuit derivation of nonsalient pole WFSM.

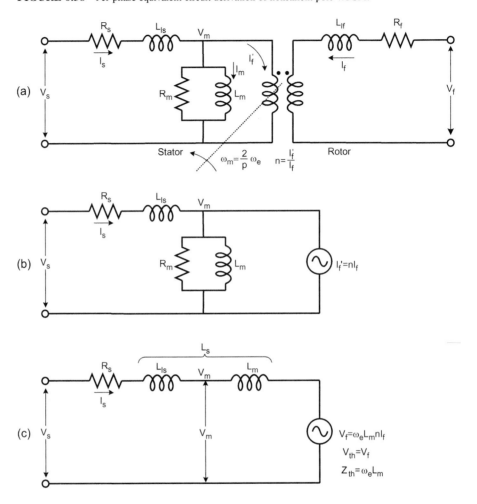

Like an induction motor, equivalent circuit analysis is a good method for studying the performance of synchronous machines at steady state. Part (a) of the figure shows a transformer-like per-phase rms equivalent circuit, where the field circuit with current I_f is rotating at speed ω_m and is coupling to the stator. Part (b) shows the equivalent circuit with an ac current source, where n represents the ratio of rms field current I_f' to the actual dc field current I_f. The power fed at the stator input flows to the shaft through the airgap after the stator resistance (R_s) loss and iron loss (R_m). If R_m is neglected, the Thevenin equivalent circuit is shown in (c), where the voltage $V_f = \omega_e L_m n I_f = \omega_e \psi_f$ is the speed emf, and $L_s = L_{ls} + L_m$ is the synchronous inductance of the machine. Since the magnetizing flux linkage $\omega_m = L_m I_m$ is nearly constant for the constant V_m, the excessive field current will cause leading reactive current at the machine terminal.

FIGURE 6.39 Phasor diagram of nonsalient pole machine: (a) motoring mode and (b) generating mode.

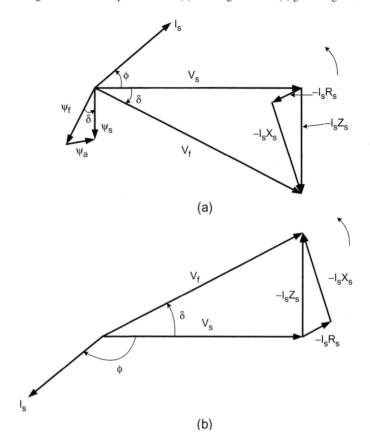

(a)

(b)

The figure shows the rms phasor diagram that corresponds to the equivalent circuit of Figure 6.38(c). The motoring mode in the upper figure is considered at leading power factor, where V_s is the terminal voltage. The counter or speed emf V_f is constructed by subtracting the R_s and X_s (synchronous reactance) drops. The R_s drop is very small for a large machine, and is often neglected. The flux linkage phasor diagram is added on the same figure, where $V_f = \omega_e \psi_f$, $V_s = \omega_e \psi_s$, and $I_s L_s = \psi_a$, neglecting the R_s drop. Note that the armature reaction flux phasor ψ_a adds to the field flux ψ_f to generate the stator flux ψ_s. The voltages V_s and V_f are perpendicular to the respective fluxes. The generating mode phasor diagram is shown in the lower figure, where the direction of I_s is reversed. The flux phasor diagram can be easily constructed on it.

FIGURE 6.40   Phasor diagram of salient pole machine: (a) motoring mode and (b) generating mode.

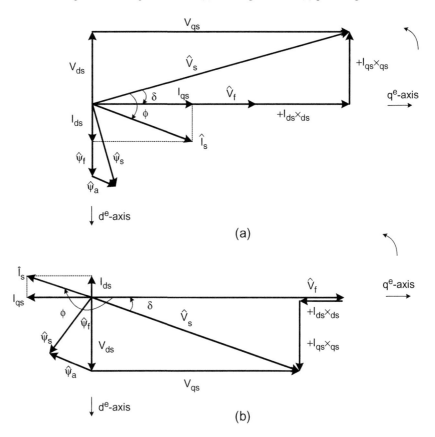

(a)

(b)

In the salient pole machine shown in Figure 6.35(a), the nonsymmetry in the airgap in the d^e and q^e axes makes the magnetizing or synchronous reactances unsymmetrical, i.e., $X_{dm} > X_{qm}$ or $X_{ds} > X_{qs}$, respectively; therefore, the phasor diagrams shown in Figure 6.39 are not valid. This figure shows the phasor diagrams in motoring and generating modes, where the phasor amplitudes are given by sinusoid peak values ($\sqrt{2}$ rms value), and resolved into d^e and q^e components. Again, the stator resistance R_s is neglected. The speed emf $\hat{V}_f$ is aligned to the q^e axis, whereas the corresponding field flux $\hat{\psi}_f$ is aligned to the d^e axis. Phase voltage $\hat{V}_s$ and phase current $\hat{I}_s$ are resolved into d^e and q^e components, and voltage phasor diagrams are drawn considering the reactive drops. In the figure, the motoring mode phasor diagram is drawn at lagging DPF, which makes $\hat{V}_s > \hat{V}_f$ or $\hat{\psi}_s > \hat{\psi}_f$, whereas in the generating mode or $\hat{V}_s < \hat{V}_f$ or $\hat{\psi}_s < \hat{\psi}_f$ because the power factor is leading. The figure can be defined as a vector diagram, where the rms phasor diagram is modified by the respective peak values of the variables.

FIGURE 6.41 Torque-δ angle characteristics of synchronous machine: (a) nonsalient pole and (b) salient pole.

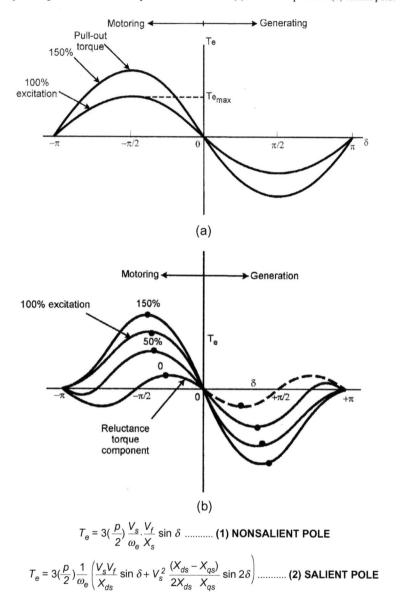

(a)

(b)

$$T_e = 3\left(\frac{p}{2}\right)\frac{V_s}{\omega_e}\cdot\frac{V_f}{X_s}\sin\delta \text{ .... (1) NONSALIENT POLE}$$

$$T_e = 3\left(\frac{p}{2}\right)\frac{1}{\omega_e}\left(\frac{V_s V_f}{X_{ds}}\sin\delta + V_s^2\frac{(X_{ds}-X_{qs})}{2X_{ds}\,X_{qs}}\sin 2\delta\right) \text{ .... (2) SALIENT POLE}$$

The figure shows the plot of developed torque with torque angle δ for nonsalient pole as well as salient pole machines. The corresponding equations are given at the bottom. The equations can also be written in terms of fluxes, where $\psi_s = V_s/\omega_e$ and $\psi_f = V_f/\omega_e$.

Note that if V_s/ω_e is maintained constant, for a fixed excitation, the torque curve is a function of angle δ only. In motoring mode, angle δ is positive, whereas in generating mode, it is negative. The stability of operation in motoring and generating modes is limited in the range $-\pi/2 < \delta < +\pi/2$ for nonsalient pole machines, whereas the range varies for the salient pole machines, as shown in the figure. The maximum torque can be varied by machine field excitation control. In the salient pole machine, the reluctance torque adds with the excitation component to increase the maximum torque as shown. The reluctance torque component in Eq. (2) is valid for the SyRM shown in Figure 6.35(b).

FIGURE 6.42 Dynamic (d^e-q^e) equivalent circuits of WFSM (a) q^e-axis circuit and (b) d^e-axis circuit.

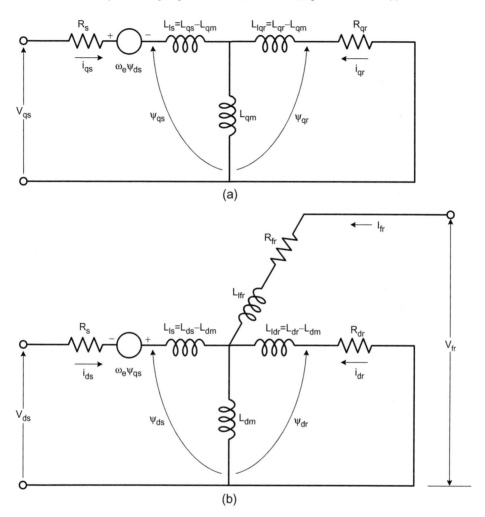

The dynamic equivalent circuits for a salient pole WFSM in synchronous reference frame, which were originally developed by Park, are shown in this figure. The machine is considered with a damper winding. The derivation of these circuits follows the same principle as that for the induction motor shown in Figure 6.20. The damper winding is equivalent to the cage winding of the induction motor, and it is resolved into dr and qr components. Included in this figure is the field excitation circuit, which is active in the d^e axis only, where the field circuit parameters V_{fr}, I_{fr}, R_{fr}, and L_{lfr} are referred to the stator side. The field current I_{fr} flows through magnetizing inductance L_{dm} to create the field flux ψ_f on the d^e axis. Because $\omega_r = \omega_e$, there is no emf on the rotor side. If the damper winding is absent, the rotor circuits are open, i.e., $i_{dr} = i_{qr} = 0$.

FIGURE 6.43 Synchronous frame dynamic model equations of WFSM.

$$
\begin{bmatrix} v_{qs} \\ v_{ds} \\ 0 \\ 0 \\ v_{fr} \end{bmatrix} = \begin{bmatrix} R_s+SL_{qs} & \omega_e L_{qs} & SL_{qm} & \omega_e L_{dm} & \omega_e L_{dm} \\ -\omega_e L_{qs} & R_s+SL_{ds} & -\omega_e L_{qm} & SL_{dm} & SL_{dm} \\ SL_{qm} & 0 & R_{qr}+SL_{qr} & 0 & 0 \\ 0 & SL_{dm} & 0 & R_{dr}+SL_{dr} & SL_{dm} \\ 0 & SL_{dm} & 0 & SL_{dm} & R_{fr}+S(L_{lfr}+L_{dm}) \end{bmatrix} \begin{bmatrix} i_{qs} \\ i_{ds} \\ i_{qr} \\ i_{dr} \\ I_{fr} \end{bmatrix} \quad \text{......... (1)}
$$

$$
T_e = \frac{3}{2}(\frac{p}{2})(\psi_{ds} i_{qs} - \psi_{qs} i_{ds}) \text{......... (2)}
$$

$$
T_e - T_L = \frac{2}{p} J \frac{d\omega_e}{dt} \text{......... (3)}
$$

The equivalent circuits in Figure 6.42 have altogether five loops; therefore, the electrical dynamics of the machine are 5th order, as shown here in matrix form by Eq. (1), where S = Laplace operator. The developed torque is given by Eq. (2), and Eq. (3) shows the mechanical dynamics with pure inertia load. Therefore, electromechanical dynamics are nonlinear and is of 6th order, i.e., one order higher than that of an induction motor. The electrical dynamics can be easily expressed in state variable form $dX/dt = AX + BV$, where the states are $X = \begin{bmatrix} i_{qs} & i_{ds} & i_{qr} & i_{dr} & I_{fr} \end{bmatrix}^T$ and input signals are $V = \begin{bmatrix} v_{qs} & v_{ds} & 0 & 0 & v_{fr} \end{bmatrix}^T$. The electrical model can be combined with Eqs. (2) and (3) for computer simulation studies. The steady-state electrical model of the machine can be derived by substituting $S = 0$. The phasor diagram in Figure 6.40 can be shown basically as a vector diagram at steady state in synchronous reference frame, i.e., rotates at speed ω_e with respect to the stator. Note that if the field current is supplied from an ideal current source ($Z_{th} = \infty$), the dynamics of the field circuit disappear and the resulting loop dynamics are of 4th order.

FIGURE 6.44 Synchronous frame (d^e-q^e) dynamic equivalent circuits for an IPM machine.

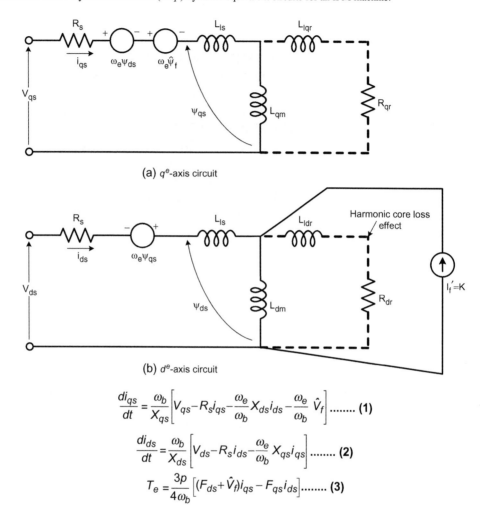

(a) q^e-axis circuit

(b) d^e-axis circuit

$$\frac{di_{qs}}{dt} = \frac{\omega_b}{X_{qs}}\left[V_{qs} - R_s i_{qs} - \frac{\omega_e}{\omega_b}X_{ds}i_{ds} - \frac{\omega_e}{\omega_b}\hat{V}_f\right] \quad \text{........ (1)}$$

$$\frac{di_{ds}}{dt} = \frac{\omega_b}{X_{ds}}\left[V_{ds} - R_s i_{ds} - \frac{\omega_e}{\omega_b}X_{qs}i_{qs}\right] \quad \text{........ (2)}$$

$$T_e = \frac{3p}{4\omega_b}\left[(F_{ds} + \hat{V}_f)i_{qs} - F_{qs}i_{ds}\right] \quad \text{........ (3)}$$

The IPM synchronous machine equivalent circuits shown in this figure are the same as the WFSM circuits shown in Figure 6.42 except the field current I_f, i.e., the flux linkage $\hat{\psi}_f = L_{dm}I_f$, can be considered constant. Therefore, the dynamics of the field circuit do not appear. A variable-speed IPM machine does not have damper winding, because the loss in the winding due to inverter-fed harmonics is large. However, the rotor equivalent circuits are shown by dotted lines because of some core loss. Note that ψ_{ds} ($= L_{ds} i_{ds}$) and $\hat{\psi}_f$ in the d^e axis induce the corresponding voltages on the q^e axis. If the core loss, i.e., the equivalent damper winding, is ignored, the electrical dynamics are of 2nd order and is given here by Eqs. (1) and (2), where ω_b is the base frequency and all the reactances (X) correspond to ω_b frequency. The torque equation [Eq. (3)] is given with the usual symbols. The same equivalent circuits are also valid for nonsalient pole sinusoidal SPM machines except the d^e and q^e circuit parameters are equal.

FIGURE 6.45   Comparison of induction and synchronous machines (multi-MW size).

	IM	WFSM
1. EFFICIENCY (INCLUDING EXCITATION):	93.9%	95.5%
2. DISPLACEMENT FACTOR (AT RATED CONDITION):	0.89	1.0
3. MAXIMUM OUTPUT (AT 200% SPEED):	240%	240%
4. MOMENT OF INERTIA:	134%	100%
5. TOTAL MOTOR WEIGHT:	101%	100%
6. TORQUE RESPONSE TIME:	<10 ms	<10 ms
7. FLUX TIME CONSTANT:	3.0 s	0.35 s
8. CONVERTER kVA:	354%	258%
9. EXCITATION RECTIFIER kVA (PEAK LOAD):	—	10%

The figure gives numerical comparisons of IMs and WFSMs for particular candidate machines in heavy reversing metal rolling mill drives [9] using a 36-thyristor phase-controlled cycloconverter. Both machines have a rating of 6 MW, 60/120 rpm, i.e., the maximum speed is twice the base speed. Both machines have the same peak power rating (240% of the rated power) at 200% speed. The efficiency is higher in WFSMs by 1.6% mainly because the loss is less with direct rotor feeding of excitation. The DPF in WFSMs can be easily programmed to unity by excitation control. However, the best DPF of IM at the rated condition is 0.89 because lagging excitation current should be fed from the stator side. This also causes larger stator and rotor diameter with the result that inertia is higher for IMs. For transient response, machine inertia is important. However, the total weight of IMs is slightly larger than that of WFSMs. Brush excitation in WFSMs is used instead of brushless excitation because of faster transient response of field current. The maximum kVA demand at 200% speed is considerably higher for IMs because of higher voltage and current loading than WFSMs. However, WFSMs require a separate excitation rectifier with 10% kVA rating at peak load. Both machines have comparable torque response times, although the flux time constant is much faster for WFSMs. As a result of the overall comparison, the WFSM was selected in this application in spite of its higher cost.

FIGURE 6.46 (a) Trapezoidal SPM machine cross section and (b) stator phase voltage and current waves.

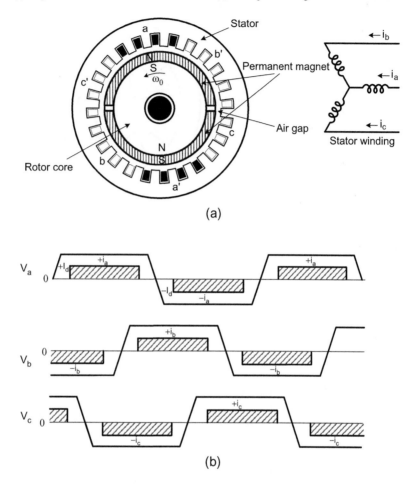

(a)

(b)

A trapezoidal SPM machine, unlike a sinusoidal SPM machine, has concentrated (nonsinusoidal) phase windings in the stator. For the three-phase, two-pole, wye-connected machine shown in this figure, the phase a winding distribution, for example, is shown by black shading in the slots. If the rotor is rotated by a prime mover, the induced trapezoidal phase voltages in the stator are shown in the figure. An inverter in the front end generally establishes $120°$ pulse phase current waves symmetrically, as shown. The machine along with the inverter and position sensor to establish the current wave are defined as a brushless dc motor (BLDM or BLDC), which will be discussed in Chapter 8. Trapezoidal machines have the advantages of simple and inexpensive construction and higher power density compared to sinusoidal machines. These machines are widely used in the low-power range.

FIGURE 6.47   (a) Cross section of switched reluctance machine (SRM) and (b) inductance profile and phase current waves.

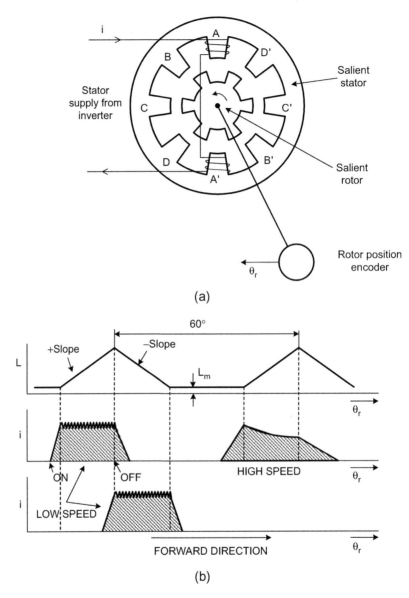

(a)

(b)

The switched reluctance motor shown in this figure is often called a variable reluctance machine (VRM) because there is reluctance variation in the stator as well as in the rotor. The two classes of VRM are SRMs and stepper motors. The SRM shown in the figure has four stator pole pairs with four-phase concentrated winding and six rotor poles

(8/6 machine). The rotor does not have any winding, nor any permanent magnet. As the stator pole pair is excited by the current, it produces magnetic flux, and the corresponding magnetic force or torque tries to align the rotor pole pair. All the stator phases are excited sequentially by a converter to produce unidirectional torque. The inductance profile L seen by the stator winding as the rotor pole-pair approaches, aligns, and moves away is shown in the figure. For the forward direction of the rotor as indicated, the positive torque is developed when the current pulse is established at the positive slope of L, whereas regenerative braking torque is developed by current in the negative slope of L. The construction of the machines is simple and inexpensive, and great effort is being made to commercialize them. However, the disadvantages are large pulsating torque and acoustic noise in the machine. SRM drives are discussed further in Chapter 8.

FIGURE 6.48   Three-phase, three-stack variable reluctance stepper motor construction (Warner Electric) [7].

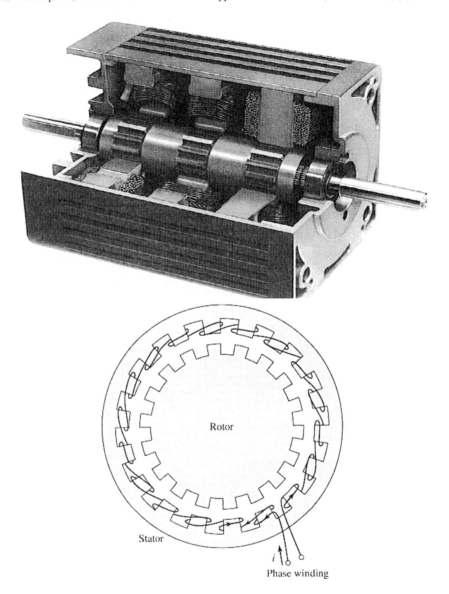

A stepper or stepping motor is a type of VRM, but it is not generally used for variable-speed applications. It is described briefly for completeness of the subject. Basically, it is a digital stepping-type synchronous motor, where an input pulse will move the rotor by a step. The stepping angle may be 1.8°, 3.6°, 7.2°, ..., 45°, 90°, etc., corresponding to

200, 100, 50, ..., 8, and 4 steps per revolution, respectively. Stepper motors are normally used in digital control systems, where a train of pulses in an open-loop manner turn the motor shaft by a definite angle. Typical applications include printers, plotters, disk drives, and CD players. Stepper motors are classified as variable reluctance, permanent magnet, and hybrid (variable reluctance with PM) types. The motor can be a single-stack or multi-stack type. The upper figure shows the construction of a three-stack, three-phase variable reluctance type motor, where the stack can be identified as a toothed wheel. The lower figure shows the details of one stack with the corresponding stator winding. When current i excites the stator winding, the rotor poles become aligned with the stator poles. Consider in the upper figure that all the three rotor poles are aligned, but the stator poles from the left to the right are offset in angular displacement by one-third of the pole pitch (20°). By successively exciting the phase windings, the rotor will move by 6.7° in each step. Increasing the number of poles, stacks, and phases will increase the angle resolution.

FIGURE 6.49   Advances and trends in electrical machines.

- MACHINE EVOLUTION HAS BEEN SLOW AND SUSTAINED OVER 100 YEARS
- ADVANCED CAD PROGRAMS AND IMPROVED MATERIALS HAVE CONTRIBUTED TO LOWER COST, HIGHER EFFICIENCY, IMPROVED RELIABILITY AND POWER DENSITY
- DC MACHINES WILL TEND TO BE OBSOLETE IN THE FUTURE
- CAGE-TYPE INDUCTION MOTORS REMAIN INDUSTRY'S WORKHORSE IN WIDE POWER RANGE.
- WFSM REMAINS POPULAR IN VERY HIGH POWER APPLICATIONS
- PM SYNCHRONOUS MACHINES ARE EFFICIENT BUT AT A HIGHER COST—THEY ARE SUPERIOR TO INDUCTION MACHINES IN LIFE CYCLE COST
- MOST MACHINES (FOR CONSTANT OR VARIABLE SPEED DRIVES) WILL HAVE FRONT-END CONVERTERS IN THE LONG RUN
- INTELLIGENT MACHINES WITH INTEGRATED CONVERTER AND CONTROLLERS LOOK VERY PROMISING IN THE FUTURE

The figure summarizes the advances and trends of electrical machines in the perspective of variable speed drives. Electrical machines are very complex electrically, magnetically, and thermally. Since their invention more than 100 years ago, technological evolution has been slow and steady, which is evident by the large number of publications even today. The primitive machines were bulky and had poor efficiency. The invention of structural, insulating, ferro-magnetic, and magnetic materials along with the advanced CAD programs, such as finite element analysis technique, has made the modern machines so optimal in size and performance. Historically, dc machines have been used for variable speed, whereas ac machines, particularly induction motors, for constant speed applications. Although a large number of dc drives are being used today, technology advancement in power electronics, advanced control, signal estimation, DSP, and ASIC has made ac drives economical and superior in performance, and their applications are fast increasing. Induction motor applications are very common because of their ruggedness and reliability. However, efficiency of these machines is inferior to synchronous machines, particularly PM machines. With the trend of decreasing cost of converters and their importance in machines, it is expected that in the future, converters will be an integral part of all machines, and machines will be integrally built with converters and control at least in the low to medium power range in so-called "intelligent machines." Further discussion on machines will be given in Chapters 7 and 8.

Summary

In this chapter, the different types of machines generally used for variable-speed applications are discussed. The dc machine features are touched on at the beginning. Converters and control for dc drives were covered in Chapter 3. The principal focus in this chapter is on ac machines, particularly induction and synchronous motors with radial geometry, although axial and linear machines are important in many applications. Knowledge of the principles and characteristics of machines is very important to understand their use in drives, which are treated in later chapters. Unlike fixed-speed machines, converter-fed machines have special characteristics that are important for drives. Any machine can be operated as a motor or generator, and the regenerative braking mode of operation of a drive helps energy conservation. Induction motors, particularly the cage type, have been the principal workhorse for most industrial applications in constant and variable speeds that range from FHP to multi-megawatt. Of course, the traditional variable-speed dc drives are still common today. Single-phase motors are normally FHP and are used in both constant- and variable-speed applications. The efficiency of these machines is low but the cost is also low. Synchronous machines, particularly the PM type, have improved efficiency, but they are more expensive. Recently, the cost of high-energy NdFeB permanent magnets is decreasing, thus promoting applications of PMSM. Low-power brushless dc drives have been used extensively. The dynamic model of ac machines is complex, but it is important when the machine is controlled in a feedback loop. The variation of machine parameters (resistances and inductances) causes additional problems in drives. The variable reluctance SRM has the simplicity of construction, but the disadvantages are pulsating torque and acoustic noise. This machine has been extensively discussed in the literature, but its applications are very few. Stepper motors are not suitable for variable-speed applications, and they are treated briefly for completeness of the subject. Selection of a particular machine for a drive system depends on trade-off considerations of many factors, such as cost, efficiency, power factor, reliability, ruggedness, rotor inertia, quietness of operation, maintainability, etc. The machines will be further discussed in Chapters 7 and 8. In fact, electrical machines constitute a vast subject, and their detailed discussion is beyond the scope of this chapter.

References

[1]   P. C. Sen, *Principles of Electrical Machines and Power Electronics*, 2nd ed., Wiley, New York, 1997.

[2]   B. K. Bose, *Modern Power Electronics and AC Drives*, Prentice Hall, Upper Saddle River, NJ, 2002.

[3]   G. R. Slemon, "Electrical machines for variable-frequency drives," *Proc. IEEE*, vol. 82, pp. 1123–1139, August 1994.

[4]   R. H. Park, "Two-reaction theory of synchronous machines—generalized method of analysis—Part 1," *AIEE Trans.*, vol. 48, pp. 716–727, July 1929.

[5]   H. C. Stanley, "An analysis of induction motors," *AIEE Trans.*, vol. 57(supplement), pp. 751–755, 1938.

[6]   G. Kron, *Equivalent Circuits of Electric Machinery*, John Wiley, New York, 1951.

[7]   A. E. Fitzgerald, C. Kingsley, and S. D. Umans, *Electric Machinery*, 6th ed., McGraw Hill, New York, 2003.

[8]   S. D. T. Robertson and K. M. Hebber, "Torque pulsation in induction motors with inverter drives," *IEEE Trans. Ind. Appl.*, vol. 7, pp. 318–323, March/April 1971.

[9] R. Hagmann, "AC cycloconverter drives for cold and hot rolling mill applications," *IEEE IAS Meeting Conference Record,* pp. 1134–1140, 1991.

[10] B. K. Bose, "Power electronics and motion control—technology status and recent trends," *IEEE Trans. Ind. Appl.,* vol. 29, pp. 902–909, September/October 1993.

[11] C. M. Ong, *Dynamic Simulation of Electric Machinery,* Prentice Hall, Upper Saddle River, NJ, 1998.

[12] R. Krishnan, *Electric Motor Drives,* Prentice Hall, Upper Saddle River, NJ, 2001.

[13] G. R. Slemon, *Electric Machines and Drives,* Addison Wesley, Reading, MA, 1992.

CHAPTER 7

Induction Motor Drives

FIGURE 7.1   Why use variable-frequency, variable-speed ac drives?

- INDUSTRIAL PROCESS CONTROL NEEDS VARIABLE SPEED
- ENERGY SAVINGS REALIZED IN VARIABLE-FLOW CONTROL APPLICATIONS
- SUPERIORITY OF AC DRIVES OVER VARIABLE-SPEED DC DRIVES
- MACHINE MATERIAL SAVING AT HIGHER FREQUENCY
- HIGHER RANGE OF SPEED CONTROL
- PRODUCTIVITY AND PRODUCT QUALITY CAN BE IMPROVED BY HIGH PERFORMANCE CONTROL
- VARIABLE-FREQUENCY SOFT STARTING OF CONSTANT-SPEED MOTOR
- SHIELDING MACHINE FROM LINE TRANSIENTS AND POWER QUALITY PROBLEMS
- INTEGRATED INTELLIGENT MACHINE OF THE FUTURE

Speed control of induction motors by means of variable-voltage, constant-frequency and variable-voltage, variable-frequency drives was introduced in Chapter 6. By far, the majority of variable-speed drives for industrial process control uses the latter method. Traditionally, dc drives have been used, but recent technological advancements are making ac drives (particularly induction motor drives) more popular. AC machines are generally cheaper, smaller, and more rugged and reliable with the absence of commutators and brushes, but the converter costs may be somewhat higher. Overall, ac drives are much superior to dc drives, and it appears that eventually dc drives will be totally obsolete. One of the motivations for using variable-speed drives in flow control is the energy savings, which will be described later. Higher frequency operation of machines through an inverter permits machine size reductions because of the reduced volume of magnetics, but the loss density becomes higher. In modern factories, variable-frequency drives with high performance control permit high-volume, automated production with superior product quality. Even for constant-speed applications, the variable-frequency converter can permit soft starting. Variable-frequency soft starting has the advantages of high starting torque, less harmonic distortion, and flux-programming efficiency improvements compared to phase-control-type starting. With the trend toward reduced costs and sizes of converters, it is expected that eventually the majority of machines—at least those in the low power end—will be built with integrated intelligent converters.

FIGURE 7.2   Application examples of variable-frequency drives.

- PUMPS, BLOWERS, AND COMPRESSORS
- PAPER AND TEXTILE MILLS
- TRAM, SUBWAY, AND LOCOMOTIVE PROPULSION
- ELECTRIC AND HYBRID VEHICLES
- ELEVATORS
- METAL ROLLING AND CEMENT MILLS
- ENGINE STARTER/GENERATORS
- HOME APPLIANCES
- MACHINE TOOLS AND ROBOTICS
- VARIABLE-SPEED AC/HEAT PUMPS
- VARIABLE-SPEED WIND-ELECTRIC GENERATION
- SHIP PROPULSION
- COMPUTER DISKS

The use of variable-frequency ac drives has recently proliferated in industrial, commercial, residential, military, aerospace, and transportation applications because of recent technological advancements that have resulted in reduced costs and better performance. Example applications are given here. Either an induction or synchronous motor can be used although the use of the former is more common. The larger segment of the applications is in pump and fan-type drives. Modern transportation drives invariably use variable-frequency speed control. For subway drives, the traditional resistance-controlled dc motor is not efficient. Of course, chopper-fed dc drives can be used to improve efficiency. Modern electric and hybrid vehicles invariably use ac drives with battery power supply. Variable-frequency starters are used to start the aircraft engine and then the same machine acts as the generator for VSCF (variable-speed, constant-frequency) power generation. The future trend in automobiles is also similar. Home appliances such as blenders, mixers, and vacuum cleaners also use variable-speed drives, but they may not always be variable-frequency drives. Renewable wind energy systems use variable-speed turbine-generators to capture more energy. Modern ships have tended to use electric propulsion mainly for saving of fuel in diesel engines. Variable-speed air conditioners and heat pumps (which are very common in Japan) save energy and improve comfort by load-proportional cooling.

FIGURE 7.3   Comparison of power consumption in variable-speed and constant-speed drives
for flow control.

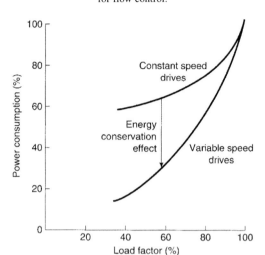

Many industrial applications require variable flow control of fluid (air, chemical gases,
water, and liquid chemicals). The traditional method of such flow control is to use an
induction motor at constant speed with 60/50-Hz power supply, and then control the
flow by means of a throttle. The efficiency of the control is poor because of the energy
loss due to vortex in the fluid. This figure shows the power consumption (as percent of
full power) with the loading factor (also as percent). Using a variable-frequency drive
to control the fluid flow with a fully open throttle saves a considerable amount of power
as shown in the figure. For example, with 60% loading (i.e., 60% flow), the efficiency
improvement can be as high as 35%. Because most of the drives operate at part load
most of the time, the accumulated energy saving, or the corresponding financial benefit,
may be substantial over a prolonged period of time. Note that the payback period for
the additional cost of power electronics is small, particularly where the electricity cost
is high (such as Japan). Additional energy savings are possible with high-performance
drives, which will be discussed later. Because this type of fluid flow control is common
in industry, widespread application of variable-frequency drives with power electronics
in this area can help in large energy conservation.

FIGURE 7.4   Principal classes of induction motor drives.

- STATOR VOLTAGE CONTROL AT CONSTANT FREQUENCY
- VOLTAGE-FED PWM INVERTER DRIVE
- CURRENT-FED INVERTER DRIVE (SIX-STEP OR PWM)
- CYCLOCONVERTER DRIVE
- SLIP POWER RECOVERY DRIVE
  - STATIC KRAMER DRIVE
  - STATIC SCHERBIUS DRIVE

The figure shows the principal classes of induction motor drives, which include both cage-type and wound-rotor machines. By far, the majority of applications uses cage-type machines. Stator voltage control at constant frequency is mainly used for low-power, low-performance single-phase machines. This topic was covered in Chapter 6. Again, the largest segment of applications uses PWM voltage-fed inverters, which cover the power range from FHP (fractional hp) to multi-megawatts. Voltage-fed inverters were discussed in Chapter 4. The current-fed inverter drives are used in higher power ranges, and the machine current waveforms can be either multistepped or PWM type. This class of converters was covered in Chapter 5. Phase-controlled cycloconverter drives are also used normally in the multi-megawatt range. Cycloconverters were covered in Chapter 3. There is another class of drives that uses wound-rotor induction motors only. In this type, the slip power in the rotor winding is retrieved through slip rings and controlled to control the speed. Generally, there are two classes of slip power controlled drives as indicated. This type of drive is of high power and is used in limited speed range applications. An engineer selecting a drive for a certain application should analyze all trade-offs. The control and signal processing for all classes of drives, except that with stator voltage control, will be further discussed in detail in this chapter.

FIGURE 7.5 General design considerations.

- WHAT APPLICATION?
- POSITION, SPEED, OR TORQUE CONTROL IN OUTER LOOP?
- SINGLE OR MULTIMOTOR DRIVE?
- HORSE POWER?
- POWER SUPPLY?
- ONE-, TWO-, OR FOUR-QUADRANT DRIVE?
- TORQUE AND SPEED RANGE?
- PRECISION AND ROBUSTNESS OF CONTROL?
- LINE HARMONICS AND POWER FACTOR?
- COST AND EFFICIENCY CONSIDERATIONS?
- NEW OR RETROFIT APPLICATION?
- INDOOR OR OUTDOOR INSTALLATION?
- RELIABILITY AND MAINTAINABILITY CONSIDERATIONS?

Once the application of the drive is identified, the next step is collection of all the desired performance features in detail. For example, EV/HV drives normally require torque control by the driver, whereas industrial drives generally use speed control. Certain applications, such as robot and satellite dish drives, require position control. Note that torque and speed control loops may be embedded in the position control loop. Subways and locomotives normally use multiple machines on the same inverter, but the majority of applications uses a single-motor drive. If the power supply is ac, voltage, frequency, and number of phases are important. Similarly, dc supply voltage is important for traction drive. Pumps and fans normally use one-quadrant control, but two-quadrant control with dynamic braking can also be used. Rolling mills, EV/HV, electric locomotives, elevators, etc., use four-quadrant speed control. In all applications, the evaluation of torque, speed, and power range is important. The cost is the most dominant factor, but its trade-off with performance, reliability, and maintainability should be evaluated. For military and aerospace applications, reliability cannot be sacrificed for cost. Some applications are retrofit, i.e., the present machine, power supply, and switchgears are to be used in the drive. For example, currently, many fixed-speed pumps and blowers are being considered for replacement with variable-speed drives to save energy (see Figure 7.3).

FIGURE 7.6   Features of cage-type induction motor drives.

- CHEAP AND ROBUST MACHINE
- HIGH-SPEED FIELD-WEAKENING MODE CONTROL POSSIBLE
- CONTROL MAY BE COMPLEX
- LOWER EFFICIENCY DUE TO ROTOR LOSSES
- DOES NOT NEED ABSOLUTE POSITION ENCODER IN CLOSED-LOOP POSITION CONTROL
- POSSIBILITY OF EFFICIENCY IMPROVEMENT BY LIGHT LOAD FLUX PROGRAMMING
- CONVERTER SUPPLIES MACHINE EXCITATION – HIGHER CONVERTER RATING
- ROTOR INERTIA MAY BE HIGH – DISADVANTAGE FOR SERVO DRIVES
- MINIMAL PULSATING BRAKING TORQUE AT CONVERTER FAULT

The figure highlights the favorable and unfavorable features of induction motor drives. One reason for the popularity of cage-type induction motors is that they are cheap and rugged. However, their efficiency is somewhat inferior because of rotor copper losses. The drive can be designed easily for an extended speed field-weakening region, which is important in many applications, such as traction and spindle drives. For closed-loop speed or position control, the machine needs only incremental speed encoder (usually optical type). The induction motor drive control may be simple or complex depending on the performance needed. The front-end inverter supplies the machine excitation, which can be controlled to control the airgap flux at light load, resulting in improved efficiency. This will be discussed later. The machine has somewhat higher rotor inertia (J) compared to a high-energy magnet synchronous motor, and this is a disadvantage in a closed-loop speed or position control system. Because excitation is supplied by the inverter, the inverter KVA rating is somewhat higher. For an inverter fault, the machine counter emf tends to decrease rapidly, thus decreasing the possibility of large fault current and the resulting pulsating braking torque.

FIGURE 7.7   Steps in drive development and design.

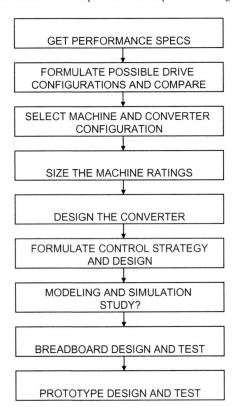

Once the application of a drive is identified, the next is to collect all the needed performance specs as mentioned earlier. At this point, it is desirable to consider alternate drive types and compare their cost and performance. Both induction and synchronous machines (particularly PM type) should be considered with possible converter topologies and control techniques. Based on the trade-off study, a final drive configuration should be selected. Voltage-fed PWM converters are usually preferred. Digital control is invariably used nowadays. The possibility of using an off-the-shelf drive from a product vendor should be considered. Often, the vendors can satisfy the drive requirement with some small changes in the design. Otherwise, custom design in detail is needed. Generally, a machine is sized first and then the converters are designed and control strategy formulated. For a new and complex drive, it is desirable to validate the system by means of a simulation study, followed by model and prototype design and testing.

FIGURE 7.8 Control of induction motor drives.

- VARIABLE-VOLTAGE, CONSTANT-FREQUENCY (VVCF) CONTROL
- VARIABLE-VOLTAGE, VARIABLE-FREQUENCY(VVVF) CONTROL
- VARIABLE-CURRENT, VARIABLE-FREQUENCY (VCVF) CONTROL
- SIMPLE AND LOW-PERFORMANCE OPEN-LOOP CONTROL
- SCALAR CONTROL TECHNIQUES (OPEN OR CLOSE-LOOP)
- VECTOR OR FIELD-ORIENTED CONTROL (CLOSE-LOOP) TECHNIQUES
- DIRECT TORQUE AND FLUX CONTROL (DTC or DTFC)
- CLOSE–LOOP CONTROL COMPLEXITY:
  - NONLINEAR MULTIVARIABLE DISCRETE TIME SYSTEM – PLANT PARAMETER VARIATION
  - DIFFICULTY OF PROCESSING FEEDBACK SIGNALS

The figure summarizes the principal control techniques of cage-type induction motors using voltage-fed or current-fed converters. More advanced control techniques will be discussed later. The principal control objectives are that the drive should follow the desired command as closely as possible, and any type of external disturbances should be rejected. The VVCF and VVVF control principles were discussed in Chapter 6. With a voltage-fed PWM inverter, both voltage and frequency can be controlled to control the machine flux like that of a separately excited dc motor. The same is possible with a current-fed inverter, where machine current and frequency are controlled by the inverter. Indirectly, it is equivalent to a VVVF drive because machine counter emf varies directly with frequency with the flux remaining constant or programmable. In general, control can be classified as the open-loop or closed-loop type. Open-loop volts/Hz control with a voltage-fed PWM inverter is the most simple and commonly used control in industrial applications. However, the performance of this type of drive is low. Another type of control classification is scalar and vector (often called field-oriented) control. Open-loop volts/Hz control and several other closed-loop controls fall under the category of scalar control. Vector control (which is closed loop) results in a high-performance drive like that of a separately excited dc motor. Recently, DTC control has emerged and is somewhat competitive to vector control, but its performance is inferior. Any type of closed-loop control, particularly vector control, is complex because of the complexity of the machine model and its parameter variation problem. In addition, processing of machine feedback signals is complex because of variable frequency and variable magnitude, and the signal is distorted with harmonics.

FIGURE 7.9   Estimation of drive signals and machine parameters.

- SIGNALS
    - MACHINE VOLTAGE
    - FLUX
    - TORQUE
    - SPEED
    - POSITION
    - ACTIVE POWER
    - REACTIVE POWER
    - POWER FACTOR OR DPF
    - COPPER AND IRON LOSSES
    - EFFICIENCY
- PARAMETERS
    - STATOR RESISTANCE
    - ROTOR RESISTANCE
    - MAGNETIZING INDUCTANCE
    - STATOR AND ROTOR LEAKAGE INDUCTANCES

In a drive system, a number of sensors can be used to measure the signals listed in this figure. These signals are useful for feedback control, monitoring, and diagnostic purposes. Some sensors are expensive, unreliable, and may result in drift problems. The modern trend is to estimate as many signals as possible with the help of a powerful DSP (digital signal processor) in "sensorless" drives. The current and voltage sensors are most commonly used. The current signal is very basic for protection of the system. Knowing the dc link voltage and PWM pattern of a VFI, the machine terminal voltage can be estimated. Similarly, knowing the dc link current and switching pattern of a CFI, machine current can be estimated. Knowing the mathematical expression (or model) of all the other signals and drive system parameters, a DSP can calculate them from the basic voltage and current signals. High-performance feedback control of a drive, for example, may need machine voltage, flux, torque, and speed (or position) signals. Some of these computation techniques are very complex and constitute modern R&D topics. There is also the problem of machine parameter variation in the drive running condition. Correct estimation of the parameters is needed for accurate signal estimation. Also, because the machine is within the feedback loop, parameter variation will make the control inferior and may cause stability problems also. Accurate parameter estimation is a challenging research topic in literature.

FIGURE 7.10   Open-loop volts/Hz control with PWM voltage-fed converter.

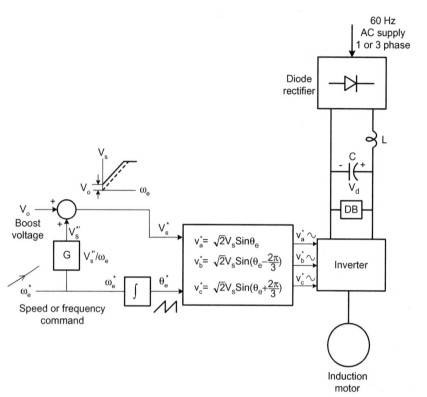

The figure shows simple open-loop volts/Hz control of an induction motor, which is very commonly used in industry. Feedback signal processing complexity and stability problems are avoided at the cost of inferior performance. In the front end, a simple diode rectifier maintains the dc link voltage V_d. The filter inductance L is often omitted. The speed or frequency is the command signal and the proportional voltage signal $V_s^{*'}$ is derived from it so that the airgap flux remains constant. A boost voltage is added to this signal so that flux does not decrease at low frequency (see Figure 6.12). The command sinusoidal phase voltages are calculated from the voltage magnitude and angle command signal for the PWM inverter, as shown in the figure. The drive can be accelerated or decelerated by slowly ramping the speed command signal. The motor speed can be reversed by reversing the phase sequence of inverter output. While decelerating the drive, the motor acts as a generator, and electrical braking power is dissipated in the dynamic brake (DB) resistor. The DB is basically a dc-dc converter that keeps the dc link voltage constant during generation. As the command speed exceeds the base speed, the supply voltage V_s saturates, i.e., the proportionality with frequency is lost, and the drive enters the field-weakening region. In this condition, developed torque will decrease because of flux reduction.

FIGURE 7.11   Features of volts/Hz controlled drive.

- SIMPLE AND INEXPENSIVE – VERY POPULAR
- NO NEED FOR FEEDBACK SENSORS
- COMPLEXITY OF FEEDBACK SIGNAL PROCESSING IS AVOIDED
  - WIDE FREQUENCY VARIATION
  - WIDE MAGNITUDE VARIATION
  - COMPLEX HARMONICS
  - PHASE UNBALANCE
- TWO-QUADRANT OPERATION
- CONSTANT-TORQUE AND FIELD-WEAKENING
  MODES OF SPEED CONTROL
- DRIFT OF SPEED WITH LOAD TORQUE VARIATION
- DRIFT OF SPEED AND FLUX WITH SUPPLY VOLTAGE VARIATION
- POOR SYSTEM STABILITY
- SLUGGISH SYSTEM RESPONSE
- MULTI-MOTOR OPERATION POSSIBLE

Open-loop volts/Hz control is simple, inexpensive, and extensively used in industry (for example, for speed control of pumps and fans). In a variable-frequency drive, the feedback signals vary throughout a wide frequency range, and harmonics in these signals may be complex. The complexity of processing these signals is avoided. Besides, there is the problem of machine parameter variation. However, performance of the drives is poor. For example, a drop in the supply voltage decreases the airgap flux that reduces torque and speed. Similarly, higher load torque reduces the speed (see Figure 7.14). One of the features of open-loop control is that it can easily control the speed of multiple machines operating in parallel as shown in the next figure.

FIGURE 7.12   Multimotor subway drive system.

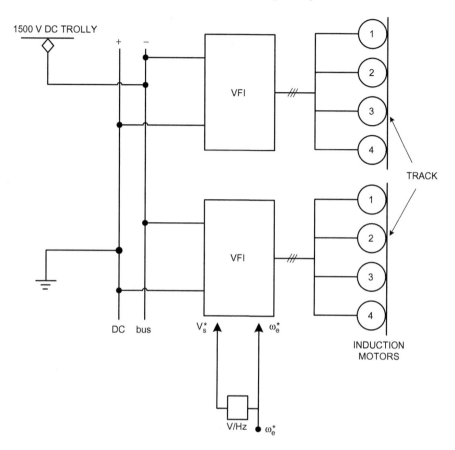

The figure shows a typical multimotor subway drive system with induction motors. The dc trolly supply is shown as 1500 V. Traditionally, it was 600 V, but has been gradually raised as high-voltage self-controlled devices (GTO, IGBT) have become available. Two voltage-fed PWM inverters are shown on the dc bus, and each inverter has four parallel machines (see also Figure 4.49). Each machine normally drives an axle with two wheels. The inverters can be two level or three level depending on the trolly voltage and device voltage ratings. Parallel machines with a single inverter are more economical than an individual inverter drive. Normally, open-loop volts/Hz control is used so that all the machines run at identical frequency and voltage. Recently, sophisticated vector control (described later) has been used in the system. For parallel operation with equal developed torque, it is desirable for all four machines to be of equal rating and have matched characteristics. Again, because all wheels run on the same track, their linear speed is constrained to be identical. This demands that all wheel diameters should be identical so that angular speed (rpm) of all the parallel machines remains the same. Matching of machines and wheel diameters is a problem that is discussed in Figure 7.13.

FIGURE 7.13   Multimotor drive with machine and wheel mismatching problems.

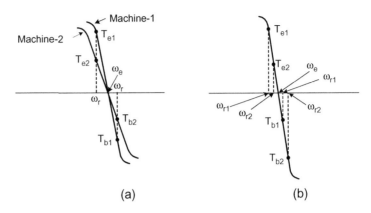

(a)   (b)

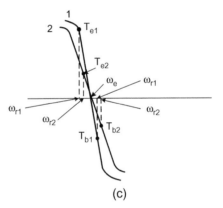

(c)

Let us consider the problems of mismatching of machines and wheel diameters in the subway drive shown in Figure 7.12. For simplicity, we will consider only two parallel machines. Figure 7.13(a) considers that the machines are mismatched with nonidentical torque-speed characteristics but the wheel diameters are equal. Equal wheel diameter means the speed will always be identical, i.e., $\omega_{r1} = \omega_{r2} = \omega_r$. Therefore, slip S is identical. In motoring mode, developed torques are T_{e1} and T_{e2} for machines 1 and 2, respectively. With excessive torque sharing of machine 1, it may become unstable or its wheels may slip, transferring all torque to machine 2. When the drive goes into regeneration, the corresponding torque sharings are T_{b1} and T_{b2} as shown. In (b), the machines are considered identical, but the wheels of machine 2 are shorter so that $\omega_{r2} > \omega_{r1}$. In that case, $T_{e1} > T_{e2}$ as before, but in regenerative mode, $T_{b2} > T_{b1}$. It is possible for one motor to be in motoring mode while the other is in braking mode. In (c), mismatching of both machines and wheel diameters is considered. As shown, in motoring mode, machine 1 will share larger torque, but in braking mode, machine 2, which was sharing a larger torque in (b), will tend to equalize. Traction authorities try to keep close tolerance in the wheel diameters and parallel machines.

FIGURE 7.14   Closed-loop slip-controlled drive with volts/Hz control.

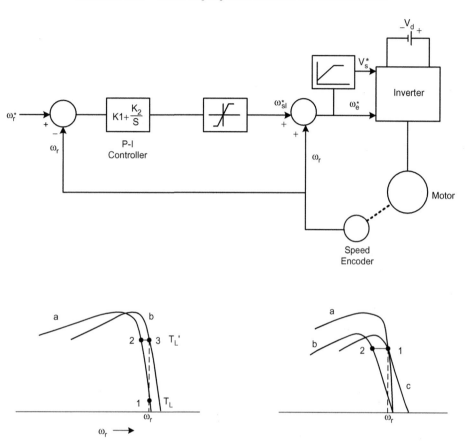

Closed-loop speed control with slip regulation adds some performance improvement to open-loop volts/Hz control, as explained in this figure. Here, the motor speed is compared with the command speed, and the error generates the slip frequency (ω_{sl}^{*}) command through a P-I compensator and limiter. The slip is added to the feedback speed to generate the frequency and voltage command as shown. Because slip is proportional to torque at constant airgap flux, the scheme can be considered as torque control within a speed control loop. The machine can accelerate/decelerate within the slip limit (i.e., the current limit). The scheme compensates speed drift for supply voltage and load torque variation, as explained in the lower figures. As T_L increases from point 1 to 2 on curve a (left), ω_r will tend to decrease, but it will be compensated by increasing the frequency as shown by curve b. Similarly, if the supply voltage drops at constant T_L (right), operating point 1 on curve a will tend to drift to point 2 on curve b, and speed will tend to decrease. However, the speed control loop will increment the frequency, and the speed drop will be restored as shown on curve c. The flux drop can be compensated by an independent flux control loop that corrects the voltage command restoring the torque-slip sensitivity.

FIGURE 7.15   (a) Basic current-fed inverter drive configuration and (b) acceleration/deceleration characteristics.

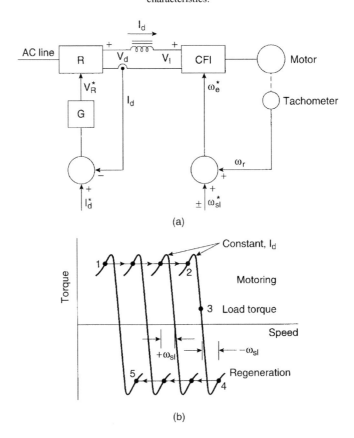

(a)

(b)

So far, we have discussed scalar control of voltage-fed inverter drives. In a current-fed inverter drive, as shown in this figure, simple open-loop control, unfortunately, is not possible. The minimum control needed, as shown, is the closed-loop dc link current control for the front-end rectifier and slip control of the inverter. Note that both I_d and ω_{sl} can control torque as well as flux (see Figure 6.13). In (b), the drive acceleration/deceleration characteristics at constant slip but limiting current, which indirectly maintains constant airgap flux, are explained. Point 3 is shown as a steady-state operating point $(T_e = T_L)$ at reduced slip but rated current. The disadvantage of such operation is heavy saturation of flux as explained in Figure 6.13. The steady-state flux can be maintained at the rated value by programming I_d as a function of slip while controlling torque by I_d, which will be explained later.

FIGURE 7.16 Current-fed inverter drive with volts/Hz control.

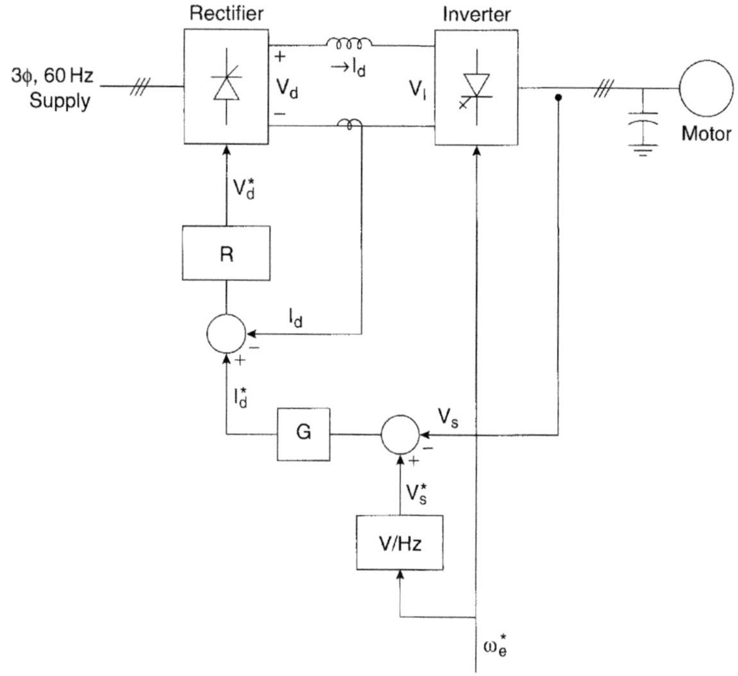

A current-fed converter system can be controlled with volts/Hz control like a voltage-fed converter system as shown in this figure. The frequency of the self-controlled GTO inverter is controlled directly. Because of self-control, a commutating capacitor bank is shown at the machine terminal. The machine terminal voltage is compared with the command voltage generated from the frequency command. The loop error controls the dc current I_d generated by the rectifier to regulate the machine voltage. Note that the supply voltage fluctuation does not affect the machine performance because of the closed-loop voltage control. In addition, the inherent regenerative property of current-fed converter systems is retained here. Otherwise, the inherent demerits of volts/Hz control, such as instability as a result of a sudden frequency change and speed drift if the load torque changes, are retained here.

FIGURE 7.17   Current-fed inverter drive with speed and flux control.

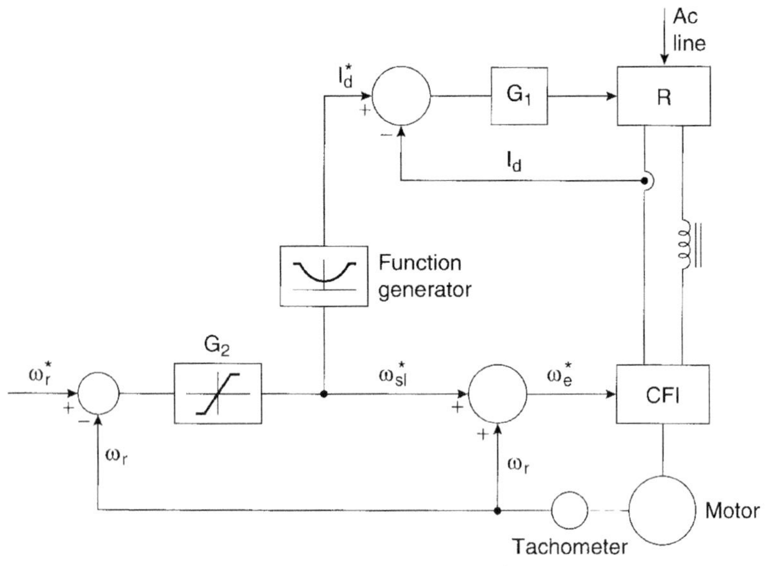

In both voltage-fed and current-fed inverter drives, it is desirable to control the machine flux at the rated value like a separately excited dc motor. This figure shows a closed-loop speed-controlled current-fed drive with indirect flux control. The speed loop generates the slip command through a limiter because slip is proportional to torque at the rated flux. The dc link current is controlled by the slip command through a function generator to maintain the flux constant. The function generator can be programmed analytically from the machine equivalent circuit (Figure 6.13) and verified by test. When the drive is accelerated or decelerated at the limit slip, I_d goes to the rated value to generate the rated flux and torque. At zero torque, slip is zero and the minimum I_d corresponds to the magnetizing current for rated flux. However, without direct flux control loop, a small variation of flux is possible because of machine parameter variations.

FIGURE 7.18 (a) Efficiency improvement by flux program and (b) volts/Hz control with voltage program for optimal efficiency with pump load.

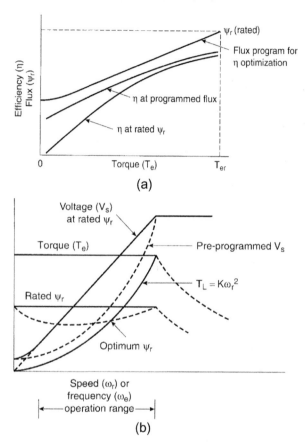

(a)

(b)

Machines are normally operated at rated flux irrespective of load torque and speed so that the transient response is fast and developed torque/amp is high. With rated flux at a light load condition, core loss is excessive, resulting in poor drive efficiency. If the flux is reduced from the rated value, core loss decreases, but the machine and converter copper losses increase. However, the total loss decreases, resulting in efficiency improvements. Figure 7.18(a) shows efficiency curves at rated flux and optimum efficiency programming flux, respectively, at variable torque but constant speed. Because core loss is a function of flux and frequency, the optimum flux program will be different at different speed. Part (b) shows a volts/Hz program for optimum flux program control for a pump-type drive, where the load torque as a function of speed is known. At each speed, the optimum flux is shown for the given load torque condition. This flux determines the supply voltage. Commercial off-the-shelf drives are available with a selectable volts/Hz program for efficiency optimization control. A more advanced method of efficiency optimization control based on an online search will be discussed in Chapter 10.

FIGURE 7.19   Direct torque and flux control (DTC).

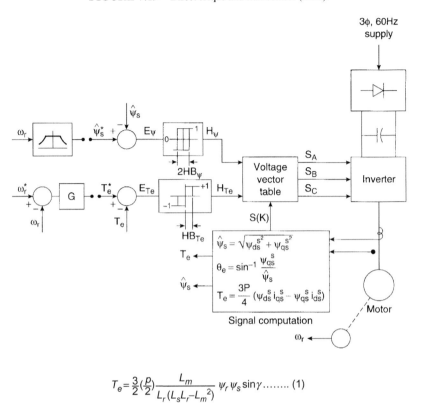

$$T_e = \frac{3}{2}\left(\frac{P}{2}\right)\frac{L_m}{L_r(L_sL_r - L_m^2)}\,\psi_r\psi_s\sin\gamma\ \dots\dots\ (1)$$

An advanced scalar control technique based on direct torque and flux control (known as DTFC or DTC) was introduced in 1985 [5], and was recently developed as a product by a large company. The strategy of DTC control is shown in this figure, and its control principle will be explained in the next two figures. Basically, it uses torque and stator flux control loops, where the feedback signals are estimated from the machine terminal voltages and currents. The torque command can be generated by the speed loop as shown. The loop errors are processed through hysteresis bands and fed to a voltage vector look-up table. The flux loop has outputs +1 and −1, whereas the torque loop has three outputs, +1, 0, and −1 as shown. The inverter voltage vector table also gets the information about the location of the stator flux vector $\overline{\psi}_s$ (Figure 7.20). From the three inputs, the voltage vector table selects an appropriate voltage vector to control the PWM inverter switches. The control strategy is based on the torque equation shown at the bottom of the figure [2], where ψ_s and ψ_r are the stator and rotor fluxes, respectively, and γ is the angle between them. Note that the control does not use any PWM algorithm nor any feedback current signal. It can be used for wide speed ranges including the field-weakening region but excludes the region close to zero speed. DTC control has been used widely, for example, in pump and compressor drives as improvement of open-loop volts/Hz control.

FIGURE 7.20   (a) Stator flux vector trajectory in DTC control and (b) control of voltage vectors to control flux trajectory.

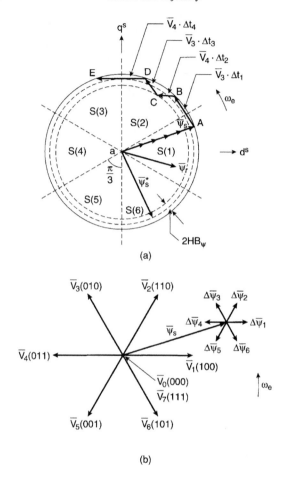

(a)

(b)

The stator flux vector $\overline{\psi}_s$ rotates in a circular orbit within a hysteresis band covering the six sectors as shown in part (a). The six active voltage vectors and the two zero vectors of the inverter controlled by the look-up table are shown in the lower figure. If a voltage vector is applied to the inverter for time Δt, the corresponding flux change is given by the relation $\Delta \overline{\psi}_s = \overline{V}_s \cdot \Delta t$. The flux increment vector for each voltage vector is indicated in (b). The flux is initially established at zero frequency in the radial trajectory aA. With the rated flux, command torque is applied and the flux vector starts rotating in the counter-clockwise direction within the hysteresis band depending on the selected voltage vector. The flux is altered in the radial direction due to flux loop error, whereas the torque is altered by tangential movement of the flux vector. Note that ψ_s moves in a jerky manner at γ angle ahead (for $+T_e$) of rotor flux $\overline{\psi}_r$, which has smooth rotation. The jerky variation of stator flux and γ angle introduces the torque ripple. Note that the lowest speed is restricted because of the difficulty of voltage model flux estimation at low frequency.

FIGURE 7.21   (a) Switching table of voltage vectors and (b) flux and torque sensitivity by voltage vectors.

H_ψ	H_{Te}	S(1)	S(2)	S(3)	S(4)	S(5)	S(6)
1	1	V_2	V_3	V_4	V_5	V_6	V_1
	0	V_0	V_7	V_0	V_7	V_0	V_7
	−1	V_6	V_1	V_2	V_3	V_4	V_5
−1	1	(V_3)	(V_4)	V_5	V_6	V_1	V_2
	0	V_7	V_0	V_7	V_0	V_7	V_0
	−1	V_5	V_6	V_1	V_2	V_3	V_4

(a)

Voltage vector	V_1	V_2	V_3	V_4	V_5	V_6	V_0 or V_7
ψ_s	↑	↑	↓	↓	↓	↑	0
T_e	↓	↑	↑	↑	↓	↓	↓

(b)

The voltage vector look-up table for DTC control is shown in (a) [6] for the three inputs $[H_\psi, H_{Te}$, and $S(K)]$. The flux trajectory segments AB, BC, CD, and DE by the respective voltage vectors V_3, V_4, V_3, and V_4 are shown in Figure 7.20(a). For example, if $H_\psi = -1$, $H_{Te} = 1$, and $S(K) = S(2)$, vector V_4 will be selected to describe the BC trajectory because at point B, the flux is too high and torque is too low. At point C, $H_\psi = +1$ and $H_{Te} = +1$, and this will generate V_3 vector from the table. The lower table summarizes the flux and torque sensitivity and direction for applying a voltage vector for the flux location shown in Figure 7.20(b). The flux can be increased by V_1, V_2, and V_6, whereas it can be decreased by V_3, V_4, and V_5. The zero vector short circuits the machine terminal and keeps the flux and torque essentially unchanged.

FIGURE 7.22   Salient features of DTC control.

- SIMPLE DIRECT CONTROL OF TORQUE AND STATOR
  FLUX BY SELECTION OF A VOLTAGE VECTOR
- BASICALLY ADVANCED SCALAR CONTROL
- SOMEWHAT ANALOGOUS TO HYSTERESIS BAND CURRENT
  CONTROL PWM
- NO FEEDBACK CURRENT CONTROL
- NO TRADITIONAL PWM TECHNIQUE
- RIPPLE IN CURRENT, TORQUE, AND FLUX
- LIMITATION OF MINIMUM SPEED (INTEGRATION AND STATOR
  RESISTANCE VARIATION PROBLEMS)
- HAS BEEN APPLIED TO PUMPS, FANS, EXTRUDERS, ETC.,
  AS IMPROVEMENT OF VOLTS/HZ CONTROL

FIGURE 7.22   Salient features of DTC control.

The figure summarizes the essential features of DTC control. Basically, the control is accomplished by simple but advanced scalar control of torque and stator flux by hysteresis-band feedback loops. There is no feedback current control although current sensors are essential for protection. Note that no traditional SPWM or SVM technique is used as in other drives. The indirect PWM control is due to voltage vector selection from the look-up table to constrain the flux within the hysteresis band. Similar to hysteresis band (HB) current control, there will be ripple in current, flux, and torque. The current ripple will give additional harmonic loss, and torque ripple will try to induce speed ripple in a low inertia system. In recent years, the simple HB-based DTC control has been modified by fuzzy and neuro-fuzzy control in inner loops with SVM control of the inverter. Multiple inverter vector selection in SVM within a sample time smooths current, flux, and torque. However, with the added complexity, the simplicity of DTC control is lost. DTC control can be applied to PM synchronous motor drives also.

FIGURE 7.23   Commercial DTC-controlled induction motor drive (ACS1000).

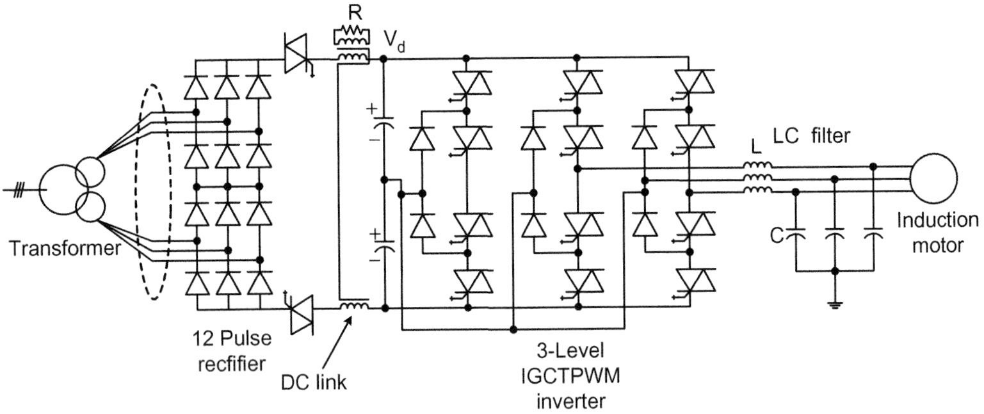

Recently, ABB commercially introduced a medium-voltage (ACS1000) induction motor drive family that uses DTC control [8], and its power circuit is shown in this figure. It is, in fact, an update of the ACS600 drive with DTC control that was introduced a few years earlier. The power range of the drive is from 315 kW to 5.0 MW (400 to 6500 hp) with input supply voltages that include 2.3, 3.3, and 4.0 kV. Typical applications are fans, pumps, compressors, conveyers, and extruders, where the performance has been improved over the traditional volts/Hz control. The front end of the converter system shows a 12-pulse diode rectifier with dc link LC filter, which is supplied by a transformer with a dual, secondary delta-wye connection. The resulting 12-step line current wave makes the line power factor high (0.95) and satisfies the IEEE 519-1992 harmonic standards. Optionally, a 24-pulse quad-bridge rectifier with a quad secondary transformer can be used for further improvement of line current harmonics. A three-level diode-clamped PWM inverter with snubberless IGCTs is used to generate the VVVF power supply for the motor, which can be of the standard or high-efficiency type. The integrated diode with the IGCT is indicated in the figure. An IGCT is used in both positive and negative buses of the dc link for protection in case of inverter or capacitor faults. The common mode inductors in the dc link with secondary damping resistors permit initial soft charging of the capacitors, which also limits the common mode leakage current with adequate damping of oscillation. The drive uses a three-phase sine wave LC filter with damping resistor (not shown) at the machine terminal with grounding to prevent bearing current and voltage boost at the machine terminal due to long cables. It also makes the machine current sinusoidal to improve its efficiency. The drive can have torque or speed control in the outer loop without any speed sensor. Light load flux programming efficiency enhancement for pump or fan drives can be easily incorporated. (ACS1000 has recently been added with ACS5000 and ACS6000 families) [abb.com].

FIGURE 7.24   Features of ACS1000 drive system.

- WORLD'S FIRST DTC-CONTROLLED INDUCTION MOTOR DRIVE
- SPECS: POWER: 315–5000 kW (AIR OR WATER COOLED)
  OUTPUT VOLTAGE: 0–2.3 kV, 0–3.3 kV, 0–4.16 kV
  OUTPUT FREQUENCY: 0–66 Hz (OPTIONALLY 200 Hz)
  LINE DPF: 0.97
  LINE PF: 0.95
- THREE-LEVEL SINGLE DEVICE IGCT INVERTER WITH
  INTEGRATED INVERSE DIODE-SNUBBERLESS
- SCALAR CONTROL – PERFORMANCE ENHANCEMENT OVER
  VOLTS/Hz CONTROL
- 12-PULSE DIODE RECTIFIER (OPTIONALLY 24-PULSE)
- CAPACITOR AND INVERTER FAULT PROTECTION BY IGCT
- MACHINE TERMINAL LC FILTER
  – SINUSOIDAL MACHINE CURRENT
  – NO BEARING CURRENT
  – NO VOLTAGE BOOST
- SPLIT DC CHOKE – LIMITS COMMON MODE CURRENT
  HIGH-INPUT PF
- LINE POWER LOSS RIDE-THROUGH CAPABILITY
- FLUX PROGRAM EFFICIENCY OPTIMIZATION

The figure summarizes the essential features of the ACS1000 drive. As indicated in the specs, the drive is available in wide power and voltage ranges. In lower power, the inverter is air cooled, whereas at high power it is water cooled. Replacing the traditional GTOs by an IGCT (or IGBT) permits higher PWM frequency and snubberless operation. Although the machine LC filter is somewhat bulky, the usual disadvantages of inverter hard switching are eliminated and harmonic losses are eliminated. The drive has ride-through capability in case of a line black-out or brown-out. In this mode, the machine kinetic energy is used to maintain the dc link voltage.

FIGURE 7.25   Vector control analogy with separately excited dc motor drive.

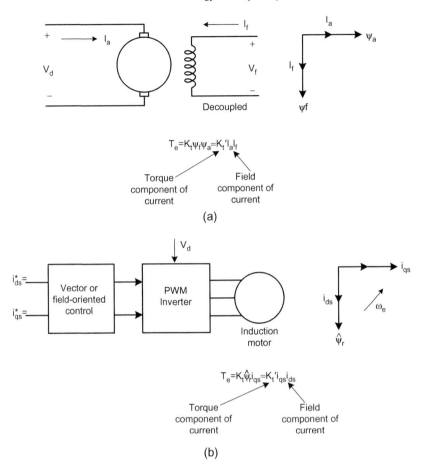

(a)

(b)

The scalar control methods of voltage-fed and current-fed inverters discussed so far are simple to implement but have the disadvantage of sluggish control response because of the inherent coupling effect in the machine. This problem is overcome in vector or field-oriented control as explained in this figure. Basically, in vector control, an induction motor is controlled like a separately excited dc motor. In a dc motor, the field flux ψ_f and armature flux ψ_a, established by the respective field current I_f and armature or torque component of current I_a, are orthogonal in space so that when torque is controlled by I_a, the field flux is not affected, thus giving fast torque response. Similarly, in induction motor vector control, the synchronous reference frame currents i_{ds} and i_{qs} are analogous to I_f and I_a, respectively, and i_{ds} is oriented in the direction of rotor flux ψ_r (defined as ψ_r orientation). Note that ψ_r is used instead of ψ_m or ψ_s because with ψ_r orientation true decoupling is obtained. However, if leakage inductances are neglected, $\psi_r = \psi_m = \psi_s$. Therefore, when torque is controlled by i_{qs}, the rotor flux is not affected thus giving fast dc motor-like torque response. The drive dynamic model also becomes simple like that of a dc machine because of decoupling vector control. Vector control has brought a renaissance to the control of ac drives (both induction and synchronous machines) since its invention at the beginning of the 1970s.

FIGURE 7.26   (a, b) Vector control principle by phasor diagram and (c) vector control implementation.

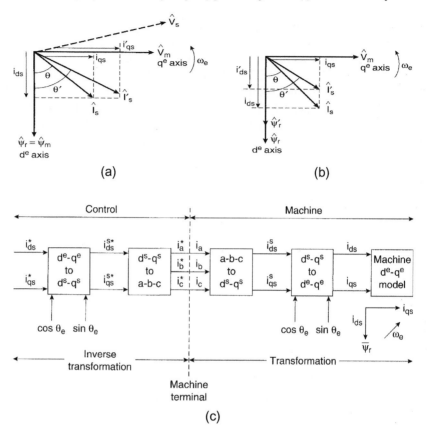

In a vector-controlled drive, the machine stator current vector $\bar{I}_s$ (or $\hat{I}_s$) has two components: i_{ds} or flux component and i_{qs} or torque component, as shown in the phasor diagram. These current components are to be controlled independently, as in a dc machine, to control the flux and torque, respectively. The i_{ds} is oriented in the direction of $\hat{\psi}_r$ (or $\hat{\psi}_m$ if L_{lr} is neglected), and i_{qs} is oriented orthogonally to it. If i_{qs} is increased to i'_{qs}, the stator current $\hat{I}_s$ changes $\hat{I}'_s$ as shown. Similarly, if i_{ds} is decreased to i'_{ds}, the corresponding change of $\hat{I}_s$ is also shown. The actual implementation principle of vector control is shown in the lower figure. The machine model is shown in a synchronous frame at the right, and the two front-end conversions of phase currents in a stationary frame are also shown. The controller should make the two inverse transformations, where the unit vector $\cos\theta_e$ and $\sin\theta_e$ in the controller should ensure correct alignment of i_{ds} in the direction of $\hat{\psi}_r$ and i_{qs} at 90° ahead of it. Obviously, the unit vector is the key element for vector control. Note that the inverter's dynamics and its transfer characteristics, if any, have been neglected. There are two methods of vector control depending on the derivation of the unit vector. These are the direct (or feedback) method and indirect (or feedforward) method.

FIGURE 7.27   Induction motor flux estimation methods.

- FLUX COILS IN AIRGAP
- HALL SENSORS IN AIRGAP
- OPEN-LOOP VOLTAGE MODEL
- CASCADED LOW-PASS FILTER VOLTAGE MODEL
- OPEN-LOOP CURRENT MODEL (OR BLASCHKE EQUATION)
- INTEGRATION OF VOLTAGE AND CURRENT MODELS
- MODEL REFERENCING ADAPTIVE PRINCIPLE
- SPEED ADAPTIVE FLUX OBSERVER

In a scalar or vector control method, the machine flux is normally maintained at a pre-determined value (as in a dc motor) by feedback control. In vector control, the flux vector signal (magnitude as well as direction) is required. The feedback flux signal (scalar or vector) can be obtained with the help of sensors, or estimated from the machine terminal voltage and current signals. Modern drives invariably use the latter method. This figure summarizes the different methods for estimation of flux. The flux coils and Hall sensors are basically sensor-based methods. In the former method, flux coils are mounted in the airgap direct (d^s) and quadrature (q^s) axes. The induced voltages in these coils ($v_{dm}{}^s$, $v_{qm}{}^s$) are then integrated and processed further to obtain the rotor flux vector components ($\psi_{dr}{}^s$, $\psi_{qr}{}^s$). The magnitude (peak value of sinusoid) can be calculated by $\hat{\psi}_r = \sqrt{\psi_{dr}{}^{s^2} + \psi_{qr}{}^{s^2}}$. Apart from the mounting difficulty, at very low frequency, integration of very small signals becomes difficult due to a drift problem. Hall sensors are also mounted in the airgap like flux coils, but here the accuracy becomes difficult due to temperature drift. These two methods are now practically obsolete. The voltage model depends on machine terminal voltages and currents, whereas the current model depends on the currents and speed signal. In the hybrid model, the voltage and current models are blended. The closed-loop observer methods are more sophisticated. The voltage and current models will be described next, and the remaining methods will be discussed later.

FIGURE 7.28   Voltage model feedback signal estimation of induction motor.

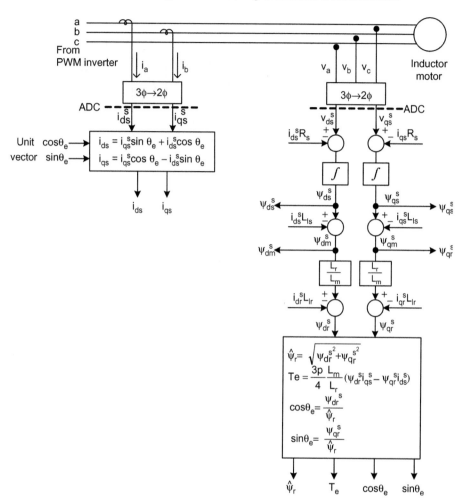

The figure shows the calculation of various feedback signals with the help of DSP. The derivation of all equations was given in Chapter 6. For an isolated neutral machine, only two current sensors are needed. The analog voltage and current signals are filtered and converted to two-phase signals by op amps, and then converted to digital signals by A/D converters. The block diagram indicates that the vector components of stator flux (ψ_{ds}^s, ψ_{qs}^s), airgap flux (ψ_{dm}^s, ψ_{qm}^s), rotor flux (ψ_{dr}^s, ψ_{qs}^s), torque (T_e), and unit vector ($\cos\theta_e$, $\sin\theta_e$) signals can be calculated. Then, using the unit vector, synchronous frame stator currents (i_{ds}, i_{qs}) can be calculated. The accuracy of the estimated signals depends on the machine parameters, which vary during machine operation. At very low speed, ideal integration becomes difficult because the voltage signals are very low at low frequency and sensor dc offsets tend to build up.

FIGURE 7.29   Current model feedback signal estimation.

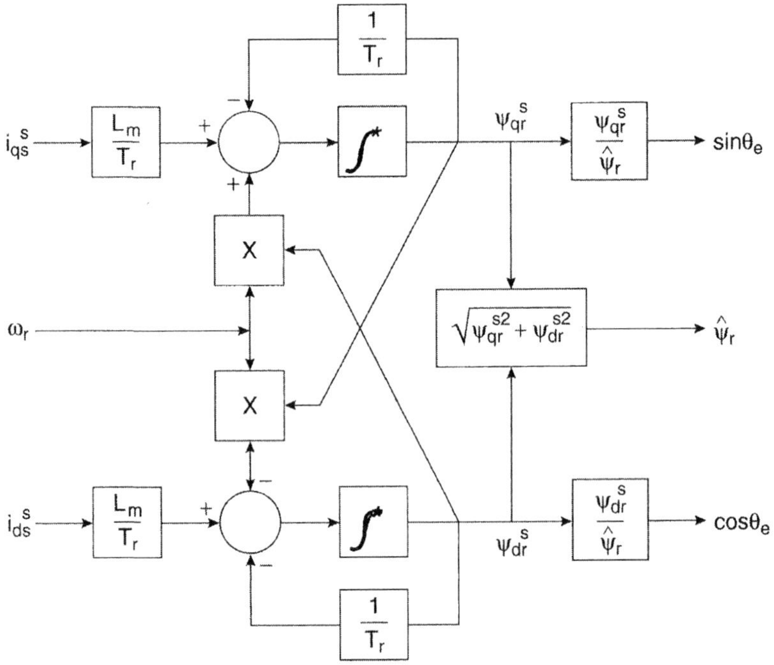

$$\frac{d\psi_{dr}{}^s}{dt} = \frac{L_m}{T_r} i_{ds}{}^s - \omega_r \psi_{qr}{}^s - \frac{1}{T_r} \psi_{dr}{}^s \ldots\ldots (1)$$

$$\frac{d\psi_{qr}{}^s}{dt} = \frac{L_m}{T_r} i_{qs}{}^s + \omega_r \psi_{dr}{}^s - \frac{1}{T_r} \psi_{qr}{}^s \ldots\ldots (2)$$

The current model signal estimation was first proposed by Blaschke (the inventor of direct vector control) and, therefore, the equation is known as the Blaschke equation. Basically, it uses the stator currents ($i_{ds}{}^s$, $i_{qs}{}^s$) and speed (ω_r) signals to estimate the rotor flux vector ($\psi_{qr}{}^s$, $\psi_{dr}{}^s$) by solving Eqs. (1) and (2) in real time. These equations are the rotor circuit equations of stationary frame equivalent circuits given in Fig. 6.26 [2], where the rotor time constant $T_r = L_r/R_r$ and the rotor currents are replaced by the stator currents. The block diagram for solving these equations is shown in the figure. Although the flux vector estimation requires the speed signal, the advantage is that ideal integration is not required and the model works well from zero speed. However, one problem of the model is that it is dependent on a rotor time constant (T_r) which varies widely primarily due to temperature variations of R_r. Note that the voltage model has good accuracy at higher frequencies, whereas the current model is good at low frequencies. These two models can be blended into a hybrid model to cover the whole frequency range.

FIGURE 7.30 Direct or feedback vector control with rotor flux (ψ_r) orientation.

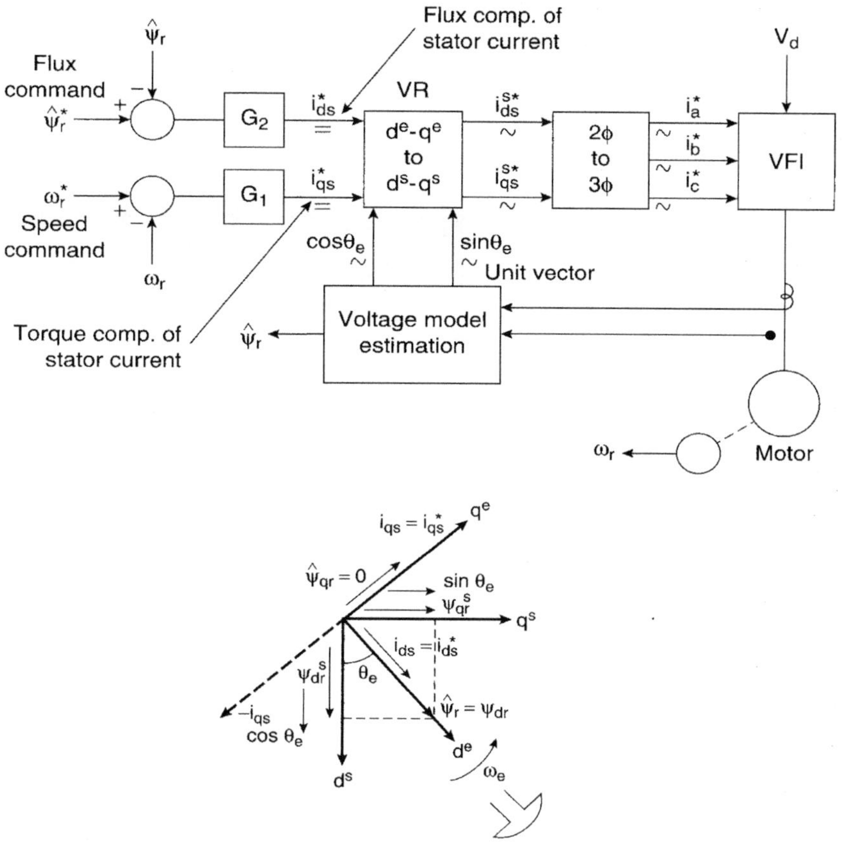

The figure shows the block diagram of direct vector control (DVC) with rotor flux orientation, where $i_{ds}{}^*$ is the flux component of stator current and $i_{qs}{}^*$ is the torque component of stator current. The vector rotation (VR) and $2\varphi/3\varphi$ transformation in the forward direction, as explained in Figure 7.26, are indicated in the figure. The $i_{ds}{}^*$ signal is obtained from a flux control loop, whereas the $i_{qs}{}^*$ signal is obtained from the speed control loop. An additional torque control loop can be added within the speed loop, if desired. The unit vector and $\hat{\psi}_r$ signals are calculated from the voltage model estimation shown in Figure 7.28. The alignment of i_{ds} in the direction of $\hat{\psi}_r$ and i_{qs} perpendicular to it is explained in the phasor diagram, where the d^e-q^e axes rotate counterclockwise at speed ω_e with respect to the d^s-q^s axes. The unit vectors $\cos\theta_e$ and $\sin\theta_e$ remain aligned with the signals $\psi_{dr}{}^s$ and $\psi_{qr}{}^s$, respectively. Note that if $i_{qs}{}^*$ is negative for regeneration or the reverse speed, its direction is reversed as shown. Note also the self-control nature, where the inverter frequency as well as the phase for the control are generated by feedback with the help of the unit vector. The current model can also be used to estimate the flux and unit vector.

FIGURE 7.31   Indirect vector control phasor diagram and equations.

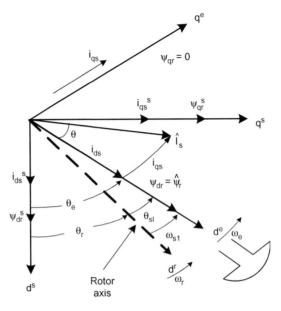

$$\frac{d\psi_{qr}}{dt} + \frac{R_r}{L_r}\psi_{qr} - \frac{L_m}{L_r}R_r i_{qs} + \omega_{sl}\psi_{dr} = 0 \ldots\ldots(1)$$

$$\frac{d\psi_{dr}}{dt} + \frac{R_r}{L_r}\psi_{dr} - \frac{L_m}{L_r}R_r i_{ds} - \omega_{sl}\psi_{qr} = 0 \ldots\ldots(2)$$

$$\omega_{sl} = \frac{L_m}{\hat{\psi}_r} \cdot \frac{R_r}{L_r} \, i_{qs} \ldots\ldots(3)$$

$$\hat{\psi}_r = L_m i_{ds} \ldots\ldots(4)$$

Indirect vector control (IVC) is essentially the same as direct vector control except the unit vector ($\cos\theta_e$, $\sin\theta_e$) is generated in feedforward manner as explained in this phasor diagram. The rotor d^r-q^r axes fixed on the rotor rotate at speed ω_r, whereas the synchronously rotating d^e-q^e axes are at slip angle θ_{sl} ahead of it ($+\omega_{sl}$) so that $\theta_e = \theta_r + \theta_{sl}$, and ψ_r is oriented at d^e axis as before. If torque is negative, d^e axis falls behind d^r because ω_{sl} (i.e., θ_{sl}) is negative. The derivation of IVC control equations is explained below the figure. The rotor circuit equations from d^e-q^e circuits (Figure 6.20) can be written. Then, by eliminating the rotor currents, Eqs. (1) and (2) can be derived [2] relating rotor fluxes with stator currents. Substituting the conditions $\psi_{qr} = 0$ and $d\psi_{qr}/dt = 0$ for decoupled control and $\psi_{dr}/dt = 0$ for constant flux, Eqs. (3) and (4) can be derived. Equation (3) indicates how the control slip command ω_{sl} can be derived in feedforward manner from the control current i_{qs}, whereas Eq. (4) shows that rotor flux is a function of i_{ds} in the steady-state condition.

FIGURE 7.32 Indirect or feedforward vector control with rotor flux (ψ_r) orientation.

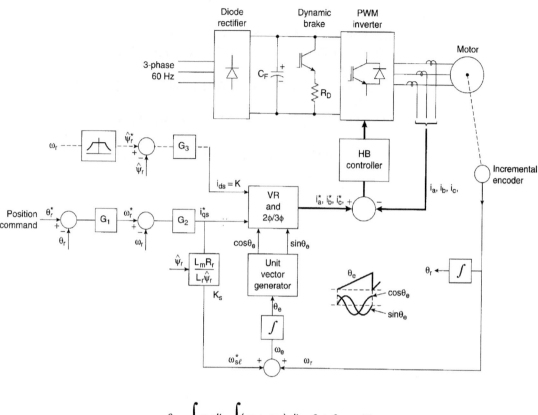

$$\theta_e = \int \omega_e \, dt = \int (\omega_r + \omega_{sl}) \, dt = \theta_r + \theta_{sl} \, ...(1)$$

In indirect vector control, the slip command signal ω_{sl}^* is derived from the command i_{qs}^* through the slip gain (K_s), as discussed earlier. This signal is then added to the speed signal, integrated, and then the unit vector components are derived as shown in this figure. Thus, the rotor pole and the corresponding current i_{ds} are held ahead of the rotor d^r axis at correct angle as indicated in Figure 7.31. The speed signal is obtained from an incremental encoder. In the constant torque region, the rated flux is generated by constant i_{ds} command. For closed-loop flux control in both constant-torque and field-weakening regions, i_{ds} can be controlled within the programmed flux control loop so that the inverter always operates in PWM mode. The loss of flux in the field-weakening region causes some loss of torque from that of the square-wave mode, but fast vector control response is retained. The figure shows a position servo, where speed command is generated from the position loop. For simplicity, a hysteresis-band current control is shown, although other types of PWM are entirely possible. The control operates smoothly from zero speed. The IVC is used widely in industry. However, one disadvantage is that the slip gain parameters, particularly R_r, vary widely with temperature, causing a coupling effect that deteriorates transient response and affects the flux and torque transfer characteristics.

FIGURE 7.33   Indirect vector control tuning methods.

- Initial Tuning:
  - USE NOMINAL MACHINE PARAMETERS
  - AUTOMATED PARAMETER MEASUREMENT
- Online Tuning:
  - INJECT PRB SIGNAL IN d^s-AXIS AND OBSERVE IN q^s-AXIS
  - EXTENDED KALMAN FILTER (EKF) PARAMETER ESTIMATION
  - DIRECT SOLUTION OF MACHINE d^s-q^s MODEL
  - MRAC TUNING
  - FUZZY-MRAC TUNING

Parameter variation (particularly the rotor resistance) is an important problem in indirect vector control, and this figure summarizes the possible methods for tuning it. Initial tuning can be done or tuning can be done in the running condition (note that parameters change in this condition). The machine equivalent circuit parameters can be determined by automated measurement [9], where signals are injected by the inverter and the parameters are calculated from the response with the help of DSP. A square-wave $i_{qs}{}^*$ command signal can be applied and slip gain K_s can be modified until the correct torque wave is obtained. A PSB (pseudo-random binary) signal can be injected in the d^s-axis and the response can be measured in the q^s-axis [10] to detect the coupling effect. Extended Kalman filter [11] is a complex observer method of parameter estimation. The complex machine model can be solved online to determine the parameter values. Model referencing adaptive control (MRAC) [12, 13] methods are also used for slip gain tuning. Most of these continuous tuning methods are complex and time consuming in terms of DSP implementation. Most IVC drives operate with some amount of detuning, which gives the resulting coupling effect. Note that in the DVC method, no serious parameter variation problem arises if the speed does not fall too low in the voltage model feedback estimation. Otherwise, near zero speed, the current model estimation must be used, which gives a similar coupling problem.

FIGURE 7.34   Slip gain tuning of indirect vector drive by MRAC principle.

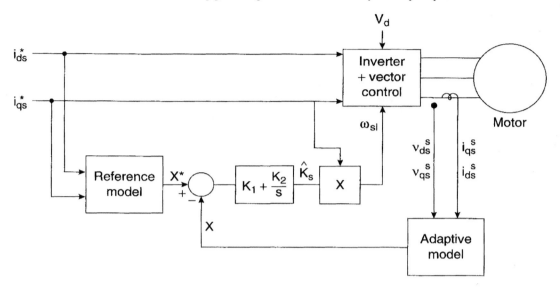

$$X^* = T_e^* = \frac{3}{2}(\frac{p}{2})\frac{L_m}{L_r}\hat{\psi}_r^* i_{qs}^* = Ki_{ds}^* i_{qs}^* \ldots\ldots\ldots (1)$$

$$X = T_e = \frac{3}{2}(\frac{p}{2})(\psi_{ds}^{\ s} i_{qs}^{\ s} - \psi_{qs}^{\ s} i_{ds}^{\ s}) \ldots\ldots\ldots (2)$$

A model referencing adaptive control (MRAC) principle for slip gain tuning [12] of an IVC drive is illustrated in this figure. In this case, a reference model that satisfies the tuned condition of the drive is solved to generate a variable that is compared with the same variable estimated from the actual operating condition. The loop error modifies slip gain K_s until the computed variable matches the reference variable. A number of variables can be chosen for MRAC tuning, but in this case, the torque (T_e) is chosen. The expression of T_e^* in the tuned condition is given by Eq. (1), where $\hat{\psi}_r = L_m i_{ds}$. This means that T_e^* can be calculated from the command values of i_{ds}^* and i_{qs}^*. The actual torque with detuning is computed by Eq. (2). When K_s is properly tuned, the developed torque matches the ideal tuned torque. There is a small parameter variation problem for T_e^* generation because it is a function of L_m and L_r values. In addition, at low speed, T_e estimation will be affected by stator resistance (R_s) variation.

FIGURE 7.35 Synchronous current control of vector drive.

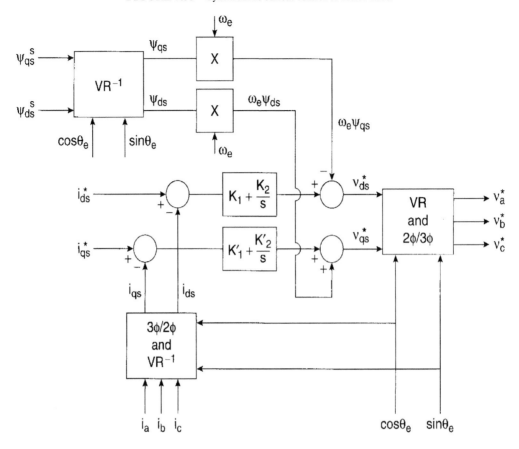

In both DVC and IVC, the current control is invariably used. The hysteresis-band current control is simple, but does not give optimal harmonic performance. Besides, there is a current tracking problem (magnitude and phase) in the quasi-PWM field-weakening region. Therefore, voltage PWM is usually used within current control loops. Either the SPWM or SVM technique can be used, but SVM is preferred. Synchronous current control (also called dc current control), described in this figure, gives the best performance. The synchronous frame command currents ($i_{ds}{}^*$, $i_{qs}{}^*$) are compared with the corresponding calculated values, and then the sinusoidal command voltage waves ($v_a{}^*$, $v_b{}^*$, $v_c{}^*$) are generated through P-I control, vector rotation, and $2\varphi/3\varphi$ transformation, as shown in the figure. The unit vector components can be generated either by the voltage model (Figure 7.28) or current model (Figure 7.29). A small amount of coupling is introduced because of P-I control. A calculated counter emf signal ($\omega_e\psi_{qs}$, $\omega_e\psi_{ds}$) is injected in the respective loop in a feedforward manner to improve the loop response. Because of dc current balancing by P-I control, the scheme operates well in both constant-torque and field-weakening regions.

FIGURE 7.36 Salient features of vector control.

- INDUCTION MOTOR IS CONTROLLED LIKE A SEPARATELY EXCITED DC MOTOR
- MACHINE IS SELF-CONTROLLED – CANNOT BECOME UNSTABLE AS WITH SCALAR CONTROL
- VC GENERALLY COVERS ZERO SPEED TO FIELD-WEAKENING REGION
- INDIRECT VC IS MORE PARAMETER SENSITIVE, BUT EASILY COVERS ZERO SPEED
- EASY HARMONIC-INSENSITIVE UNIT VECTOR ESTIMATION IN INDIRECT VC
- DIRECT VC IS MORE PARAMETER SENSITIVE IF ZERO SPEED COVERAGE IS NEEDED
- INHERENTLY FOUR-QUADRANT SPEED CONTROL
- CONTROL IS COMPLEX – NEEDS DSP
- IDEAL VECTOR CONTROL IS IMPOSSIBLE

The figure summarizes the salient features of vector control (compared to scalar control), and also illustrates the main differences between IVC and DVC. The most important feature of VC is that an induction motor is controlled like a separately excited dc machine in all operating conditions. Of course, ideal decoupling control is not possible because of machine parameter variations and discrete time operation of the PWM inverter and DSP-based controller. The complex machine dynamics become simple like a dc machine, and there is no stability problem like that in scalar control. The machine is operated within a stator current limit that maintains the limit of torque and slip automatically. Both DVC and IVC can be operated from zero speed to field-weakening regions. However, rotor resistance (R_r) variation causes coupling problems for IVC at any speed and DVC in low speed. A motoring vector drive with transition to negative speed command goes into the regenerative braking mode first; then at zero speed, the supply sequence is automatically reversed to give speed reversal.

FIGURE 7.37 Phasor diagram for stator flux (ψ_s) oriented vector control.

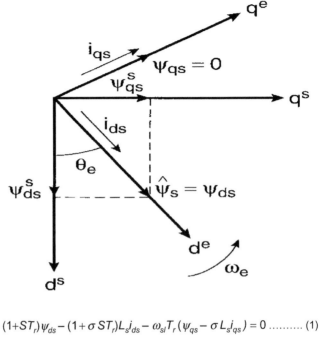

$$(1+ST_r)\psi_{ds} - (1+\sigma ST_r)L_s i_{ds} - \omega_{sl}T_r(\psi_{qs} - \sigma L_s i_{qs}) = 0 \ \ldots\ldots (1)$$

$$(1+ST_r)\psi_{qs} - (1+\sigma ST_r)L_s i_{qs} + \omega_{sl}T_r(\psi_{ds} - \sigma L_s i_{ds}) = 0 \ \ldots\ldots (2)$$

$$\omega_{sl} = \frac{(1+\sigma ST_r)L_s i_{qs}}{T_r(\psi_{ds} - \sigma L_s i_{ds})} \ \ldots\ldots\ldots (3)$$

$$i_{dq} = \frac{\sigma L_s i_{qs}^2}{(\psi_{ds} - \sigma L_s i_{ds})} \ \ldots\ldots (4)$$

So far, we have discussed a rotor flux (ψ_r) oriented vector drive in order to accomplish decoupling control. Stator (ψ_s) or airgap flux (ψ_m) oriented vector control is also possible. The phasor diagram in this figure explains ψ_s oriented vector control, where i_{ds} is oriented in the direction of ψ_s and i_{qs} is perpendicular to it. One advantage of ψ_s orientation is that the estimation of the flux and the corresponding unit vector is more accurate because only stator resistance variation affects the accuracy. Compensation of R_s due to temperature variation is somewhat easy. However, the problem is that ψ_s orientation introduces a coupling effect, which requires special decoupling compensation. The derivation for the compensation is shown in the figure [14]. Equations (1) and (2) are obtained from d^e-q^e rotor circuits, where $\sigma = 1 - L_m^2/L_s L_r$. Substituting $\psi_{qs} = 0$ and manipulating for decoupling control, the slip ω_{sl} and decoupling current i_{dq} expressions can be derived [2].

FIGURE 7.38 Stator flux-oriented vector control block diagram with estimation equations.

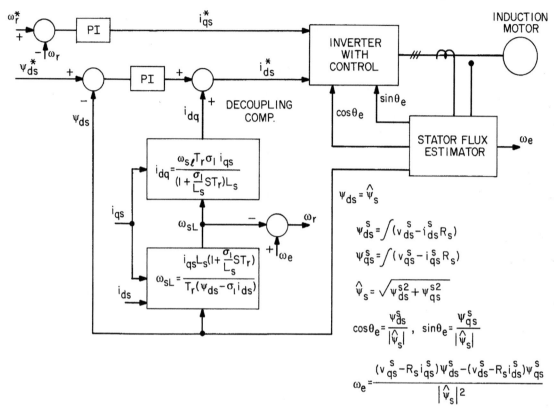

The figure shows the block diagram for stator flux-oriented direct vector control, which also includes the estimation equations for the flux vector ($\psi_{ds}{}^s$, $\psi_{qs}{}^s$), unit vector ($\cos\theta_e \cdot \sin\theta_e$), and frequency ($\omega_e$). The speed control loop generates the $i_{qs}{}^*$ command, as usual. Command $\psi_{ds}{}^*$ (constant or programmed with speed for field-weakening control) is compared with the corresponding estimated value, which then generates the $i_{ds}{}^*$ command after adding the decoupling compensation current i_{dq} in the control loop as shown. The feedback signals ψ_{ds}, i_{qs}, and i_{ds} can be generated from the stationary frame signals with the help of unit vector $\cos\theta_e$ and $\sin\theta_e$. Note that i_{dq} is not directly a function of ω_{sl} (see Figure 7.37) as indicated in this figure. However, the slip signal estimation is isolated with the objective of speed estimation by the relation $\omega_r = \omega_e - \omega_{sl}$. The signal can be estimated by differentiating $\tan\theta_e = \psi_{ds}{}^s/\psi_{qs}{}^s$ and substituting the relation $\omega_e = d\theta_e/dt$. If speed estimation is not desired, the i_{dq} expression can be used directly. Control is difficult in the very low speed range because of integration problems. Machine parameter variations will result in some inaccuracy in the estimation of the i_{dq} and ω_r signals, but the effect on the i_{dq} is not very dominant because it is within a feedback loop.

FIGURE 7.39   Vector control of line-side PWM converter.

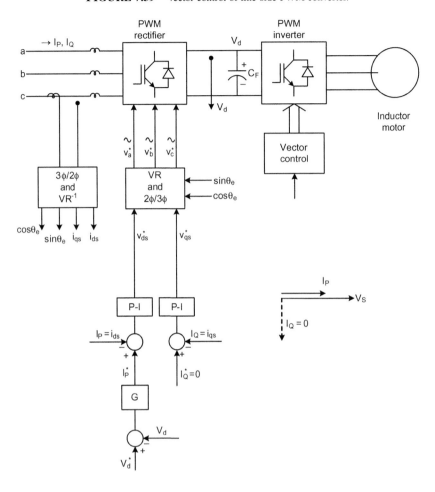

In a two-sided PWM converter system, the line-side converter must maintain the required dc link voltage V_d by balancing the power flow and sinusoidal line current at the desired DPF. One simple method of current control at unity DPF is to generate the sinusoidal phase current command waves such that they are co-phasal with the sensed phase voltage waves within the V_d control loop, and then use HB current control PWM. More advanced DVC is shown in this figure and can provide arbitrary DPF control. The V_d control loop generates the active line current command $I_P{}^*$, which is in phase with the line voltage. In this case, the current i_{ds}, which is aligned with the d^e axis (see Figure 7.40), is the same as I_P, whereas the leading reactive current I_Q, which is aligned with the q^e axis is i_{qs}, as indicated in the figure. The unit vector ($\cos\theta_e$, $\sin\theta_e$), generated from the line voltage, performs the vector rotation for correct alignment so that v_{ds} and v_{qs} control I_P and I_Q, respectively. Although $i_Q{}^*$ is shown as zero, it can be leading (+) or lagging (−). The line-side filter inductance will introduce a coupling effect, but it is neglected.

FIGURE 7.40   Phasor diagram and equations for line-side converter vector control.

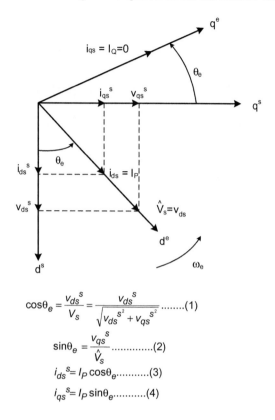

$$\cos\theta_e = \frac{v_{ds}{}^s}{V_s} = \frac{v_{ds}{}^s}{\sqrt{v_{ds}{}^{s^2} + v_{qs}{}^{s^2}}} \quad \text{........(1)}$$

$$\sin\theta_e = \frac{v_{qs}{}^s}{\hat{V}_s} \quad \text{.............(2)}$$

$$i_{ds}{}^S = I_P \cos\theta_e \text{..........(3)}$$

$$i_{qs}{}^S = I_P \sin\theta_e \text{..........(4)}$$

The figure explains the unit vector generation principle and direct vector control of Figure 7.39 with the help of a phasor diagram. In the figure, current i_{ds} is in phase with $\hat{V}_s$, where $\hat{V}_s$ = line voltage vector. It is aligned with the d^e axis as mentioned earlier, and the leading reactive current i_{qs} is aligned with the q^e axis. The stationary frame line phase voltages are converted to $v_{ds}{}^s$, $v_{qs}{}^s$, by $3\varphi/2\varphi$ transformation, and then unit vector is generated by Eqs. (1) and (2). This unit vector generation and its use in VR ensures that, in the time domain (stationary frame), $\cos\theta_e$, $v_{ds}{}^s$, and $i_{ds}{}^s$ are in the same phase and aligned with the d^s axis. Similarly, $\sin\theta_e$, $v_{qs}{}^s$, and $i_{qs}{}^s$ are also co-phasal and aligned with the q^s axis, thus ensuring correct phase alignment. Equations (3) and (4) are shown for $I_Q = 0$, i.e., unity DPF condition. The unit vector is also used to calculate i_{ds} and i_{qs} from line currents that are required for feedback current control loops.

FIGURE 7.41   Vector control of current-fed inverter drive.

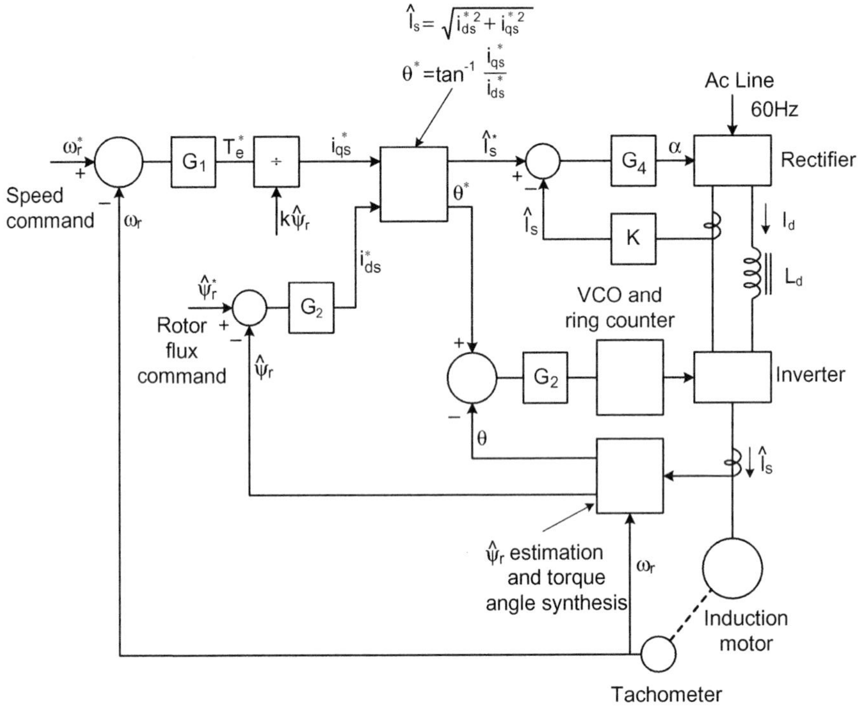

$$i_{qs} = i_{qs}{}^s \cos\theta_e - i_{ds}{}^s \sin\theta_e \quad\ldots\ldots\ldots (1)$$

$$i_{ds} = i_{qs}{}^s \sin\theta_e + i_{ds}{}^s \cos\theta_e \quad\ldots\ldots\ldots (2)$$

$$\angle\theta = \tan^{-1}\frac{i_{qs}}{i_{ds}} \quad\ldots\ldots\ldots\ldots (3)$$

Vector control can also be applied to current-fed inverter drive to get a fast response, as shown in this figure. The inverter can be of the six-step or PWM type, and the phase-controlled rectifier generates the current source for the inverter. The drive operates with rotor flux orientation, where $\hat{\psi}_r$ and torque angle θ signals are estimated from the current model (Figure 7.31) and Eqs. (1)–(3). The speed control loop generates the torque command, which is divided by the flux to get the command current $i_{qs}{}^*$. The flux loop correspondingly generates the $i_{ds}{}^*$ current. These signals are then converted to polar form, where $\hat{I}_s{}^*$ is the stator current (peak of sine wave current) and is the torque angle. The rectifier is controlled to generate I_d, i.e., $\hat{I}_s$, whereas the inverter is controlled by the PLL (phase-locked-loop) principle so that the actual feedback angle θ matches with the command value. Thus, the stator current is positioned at the correct angle to get vector control. The vector control is sluggish in a current-fed drive because of the inherent delay in current generation.

FIGURE 7.42   Salient features of slip power recovery drives.

- REQUIRES EXPENSIVE DOUBLY FED WRIM – DISADVANTAGES OF SLIP RINGS AND BRUSHES
- CONTROL OF SLIP POWER TO CONTROL TORQUE AND SPEED
- ECONOMICAL CONVERTER FOR LIMITED SPEED RANGE APPLICATIONS
  - LARGE CAPACITY PUMPS AND FANS
  - VARIABLE-SPEED WIND GENERATION
  - FLYWHEEL ENERGY STORAGE
  - VSCF SYSTEM, ETC.
- FAST DC MACHINE-LIKE TRANSIENT RESPONSE
- MACHINE FLUX CANNOT BE CONTROLLED
- NEEDS SEPARATE STARTING
- ONE- OR TWO-QUADRANT DRIVE WITH NONREVERSABLE SPEED
- LOW LINE POWER FACTOR

The slip power of a doubly fed wound-rotor induction machine (WRIM) can be controlled to control speed and torque. A simple and primitive method of speed control for this type of motor is to vary the rotor circuit rheostat mechanically. All the power in the rotor circuit (slip power) is wasted in this case. However, the advantages are that the machine can be started smoothly with maximum torque with no in-rush current and no line harmonics, and a high line power factor can be obtained. Instead of wasting the slip power, it can be controlled by a converter and fed back to the line to improve the efficiency. For limited range speed control near the machine's synchronous speed, the converter size is reduced, thus offsetting the disadvantages of WRIM. A number of limitations of this drive are poor line power factor and harmonics, nonreversible speed control at only rated flux, and the need for a separate starter. For variable-speed pumps and fans within typically a 2:1 speed control range, this type of drive with a high power rating has been widely used in industry. The two main classes of this drive are static Kramer drives and static Scherbius drives.

FIGURE 7.43 Static Kramer drive.

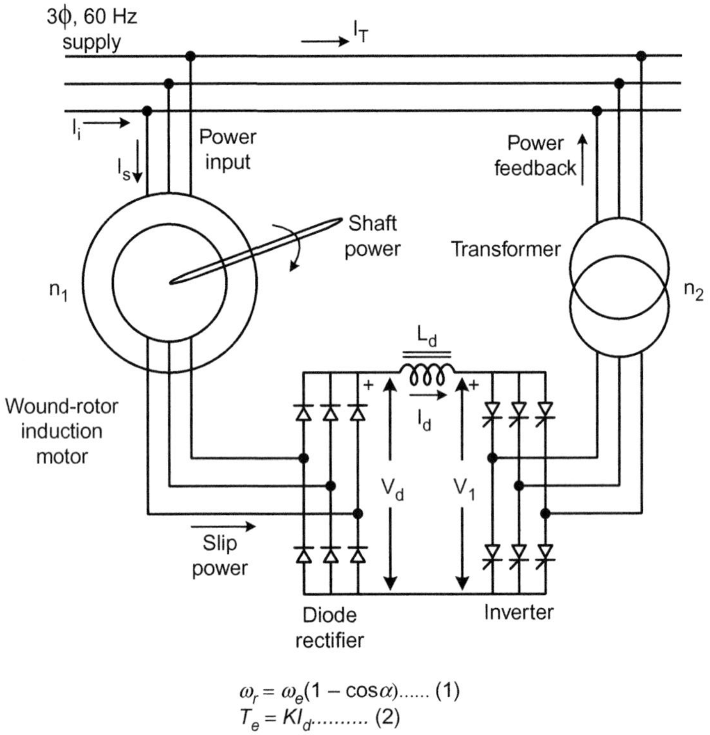

$$\omega_r = \omega_e(1 - \cos\alpha)\ldots\ldots (1)$$
$$T_e = KI_d \ldots\ldots\ldots (2)$$

In a static Kramer drive, shown in this figure, the rotor circuit slip power is converted to dc by the diode rectifier and then inverted to ac by a current-fed phase-controlled line-commutated inverter. The inverted 60-Hz power is then fed back to the line through a transformer. The machine always operates at subsynchronous speed with the airgap flux remaining constant. Speed reversal and regenerative braking are not possible. However, these are not disadvantages for some applications, such as when used in underground waste water pumps and fans. The speed is controlled by inverter firing angle α. The machine torque can be shown to be proportional to dc link current I_d. As speed decreases from synchronous speed, dc link voltage V_d increases, but at steady state $V_d = V_I$. Therefore, as V_I is increased by decreasing the inverter firing angle, speed increases. Because a larger firing angle reduces the line DPF, the transformer reduces the line voltage to decrease the firing angle control range for the desired range of speed control. The converter is economical on the rotor side because it handles the slip power only. However, with a larger range of speed control, the converter rating should be high. The drive has to be started separately (such as resistance switching) and brought up to the lowest speed before switching in the converter. The line PF is low and harmonic currents are introduced on the inverter side. The importance of the Kramer drive has been reduced in recent years.

FIGURE 7.44   Static Kramer drive control block diagram.

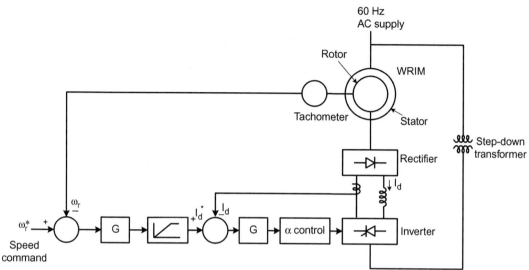

A closed-loop one-quadrant speed control system for a static Kramer drive is shown in this figure. The drive has the characteristic of a separately excited dc motor and, there-fore, the control strategy is similar to that of a phase-controlled rectifier dc drive. With constant airgap flux, the torque is proportional to dc link current I_d since the rotor cur-rent fundamental component is aligned at 90° with respect to the flux. Current I_d is con-trolled in the inner feedback loop as shown. If the command speed is increased by a step, the motor accelerates at constant developed torque corresponding to the I_d limit. The inverter firing angle α initially decreases fast to establish I_d and then decreases gradually as speed increases to track with rectifier voltage V_d. At steady state, I_d will settle down to correspond to the load torque. If speed command is decreased by a step, I_d goes to zero, and the machine slows down by inherent load torque. During deceleration, α angle increases continuously so that the dc link voltage V_I balances with V_d.

FIGURE 7.45   Static Scherbius drive with 18-thyristor cycloconverter.

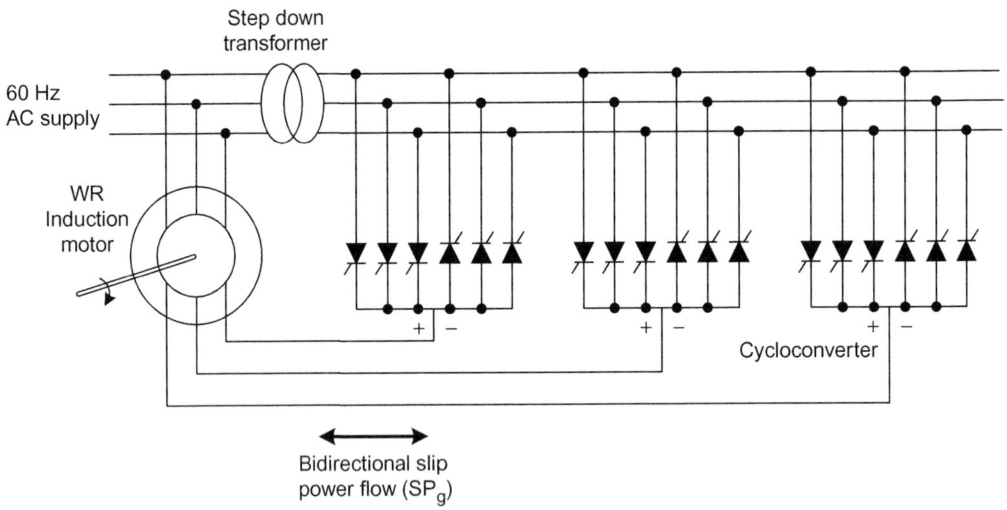

In a Scherbius drive, as shown in this figure, the converter system in the Kramer drive is replaced by a phase-controlled cycloconverter (CCV) so that slip power can be controlled to flow in either direction. With bidirectional power flow capability, the drive cannot only be controlled for motoring and regeneration, but for subsynchronous as well as supersynchronous speed regions as well. The range of speed, however, typically remains limited within ±50% of synchronous speed. The speed reversal is not possible (as is Kramer drive) because it requires reversal of stator supply voltage phase sequence. The CCV is expensive and has control complexity, but the advantages are a near-sinusoidal rotor current that gives reduced harmonic loss and a machine overexcitation capability that permits leading power factor operation on the stator side. In fact, CCV's input lagging DPF can be canceled by machine leading DPF so that the line PF can be unity or leading. In addition, true synchronous speed operation of the drive is possible when the CCV operates as a rectifier to generate dc excitation current for the machine. During the drive operation, the CCV output frequency and phase should closely track those of the rotor output. The step-down transformer at the input reduces the input voltage so that CCV can operate at the best DPF for the speed range of operation. Unfortunately, like a Kramer drive, the drive also requires separate resistive starting.

FIGURE 7.46   Modes of operation of a static Scherbius drive.

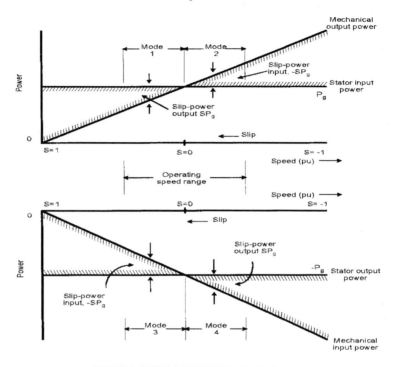

MODE 1: SUBSYNCHRONOUS MOTORING
MODE 2: SUPERSYNCHRONOUS MOTORING
MODE 3: SUBSYNCHRONOUS REGENERATION
MODE 4: SUPERSYNCHRONOUS REGENERATION

The figure explains the four modes of operation of a Scherbius drive, and the corresponding power distribution vs. slip power in sub/supersynchronous speed ranges. For simplicity, it is assumed that the shaft torque is constant at the rated value, which is positive in motoring but negative in regeneration. Mode 1 is identical to that of a Kramer drive. The stator input or airgap power P_g is positive and remains constant, and the positive slip power SP_g, which is proportional to slip, is returned back to the line through the CCV. Therefore, the line supplies the mechanical output power $P_m = (1 - S)P_g$ consumed by the shaft. The rotor speed corresponds to $\omega_e - \omega_{sl}$, i.e., $(1 - S)\omega_e$. In mode 3, i.e., the regenerative braking mode, the electrical energy is pumped out of the stator. The mechanical power input to the shaft $P_m = (1 - S)P_g$ is added with the slip power SP_g so that the total airgap power is P_g. Therefore, the line power is the same as shaft power. In mode 2, the slip is negative and slip power is absorbed by the rotor. Therefore, the total mechanical power output, i.e., the line power input is $(1 + S)P_g$, which consists of slip power that is added with the airgap power. In mode 4, the shaft input power $(1 + S)P_g$ is subtracted by the slip power SP_g so that the airgap power is P_g. The airgap power adds to the slip power to constitute line power, which is $(1 + S)P_g$. The bidirectional slip power flow is also possible by a self-commutated voltage-fed (see Figure 7.50) or current-fed converter system.

FIGURE 7.47   Vector control of cycloconverter-based Scherbius drive.

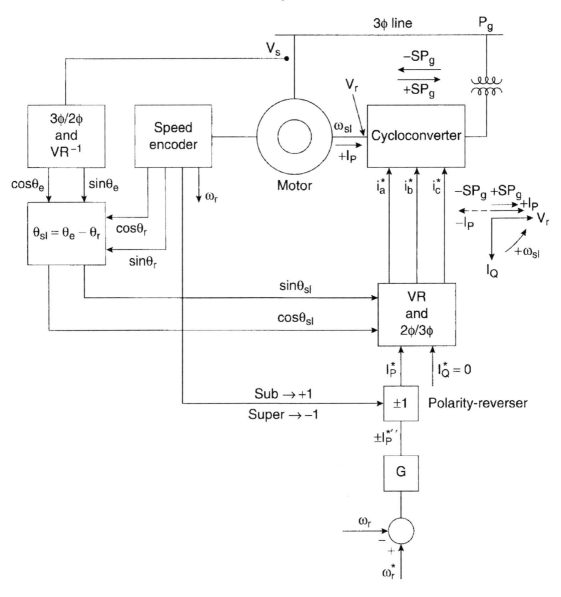

$$\cos\theta_{sl} = \cos(\theta_e - \theta_r) = \cos\theta_e \cos\theta_r + \sin\theta_e \sin\theta_r \ \text{.......} \ (1)$$
$$\sin\theta_{sl} = \sin(\theta_e - \theta_r) = \sin\theta_e \cos\theta_r - \cos\theta_e \sin\theta_r \ \text{.......} \ (2)$$

The principle of vector control discussed so far can also be extended to cycloconverter drives, where the CCV can be on either the stator or rotor side. This figure shows vector control for a Scherbius drive and includes an explanatory phasor diagram. In the speed-controlled system, the CCV should send slip energy ($+SP_g$) to the line in the subsynchronous motoring and supersynchronous regeneration modes, whereas it should feed slip energy ($-SP_g$) to the rotor in subsynchronous regeneration and supersynchronous motoring modes. The CCV active current I_P is in phase with V_r (rotor voltage at slip frequency) for $+SP_g$, whereas it is negative for $-SP_g$ as shown in the phasor diagram. The reactive current I_Q is perpendicular to V_r, but it is shown as zero in the figure. The speed control loop generates the I_P^* command through a polarity reverser, which becomes negative at supersynchronous speed. The unit vector signal at slip frequency is derived from the stator voltage and rotor encoder signals as shown. It converts I_P^* to the CCV phase currents to satisfy the phasor diagram. At true synchronous speed, $\omega_{sl} = 0$, which means that CCV generates dc current excitation for the machine.

FIGURE 7.48   400-MW Scherbius drive for variable-speed hydro generator and pump storage system.

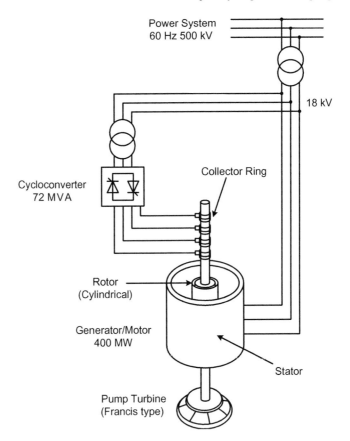

Hitachi recently installed a variable-speed hydro pump [15] generator at the Ohkawachi plant under Kansai Electric Power Co. in Japan. Normally, a synchronous machine at constant speed is used in such a system. In a hydro system, the water head in the reservoir normally varies widely. It can be shown that with a variable head, if the constant-speed machine is replaced by an adjustable-speed drive system, improved efficiency is obtained. The Kansai hydro unit is basically a 400-MW Scherbius drive with speed control as shown in this figure. The utility supply of 500 kV at 60 Hz is stepped down for the drive. In the daytime, the machine runs as a generator feeding electricity to the grid. In generating mode, the speed varies from subsynchronous to supersynchronous mode as the head increases. At night, when utility power demand is low and surplus grid power is available, the machine runs at variable speed to pump water from the tail to the head storage. In pump mode, it also runs from sub- to supersynchronous speed with the increase of head. Toshiba recently built a large flywheel energy storage system that uses the same principle where the water turbine is replaced by a flywheel [3].

FIGURE 7.49   Salient features of a 400-MW Scherbius drive.

- WORLD'S FIRST AND ONLY VARIABLE-SPEED HYDRO
  PUMP/GENERATOR IN OHKAWACHI PLANT OF
  KANSAI ELECTRIC POWER CO.
- 400-MW SCHERBIUS DRIVE WITH SLIP POWER CONTROL
- 3.0% EFFICIENCY IMPROVEMENT WITH VARIABLE HEAD
- THYRISTOR CYCLOCONVERTER:
  - NONCIRCULATING MODE
  - – 5.0 Hz TO +5.0 Hz FREQUENCY VARIATION
  - 12-PULSE, 72 MVA
- INDUCTION MACHINE:
  - 20 POLE
  - 330 RPM TO 390 RPM (SYNC. SPEED = 360 RPM)
  - LEADING/LAGGING STATOR CURRENT
- POWER SYSTEM: 500 kV, 60 Hz, LEADING/LAGGING DPF.

The salient features of the pump/generator system shown in Figure 7.48 are summarized here. A 72-MVA phase-controlled thyristor CCV is used for control of slip power circulation. The 72-thyristor, 12-pulse CCV feeds the three-phase rotor winding with a neutral wire. It operates in blocking mode within the frequency range of −5.0 to +5.0 Hz. For the nominal 60-Hz, 20-pole machine, the corresponding speed range is 330 rpm (−8.3%) to 390 rpm (+8.3%) with the synchronous speed at 360 rpm. The additional transformer in series with the CCV gives phase-splitting for 12-pulse operation, decreases the CCV power rating by decreasing the voltage, and improves the system power factor. As explained earlier, CCVs always run at lagging DPF, but the machine can run at leading DPF if additional excitation current is fed to the rotor. The total DPF of the system is controllable at unity, leading, or lagging. Although the equipment is expensive, the extra cost is soon recovered because of the overall 3% improvement in efficiency.

FIGURE 7.50   Scherbius drive with two-sided voltage-fed PWM converter.

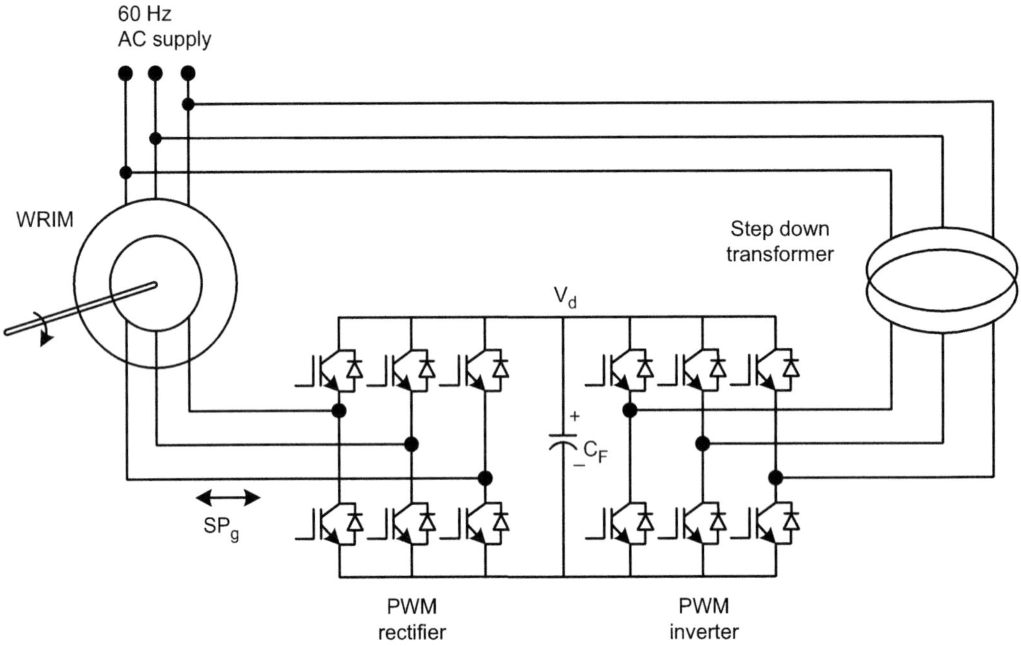

The CCV-based Scherbius drive system discussed in earlier figures can offer improved performance if the CCV is replaced by the two-sided PWM voltage-fed converter system shown in this figure. The slip power can be controlled to flow in either direction and vector control can be easily applied to both converters. DC link voltage V_d should be sufficiently high so that both the line-side and machine-side converters can always operate in undermodulation mode (buck mode with respect to the dc link) to fabricate sinusoidal machine and line currents. The PWM rectifier can easily track the variable-frequency, variable-magnitude slip voltage including the ideal dc condition at synchronous speed. This is basically dc-dc converter mode operation of the rectifier. The line-side step-down transformer helps keep the converter rating low with reasonably low V_d. Note that the machine stator-side DPF is always lagging, but the line-side converter can be controlled to be leading so that the total DPF can be maintained at unity.

FIGURE 7.51 Wind generation systems.

- TRADITIONALLY VARIABLE-PITCH, CONSTANT-SPEED TURBINE
- RECENT USE OF VARIABLE-SPEED GENERATION SYSTEM
  WITH POWER ELECTRONICS
  - LARGER ENERGY CAPTURE
  - LIFE-CYCLE COST IS LOWER
- POSSIBLE GENERATOR CONVERTER SYSTEMS
  - DOUBLY FED INDUCTION GENERATOR WITH SLIP POWER
    CONTROL
  - CAGE-TYPE INDUCTION GENERATOR WITH
    SHUNT PASSIVE OR ACTIVE SVG
    CASCADED VOLTAGE-FED CONVERTERS
    CASCADED CURRENT-FED CONVERTERS
    CYCLOCONVERTERS
  - SYNCHRONOUS GENERATOR WITH
    CASCADED VOLTAGE-FED CONVERTERS
    CASCADED CURRENT-FED CONVERTERS
    CYCLOCONVERTERS

Considering the recent importance of wind power generation, we will include the subject in this chapter. In a wind-electric generation system, the wind turbine, which is the prime mover, is coupled to a machine that generates electric power. The power is then fed to a utility grid, normally through a power electronic system. The wind turbine can operate at constant speed with variable-pitch angle control that permits the machine to generate power directly at 60/50 Hz and tie to the grid. A variable-speed wind generation system generates variable-frequency, variable-voltage power that is converted to constant frequency, constant voltage before connecting to the grid. Although the latter system is expensive, the energy capture is large enough that it makes the life-cycle costs lower. Recent technological advances in wind turbines, power electronics, and ac drives make the constant-speed system practically obsolete. A wind generation system can be looked on as a regenerative fan or blower drive. Therefore, many drive configurations with induction and synchronous machines are possible. The Scherbius drive with CCV or two-sided PWM voltage-fed converters, as discussed earlier, can be used within limited speed range. Also, a standard cage or synchronous machine can be used with voltage-fed converters, current-fed converters, or CCVs. The drive operates in the continuous regeneration mode, although sometimes start-up motoring power is also needed.

FIGURE 7.52   Characteristics of vertical and horizontal axis wind turbines [32].

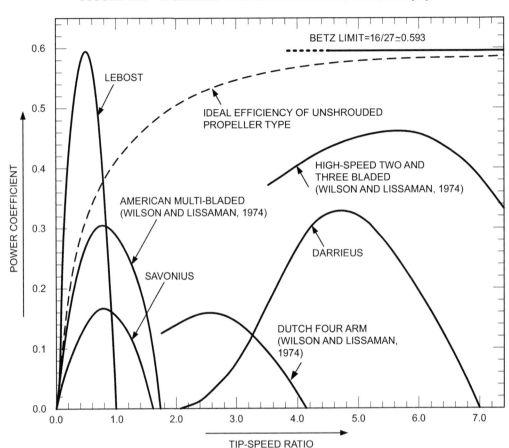

The two classes of wind turbines are vertical axis wind turbines (VAWTs) and horizontal axis wind turbines (HAWTs). The VAWT or Darrieus type, popularly called *egg beater* due to the blade configuration, has the advantages that it can be installed on the ground and can accept wind from any direction without any special yaw mechanism and, therefore, is preferred for high power. The disadvantages are that it is not self-starting and there is a large pulsating torque. The HAWT does not have some of these disadvantages, but the turbine and machine need to be mounted at high altitude. This figure summarizes the power coefficient (C_p) vs. tip-speed ratio (TSR) of various types of vertical and horizontal axis wind turbines available today. The TSR is the ratio of turbine peripheral speed to the wind velocity (V_w), and C_p is the index of output power. The figure shows that for a particular turbine, the maximum C_p is a function of TSR. This means that to extract maximum power from the turbine, the turbine speed must be varied with wind velocity to maintain the optimum TSR.

FIGURE 7.53 Family of torque-speed characteristics for a variable-speed wind turbine at various wind velocities.

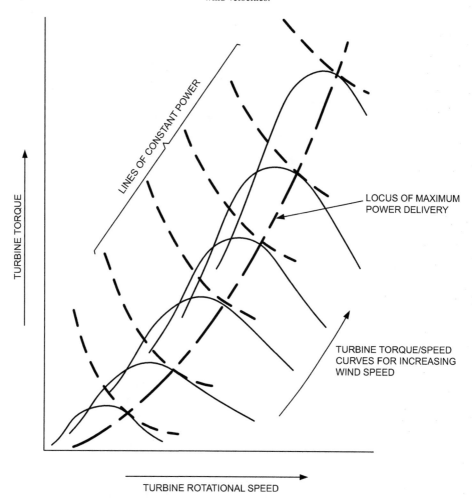

The figure shows the wind turbine developed torque as a function of its rotational speed at different wind velocities. If the wind velocity remains constant and turbine speed is increased from a low value, the developed torque increases, reaches a peak value, and then decreases. Superimposed on the family of curves is a set of peak constant power curves (dotted) that are tangential to the respective torque-speed curves. This indicates that the peak power point, where the turbine operates at its highest aerodynamic efficiency, is slightly to the right of the peak torque point. For a particular wind velocity, the turbine speed is varied so that it can deliver the maximum power. Because the torque-speed characteristics of a wind generation system are analogous to those of a motor-blower system, except that the turbine runs in the reverse direction, the torque follows the square-law characteristics ($T_e = K \cdot \omega_r^2$) and the output power follows the cube law ($P_0 = K_1 \cdot \omega_r^3$), as shown in the figure.

FIGURE 7.54   Wind power induction generator system with SVG and diode rectifier.

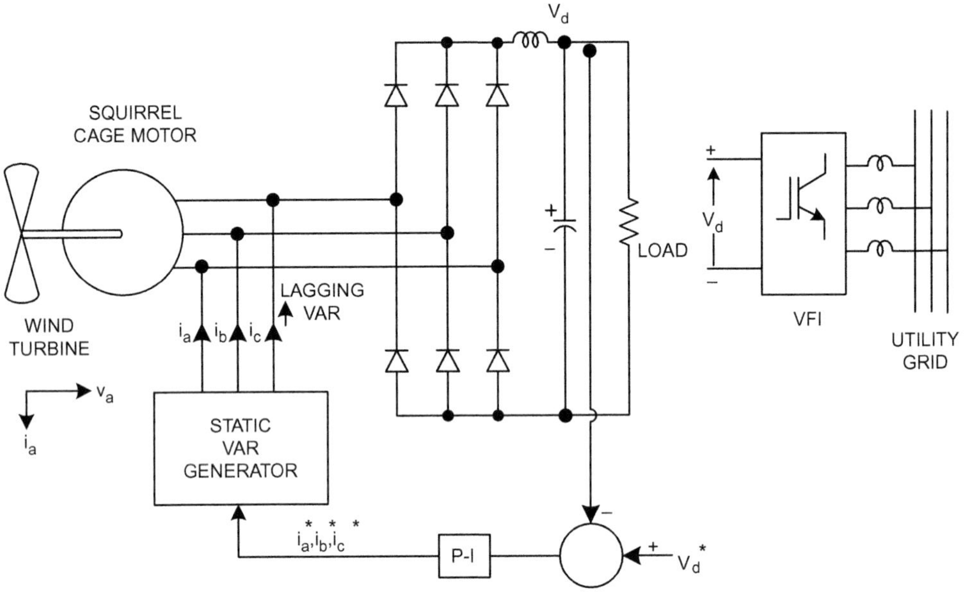

The figure shows a simple wind generation system using a cage-type induction generator, static VAR generator (SVG), and diode rectifier in the power circuit. One advantage compared to a traditional synchronous generator is that the machine is simple and does not require any current control in the rotor circuit. However, the lagging excitation current for the machine is to be supplied in the stator circuit by an SVG (see Figure 4.57). Basically, the SVG acts as a three-phase variable capacitor bank that generates lagging excitation current for the machine to regulate the airgap flux (often called a self-excited induction generator, SEIG). The SVG can be replaced by a thyristor-controlled reactor/capacitor system. The machine operates at supersynchronous speed (negative slip) to function as a generator. The shaft torque determines the slip frequency and the stator frequency is determined from the relation $\omega_e = \omega_r - \omega_{sl} (\omega_r > \omega_e)$. Therefore, with variation of the rotor speed, the stator frequency can be varied. The machine excitation current is varied to regulate the machine terminal voltage, i.e., the output dc voltage V_d. The stand-alone generator can feed the dc load as shown, or can be inverted to 60/50-Hz ac and fed to the utility grid as indicated. The system has several limitations. The machine cannot generate without initial excitation. The SVG's precharged capacitor supplies the initial excitation. The generated power is not optimum because the turbine speed is not tracked with the wind velocity. The turbine speed range is limited because at low speed it is difficult to regulate the voltage V_d because of flux saturation. There is of course the diode rectifier ripple current in the machine that gives high harmonic copper losses. Note that the static Scherbius drive system with cycloconverter (Figure 7.47) or with a two-sided PWM converter (Figure 7.50) can be operated in a wind generation system, where the machine always operates in the regenerative braking mode. The motor speed command ω_r^* in Figure 7.47 should track wind velocity to maximize the power output.

FIGURE 7.55   Optimum wind power generation system with 2-sided PWM converters.

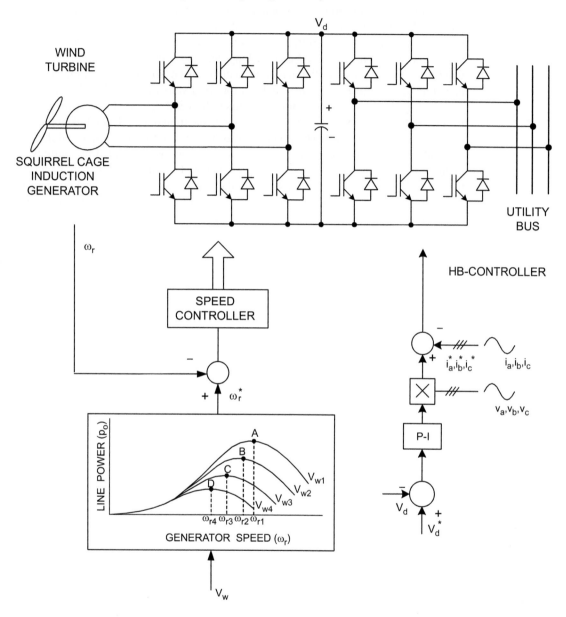

An optimum wind power generation system with a cage-type machine and two-sided PWM voltage-fed converter is shown in this figure [18]. The machine excitation is supplied by the PWM rectifier, where the excitation current i_{ds} maintains the flux constant. Currents i_{ds} and i_{qs} are controlled (not shown here, but discussed in Chapter 10) by vector control within the speed control loop, and the speed is programmed with the wind velocity V_W to extract the maximum output power. This means that the optimum operating speeds are ω_{r1}, ω_{r2}, ω_{r3}, and ω_{r4} for wind velocities V_{w1}, V_{w2}, V_{w3}, and V_{w4}, respectively, as shown in the figure. Of course, the wind velocity requires monitoring for the control. The line-side converter is responsible to maintain the dc link voltage V_d constant as shown. The line phase voltage waves v_a, v_b, and v_c are sensed, and the corresponding co-phasal line current commands i_a^*, i_b^*, i_c^* are generated by multiplying them by the output of the V_d control loop as shown. The phase currents are then controlled by HB PWM current control. When the turbine output power tends to increase the dc link voltage, the line currents tend to increase so that a balance is maintained between the line output power and turbine power.

FIGURE 7.56   Vector-controlled induction motor drive with two-sided, three-level PWM converters.

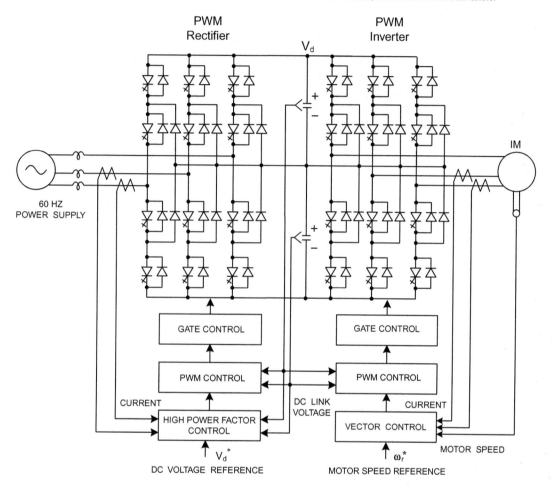

The figure shows the block diagram of a vector-controlled drive with a three-level GTO-based rectifier-inverter system. The usual advantages of a two-sided PWM converter were discussed in Chapter 4. The four-quadrant multi-megawatt drive system can be used, for example, in rolling mill applications, replacing the traditional cycloconverter drive. The machine uses rotor flux-oriented indirect vector control with speed control in the outer loop, which uses a discrete speed encoder. Synchronous current control with space vector PWM is used in the inner loop. The closed-loop rotor flux control is programmed with speed (not shown) so that the drive can operate smoothly from zero speed to field-weakening region. The PWM rectifier is controlled in direct vector control mode. It maintains constant dc link voltage and input power factor unity irrespective of motoring or regeneration mode of the drive. In the inner loop, the rectifier also uses synchronous current control with SVM algorithm. There is often feedforward control from the inverter to the rectifier to prevent dc link voltage fluctuation with fast loading of the machine.

FIGURE 7.57   Induction motor speed estimation methods.

- SLIP AND STATOR FREQUENCY ESTIMATION FROM MACHINE VOLTAGE AND CURRENT SIGNALS ($\omega_r = \omega_e - \omega_{sl}$)
- ESTIMATION FROM SLOT HARMONIC VOLTAGES
- DIRECT ESTIMATION FROM MACHINE DYNAMIC MODEL
- KALMAN FILTER METHOD
- MODEL REFERENCING ADAPTIVE CONTROL (MRAC)
- SPEED ADAPTIVE FLUX OBSERVER
- AUXILIARY SIGNAL INJECTION IN SALIENT ROTOR

An incremental speed signal for an induction motor is essential for closed-loop speed control of scalar or vector drives. The signal is also needed for indirect vector control, and direct vector control if speed control is necessary from zero speed. A physical speed encoder (typically of the optical type) mounted on the shaft adds cost and reliability problems to the drive, in addition to the need for a shaft extension for mounting it. In modern speed sensorless vector control, precision speed estimation from the machine terminal voltages and currents with the help of DSP is an important topic of research. Various estimation techniques have been proposed that are summarized here [19]. One method that was mentioned before is to calculate the ω_r signal from the ω_{sl} and ω_e expressions shown in Figure 7.38. The problem is that ω_{sl} is a very small signal, and the estimation becomes particularly difficult at low speed mainly due to parameter variations. Another problem is integration at very low frequency. The slot harmonic method is possibly the simplest for speed estimation. The slots on the rotor surface provide reluctance modulation that produces space harmonics in the airgap flux. Therefore, the machine counter emf waves will contain ripple voltage, the frequency and magnitude of which are proportional to speed. The speed can be identified by separating the ripple frequency signal through a signal processing circuit. Due to the finite number of rotor slots and small reluctance variation, the estimation at low speed becomes difficult. The remaining methods will be discussed in the following pages. However, accurate speed estimation, particularly near zero speed, remains a challenge.

FIGURE 7.58 Speed estimation from dynamic d^s-q^s model.

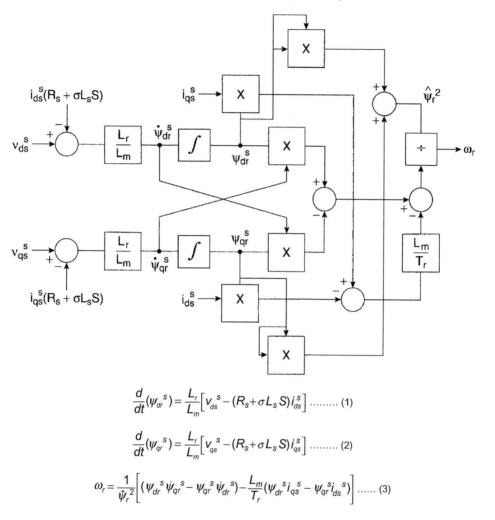

$$\frac{d}{dt}(\psi_{dr}^{\ s}) = \frac{L_r}{L_m}\left[v_{ds}^{\ s} - (R_s + \sigma L_s S)i_{ds}^{\ s}\right] \ \text{......... (1)}$$

$$\frac{d}{dt}(\psi_{qr}^{\ s}) = \frac{L_r}{L_m}\left[v_{qs}^{\ s} - (R_s + \sigma L_s S)i_{qs}^{\ s}\right] \ \text{......... (2)}$$

$$\omega_r = \frac{1}{\hat{\psi}_r^{\ 2}}\left[(\psi_{dr}^{\ s}\dot{\psi}_{qr}^{\ s} - \psi_{qr}^{\ s}\dot{\psi}_{dr}^{\ s}) - \frac{L_m}{T_r}(\psi_{dr}^{\ s}i_{qs}^{\ s} - \psi_{qr}^{\ s}i_{ds}^{\ s})\right] \ \text{...... (3)}$$

The machine stationary frame (d^s-q^s) equations, shown in Figure 6.25, contain speed (ω_r) as a variable that can be solved from the known values of $v_{ds}^{\ s}$, $v_{qs}^{\ s}$, $i_{ds}^{\ s}$, and $i_{qs}^{\ s}$ of an operating machine. The simplified forms of equations that are actually solved in real time for speed estimation and the corresponding block diagrams are shown in this figure [20]. Equations (1) and (2) essentially relate to voltage model rotor flux vector estimation (see Figure 7.28), where $\sigma = 1 - L_m^2/L_r L_s$ and $S = d/dt$. These equations are derived from the stator equations and express the rotor fluxes in terms of stator voltages and currents. Equation (3) is derived from rotor circuits, where frequency ω_e is eliminated and expressed in terms of rotor fluxes and their derivatives. Equations (1) and (2) generate the rotor fluxes, which are then substituted in Eq. (3) to solve the speed. Obviously, the model is complex and highly parameter dependent. Therefore, the accuracy of estimation is expected to be poor, particularly at low speed.

FIGURE 7.59 Speed estimation by MRAC principle.

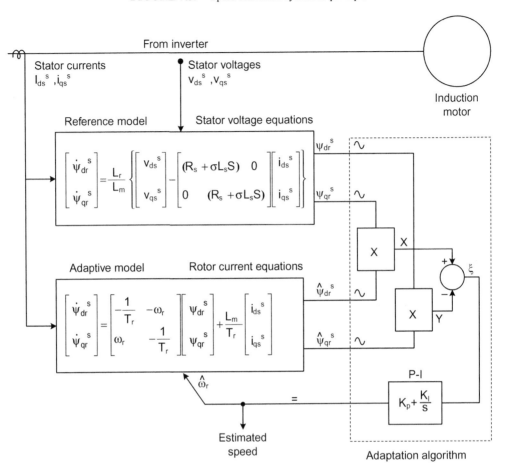

Speed estimation is possible by the principle of model referencing adaptive control (MRAC) [20], where the output of a reference model is compared with the output of an adaptive model. The adaptive model parameter is adjusted until the error between the two models vanishes to zero. This figure shows the voltage model stator equations of rotor fluxes as the reference model, and the adaptive model is the rotor current model equations of rotor fluxes. The reference model calculates $\psi_{qr}{}^s$ and $\psi_{dr}{}^s$ from the stator voltages and currents, and then compares them with the estimated values of $\hat{\psi}_{qr}{}^s$ and $\hat{\psi}_{dr}{}^s$ of the adaptive model. Note that the adaptive model parameter matrix contains speed ω_r as the unknown coefficient. This coefficient is tuned by an adaptation algorithm as shown until the output of the two models match, i.e., $\xi = 0$. The details of the adaptation algorithm are shown in the dashed box. The scheme has better accuracy than Figure 7.58. However, solving the reference model is difficult at low speed because of pure integration. There are, of course, parameter variation problems that reduce accuracy at low speed.

FIGURE 7.60   Speed estimation by speed-adaptive flux observer.

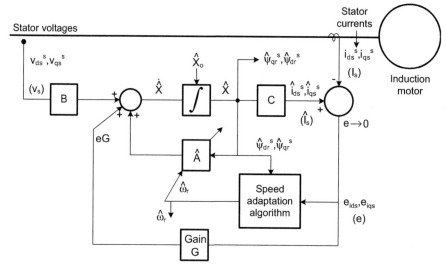

OBSERVER EQUATION : $\hat{X} = A\hat{X} + BV_s + GAIN(\hat{I}_s - I_s)$

where:

$$\hat{X} = \left[\, i_{qs}{}^s \, i_{ds}{}^s \, \psi_{qr}{}^s \, \psi_{dr}{}^s \,\right]^T$$

$$V_s = \left[\, v_{qs}{}^s \, v_{ds}{}^s \, 0 \, 0 \,\right]^T$$

$$I_s = \left[\, i_{qs}{}^s \, i_{ds}{}^s \, 0 \, 0 \,\right]^T$$

The speed estimation by an observer technique [21] is shown in this figure. An observer is basically an estimator that uses the plant model (full or partial) and a feedback loop with measured plant variables. In this case, the full plant d^s-q^s model is used with the state variables of rotor fluxes and stator currents. The model is solved in real time with machine terminal voltages ($v_{ds}{}^s$, $v_{qs}{}^s$), and the resulting current signals ($i_{ds}{}^s$, $i_{qs}{}^s$) are iso-lated by the output matrix C for comparison with the actual machine currents as shown in the figure. The speed signal ω_r is buried in the parameter matrix $\hat{A}$. If the speed signal is known and the machine parameters are correct, all the state variables $\hat{X}$ could be solved accurately. With an incorrect speed signal, there will be deviation between actual and estimated currents. The resulting error matrix e is multiplied by the gain matrix G and the signal eG is injected as shown so that $e \to 0$. The observer equation is shown below the figure. There is in addition the speed adaptation for matrix $\hat{A}$ following the *P-I* type algorithm as follows:

$$\hat{\omega}_r = K_P(e_{ids}\hat{\psi}_{qr}{}^s - e_{iqs}\hat{\psi}_{qr}{}^s) + K_I\int(e_{ids}\hat{\psi}_{qr}{}^s - e_{iqs}\hat{\psi}_{dr}{}^s)dt$$

where K_P and K_I are the gains. The speed estimation accuracy is improved by the observer. But the parameter variation effect remains dominant near zero speed. The scheme has been used in commercial drives.

FIGURE 7.61   Speed estimation by extended Kalman filter (EKF) principle.

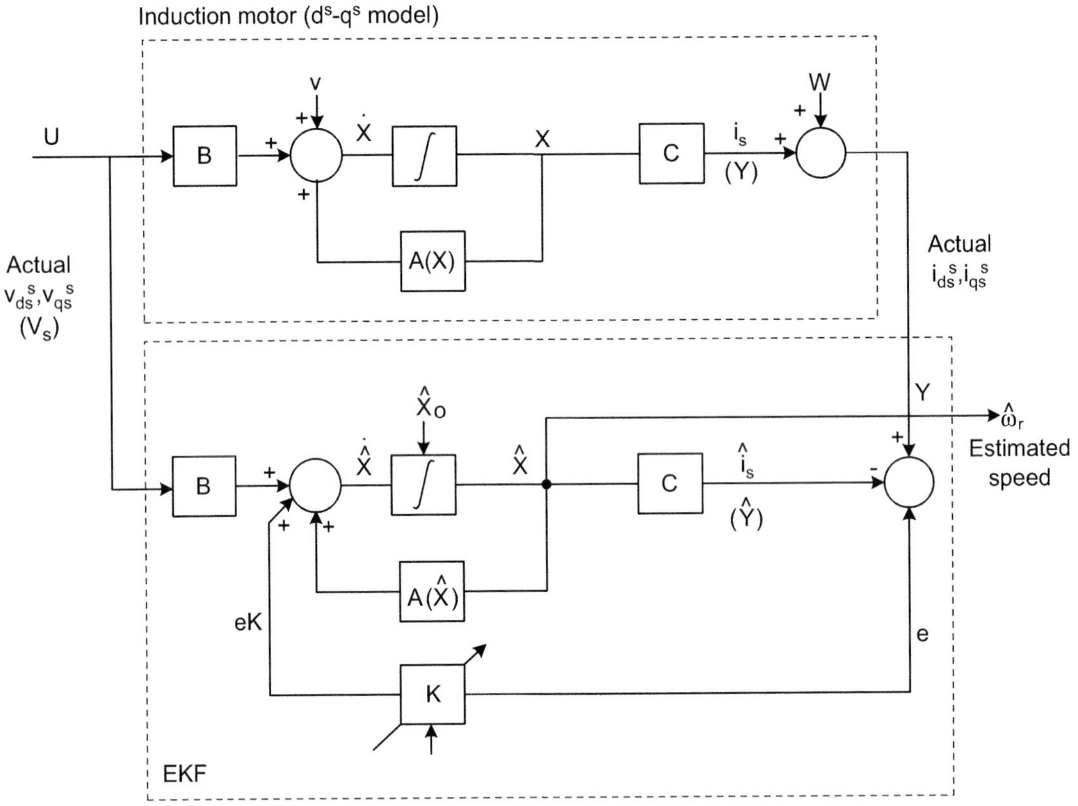

$$\frac{dX}{dt} = A(X)X + BV_s \dots\dots(1)$$

$$Y = CX \dots\dots\dots(2)$$

The extended Kalman filter (EKF) is basically a full-order stochastic (corrupted by noise) observer for recursive estimation of state in a nonlinear dynamic system. A deterministic (free from noise) observer was illustrated in Figure 7.60. The EKF algorithm is also useful for estimation of the machine parameters or parameters with the states. This figure shows the block diagram for EKF-based estimation of speed $\hat{\omega}_r$ [11, 22]. The upper part of the figure shows the block diagram of an actual machine d^s-q^s model, and the lower part shows the corresponding block diagram of EKF. The machine model equations are shown below the figure. Note that in the EKF algorithm, speed is defined as a state as well as a parameter in matrix $A(X)$. With slowly varying speed, or with fast DSP computation, this definition is justified. With speed as a constant parameter, the model is linear and of 5th order with the states $i_{ds}{}^s$, $i_{qs}{}^s$, $\psi_{dr}{}^s$, $\psi_{qr}{}^s$, and ω_r. The model equations for EKF are to be solved in discrete form with the help of DSP. The Gaussian noise vectors V and W, shown in the machine model, are for X and Y, respectively. The statistics of noise are given by three covariance matrices Q, R, and P (not shown in the figure), where they correspond to system noise vector, measurement noise vector, and system state vector, respectively. In the recursive computational flow diagram of the EKF, these covariance matrices should be taken into consideration for computation of Kalman gain K and $\hat{\omega}_r$ in every cycle. Obviously, the computation is very complex and time consuming, so a powerful DSP is essential. The accuracy tends to be poor, particularly at low speed.

FIGURE 7.62   Speed and position estimation block diagram by auxiliary signal injection method.

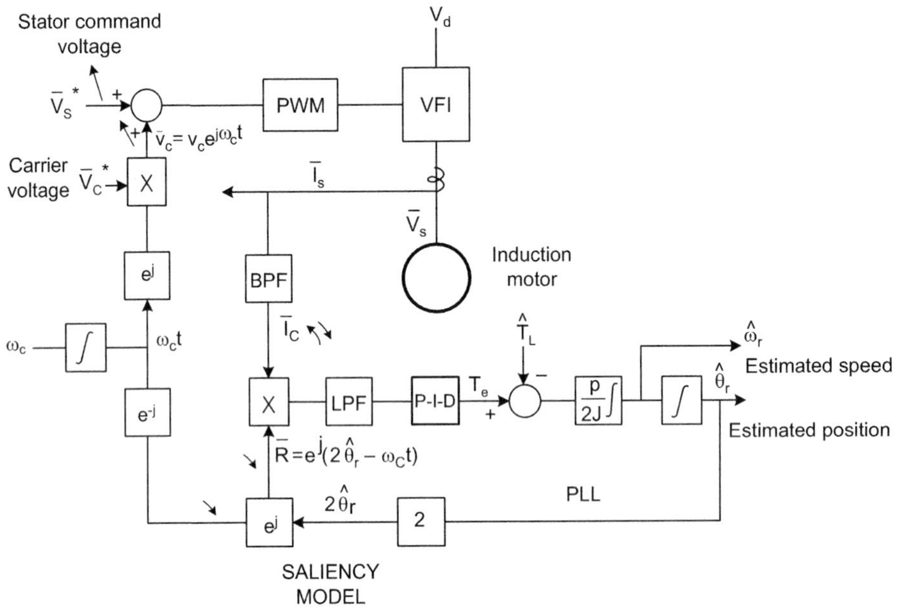

The speed estimation methods discussed so far fail at zero frequency, or dc condition when the rotor conditions become unobservable. However, it is possible to inject an auxiliary signal at carrier frequency from the stator side for a custom-designed anisotropic rotor and process the response for speed and position estimation. The anisotropic properties of the rotor can be created by periodic variation in the widths of the slot openings (causing magnetic saliency), periodic variation of the resistance of the outer conductors, depths of rotor bars, or by conductor heights of the slot openings. The block diagram for the estimation is shown in this figure [23]. The stator command voltage vector $\bar{V}_s$ is applied to the PWM controller of an inverter-fed machine as shown. A carrier frequency (typically 250-Hz) vector signal $\bar{V}_c = V_c e^{j\omega_c t}$ is generated and added to the stator voltage signal as shown. The analysis of a high-frequency machine model with magnetic saliency indicates that the resulting stator current vector $\bar{I}_s$ signal contains the fundamental component, harmonic ripple, and positive and negative sequence components of carrier frequency current. A band-pass filter (BPF) isolates the carrier frequency sequence currents $\bar{I}_c$ as indicated. The phase-locked-loop (PLL) creates the negative sequence reference vector $\bar{R} = e^{j(2\hat{\theta}_r - \omega_c t)}$ that rotates at the same frequency and phase of the negative sequence component of $\bar{I}_c$. The LPF of the PLL eliminates the oscillatory components created by the interaction of positive and negative sequence components. The LPF output flows through the P-I-D controller and then estimates the speed and position signals by solving the mechanical model with estimated load torque ($\hat{T}_L$). The factor 2 in the saliency model, as indicated, assumes that there are two saliencies per pole pair. The estimation method is obviously very complex and the possibility of a large error exists due to mechanical model parameters. In addition, having to custom design the rotor for speed estimation may not be acceptable by drive manufacturers.

FIGURE 7.63 Stator flux vector estimation by multistage programmable cascaded low-pass filters (PCLPFs).

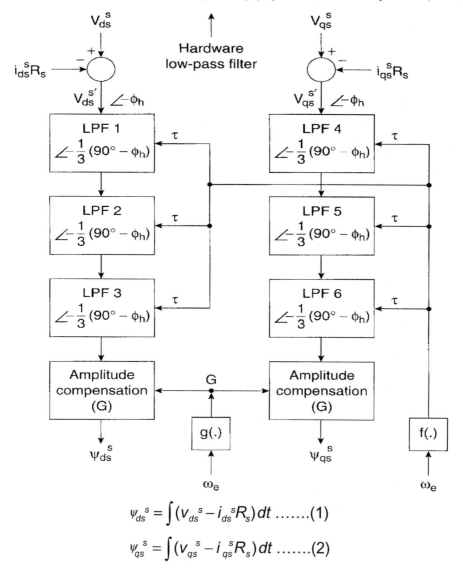

$$\psi_{ds}{}^s = \int (v_{ds}{}^s - i_{ds}{}^s R_s)\, dt \quad ....(1)$$

$$\psi_{qs}{}^s = \int (v_{qs}{}^s - i_{qs}{}^s R_s)\, dt \quad ....(2)$$

We will now discuss speed sensorless direct vector control with ψ_s orientation. It was mentioned earlier that direct vector control from zero speed requires a flux vector calculation from the current model that requires an ω_r signal, because the voltage model at low frequency has an integration problem. In fact, the voltage model can be extended to very low frequency (fraction of a hertz) if the ideal integrator is avoided and replaced by cascaded programmable (or adaptive) first-order low-pass filters (LPFs), as indicated in this figure [24]. Both the channels for computation of $\psi_{ds}^{\ s}$ and $\psi_{qs}^{\ s}$ are identical. The machine voltage and current signals are sensed, converted to d^s-q^s signals, and filtered and then the stator resistance drop is subtracted from each channel by using op amps, as indicated in the figure. The resulting voltage has lagging phase shift angle $-\varphi_h$, which increases with frequency. Each PCLPF channel is considered with three stages of LPFs, with each stage going through phase shift angle $-1/3(90° - \varphi_h)$ and some amplitude attenuation. To obtain the desired total 90° phase shift and amplitude compensation such that integration will be ideal, filter time constant τ and compensation gain G should be programmed with frequency as indicated in the figure. Note that the PCLPF compensates the varying phase shift of the hardware filter. If $\varphi_h = 0$, each LPF stage is required to give a 30° phase shift at all frequencies. The stator flux vector, calculated from the figure, can be used to calculate the unit vector and flux magnitude needed for vector control.

FIGURE 7.64   PCLPF time constant (τ) and gain (G) variation with frequency.

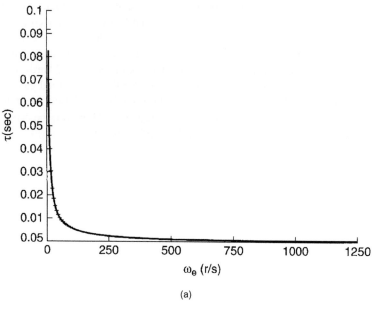

(a)

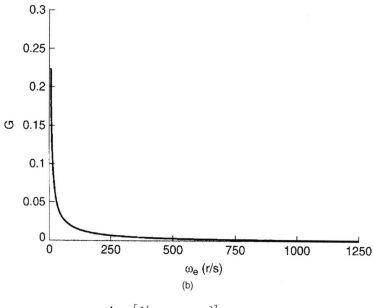

(b)

$$\tau = (\frac{1}{\omega_e})\tan\left[\frac{1}{n}\left\{\tan^{-1}(\tau_h\omega_e) + \frac{\pi}{2}\right\}\right] = f\ (.)\,\omega_e\ \ ......(1)$$

$$G = (\frac{1}{\omega_e})\sqrt{\left[1+(\tau\omega_e)^2\right]^n\left[1+(\tau_h\omega_e)^2\right]} = g(.)\,\omega_e\ \ ......(2)$$

The mathematical functions $f(\cdot)$ and $g(\cdot)$ for the parameters τ and G, respectively, from Figure 7.63 and their corresponding plots are shown in this figure, where n = number of PCLPF stages (three in this case), τ_h = hardware filter time constant (0.16 ms in this case), and ω_e = frequency. For an ordinary first-order LPF, τ is constant and, therefore, phase lag and attenuation increase with frequency. This means that for frequency-insensitive constant phase lag (with due compensation of hardware LPF lag angle), τ should decrease with ω_e as shown. Again, for the correct total gain requirement of the integrator, G also attenuates with ω_e. With the PCLPF-based integrator, vector control is valid down to a fraction of a hertz of frequency. However, the estimation is not valid at zero frequency, where both τ and G tend to be infinity.

FIGURE 7.65   Block diagram for stator flux vector estimation at start-up from current model without speed sensor.

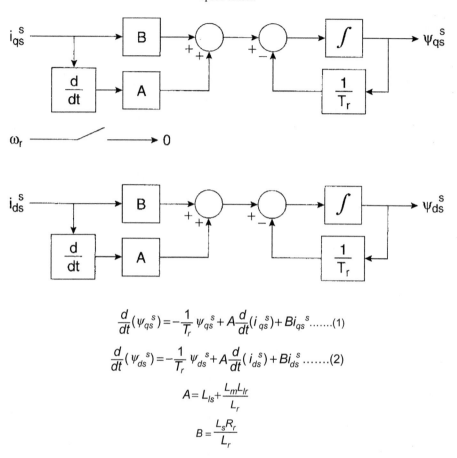

$$\frac{d}{dt}(\psi_{qs}^{s})=-\frac{1}{T_r}\psi_{qs}^{s}+A\frac{d}{dt}(i_{qs}^{s})+Bi_{qs}^{s}\ \dots\dots(1)$$

$$\frac{d}{dt}(\psi_{ds}^{s})=-\frac{1}{T_r}\psi_{ds}^{s}+A\frac{d}{dt}(i_{ds}^{s})+Bi_{ds}^{s}\ \dots\dots(2)$$

$$A=L_{ls}+\frac{L_m L_{lr}}{L_r}$$

$$B=\frac{L_s R_r}{L_r}$$

The stationary frame current model equations (or Blaschke equation) were given in Figure 7.29, where the rotor fluxes were expressed as a function of stator currents and speed. For stator flux-oriented vector control, these equations are expressed in terms of stator fluxes. They are further modified for the start-up condition by substituting $\omega_r = 0$. The resulting equations, Eqs. (1) and (2), and the corresponding block diagram are shown in the figure. Therefore, in the start-up condition for a direct vector-controlled drive, the flux vector can be estimated by this block diagram. Because the current signals are heavily filtered from harmonics, it is not difficult to solve the derivatives by DSP on an incremental basis. Note that this estimation is valid only at zero speed with finite torque (at a slip frequency), but it becomes invalid as the speed begins to develop. There is, of course, the usual parameter variation problem.

The figure shows the ψ_s-oriented direct vector control of a speed sensorless drive that is suitable for EV-type torque-controlled drive [24]. For closed-loop speed control, a feedback speed signal from an encoder is essential. The drive starts at zero speed in the Blaschke equation (BE) mode (see Figure 7.65). The usual "signal computation" is indicated in Figure 7.38. Initially, the rated dc flux is established at zero slip. The slip frequency, i.e., the torque, is developed when the torque command is applied. As the speed begins to develop, the control is transferred to the DVC (direct vector control) mode, where the stator flux vectors are computed from a PCLPF-based voltage model.

Note that PCLPF computation requires the ω_e signal, which is also dependent on the PCLPF output. The computation is done in a circulatory manner. The control is valid in constant torque as well as field-weakening regions. The control shown in the forward path consists of VR^{-1} of $(i_{ds}{}^s, i_{qs}{}^s)$ to (i_{ds}, i_{qs}), and synchronous current control. The unit vector $\cos\theta_e$, $\sin\theta_e$ is used for VR in the forward path, whereas for VR^{-1}, a modified unit vector $(\cos\theta_e, \sin\theta_e)$ is generated from the flux vector by phase compensation (advance) due to hardware filter time constant τ_h. The back-and-forth transition between the DVC mode and BE mode should be smooth [25]. The scheme is applicable with ψ_r orientation also.

FIGURE 7.67 Direct vector drive of EV with indirect vector control start-up.

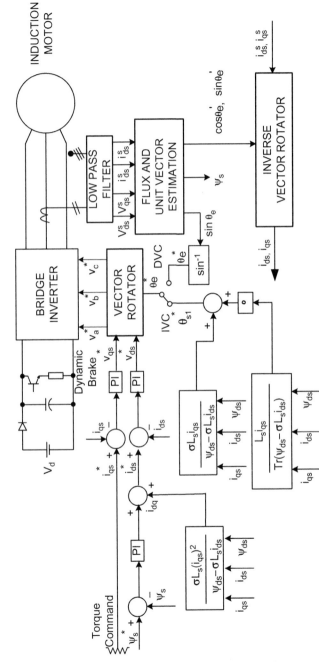

The figure shows a ψ_s-oriented vector-controlled speed sensorless drive in which both IVC and DVC have been hybridized [26]. The drive is torque controlled, and it starts at zero speed ($\omega_r = 0$) in the IVC mode as shown. The unit vector θ_e^* (in polar form) is generated as $\theta_e^* = \theta_{sl}^* \int \omega_{sl} dt$, where the slip frequency ω_{sl} is obtained from Eq. (3) of Figure 7.37. The usual coupling problem with ψ_s orientation exists, but is solved by injecting the compensating current i_{dq} in the flux loop. The drive uses synchronous current control in both modes. Vector control is not valid in the IVC mode at any finite speed. As the speed begins to develop, control is transitioned to the DVC mode, where the flux and unit vector estimation are based on the PCLPF-based estimator discussed earlier. The modified unit vector with compensation due to the hardware low-pass filter is used for converting i_{ds}^s, i_{qs}^s to i_{ds}, i_{qs} as shown. This computation as well as ψ_s synthesis are also valid in the IVC mode with reasonable accuracy when $\omega_e = \omega_{sl}$. Again, the transition between the IVC and DVC modes has to be smooth in either direction. (In a practical EV drive, the dynamic brake and series diode in dc link should be omitted.)

FIGURE 7.68 Advanced control techniques for induction motor drives.

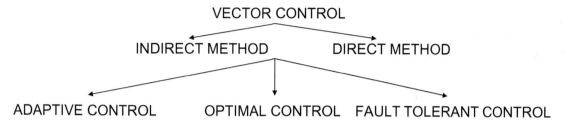

VECTOR CONTROL

INDIRECT METHOD   DIRECT METHOD

ADAPTIVE CONTROL   OPTIMAL CONTROL   FAULT TOLERANT CONTROL

- SELF-TUNING REGULATOR (STR)
- MODEL REFERENCING ADAPTIVE CONTROL (MRAC)
- SLIDING MODE OR VARIABLE STRUCTURE CONTROL (SMC or VSS)
- H-INFINITY CONTROL
- INTELLIGENT CONTROL
- EXPERT SYSTEM (ES)
- FUZZY LOGIC (FL)
- ARTIFICIAL NEURAL NETWORK (ANN)
- GENETIC ALGORITHM (GA)

A number of advanced control techniques for induction motor drive are summarized in this control tree. Most of the advanced controls use vector control in the inner loop because the drive model becomes simple and linear, like that of a separately excited dc motor drive. Adaptive control tends to adapt the drive control against plant model and parameter variations, and makes the system response robust against load disturbances. The goal is for the system response to follow the command profile irrespective of such variations. For example, in a self-tuning speed control system with variable inertia (J) load, J can be identified by the Kalman filter, and the loop gain can be tuned to compensate for the effect of this variation. Similarly, the rotor resistance (R_r) of the machine can be identified and compensated in the slip gain of an IVC system. The load torque (T_L) disturbance can be identified by a disturbance observer and feedforward compensation can be provided to counteract such effect. H-infinity control essentially operates to compensate the effects of such disturbances. In model referencing adaptive control (MRAC), a drive system, for example with variable J, can be forced to track the response of a reference model to eliminate the J variation effect. Intelligent control is based on artificial intelligence (AI) techniques. Sliding mode control will be discussed in this chapter, and intelligent control will be discussed in Chapters 10 and 11. Optimal control tends to optimize the desired objective, such as control response or system efficiency, within some constraints. Fault-tolerant control permits drive operation in a degraded mode under certain fault conditions of components instead of shutting down the system. Online fault diagnostics with a DSP are essential for such control.

FIGURE 7.69   (a) Variable structure control block diagram and (b) phase-plane portrait in negative feedback mode.

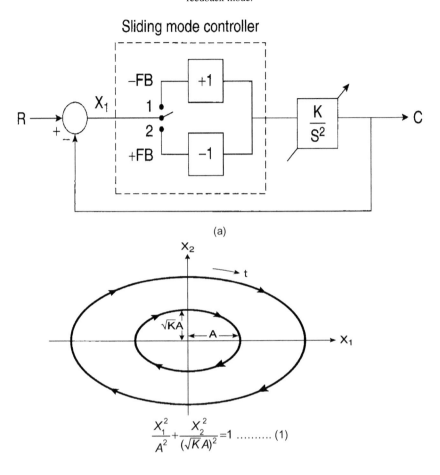

(a)

$$\frac{X_1^2}{A^2}+\frac{X_2^2}{(\sqrt{K}A)^2}=1 \ldots\ldots (1)$$

A sliding mode or variable structure control system (SMC or VSS) is basically an adaptive control that allows robust performance of the drive even with parameter variations and load disturbances. The control is nonlinear and can be applied to a linear or nonlinear plant. In SMC, the drive response is forced to track or "slide" along a predefined trajectory or "reference model" in a phase plane, but the algorithm is entirely different from MRAC. The upper figure shows the control block diagram of a second-order system with variable parameter K. The control switches between positive and negative feedback modes as shown. Apparently, in either mode, the system is unstable, but it can be made stable by a switching control algorithm. In –FB mode (position 1), the control satisfies a second-order differential equation, the solution of which is given by Eq. (1), which describes a family of ellipses in the phase plane X_2-X_1, where X_2 is the derivative of the loop error X_1. Arbitrary parameter A affects the size of the ellipse, whereas gain K changes its shape as indicated.

FIGURE 7.70   Phase plane portrait in positive feedback mode.

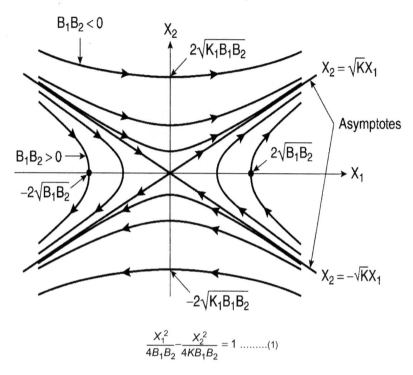

$$\frac{X_1^2}{4B_1B_2} - \frac{X_2^2}{4KB_1B_2} = 1 \ldots\ldots(1)$$

In the positive feedback mode (switch position 2) of Figure 7.69(a), the system is also described by a second-order differential equation, the solution of which is given by Eq. (1) in this figure, where the arbitrary constant B_1B_2 can be positive, negative, or zero. The equation describes a set of hyperbolas in the phase plane X_1-X_2 shown in this figure. The solution for $B_1B_2 = 0$ describes a pair of straight-line asymptotes given by the equations $X_2 = +\sqrt{K}X_1$ and $X_2 = -\sqrt{K}X_1$, which are shown in the figure. Obviously, the solution is also unstable in the positive feedback mode. If control is switched between the –FB and +FB modes, the control trajectory will be transitioned between the two phase plane portraits of Figures 7.69(b) and 7.70.

FIGURE 7.71   Sliding line control principle.

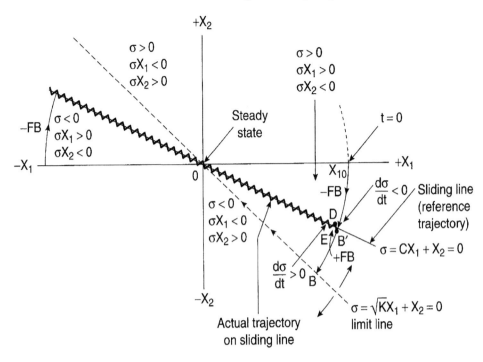

$$SLIDING\ LINE\ EQUATION:\quad \sigma = CX_1 + X_2 = 0\ \dots\dots(1)$$

$$EXISTENCE\ EQUATION:\ Lim\ \sigma\frac{d\sigma}{dt} < 0\ for\ \sigma \rightarrow 0\ \dots\dots(2)$$

The sliding line given by Eq. (1) on the phase plane X_1-X_2 is basically the reference trajectory that the SMC controller is required to follow. The negative asymptote equation given by $\sigma = \sqrt{K}X_1 + X_2 = 0$ is also included in the figure as a dashed line, where the slope $\sqrt{K} > C$. Note that the locus for $\sigma = 0$ falls on the line. Assume that, initially, the operation is at X_{10} in the –FB mode. When the elliptic trajectory crosses the sliding line and reaches B', operation switches to the +FB mode. Then at point D, it is switched back again to the –FB mode. By such back-and-forth transitions, the operation trajectory can be forced to follow the sliding line in the zigzag (chattering) path shown in the figure. Finally, steady state is reached at the origin ($X_1 = 0$ and $X_2 = 0$). In this trajectory, the system's response is guided by parameter C, but not by parameter K of the system as long as $C < \sqrt{K}$. Similarly, it can be shown that load disturbance does not affect the system. The polarity of the parameters σ, σX_1, and σX_2 that guide the switching algorithm is indicated in the figure. Evidently, as the trajectory tends to cross the sliding line from the top ($\sigma \rightarrow +0$), $d\sigma/dt < 0$. Similarly, as the trajectory is approached from the lower side ($\sigma \rightarrow -0$), $d\sigma/dt < 0$. Combining these two conditions, Eq. (2) is derived. The system parameters can vary widely. However, the existence condition must be satisfied for SMC control to be valid.

FIGURE 7.72 Block diagram for sliding mode trajectory control of an induction motor. (Courtesy of Pearson.)

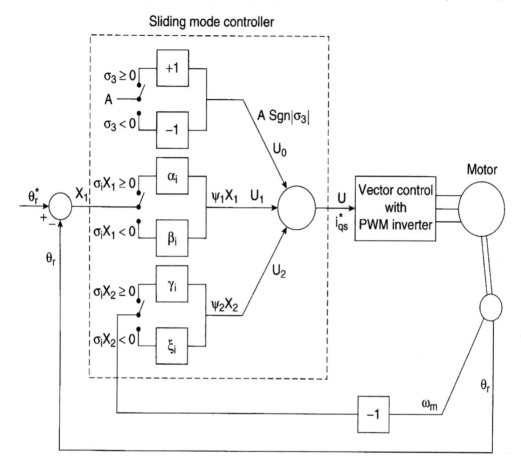

The figure shows the sliding mode position control servo for an induction motor [28] that is under vector control in the inner loop. The machine operates with rated flux, and the torque component of current (i_{qs}^{*}) comes out of the SMC. The complete sliding trajectory control is described in the next figure. The main position control loop generates the error signal X_1, which goes through SMC [represented by the single-pole, double-throw (SPDT) switch] to generate the output component U_1 as shown. The α_1 (negative) and β_1 are the general gain parameters of the main loop, and the switching criteria are indicated on the figure. The second loop is the derivative control loop that receives the input signal $X_2 = dX_1/dt = -\omega_m$ which is generated from the speed encoder signal ω_m. Although the derivative control loop is not essential, it can be shown [29] that it improves system response and permits wider plant parameter variations. The third loop, defined as the dither signal loop, receives a constant input parameter A. This loop generates an average signal (U_0) to support the load torque T_L so that steady-state error $X_1 = 0$. All the loop signals are added to generate the U or i_s^{*} signal for vector control. A DSP executes the control with small sample time.

FIGURE 7.73   Sliding mode trajectory in phase plane for an induction motor drive. (Courtesy of Pearson.)

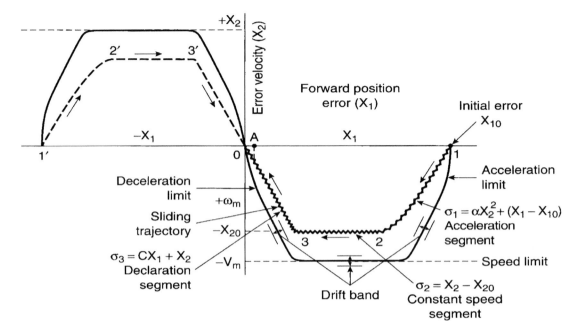

The sliding mode control of Figure 7.72 is executed during acceleration, constant speed, and deceleration modes as explained in this figure. The sliding trajectory in the phase plane X_1-X_2 in these modes is shown with their equations in the fourth quadrant (forward direction), but similar modes are also valid in the second quadrant for the reverse direction. The trajectory is defined such that optimal response is obtained, but at the same time saturation on the limit envelope is to be avoided. The limit envelope, which describes the fastest response, has a drift band due to plant parameter variations. For robustness of control, the reference trajectory should exclude the minimum limit envelope. Consider the initial positive position error X_{10} with a step position command. The control first tracks the acceleration segment, then the constant speed segment $(-X_{20})$, and finally the deceleration line before coming to steady state at the origin. The main and derivative control loops operate in all segments with different control parameters, but the dither signal is applied only in the deceleration segment so that the steady-state error is zero. The chattering in the phase plane is reflected in the time-domain response. This is one of the great disadvantages of sliding mode control. Of course, chattering can be minimized by using a higher PWM frequency with the inverter, shorter sampling time with SMC control, and shorter delay time for feedback signal computation.

FIGURE 7.74 Advances and trends in induction motor drives.

- VOLTAGE-FED CONVERTER CAGE MACHINE DRIVES ARE THE MOST COMMONLY USED INDUSTRIAL DRIVES TODAY – ALSO THE TREND FOR THE FUTURE
- FUTURE EMPHASIS ON CONVERTER AND CONTROLLER INTEGRATION WITH THE MACHINE ON THE LOWER END OF POWER – INTELLIGENT MACHINES
- OPEN-LOOP VOLTS/Hz CONTROL IS VERY POPULAR FOR GENERAL-PURPOSE INDUSTRIAL DRIVES, WHEREAS VECTOR CONTROL IS USED IN HIGH-PERFORMANCE DRIVES
- VECTOR CONTROL WILL BE UNIVERSALLY USED IN FUTURE INCREASING EMPHASIS ON VARIABLE-FREQUENCY SOFT STARTING OF CONSTANT-SPEED MOTORS
- INCREASING EMPHASIS ON SPEED SENSORLESS VECTOR AND SCALAR DRIVES – HOWEVER, PRECISION SPEED ESTIMATION, PARTICULARLY AT ZERO FREQUENCY, REMAINS A CHALLENGE
- INCREASING EMPHASIS ON ONLINE DRIVE FAULT DIAGNOSTICS AND FAULT-TOLERANT CONTROL TO IMPROVE SYSTEM RELIABILITY
- INTELLIGENT CONTROL AND ESTIMATION (DISCUSSED LATER) WITH ASIC CHIPS WILL FIND INCREASING ACCEPTANCE IN FUTURE

Some of the advances and future trends for induction motor drives are summarized here. The superiority of voltage-fed converter cage machine drives over current-fed converter and cycloconverter drive systems has been discussed. As the converter and controller size shrinks, increasingly intelligent machines with an integrated converter and controller will be available in the future. Vector-controlled drives will become the industry standard in the future because the additional software complexity of DSPs will be transparent to the user, and will only marginally increase drive costs compared to the added advantages. With the decreasing trend of cost, variable-frequency converters will be universally used in the front end of each machine even as a starter for constant-speed machines. In such systems, reduced flux light load efficiency improvements will provide an added benefit. Online drive diagnostics and fault-tolerant control will be increasingly emphasized in the future to improve system reliability even with the degraded mode of drive operation.

Summary

This chapter gives comprehensive coverage of induction motor drives, including different drive configurations, control topologies, estimation of signals, and speed sensorless control. The technology has gone through a dynamic evolution in the last two and half decades and is currently advancing with large momentum for the future. The literature in this area is very rich. Voltage-fed converter drives are most commonly used in industry in a wide power range that covers a few watts to multi-megawatts, although current-fed converter and cycloconverter drives are also used mainly in the high-power range. Cage-type machines are most common because of their simplicity, ruggedness, and cost effectiveness, but occasionally wound-rotor machines are also used in the high-power range with slip power recovery control. Doubly fed machines are expensive and have the disadvantages of slip rings and brushes. Static Kramer and static Scherbius drives have been used in industry for limited range speed control near the synchronous speed. Their converter costs are somewhat cheaper than that with cage machines, but the additional disadvantages are lagging line power factor, harmonic distortion of line current, nonreversible speed control, the need for separate starting, and, of course, a 60-Hz line transformer in the path of slip power flow. Scherbius drives with a cyclo-converter and two-sided PWM converter system can solve some of these problems. A special application of a Scherbius drive was mentioned for a 400-MW hydroelectric power plant variable-speed generator/pump storage system. Overall, slip power recovery drives will tend to become obsolete except in very special applications. Speed control of single-phase machines was discussed in Chapter 6. A variable-frequency single-phase motor drive is difficult to achieve because of its large pulsating torque. However, the variable-frequency two-phase machine drive principle is similar to a three-phase drive. Among the different scalar control techniques, open-loop volts/Hz control is simple, and such control is widely used in general industrial applications. However, the response of the drive is sluggish, can easily become unstable, and tends to be oscillatory without sufficient damping load torque. There are feedback control methods for solving these problems, but then the simplicity of the control is lost. DTC control is recently showing a lot of visibility in the family of scalar controls, but current applications are very limited. One large company is trying to promote the product vigorously. The control is somewhat simple and does not require any traditional PWM technique, but there is the problem of pulsating current and flux that introduces pulsating torque and additional harmonic loss. Many attempts have been made to overcome the demerits of DTC, which adds more control complexity and, hence, diminishes its appeal. The invention of vector or field-oriented control is a major milestone in the evolution of ac drives. It permits ac machines to be controlled like a separately excited dc motor. However, practical vector control is far from ideal. Because of the control complexity and dependence on machine parameters, the control is invariably based on a microcomputer or DSP. In fact, nowadays, all motor drives use DSP control. Among the indirect and direct vector control methods, the former is more commonly used, because the drive operates easily from zero speed to the high-speed field-weakening region with the help of a speed encoder. However, there is the problem that the slip gain (K_s) has to track with machine parameters, particularly the rotor resistance. It is extremely difficult to accurately estimate this parameter. Direct vector control requires flux vector estimation, which can be done by voltage model, current model, or PCLPF (programmable cascaded low-pass filter). Of course, flux observation in MRAC or a speed-adaptive flux observer can also be used. The voltage model flux estimation is simple but excludes the low-speed region. The rotor flux-oriented vector control gives more accurate decoupling control provided the parameters are estimated correctly. The stator flux-oriented vector control has some coupling problems (even with decoupling compensation), but its dependence on stator resistance variation can be compensated without much difficulty. Vector control generally requires a speed encoder, and currently, R&D is extensive in the literature for developing speed sensorless vector control. Various speed estimation techniques have been discussed by processing the machine terminal voltages and currents; these techniques include MRAC, speed adaptive flux observer, extended Kalman filter, and external carrier signal injection. However, machine parameter variation tries to blur the estimation accuracy, particularly near zero-frequency (zero speed and zero torque) operation. Carrier signal injection is the only viable estimation method for zero-frequency operation, but the computation complexity is enormous and the estimation accuracy is questionable. From speed estimation, zero-frequency speed estimation remains a challenge today. Fortunately, commercial vector drives are available without a speed encoder, but they exclude the very low speed region. The prime topics of research on induction motor drives today include intelligent control and estimation, on-line diagnostics, fault-tolerant control, accurate estimation of parameters, and states (particularly the speed) that cover the range down to zero frequency. Intelligent control that uses artificial intelligence techniques will be covered in Chapters 10 and 11. Fault-tolerant control remains an unexplored topic for the most part. It should be mentioned that many control and estimation concepts described here are also applicable to synchronous machine drives, which are described in the next chapter.

References

[1] B. K. Bose, "High performance control and estimation in ac drives," *IEEE IECON Conf. Rec.*, pp. 377–385, 1997.

[2] B. K. Bose, *Modern Power Electronics and AC Drives*, Prentice Hall, Upper Saddle River, NJ, 2002.

[3] H. Akagi, "The state-of-the-art of power electronics in Japan," *IEEE Trans. Power Electronics*, vol. 13, pp. 345–356, March 1998.

[4] B. K. Bose (Ed.), *Power Electronics and Variable Frequency Drives*, IEEE Press, New York, 1996.

[5] I. Takahashi and T. Noguchi, "A new quick response and high efficiency control strategy of an induction motor," *IEEE Trans. Ind. Appl.*, vol. 22, pp. 820–827, September/October 1986.

[6] G. Buja *et al.*, "Direct torque control of induction motor drives," *IEEE ISIE Conf. Rec.*, pp. TU2–TU8, 1997.

[7] S. Malik and D. Khunge, "ACS1000—world's first standard ac drive for medium voltage applications," *ABB Review*, pp. 1–11, February 1998.

[8] P. K. Steimer, J. K. Steinke, and H. E. Gruning, "A reliable, interface-friendly medium voltage drive based on the robust IGCT and DTC technologies," *IEEE IAS Annu. Meet. Conf. Rec.*, pp. 1505–1512, 1999.

[9] A. M. Khambadkone and J. Holtz, "Vector controlled induction motor drive with a self-commissioning scheme," *IEEE Trans. Ind. Elec.*, vol. 38, pp. 322–327, October 1991.

[10] R. Gabriel and W. Leonhard, "Microprocessor control of induction motor," *IEEE IAS Intl. Sem. Power Conv. Conf. Rec.*, pp. 385–396, 1982.

[11] P. Vas, *Sensorless Vector and Direct Torque Control,* Oxford University Press, New York, 1998.

[12] T. M. Rowan, R. J. Kerkman, and D. Leggate, "A simple on-line adaptation for indirect field orientation of an induction machine," *IEEE Trans. Ind. Appl.,* vol. 42, pp. 129–132, April 1995.

[13] G. C. D. Sousa, B. K. Bose, and K. S. Kim, "Fuzzy logic based on-line tuning of slip gain for an indirect vector controlled induction motor drive," *IEEE IECON Conf. Rec.,* pp. 1003–1008, 1993.

[14] X. Xu, R. De Doncker, and D. W. Novotny, "A stator flux oriented induction motor drive," *IEEE Power Elec. Spec. Conf.,* pp. 870–876, 1988.

[15] S. Mori *et al.,* "Commissioning of 400 MW adjustable speed pumped-storage system for Ohkawachi hydro power plant," *Proc. Cigre Symp.,* No. 520-04, 1995.

[16] T. Ohmae and K. Nakamura, "Hitachi's role in the area of power electronics for transportation," *IEEE-IECON Conf. Rec.,* pp. 714–718, 1993.

[17] B. K. Bose, "Recent advances and trends in power electronics and drives," *Proc. IEEE Nordic Workshop on Power Electronics and Industrial Electronics,* pp. 170–181, 1998.

[18] M. G. Simoes, B. K. Bose, and R. J. Spiegel, "Design and performance evaluation of a fuzzy logic based variable speed wind generation system," *IEEE Trans. Ind. Appl.,* vol. 33, pp. 956–965, July/August 1997.

[19] K. Rajashekara, A. Kawamura, and K. Matsuse, *Sensorless Control of AC Motor Drive,* IEEE Press, New York, 1996.

[20] C. Schauder, "Adaptive speed identification for vector control of induction motors without rotational transducers," *IEEE Trans. Ind. Appl.,* vol. 28, pp. 1054–1061, September/October 1992.

[21] H. Kubota, K. Matsuse, and T. Nakano, "DSP-based speed adaptive flux observer of induction motor," *IEEE Trans. Ind. Appl.,* vol. 29, pp. 344–348, March/April 1993.

[22] Y. R. Kim, S. K. Sul, and M. H. Park, "Speed sensorless vector control of induction motor using extended Kalman filter," *IEEE Trans. Ind. Appl.,* vol. 30, pp. 1225–1233, September/October 1994.

[23] J. Holtz, "Sensorless position control of induction motor—an emerging technology," *IEEE Trans. Indus. Elec.,* vol. 45, pp. 840–852, December 1998.

[24] B. K. Bose and N. R. Patel, "A sensorless stator flux oriented vector controlled induction motor drive with neuro-fuzzy based performance enhancement," *IEEE IAS Annu. Meet. Conf. Rec.,* pp. 393–400, 1997.

[25] T. W. Chun, M. K. Choi, and B. K. Bose, "A novel start-up scheme of stator flux oriented vector controlled induction motor drive without torque jerk," *IEEE Trans. Ind. Appl.,* vol. 39, pp. 776–782, May/June 2003.

[26] B. K. Bose and M. G. Simoes, "Speed sensorless hybrid vector controlled induction motor drive," *IEEE IAS Annu. Meet. Conf. Rec.,* pp. 137–143, 1995.

[27] Y. D. Landau, *Adaptive Control—The Model Referencing Approach,* Marcel Dekker, New York, 1979.

[28] B. K. Bose, "Sliding mode control of induction motor," *IEEE IAS Annu. Meet. Conf. Rec.,* pp. 479–486, 1985.

[29] U. Itkis, *Control Systems of Variable Structure,* Wiley, New York, 1976.

[30] B. K. Bose, "Energy, environment, and advances in power electronics," *IEEE Trans. Power Electronics,* vol. 15, pp. 688–701, July 2000.

[31] G. S. Buja and M. P. Kazmierkowski, "Direct torque control of PWM inverter-fed ac motors—a survey," *IEEE IE Trans.,* vol. 51, pp. 744–757, August 2004.

[32] R. E. Wilson and P. B. S. Lissaman, "Applied aerodynamics of wind power machines," Springfield, Va, 1974.

CHAPTER 8

Synchronous Motor Drives

FIGURE 8.1   Features of synchronous motor drives.

- DRIVES ARE MORE EFFICIENT – BUT MORE EXPENSIVE
- HIGHER POWER DENSITY AND LOWER ROTOR INERTIA, PARTICULARLY WITH HIGH-ENERGY PERMANENT MAGNET
- TRUE SPEED TRACKING IN PARALLEL MULTIPLE-MACHINE DRIVE
- POWER FACTOR CAN BE PROGRAMMABLE – LEADING, LAGGING, OR UNITY WITH FIELD CONTROL
- CONVERTER WILL BE LESS EXPENSIVE WITH UNITY POWER FACTOR
- EXCITATION CONTROL – NEEDS SEPARATE CONVERTER
- ABSOLUTE POSITION SENSOR IS MANDATORY IN ANY CLOSED-LOOP CONTROL
- CAN REPRESENT TRUE BRUSHLESS COMMUTATORLESS DC MOTOR (BLDM)
- CONVERTER FAULT CAN CAUSE DANGEROUS PULSATING TORQUE DUE TO MACHINE CEMF

Although synchronous motors, in general, are more expensive, synchronous machines have higher efficiency (with less cooling needs), particularly PM machines, which can make the life-cycle cost of the drives less depending on electricity costs. Again, the converter rating is lower with a near-unity power factor. With the present trend toward decreasing costs for NdFeB magnets, the cost of PM machines is expected to decrease substantially in the future. With the leading power factor control of WFSMs, a low-cost thyristor load-commutated inverter is even more economical than a traditional PWM inverter. Of course, an additional converter is needed for field current control of WFSM drives. For true speed tracking in multiple-motor textile mill drives, for example, PM synchronous motors (or SyRM) are essential. Another disadvantage is that an absolute position sensor is essential in most of the drives. Both induction (cage) and PM synchronous machines are brushless (also commutatorless). With vector control, the drives behave like separately excited dc motor drives, but the trapezoidal PM machines with their position sensor-controlled inverters are the most similar to separately excited dc motors (often called brushless dc motors or BLDMs). There is one potentially serious problem with synchronous machine drives. If there is a converter fault, a large nonsinusoidal fault current in the stator due to excitation CEMF can cause dangerous pulsating torque that can also cause a mechanical resonance problem. For WFSMs, the excitation should be deenergized quickly by tripping the field circuit and inserting a load resistance.

FIGURE 8.2 Principal classes of synchronous motor drives.

- VOLTAGE-FED PWM INVERTER DRIVES
    PM SINUSOIDAL MACHINES (PMSM)
    PM TRAPEZOIDAL MACHINES (BLDM)
    WOUND-FIELD SYNCHRONOUS MACHINES (WFSM)
    SYNCHRONOUS RELUCTANCE MACHINES (SyRM)
    SWITCHED RELUCTANCE MACHINE (SRM)
    STEPPER MOTORS
- CURRENT-FED SIX-STEP INVERTER DRIVES WITH LOAD COMMUTATION
    WFSM
- PHASE-CONTROLLED CYCLOCONVERTER DRIVES
    WFSM

The general classification of synchronous motor drives is similar to the classification of induction motor drives. However, there are some differences because of the features of this type of machine, as discussed in the previous figure. The most commonly used are the voltage-fed PWM inverter drives, which require a variable-frequency, variable-voltage power supply at the machine terminal. This class of drives is available in power ranges from FHP to multi-megawatts. Permanent magnet machines have constant field excitation unlike that of WFSMs. However, excitation can be controlled from the stator side by the inverter. The method of control is somewhat different for different types of machines. Switched reluctance motors are not really synchronous machines. However, they are included here for convenience. Both PWM and square-wave current-fed inverters are possible, but the former is hardly used. Large WFSMs are most commonly used with load-commutated thyristor inverters because the machines can operate at leading power factor with controllable excitation. In fact, multistepped (more than six steps) two-sided converter systems are possibly the largest drives in use today. In this case, the line-side multistepping is done by multiwinding transformers, but the machine side often uses multiphase winding instead of a separate transformer (discussed in Chapter 5). High-power cycloconverters and LCI drives have been used in ship propulsion. The recent trend is to replace cycloconverters with two-sided voltage-fed multilevel PWM converters.

FIGURE 8.3   Open-loop volts/Hz speed control of synchronous motors.

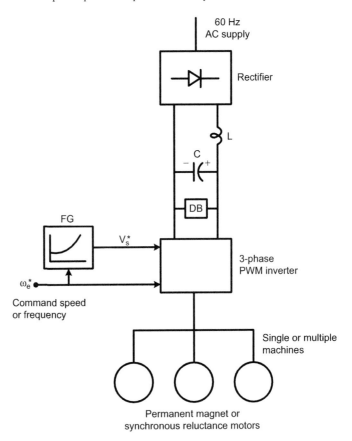

Essentially, two control techniques are used in synchronous motor drives. One is the simple open-loop volts/Hz control shown in this figure, and the other is self-control mode, which will be described later. In volts/Hz control, the motor frequency is controlled independently and the machine speed follows the frequency (zero slip). This is extremely important for parallel machines for which precision speed tracking is essential, such as in fiber-spinning mills. Either PM machines or synchronous reluctance machines can be used. The machine is usually provided with damper winding to prevent a hunting or oscillatory response. The control method is similar to that of an induction motor drive with volts/Hz control. Single or three-phase line power is rectified to dc, filtered by an L-C or C filter, and then converted to a variable-frequency, variable-voltage supply by the inverter. The frequency ω_e^* is commanded and the voltage V_s^* is slaved with it proportionately (with initial offset) so that the stator flux remains constant. A dynamic brake is shown in the dc link that absorbs the braking power. Instead of a voltage-fed inverter, a cycloconverter has also been used in a large single-machine drive. It is the simplest control scheme for synchronous motor drives.

FIGURE 8.4   Open-loop volts/Hz speed control characteristics.

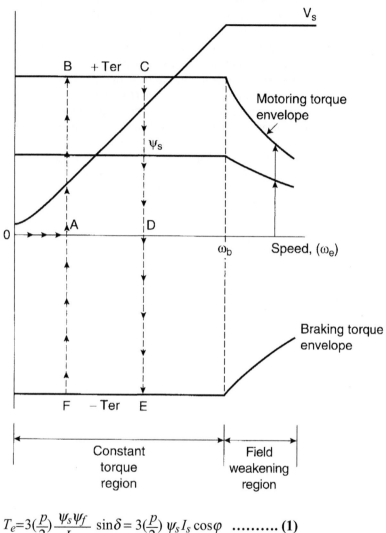

$$T_e = 3\left(\frac{p}{2}\right)\frac{\psi_s \psi_f}{L_s}\ \sin\delta = 3\left(\frac{p}{2}\right)\psi_s I_s \cos\varphi \ \ \cdots\cdots\cdots (1)$$

$$J\left(\frac{2}{p}\right)\frac{d\omega_e}{dt} = T_e - T_L \ \ \cdots\cdots\cdots (2)$$

The performance of a volts/Hz controlled drive is explained here on the torque-speed plane, and the corresponding phasor diagram is shown in the next figure. The developed torque expression for a nonsalient pole machine is shown in Eq. (1), and acceleration/deceleration capability is shown in Eq. (2). Assuming load torque $T_L = 0$, the machine can be easily started from zero speed at point O to point A by slowly increasing the ω_e^* command. At point A, T_L is gradually increased. Since $T_e = T_L$, the operating point will move vertically until the rated torque is reached at B. With increasing torque, the torque angle δ and the corresponding in-phase stator current ($I_s \cos\varphi$) increases. The rated torque T_{er} is normally due to the rated stator current and is below the limit angle $\delta = 90°$. The operating point can be changed from B to C by slowly increasing the ω_e^* command. Then, it can be brought back to D by gradually decreasing T_L. At base speed ω_b, the voltage V_s saturates, and then it enters the field-weakening mode, where the decrease of stator flux decreases the developed torque as shown. Any sudden change in ω_e^* will make the system unstable like an induction motor. The motor can be accelerated from A to D within the $d\omega_e/dt$ limit of $(p/2J)$. T_c at $T_L = 0$. At any T_L, or higher J, the acceleration limit will be lower. The deceleration from D to A within the $-T_{er}$ limit will follow a similar principle. However, with T_L, the deceleration will be faster.

FIGURE 8.5   Phasor diagram of a machine in (a) motoring mode or (b) regeneration mode.

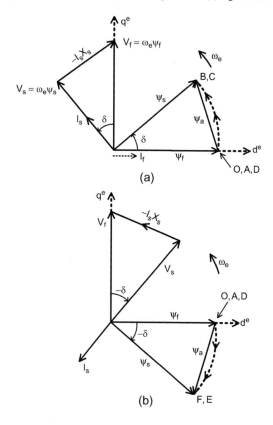

The phasor diagrams of the machine corresponding to Figures 8.3 and 8.4 are shown here assuming that the power factor is always unity. These diagrams are essentially the same as Figure 6.39, except the stator resistance R_s is neglected for simplicity. In a PM machine, the field flux ψ_f is constant and is taken as the reference phasor, and the corresponding fictitious field current I_f is shown as a dashed line. At operating point O in Figure 8.4, $\omega_e = 0$ and, correspondingly, $\psi_f = \psi_s$ and $V_f = V_s = 0$ in ideal conditions (no offset voltage). As the operating point changes to A, and then to D at zero torque, the flux ψ_s remains constant, but voltage V_f in the phasor diagram increases in the vertical direction proportional to the frequency. Both the I_s and δ angles remain at zero because torque is zero. Consider now the rise of the operating point from A to B in Figure 8.4. The magnitudes of flux ψ_s and the corresponding voltage V_s will remain constant, but will rotate counterclockwise to increase the δ and I_s magnitudes to satisfy the torque relation. The synchronous reactance drop $I_s X_s$ will appear in the voltage phasor diagram as shown. The excursion from the B to C operating points will not alter the flux phasor diagram, but the voltage phasors will increase proportional to frequency in the same direction. In regeneration mode, the operating principle is similar, but δ angle is negative and I_s is in opposite phase with V_s.

FIGURE 8.6   (a) DC motor and (b) self-controlled synchronous motor analogy.

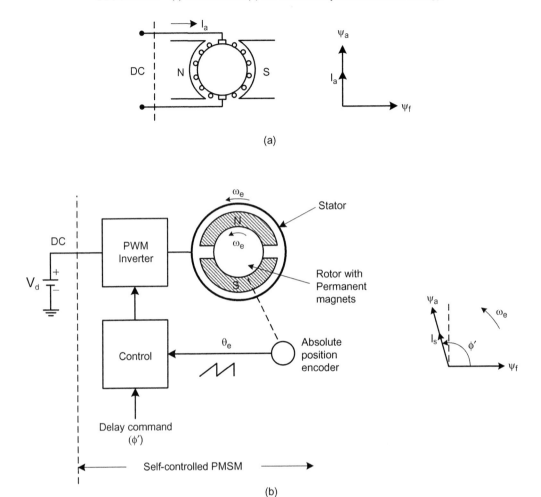

(a)

(b)

The principle of self-control in a synchronous machine and its analogy with a PM (or separately excited) dc machine is explained here. In fact, an induction motor vector drive is also a self-controlled machine, and its analogy with a dc machine was explained in Chapter 7. Consider the sinusoidal SPM machine in (b), where the voltage-fed PWM inverter generates variable-frequency, variable-voltage current-controlled power to the three-phase stator winding of the machine. The machine is self-controlled in the sense that a shaft-mounted absolute position sensor is generating the basic control signal of the inverter. The flux linkage phasor diagram is shown on the right; compare this with the dc motor phasor diagram. The sinusoidal ψ_f wave by PM is shown as the reference phasor, and it is aligned with the sawtooth wave θ_e so that the inverter frequency and phase are locked to the position sensor signal. Of course, a delay angle command in the controller can generate the adjustable φ' angle of armature MMF ψ_a phasor as shown. In a dc machine, this angle is fixed at 90° as shown. However, the main differences are that in a dc machine, these phasors are stationary, whereas in PMSM the phasor diagram rotates at synchronous speed ω_e. The angle φ' is adjustable, and the rotor and stator of the dc machine are exchanged (inside-out machine). Often such a machine is defined as a brushless dc machine (BLDM or BLDC).

FIGURE 8.7   Basic features of self-controlled synchronous machine.

- THE INVERTER, CONTROLLER, AND ABSOLUTE POSITION ENCODER ACT AS AN ELECTRONIC COMMUTATOR (EC)
- ELECTRONIC COMMUTATOR REPLACES THE MECHANICAL COMMUTATORS AND BRUSHES (MECHANICAL INVERTER) OF TRADITIONAL DC MACHINES
- THE FLUX PHASOR DIAGRAM ROTATES AT SYNCHRONOUS SPEED
- CONTROL CAN MODIFY THE ANGLE BETWEEN THE FLUX PHASORS
- BECAUSE OF SELF-CONTROL, MACHINE DOES NOT EXHIBIT THE STABILITY OR HUNTING PROBLEMS OF TRADITIONAL SYNCHRONOUS MACHINES
- THE TRANSIENT RESPONSE IS FAST – SIMILAR TO DC MACHINES
- ROTOR INERTIA IS SMALLER THAN DC MACHINES WITH HIGH-ENERGY MAGNET

Self-controlled synchronous machines have a number of advantages for which they are so popularly used in industry. Basically, they emulate dc motors, maintaining their stability and fast response characteristics. At the same time, the disadvantages that arise from their commutators and brushes—such as maintenance and reliability problems, sparking, limitations of speed and power ratings, limitation of transient current capability, difficult operation in corrosive and explosive environments, limitation of altitude, and EMI problems—are eliminated. Lower rotor inertia is a special advantage in servo-type drives. Self-controlled machines are often defined as inside-out machines, electronically commutated motors (ECMs), or brushless dc motors (BLDMs or BLDCs). A BLDM can be more expensive than a dc motor because of the extra cost of the inverter and position encoder, although the machine itself is somewhat cheaper.

FIGURE 8.8 Absolute position encoder: (a) binary-coded disk and (b) digital processing circuit.

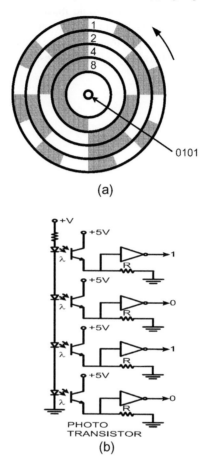

(a)

(b)

As mentioned earlier, an absolute position encoder is mandatory for a self-controlled synchronous machine drive. Sinusoidal machines require a continuous position signal at high resolution, whereas trapezoidal machines and WFSM machines with LCI require only discrete signals at certain angular intervals. Both digital- and analog-type sensors are available. This figure shows a simple digital binary-coded disk and its corresponding electronic circuit. The disk shown has four concentric rings, and each ring is binary coded, i.e., the shaded area transmits a light beam, whereas the unshaded area obstructs it. An LED generates the beam and a phototransistor detects the position by sensing the beam. The disk position shown generates the digital output 0101, i.e., the decimal count of 5. The four-ring encoder has position resolution of $2^4 = 16$, i.e., the mechanical angle resolution is $360/16 = 22.5°$. This means that the electrical angle resolution for a four-pole machine is 45°. A practical disk with 14 rings will give 14-bit position resolution, i.e., 0.04° electrical for a four-pole machine. In binary-coded disk, all the positions can change simultaneously. For this reason, a gray-coded disk is often used where only one bit change occurs per transition. The binary or gray code can easily be converted to BCD or ASCII code by a microprocessor.

FIGURE 8.9 Digital slotted disk encoder with waveforms (four-pole machine).

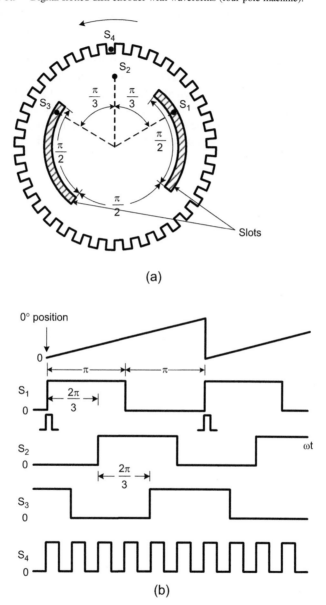

(a)

(b)

A slotted-disk-type absolute position encoder designed for a four-pole machine is shown here. The disk has four optical-type sensors as shown. Sensor S_4 is at the perimeter, where there are a large number of slots. Sensors S_1, S_2, and S_3 are in the inner radius with 60°

angular spacing as shown in the figure. There are two large slots covering a 90° angle that are located in diametrically opposite positions. The electrical signals generated from the phototransistors of all sensors are shown in (b). Basically, S_1, S_2, and S_3 generate square waves with 120° phase displacement, and S_4 generates pulses at high frequency. The encoder will be used for an LCI-WFSM drive to be described later. To understand absolute rotor position detection, consider that only S_1 and S_4 sensors are present, and the S_1 leading edge is aligned to the zero position of the rotor. A pulse at the leading edge of the S_1 wave can reset and trigger an "up" counter, which counts the pulses generated by S_4. The counter resets at 360° by another pulse of S_1. If the number of slots on the perimeter is 360, the mechanical angle resolution is 1°, i.e., 2° electrical for a four-pole machine.

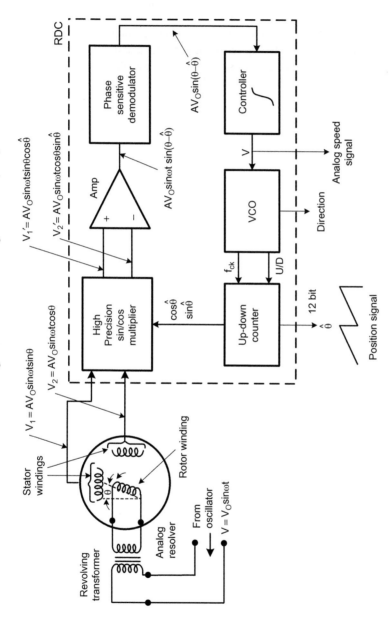

FIGURE 8.10 Analog resolver with resolver-to-digital converter (RDC).

A mechanical position encoder that is more robust and environmental contamination insensitive than an optical encoder is shown in this figure. The encoder has two parts: an analog resolver and a resolver-to-digital converter (RDC). The resolver is basically a two-phase synchro that is excited by the rotor winding, which is supplied by a carrier wave of several kilohertz frequency as indicated. Angular shaft position θ determines the stator winding outputs by amplitude-modulated carrier signals. The resolver is brushless because its rotor winding is excited by a revolving transformer whose primary receives the oscillator signal. The analog position-embedded signals in the resolver output are converted to digital signals by the RDC. The RDC is basically a closed-loop position tracking servo system. The high precision sin/cos multiplier multiplies the resolver outputs with the estimated $\cos\hat{\theta}$ and $\sin\hat{\theta}$ signals fabricated from the up/down counter output. These signals are then processed through an error amplifier, phase-sensitive demodulator, and integrating-type controller to generate the bipolar analog speed signal V. The polarity of V gives the direction signal. This signal is then converted to a position signal through a VCO and up/down counter as shown. The "up" counting indicates a positive direction of rotation, whereas "down" counting indicates reverse rotation. The encoder gives precision output, but is expensive compared to an optical encoder.

FIGURE 8.11   (a) Vector control of sinusoidal SPM machine and (b) phasor diagram.

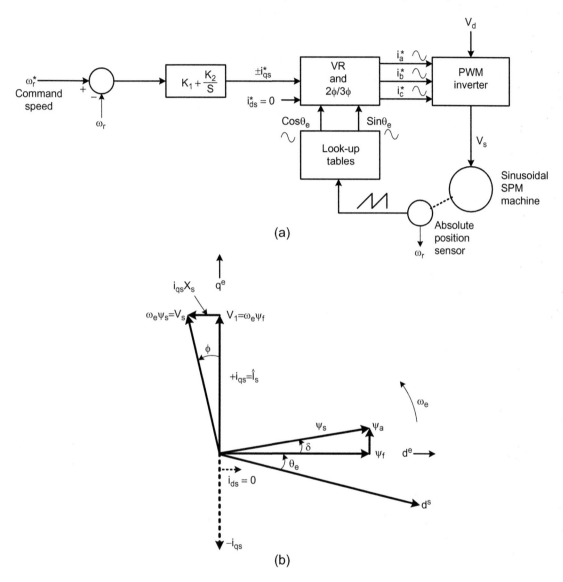

(a)

(b)

Let us now consider vector control of a sinusoidal SPM machine using the absolute position encoder just described. The phasor diagram for the control is shown in (b), neglecting the small stator resistance (R_s). The machine is nonsalient pole with a large effective airgap that makes the armature reaction flux $\psi_a = L_s I_s$ small, i.e., $\psi_s \approx \psi_m \approx \psi_f$. For best torque sensitivity with stator current, we make $i_{ds} = 0$ and $\hat{I}_s = i_{qs}$, where $\hat{I}_s$ represents the magnitude of the current vector. In vector control, i_{ds} is oriented in the direction of magnet flux ψ_f, whereas i_{qs} is oriented in the direction of induced voltage V_f. Stator voltage V_s slightly leads the V_f phasor because of the small reactance drop. In the vector control diagram, the speed control loop generates the $\pm i_{qs}{}^*$ signal, which can produce motoring or regenerating torque. The unit vector signal is generated directly from the position sensor because of its alignment with flux ψ_f. The unit vector in the vector rotator (VR) gives the desired alignment of the commanded current signals. Note the analogy with the induction motor vector control shown in Figure 7.32. In this figure, the slip (ω_{sl}) signal and magnetizing current (i_{ds}) are zero, and the unit vector is locked with the magnet pole position. The control is valid in the constant torque region.

FIGURE 8.12 SPM machine speed control in field-weakening region: (a) phasor diagram and (b) torque-speed curve.

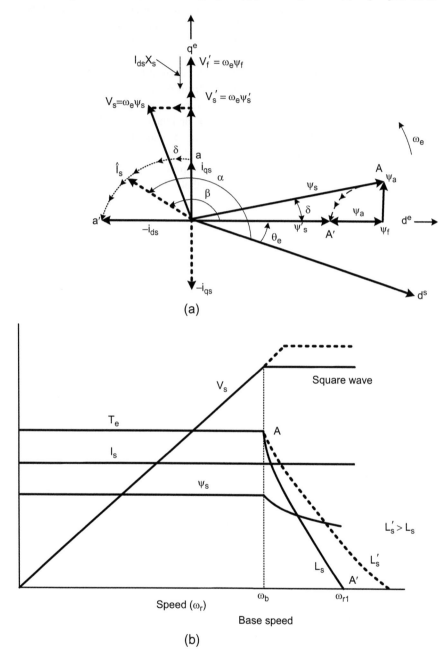

(a)

(b)

The speed of a sinusoidal SPM machine can be controlled beyond the constant torque region as explained in this figure. However, this range is small because of low synchronous inductance (L_s). At the base speed (ω_b) of the constant torque region, supply voltage V_s will saturate. Because $V_s = \omega_e \psi_s$, ψ_s is to be weakened by demagnetizing current $-i_{ds}$ so that the stator current is controllable at higher field-weakening speed. In the phasor diagram, if the rated stator current $\hat{I}_s$ is rotated counterclockwise from a to a', the constant ψ_a phasor will rotate from A to A' injecting more demagnetizing current, and correspondingly decreasing the torque component of current at higher speed. At point A', the machine speed will rise to ω_{r1} and the torque will fall to zero. With torque angle $\delta = 0$, both ψ_s and ψ_f phasors will be in the same phase, and the corresponding voltages will also be cophasal. It is also obvious that with constant ψ_f, V_f will increase proportionally with ω_e, and the overexcited machine will give leading DPF at the machine terminal. The torque-speed curve shows that higher L_s will increase the field-weakening region.

FIGURE 8.13   Vector control of sinusoidal SPM machine covering field-weakening region.

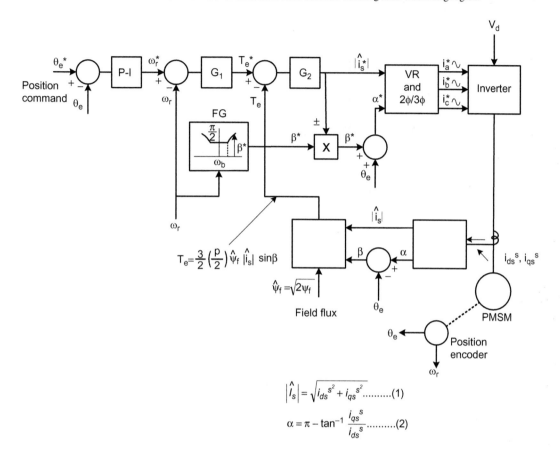

$$\left|\hat{I}_s\right| = \sqrt{i_{ds}^{s^2} + i_{qs}^{s^2}} \quad\dots\dots(1)$$

$$\alpha = \pi - \tan^{-1} \frac{i_{qs}^{s}}{i_{ds}^{s}} \quad\dots\dots(2)$$

The figure shows a block diagram for a PMSM vector drive with ψ_f orientation as before, and covers the high-speed field-weakening region. The operation principle follows the phasor diagram of Figure 8.12. The position loop feeds the speed loop, and the speed loop feeds the torque loop as usual. The torque loop generates the $\hat{I}_s$ command, the magnitude of which and the α^* angle are fed to the VR to generate the phase current commands. The θ_e angle from the position sensor and β^* from a speed-sensitive function generator (FG) are added to generate the α^* signal. The FG indicates that in either direction of rotation, $\beta = 90°$ in the constant-torque region, but can approach $180°$ in the field-weakening region. The sign of the stator current determines the sign of β^* (positive for motoring, but negative for regeneration) as indicated in the figure. The feedback signal synthesis equations from machine currents are shown below the figure and the corresponding feedback torque equation is indicated in the figure.

FIGURE 8.14   Phasor diagram of synchronous reluctance machine.

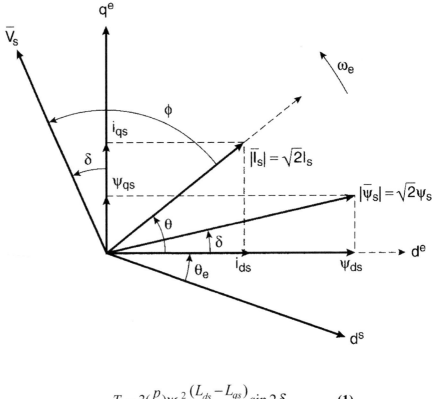

$$T_e = 3(\frac{p}{2})\psi_s{}^2 \frac{(L_{ds} - L_{qs})}{2L_{ds}L_{qs}} \sin 2\delta \ldots\ldots\ldots (1)$$

$$= \frac{3}{2}(\frac{p}{2})\frac{(L_{ds} - L_{qs})}{L_{ds}L_{qs}} \psi_{ds}\psi_{qs} \ldots\ldots\ldots (2)$$

The phasor diagram (with vectors) for a synchronous reluctance motor and the torque equations are shown in the figure, where stator resistance (R_s) is neglected. Basically, an SyRM is a salient pole ($L_{ds} \gg L_{qs}$) machine with no PM or field winding on the rotor (i.e., $\psi_f = 0$). The torque/unit volume, power factor, and efficiency are higher for a machine designed with a high saliency ratio. The open-loop volts/Hz control in Figure 8.3 can use SyRM, as indicated. A damper or cage winding is usually needed in the rotor to prevent oscillatory performance. However, with self-control no damper winding is needed. The SyRM can also be used as a line-start 60-Hz motor with cage winding like a PMSM. The developed torque due to saliency is given by Eq. (1), and it is low compared to PMSM or WFSM. The torque-speed curve in Figure 8.4 is valid here, except the torque magnitude is lower, the field-weakening range is higher, and the stability limit is reached when $\delta = 45°$ angle. The phasor diagram is similar to that of PMSM. The diagram is shown with vector variables, which means that the phasor magnitudes are equal to peak values instead of rms values. Substituting $\psi_s = \hat{\psi}_s/\sqrt{2}$, $\sin\delta = \psi_{qs}/\hat{\psi}_s$, and $\cos\delta = \psi_{ds}/\hat{\psi}_s$ in Eq. (1), Eq. (2) is obtained. Note that the ψ_f phasor and the corresponding voltage V_f are absent in the phasor diagram. The power factor of the machine is low because the stator supplies the reactive current to magnetize the rotor.

FIGURE 8.15   Self-controlled SyRM drive with constant i_{ds} control.

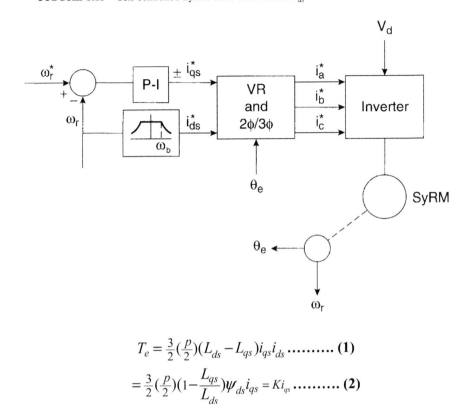

$$T_e = \frac{3}{2}\left(\frac{p}{2}\right)(L_{ds} - L_{qs})i_{qs}i_{ds} \cdots\cdots\cdots \textbf{(1)}$$

$$= \frac{3}{2}\left(\frac{p}{2}\right)\left(1 - \frac{L_{qs}}{L_{ds}}\right)\psi_{ds}i_{qs} = Ki_{qs} \cdots\cdots\cdots \textbf{(2)}$$

The figure shows the simple self-control of a SyRM drive [5] with control of i_{ds} and i_{qs} currents, as indicated in the phasor diagram shown earlier. Because $\psi_{ds} = L_{ds}i_{ds}$ and $\psi_{qs} = L_{qs}i_{qs}$, the torque equation can be written as shown in Eq. (1), which can be further modified as in Eq. (2). It is similar to the torque equation of a vector-controlled induction motor or sinusoidal PMSM drive at constant ψ_{ds}, where polarity of the torque can be changed by i_{qs} polarity. However, coefficient K in Eq. (2) is much smaller and depends on the saliency ratio. In the speed-controlled drive, current i_{ds} is constant in the constant-torque region, but is reduced for field-weakening control. Both the i_{ds}^* and i_{qs}^* currents are vector rotated by the absolute position signal θ_e shown in the figure so that i_{ds} remains oriented to the ψ_{ds} direction. Note that in the phasor diagram of Figure 8.14, as i_{qs} varies, both stator flux $(\hat{\psi}_s)$ and the torque angle (δ) will vary, causing the coupling effect. Therefore, the control cannot be defined as vector control.

FIGURE 8.16   Various self-control criteria for a SyRM machine.

(a) FASTEST TORQUE RESPONSE CONTROL

$$T_e = \frac{3}{2}(\frac{p}{2})(L_{ds} - L_{qs})i_{ds}i_{qs} \cdot \frac{\hat{\psi}_s^2}{(L_{ds}^2 i_{ds}^2 + L_{qs}^2 i_{qs}^2)} \quad \text{......(1)}$$

$$= \frac{3}{2}(\frac{p}{2})\frac{(L_{ds} - L_{qs})\hat{\psi}_s^2 \tan\theta}{(L_{ds}^2 + L_{qs}^2 \tan^2\theta)} \quad \text{.....................(2)}$$

$$dT_e/d\theta \text{ or } dT_e/d(\tan\theta) = 0 \text{ gives } \tan\theta = \frac{L_{ds}}{L_{qs}} = \frac{i_{qs}}{i_{ds}} \text{.......(3)}$$

(b) MAXIMUM TORQUE/AMP CONTROL

$$T_e = \frac{3}{2}(\frac{p}{2})(L_{ds} - L_{qs})(\hat{I}_s \cos\theta)(\hat{I}_s \sin\theta) \quad \text{...............(4)}$$

$$= \frac{3}{2}(\frac{p}{2})(L_{ds} - L_{qs})\hat{I}_s^2 \frac{\sin 2\theta}{2} \quad \text{..................(5)}$$

$$\theta = 45^o$$

(c) MAXIMUM POWER FACTOR CONTROL

$$\cos\varphi = \frac{T_e\omega_m}{3/2.\hat{V}_s\hat{I}_s} = \frac{\left[\frac{3}{2}(\frac{p}{2})(L_{ds} - L_{qs})i_{qs}i_{ds}\right].\omega_e\frac{2}{p}}{3/2\left[\hat{\psi}_s\omega_e\right]\sqrt{i_{ds}^2 + i_{qs}^2}} \quad \text{........(6)}$$

$$= \frac{(L_{ds}/L_{qs} - 1)}{\left[(L_{ds}/L_{qs})^2 + \tan^2\theta\right]^{1/2}((1/\tan^2\theta)+1)^{1/2}} \quad \text{......(7)}$$

$$\tan\theta = \sqrt{\frac{L_{ds}}{L_{qs}}}$$

The equations for various self-control criteria for a synchronous reluctance machine [5] are shown here. They all relate to the phasor diagram shown in Figure 8.14. The basic torque expression [Eq. (1)] of Figure 8.15 is modified to derive the criteria for fastest torque response and maximum torque/amp relations shown here. In the fastest response criteria, T_e is expressed in terms of stator flux ψ_s and the angle θ, where the latter is defined by $\tan\theta = i_{qs}/i_{ds}$. Equation (2) is differentiated with respect to $\tan\theta$ to determine that the fastest response is possible when $\tan\theta = L_{ds}/L_{qs}$. This means that $\psi_{ds} = \psi_{qs}$, i.e., the ψ_s line is always oriented at $\delta = 45°$. In the maximum T_e/I_s control, T_e is expressed as a function of I_s and θ as shown in Eq. (5). This indicates that maximum T_e/I_s is obtained when $\theta = 45°$; i.e., to satisfy this criteria, the $i_{ds} = i_{qs}$ condition should be maintained. To get the best terminal power factor (or DPF) irrespective of the former two criteria, the power factor is expressed as function of θ in Eq. (7) by substituting $\tan\theta = i_{qs}/i_{ds}$ in Eq. (6). To minimize the denominator of Eq. (7), it is differentiated with respect to $\tan\theta$ to derive the criteria $\tan\theta = \sqrt{L_{ds}/L_{qs}}$.

FIGURE 8.17 Control diagram of SyRM at constant θ angle with various criteria.

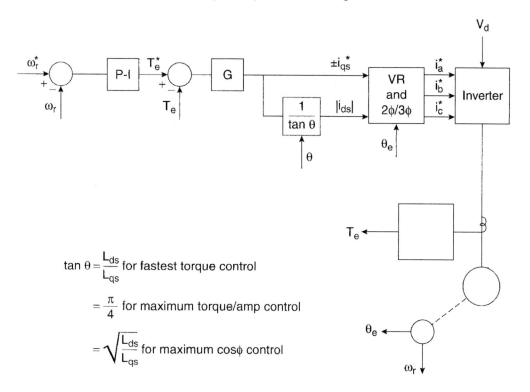

$$\tan\theta = \frac{L_{ds}}{L_{qs}} \text{ for fastest torque control}$$

$$= \frac{\pi}{4} \text{ for maximum torque/amp control}$$

$$= \sqrt{\frac{L_{ds}}{L_{qs}}} \text{ for maximum } \cos\phi \text{ control}$$

The control block diagram for a SyRM based on the various criteria discussed in Figure 8.16 is shown in this figure, where θ is the orientation angle of the stator current with respect to the d^e axis as shown in Figure 8.14. As usual, the torque component of current i_{qs} is derived from the torque control loop, which in turn is derived from the speed loop as shown. Because $\tan\theta = i_{qs}/i_{ds}$, signal i_{ds} is derived from i_{qs}. Both i_{qs} and i_{ds} are then vector rotated (VR) with the help of absolute position encoder signal θ_e to establish the phase current commands. If, for example, for a particular machine, $L_{ds}/L_{qs} = 2.63$, $\theta = 69.2°$ for fastest torque control, where the stator flux angle δ remains oriented at fixed 45°. For maximum torque/amp control, the stator current remains oriented at 45° irrespective of the L_{ds}/L_{qs} ratio. For maximum DPF at the machine terminal, $\tan\theta$ remains at $\sqrt{2.63}$, i.e., $\theta = 58.3°$. Note that although the control is frequently defined as vector control, it is not really vector control because there is no fixed orientation angle of currents with respect to the stator flux ψ_s.

FIGURE 8.18   Constant-torque loci of an IPM machine in the $i_{ds}(pu)$-$i_{qs}(pu)$ plane and corresponding maximum torque/amp curve.

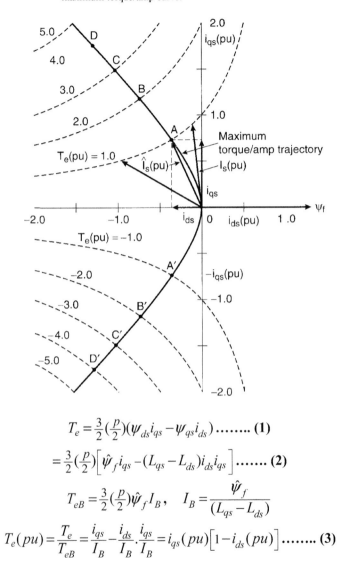

$$T_e = \tfrac{3}{2}(\tfrac{p}{2})(\psi_{ds} i_{qs} - \psi_{qs} i_{ds}) \text{........ (1)}$$

$$= \tfrac{3}{2}(\tfrac{p}{2})\Big[\hat{\psi}_f i_{qs} - (L_{qs} - L_{ds}) i_{ds} i_{qs} \Big] \text{....... (2)}$$

$$T_{eB} = \tfrac{3}{2}(\tfrac{p}{2})\hat{\psi}_f I_B, \quad I_B = \frac{\hat{\psi}_f}{(L_{qs} - L_{ds})}$$

$$T_e(pu) = \frac{T_e}{T_{eB}} = \frac{i_{qs}}{I_B} - \frac{i_{ds}}{I_B} \cdot \frac{i_{qs}}{I_B} = i_{qs}(pu)\Big[1 - i_{ds}(pu)\Big] \text{........ (3)}$$

The sinusoidal interior PM (IPM) machine is a salient-pole machine like a SyRM or WFSM, but it is characterized by $L_{qs} \gg L_{ds}$. This figure shows the plot of maximum torque/amp curve on a per-unit basis. This operation mode has the advantage of machine loss minimization but at the penalty of nonoptimum torque response. General torque Eq. (1) can be modified to Eq. (2) by substituting $\psi_{ds} = \hat{\psi}_f + L_{ds}i_{ds}$ and $\psi_{qs} = L_{qs}i_{qs}$, where $\hat{\psi}_f$ is the magnet flux on the d^e axis. Defining the base torque (T_{eB}) and base current (I_B) as indicated, and substituting in Eq. (2), Eq. (3) is derived. The figure shows constant motoring and braking torque profiles as a function of $i_{ds}(pu)$ and $i_{qs}(pu)$, and the corresponding maximum torque/amp curve is superimposed. Points A, B, C, D, A′, B′, etc., indicate the operating points, where $\hat{I}_s(pu)$ is minimum. Note that the polarity of $i_{qs}(pu)$ determines the torque polarity, but $i_{ds}(pu)$ polarity is always negative, which ensures positive contribution of reluctance torque on the field torque in Eq. (2).

FIGURE 8.19   (a) $i_{ds}(pu)$ and $i_{qs}(pu)$ curves for $T_e(pu)_{max}$/amp in motoring mode and (b) control of IPM machine in maximum torque/amp mode.

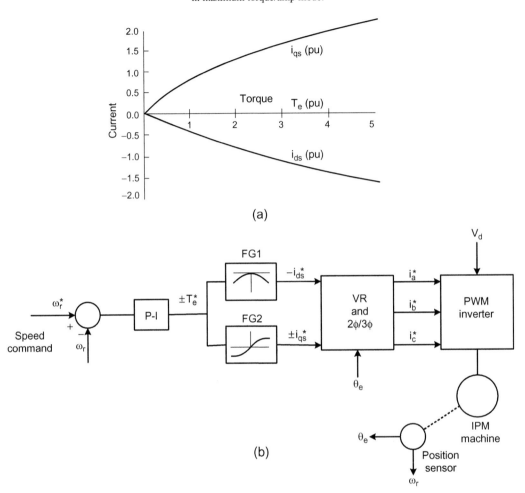

(a)

(b)

Figure 8.19(a) shows the optimum $i_{ds}(pu)$ and $i_{qs}(pu)$ profiles in the motoring mode. They have been derived from the previous figure. These functions are utilized to develop the control strategy shown in (b) [6]. The speed control loop generates the torque command T_e^* as usual. Command currents i_{ds}^* and i_{qs}^* are then generated through the function generators FG1 and FG2, which are derived from part (a). As mentioned before, i_{ds} polarity always remains negative, but i_{qs} polarity is positive in motoring and negative in regeneration. The command currents are vector rotated by the absolute position angle θ_e, which ensures i_{ds} alignment on the d^e axis. This control algorithm is valid in the constant-torque region only. Again, the simple feed-forward open-loop current control may not be accurate due to variations in the $\hat{\psi}_f$, L_{ds}, and L_{qs} parameters.

FIGURE 8.20   Square-wave voltage limit ellipse and field-weakening mode control of an IPM machine.

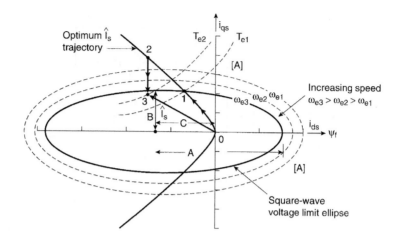

SQUARE-WAVE OPERATION:

$$\left(\frac{2V_d}{\pi}\right)^2 = v_{ds}^2 + v_{qs}^2 \quad \ldots\ldots\ldots (1)$$

$$= \omega_e^2 (L_{ds} i_{ds} + \hat{\psi}_f)^2 + \omega_e^2 (L_{qs} i_{qs})^2 \ldots\ldots (2)$$

$$\frac{(i_{ds} + \frac{\hat{\psi}_f}{L_{ds}})^2}{\left(\frac{2V_d}{\pi \omega_e L_{ds}}\right)^2} + \frac{i_{qs}^2}{\left(\frac{2V_d}{\pi \omega_e L_{qs}}\right)^2} = 1 \ \ \textbf{Or} \ \ \frac{(i_{ds} - C)^2}{A^2} + \frac{i_{qs}^2}{B^2} = 1 \ldots\ldots (3)$$

IPM machines can be operated in the extended speed field-weakening region because of a strong armature reaction effect. The machine reaches the square-wave mode at the end of the constant-torque region. At square wave, v_{ds} and v_{qs} constitute the components of fundamental peak voltage as given by $2V_d/\pi$ in Eq. (2). Equation (2) can be expressed in the form of Eq. (3), which describes an ellipse that is plotted in the figure, where A, B, and C describe the semi-major axis, semi-minor axis, and the offset, respectively. The optimum $\hat{I}_s$ trajectory is superimposed on the figure. While the boundary of the ellipse represents square-wave operation, any point inside the ellipse represents the current control mode. Note that as the speed increases, the size of the ellipse shrinks as indicated. Consider initial operating point 2 in the low-speed PWM mode. As ω_e increases and the ellipse shrinks below point 2, the i_{qs}^* can be decreased to push the operating point to 3 for the current control to remain valid. This will weaken the stator flux ψ_s in the constant-power region so as to restore current control beyond the base speed. Note that the developed torque at point 3 is higher than that in the natural crossing point.

FIGURE 8.21 Speed control of IPM machine covering the field-weakening mode.

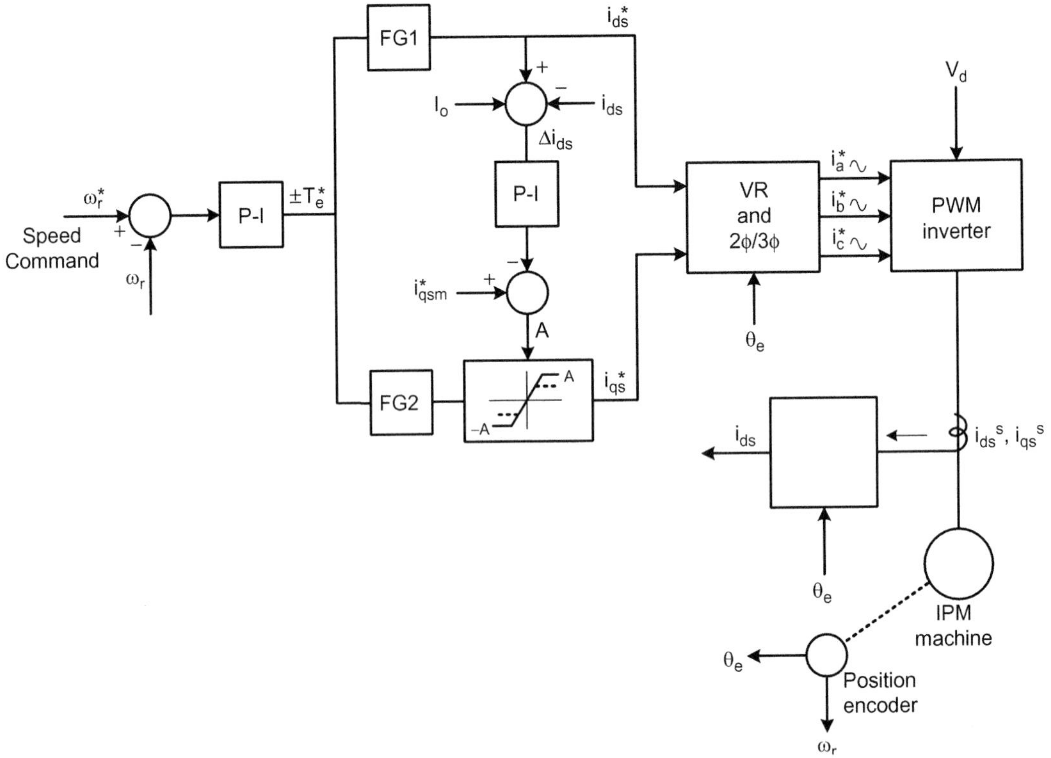

The operating principle described in the previous figure has been utilized to develop the control diagram in this figure [7], which covers both the constant-torque and field-weakening regions. Basically, it is an extension of Figure 8.19. Function generators FG1 and FG2 fabricate the i_{ds}^* and i_{qs}^* functions, as usual. The i_{qs}^* command is taken through a controllable limit A as shown. In the constant-torque region, actual i_{ds} matches with the i_{ds}^* command and, therefore, $\Delta i_{ds} = 0$, i.e., $A = i_{qsm}^*$. However, as the i_{ds} controller saturates, ΔI_{ds} builds up and pushes the value of A smaller through a *P-I* controller to force the current control mode. For negative torque, the current i_{qs} is negative and the operating point is symmetrically on the lower side of the ellipse. Again, the control is basically scalar and the problem of parameter variation exists.

FIGURE 8.22 Features of DOE electric-vehicle drive system using IPM machine (ETXII).

- IPM MACHINE RATINGS:
  - 70 HP, 4-POLE, WYE-CONNECTED Nd-Fe-B (CRUMAX 30A)
  - BASE SPEED (ω_b)= 710.48 elec rad/sec (3394 RPM)
  - MAXIMUM SPEED = 13,750 RPM
  - CROSSOVER SPEED = 5044 RPM
  - MAGNET FLUX ($\omega_b\psi_f$)= 40.2 V (75°C)
  - RATED STATOR FLUX($\omega_b\psi_s$) = 58.5 V
- MACHINE PARAMETERS:
  R_s = 0.00443 ohm
  $\omega_b L_{ls}$ = 0.0189 ohm
  $\omega_b L_{dm}$ = 0.0785 ohm
  $\omega_b L_{qm}$ = 0.1747 ohm
- LEAD ACID BATTERY: V_B = 204 V (NOMINAL) (135–264 V VARIATION)
- STATOR FLUX-ORIENTED VECTOR CONTROL IN CONSTANT-TORQUE REGION
- SQUARE-WAVE δ-ANGLE CONTROL MODE IN FIELD-WEAKENING REGION
- HYSTERESIS BAND CURRENT CONTROL WITHIN OVERLAY CURRENT CONTROL
- LIGHT LOAD ψ_s PROGRAMMING FOR EFFICIENCY OPTIMIZATION
- MICROCOMPUTER/DSP-BASED CONTROL AND ESTIMATION

The IPM machine is a prime candidate for EV propulsion because of its higher efficiency and ability to operate in the field-weakening mode for extended speed range. These features are not possible with induction or SPM machines. This figure shows the features of a prototype EV drive sponsored by the U.S. Department of Energy [8]. The details of the drive will be described in the next few figures. The complete 70-hp drive system was initially studied by simulation, and then followed by detailed experimentation on a dynamometer in a laboratory prior to testing on the road. All the controls, feedback signal processing, and diagnostics were based on multiple DSPs.

FIGURE 8.23   Simplified schematic of drive system power circuit.

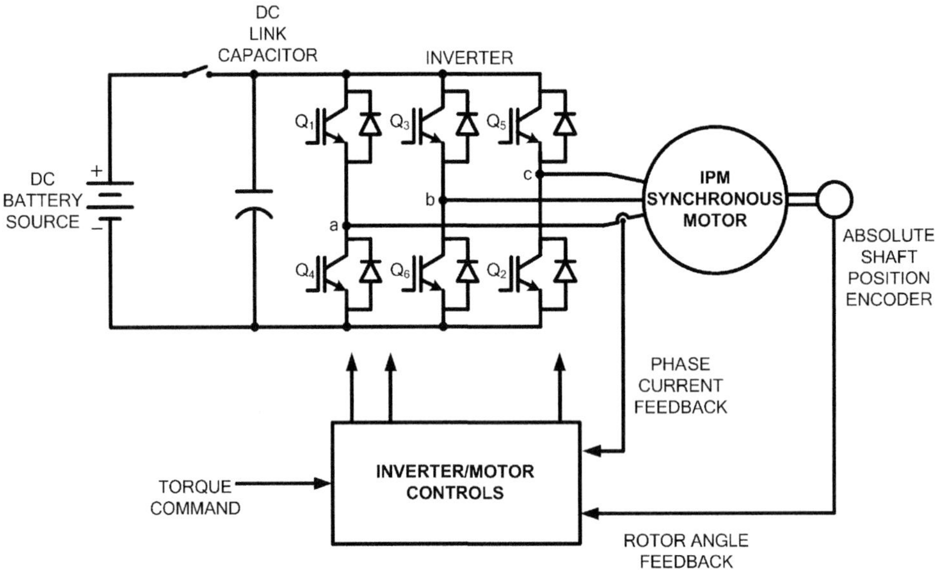

The power circuit for an EV drive along with its control block diagram are shown in this figure. (Actually, BJTs were used in the inverter.) The PWM voltage-fed inverter takes the battery-fed dc supply and converts it into variable-frequency, variable-voltage ac for the machine. In the regenerating mode, the stored mechanical energy in the vehicle is converted into electrical energy and fed back to the battery. The dc-side filter capacitor bank shunts the inverter input current harmonics from the battery to prolong the battery life. It also decouples the line leakage inductance. The IGBTs (or BJTs) are rated for 600 V, 500 A, and there are RCD snubbers across each device (not shown). As usual, the EV drive is a torque-controlled system. The absolute position encoder (analog resolver with RDC) permits the drive to be self-controlled like a dc machine from zero to the maximum speed. The machine was custom designed, and there was no damper winding on it. The microcomputer/DSP-based inverter/motor controller receives the torque command and feedback signals and generates PWM signals for the inverter.

FIGURE 8.24 (a) Torque-speed curve in motoring mode and (b) stator flux program with torque for efficiency improvement.

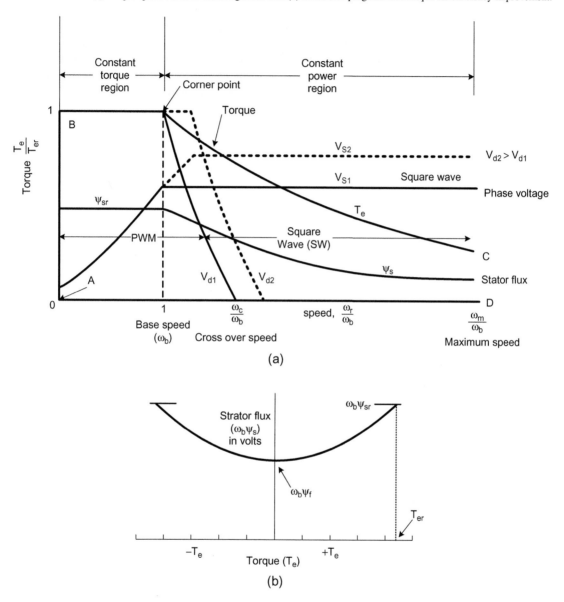

The torque-speed relation of an IPM machine in motoring mode along with phase voltage (V_s) and stator flux (ψ_s) is shown in the upper figure. The machine has two operating modes: PWM mode in the constant-torque region and square-wave mode in the constant-power field-weakening region. In PWM mode, the stator currents are controlled to operate the drive with vector control. The boundary between PWM and square-wave mode varies with battery voltage as shown in the figure. With higher V_d ($V_{d2} \gg V_{d1}$), the V_s level increases and, correspondingly, the maximum power output increases (not shown). The machine operates with programmable stator flux with torque for efficiency optimization as shown in (b). Therefore, the PWM–square-wave boundary is slanting because of the flux–speed relation $V_s = \omega_r \psi_s = \omega_b \psi_{sr} = \omega_c \psi_f$, where ψ_{sr} = stator flux at rated torque, ψ_f = magnet flux, ω_b = base speed, and ω_c = crossover speed, as indicated on the figure.

FIGURE 8.25   (a) Phasor diagram of machine and (b) reactive current as a function of active current.

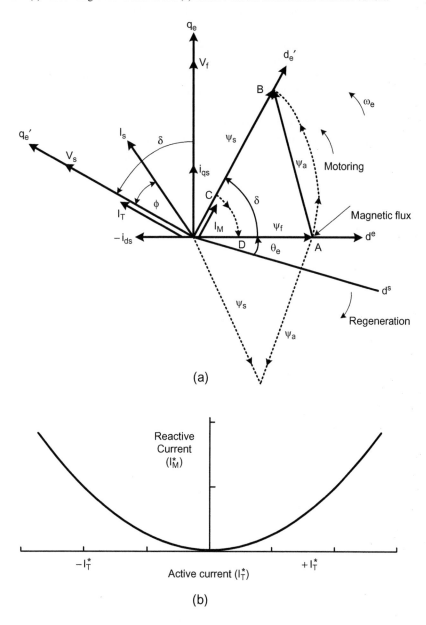

(a)

(b)

The phasor diagram of the machine in the motoring mode is shown in (a), where the torque angle δ is positive and phase angle of stator current φ is lagging. The armature reaction flux ψ_a is not aligned in the direction of I_s because of the saliency. At zero torque or current, $\psi_s = \psi_f$, and they remain aligned with the d^e axis as shown. As torque increases to the rated value, ψ_s moves along the trajectory A-B, and then in the field-weakening mode, it moves from B to C with constant δ angle. In the locus CD, the torque goes to zero at the highest speed when $\delta = 0$. Evidently, in the high-speed, low-torque, constant-power region, the current at low leading DPF will load the machine and inverter heavily. Current I_s can be resolved into I_T (torque or in-phase current) and I_M (magnetizing current) components, where they remain aligned to the newly defined $q^{e\prime}$ and $d^{e\prime}$ axes, respectively. The theoretical I_M–I_T relation from zero to the rated torque is shown in (b). In the regeneration mode, the flux phasor diagram reflects to the lower side of the d^e axis as shown.

FIGURE 8.26   Vector control block diagram of EV drive.

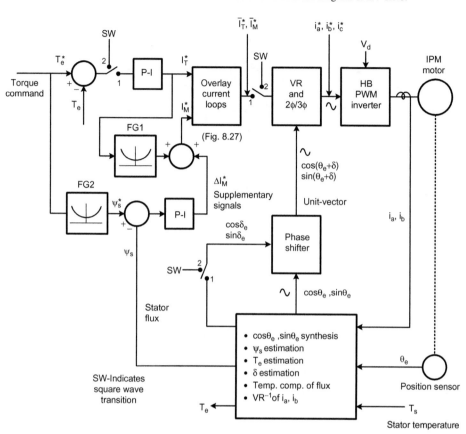

As indicated earlier, the drive system has stator flux-oriented vector control in the constant-torque region [9], where the inverter operates in PWM mode. The control block diagram satisfies the phasor diagram in the previous figure. The drive system uses a primary torque control loop with the slaved stator flux loop through the function generator of Figure 8.24(b). The figure includes some switches for transition to square-wave mode, which will be described later. The torque control loop generates the I_T^* command, whereas the flux loop generates the ΔI_M^* command as shown. The ΔI_M^* current is then added to the output of the $I_M^*-I_T^*$ function generator (see Figure 8.25) to generate the total I_M^*. These currents are then processed through overlay current loops (described later) and VR to generate the phase current commands for the inverter. In this project, HB PWM is used for simplicity, although synchronous current control could be used. The feedback signal-processing block (described later) generates the needed signals for the control system from position angle θ_e, machine terminal currents, and stator temperature (T_s). The flux ψ_f by NdFeB magnet decreases slightly with temperature, which requires correction. The unit vector signals $\cos(\theta_e + \delta)$ and $\sin(\theta_e + \delta)$ permit orientation of commanded currents on the $d^{e\prime}$-$q^{e\prime}$ axes.

FIGURE 8.27   Overlay current control of I_T^* and I_M^* with vector rotation.

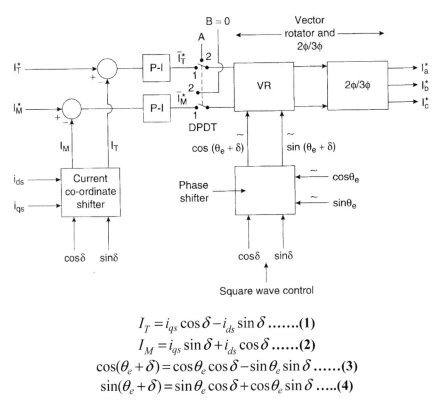

$$I_T = i_{qs}\cos\delta - i_{ds}\sin\delta \dots\dots(1)$$
$$I_M = i_{qs}\sin\delta + i_{ds}\cos\delta \dots\dots(2)$$
$$\cos(\theta_e + \delta) = \cos\theta_e\cos\delta - \sin\theta_e\sin\delta \dots\dots(3)$$
$$\sin(\theta_e + \delta) = \sin\theta_e\cos\delta + \cos\theta_e\sin\delta \dots\dots(4)$$

The overlay current control is basically redundant current control, as shown in this figure, which has the following features: (1) It permits the use of a simple HB current control PWM although it is not harmonically optimum, and (2) current control with partial saturation (overmodulation) is used, which permits the HB current controller to smoothly transition to the square-wave mode. The feedback currents are generated by the current coordinate shifter using Eqs. (1) and (2). The *P-I* control ensures tracking of command and actual currents. In the pure PWM mode, the *P-I* outputs remain the same as the command values (redundant loops). However, in overmodulation mode, the *P-I* outputs are higher than command values to force matching of command and feedback currents, until at the square-wave mode, the current control is lost and the current loops are no longer active. The equations for unit vector generation for correct VR are shown in Eqs. (3) and (4). Estimation of the δ angle along with other feedback signals will be discussed in the next figure.

FIGURE 8.28 Feedback signal estimation.

A. INVERSE VECTOR ROTATION (VR^{-1}) OF i_a, i_b:

$$i_{ds} = i_a \sin\theta_e - \frac{1}{\sqrt{3}}(i_a + 2i_b)\cos\theta_e \ldots\ldots (1)$$

$$i_{qs} = i_a \cos\theta_e + \frac{1}{\sqrt{3}}(i_a + 2i_b)\sin\theta_e \ldots\ldots (2)$$

$$i_a + i_b + i_c = 0 \ldots\ldots (3)$$

B. STATOR FLUX (ψ_s') ESTIMATION

$$\psi_s' = \sqrt{\psi_{ds}'^2 + \psi_{qs}'^2} \ldots\ldots (4)$$

$$\psi_{qs}' = L_{qs}i_{qs} = i_{qs}(L_{qs0} - f(i_{ds}, i_{qs})) \ldots\ldots (5)$$

$$\psi_{ds}' = L_{ds}i_{ds} + \psi_f = L_{ds}i_{ds} + (\psi_{f0} + f(i_{ds}, i_{qs})) \ldots\ldots (6)$$

C. TEMPERATURE CORRECTION OF ψ_s

$$\psi_{ds} = \psi_{ds}' + K_d(75° - T_R) \ldots\ldots (7)$$

$$\psi_{qs} = \psi_{qs}'\left[1 - K_q(75° - T_R)\right] \ldots\ldots (8)$$

$$T_R(n) = T_R(n-1) + \frac{1}{\tau}\left[T_s(n) - T_R(n-1)\right] \ldots (9)$$

D. TORQUE ESTIMATION

$$T_e = \frac{3}{2}(\frac{P}{2})(\psi_{ds}i_{qs} - \psi_{qs}i_{ds}) \ldots\ldots (10)$$

E. TORQUE ANGLE

$$\sin\delta = \frac{\psi_{qs}}{\hat{\psi}_s} \ldots\ldots (11)$$

$$\cos\delta = \frac{\psi_{ds}}{\hat{\psi}_s} \ldots\ldots (12)$$

The feedback signal estimation equations for the block diagram of Figure 8.26 are given here. Signals $\sin\theta_e$ and $\cos\theta_e$ can be estimated directly from the θ_e signal by the look-up table. The stator flux estimation is somewhat complex because of inductance saturation with rotor saliency, which creates complex cross-coupling effects between the d^e and q^e axes fluxes. In addition, both flux components require correction with estimated rotor temperature T_R. Estimation of T_R from stator temperature T_s can be done approximately by using a simple first-order thermal model with TC = τ. Obviously, a microcomputer is essential for such estimations.

FIGURE 8.29 Control block diagram in square-wave mode.

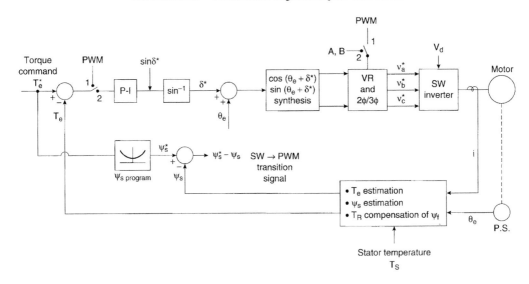

$$v_a^{*} = A\cos(\theta_e + \delta^{*})\ldots\ldots..(1)$$

$$v_b^{*} = A\cos(\theta_e + \delta^{*} - \frac{2\pi}{3})\ldots\ldots(2)$$

$$v_c^{*} = A\cos(\theta_e + \delta^{*} + \frac{2\pi}{3})\ldots\ldots.(3)$$

The scalar control block diagram of the drive in the square-wave field-weakening mode is shown [9] in this figure. Changing the control structure between PWM and square-wave modes is complex. However, the motivation for square-wave mode is to improve the drive efficiency by reducing the inverter switching loss, since the harmonic loss is low at high frequency due to high inductance of the machine. One penalty with scalar control is the sluggish response. Of course, vector control with PWM operation could have been easily extended in the field-weakening region. The basic idea of the control is to orient the V_s axis ($q^{e'}$ axis) at the desired δ angle with respect to the V_f (q^e axis) (see Figure 8.25) in a feed-forward manner. There are two switches in the figure for transition to PWM mode at low speed. The torque control loop generates the $\sin\delta^{*}$ signal, which is inverted to generate the torque angle command δ^{*}. The sign of δ^{*} is negative in the regenerative condition. This angle is added with the θ_e angle and then the look-up table generates the $\cos(\theta_e + \delta^{*})$ and $\sin(\theta_e + \delta^{*})$ signals. These are then used to vector rotate constants A and B ($= 0$) to generate the phase voltage commands by Eqs. (1), (2), and (3). Square-wave control essentially works on the right side of Figure 8.27 as indicated, where the A and B parameters are initialized at the end of PWM control. This principle will be discussed in the next figure. The flux control loop, as shown, is not valid in square wave, but the error signal is used for transition to the PWM mode.

FIGURE 8.30 Square-wave control implementation through VR and HB controller.

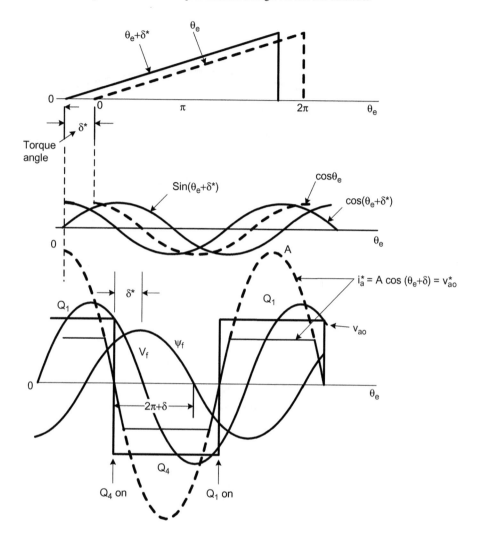

The figure shows the waveforms explaining the δ or torque angle control principle in square-wave mode. The position signal θ_e is a sawtooth wave covering a 360° angle. Addition of torque angle δ^* advances the sawtooth wave as shown. The fabrication of $\sin(\theta_e + \delta^*)$ and $\cos(\theta_e + \delta^*)$ waves is indicated in the middle of the figure. The vector rotation of A by these unit vector signals generates the fictitious current wave i_a^* as indicated. Because A is very large, it is truncated and this signal is essentially phase voltage command wave v_{ao}^* (or v_a^*). Note that this wave is at $90 + \delta°$ ahead of flux wave ψ_f as indicated in the phasor diagram of Figure 8.25. The other phase voltage waves v_{bo}^* and v_{co}^* at 120° phase differences are not shown.

FIGURE 8.31   PWM $\leftrightarrows$ square-wave transition.

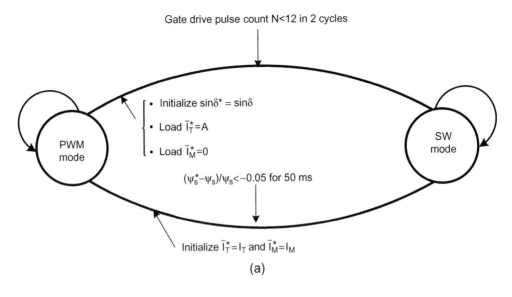

(a)

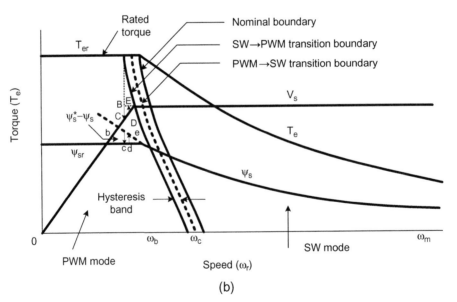

(b)

The PWM $\leftrightarrows$ SW transition is complex, and is explained by the sequence diagram in part (a) and the torque-speed curve in (b). The PWM $\leftrightarrows$ SW transition is dictated by the inverter gate drive pulse count N over two fundamental cycles. As the PWM mode saturates, N decreases, and transition occurs below a specific value of N. On the torque-speed curve, the boundary lines for the transition are shown. Both the lines with a hysteresis band lie below the nominal boundary. At transition to SW, the voltage V_s jumps up (DE), and correspondingly ψ_s jumps up (de). After the transition, the variables of the current loops are initialized as indicated in (a). The criteria for SW $\rightarrow$ PWM transition is dictated by flux loop error. When the jump occurs, V_s decreases by BC and correspondingly ψ_s decreases by bc, as shown in (b). After transition to the PWM mode, the loop currents are initialized to the estimated values. Careful design of the transition is essential to prevent any jerking in the vehicle.

FIGURE 8.32   Four-quadrant experimental performance of the drive with inertia (*J*) load (shows PWM ⇆ SW transitions).

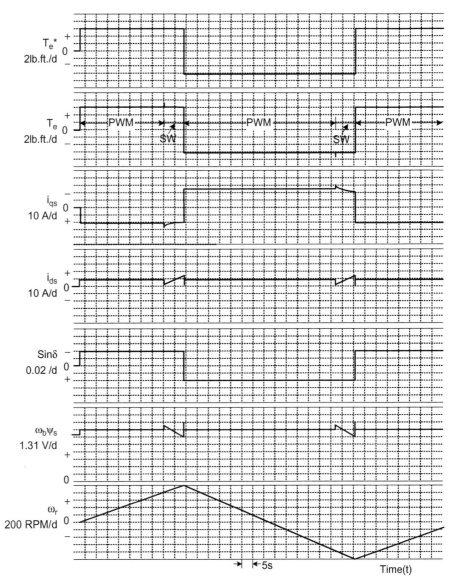

The figure shows 70-hp IPM machine drive performance with pure inertia load on a laboratory dynamometer that includes a smooth transition between the PWM and SW (square-wave) modes. All control parameters were initially set by a simulation study, but iterated during the experimental stage to obtain the best performance. The system starts in the forward direction at zero speed and constant torque. As the speed increases beyond a critical value, a transition to the SW mode occurs, causing the flux jump as shown. As torque command reverses, the drive enters the regeneration mode, but the increase in battery voltage (due to internal resistance) causes a jump to PWM mode immediately. In this transition, the flux jumps up first to satisfy the ψ_s^*-ψ_s criteria, and is then regulated by the torque command. There is some torque ripple near zero speed because of insufficient filtering of i_{ds} and i_{qs} currents. The torque ripple is also higher in the PWM overmodulation region due to an increase in harmonic currents.

FIGURE 8.33   Trapezoidal SPM machine with inverter and self-control with position sensor. (Courtesy of Pearson)

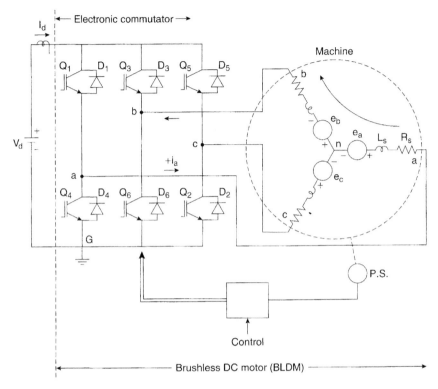

A trapezoidal SPM machine, as discussed in Chapter 6, is basically a nonsalient-pole surface PM machine with a full pitch concentrated winding that generates three-phase trapezoidal counter emf waves at the machine terminal. This figure shows a trapezoidal SPM machine in self-control mode with absolute position sensor and voltage-fed inverter. The whole configuration can be described as a brushless dc motor (BLDM), where the inverter acts as an electronic commutator, which replaces the mechanical commutators of a traditional dc machine. The drive is also called an electronically commutated motor (ECM). Note that the BLDM principle was described earlier in Figure 8.6 for a sinusoidal SPM machine. In fact, any type of synchronous machine with self-control can be defined as a BLDM. However, a trapezoidal SPM machine with self-control, as shown in this figure, most closely represents a BLDM. Therefore, this type of drive is commonly known as a BLDM. The position sensor signal is processed through a controller (explained later) to generate the inverter firing signals as shown. The dc supply at the input may be rectified dc or from a battery.

FIGURE 8.34   Stator phase CEMF and current waves showing the inverter conducting switches.

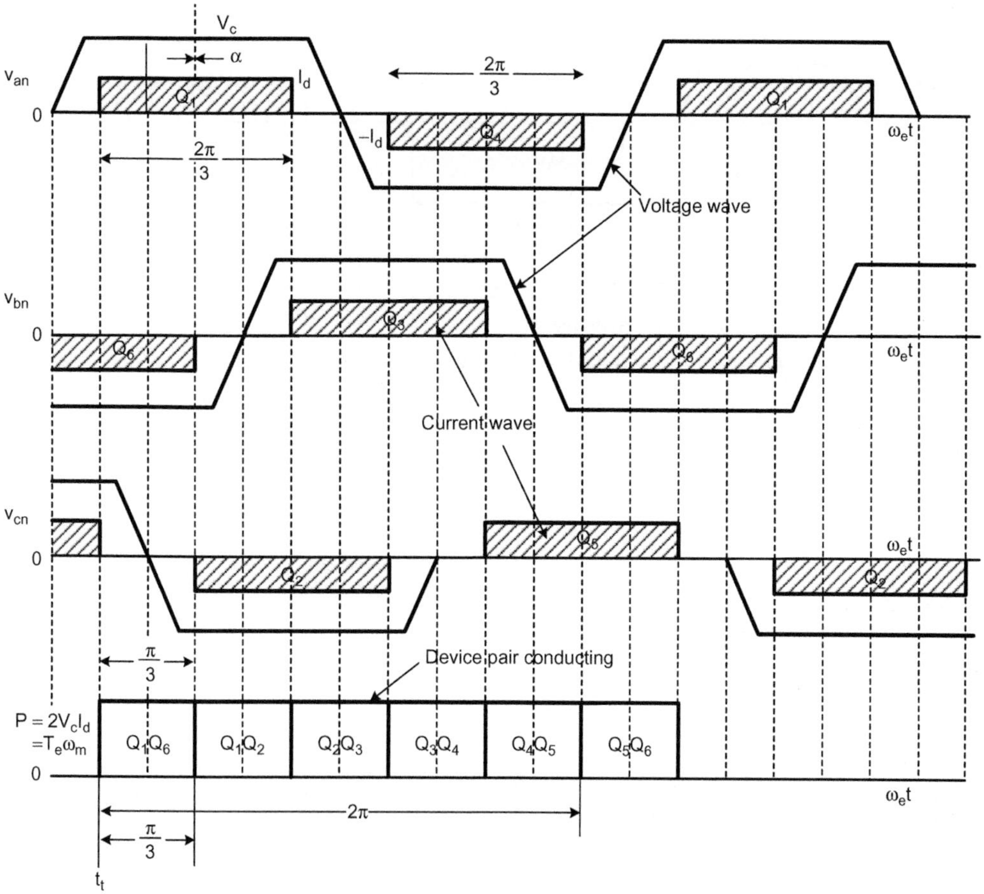

Basically, the inverter in the trapezoidal SPM can be controlled in two different modes: one is the simple 120° switch-on mode of the devices as explained in this figure, and the other is the PWM mode shown in the next figure. The inverter switches are controlled in such a way that a 120° pulse of input current I_d is placed symmetrically at the center of each phase voltage wave as shown. Because the phase position of the counter emf waves are related to the rotor position, generation of these pulses should be straightforward. Note that at any instant, two switches are on: one in the upper group and another in the lower group. The switching pattern changes every 60° as shown in the lower part. The input V_d or I_d is applied simultaneously to two phases in series, and the power supplied to the machine $(P = 2V_c I_d)$ is ideally constant as shown in the lower part of the figure. With constant power, the developed torque is constant. The supply can also be variable voltage or current source. Although it is an ac synchronous motor drive with an inverter, the unit looks like a dc motor at the input through the electronic commutator.

FIGURE 8.35   Waveforms with PWM mode current control of trapezoidal SPM machine.

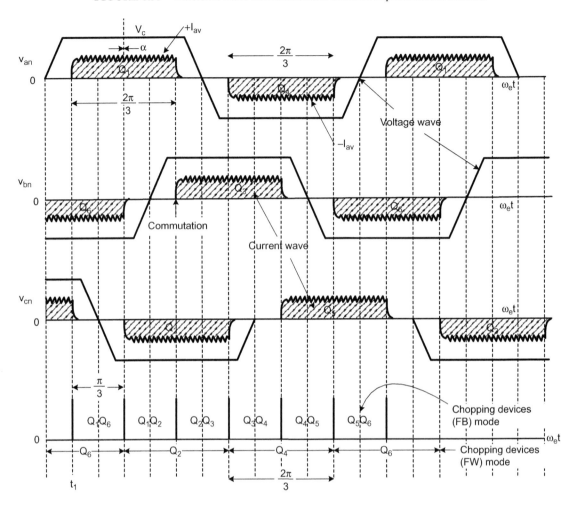

The inverter switches in Figure 8.33 can be controlled not only for the 120° switch-on commutator mode, but also for the PWM chopping mode to control voltage and current continuously across the machine. There are two PWM control modes, as shown in this figure. These are feedback (FB) and freewheeling (FW) modes. In the FB mode, the upper and lower switches are controlled simultaneously. For example, from time t_1, as Q_1Q_6 are turned on simultaneously, the phase a and b currents will build up ($V_d > 2V_c$). But, when they are turned off, the feedback currents through D_3D_4 will decay. Thus, average voltage and average current at the machine terminal are controlled by duty cycle ratio. In the FW mode, as explained in the lowest part, the upper devices (Q_1, Q_3, Q_5) are kept on sequentially for 120° in the middle of the respective positive half cycles, and the lower devices are chopped only. This results in freewheeling of machine current by shorting the phases. During commutation of adjacent phases, the finite rise and fall times of currents are indicated. PWM control can be used during initial starting of the machine (dc motor starter function) to establish the speed and counter emf, and then the 120° switch-on mode can be established. Or, continuous operation in the PWM mode will permit speed control of the motor. The machine can be regenerative by shifting the current waves by 180° (i.e., $\alpha = 180°$).

FIGURE 8.36   Torque-speed characteristics of BLDM (with trapezoidal PMSM).

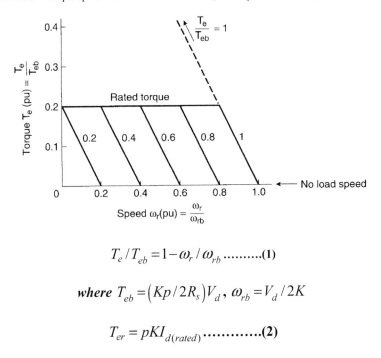

$$T_e / T_{eb} = 1 - \omega_r / \omega_{rb} \dots\dots\dots(1)$$

$$\textit{where } T_{eb} = \left(Kp / 2R_s \right) V_d, \; \omega_{rb} = V_d / 2K$$

$$T_{er} = pK I_{d(rated)} \dots\dots\dots\dots(2)$$

The torque-speed characteristics of the BLDM shown in the figure are similar to a PM dc machine. The torque-speed equation in pu form is given by Eq. (1), where K = CEMF/unit speed, p = poles, V_d = supply dc voltage, and R_s = stator resistance/phase. Assume first that the machine runs in the 120° switch-on mode with the applied voltage V_d at the machine terminal. The outermost line shows the plot of Eq. (1) in this condition. The no-load speed is maximum at zero torque, and decreases with torque (as it varies with current I_d) due to $2I_d R_s$ drop. The rated horizontal torque line shown in the figure is given by Eq. (2). The speed at any torque can be controlled linearly by varying V_d at the machine terminal by PWM control of the inverter. All torque-speed curves have identical slope as shown in the figure.

FIGURE 8.37 Closed-loop speed control of a BLDM in the PWM/FB mode.

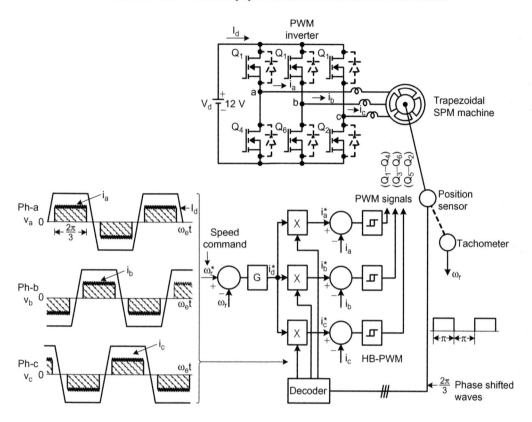

The figure shows a BLDM drive speed control system in which the inverter operates in the PWM feedback mode to control the voltage at the machine terminal. Absolute position sensors (normally three Hall sensors; see Figure 8.38) are mounted on the stator side at the edge of the rotor poles, and these generate three-phase square waves with a 120° phase shift. These waves are co-phasal with the respective phase CEMF waves. The signals are then processed through a decoder to generate 120° switch-on logic waves as shown on the left of the figure. The speed control loop generates the I_d^* current command for the phases, which are enabled by the respective phase decoder output. The output phase current commands are then controlled by the HB-PWM method. For example, $+i_a$ and $-i_b$ are enabled simultaneously. Thus, when the Q_1 and Q_6 switches are turned on simultaneously, both the phase currents increase equally. When the switches are turned off, however, the currents are fed back to the source and then decay. In motoring mode, the 120° current pulses are in phase with the phase voltage waves, but in regeneration, the phase is shifted by 180°. Because of low supply voltage, power MOSFETs are the appropriate devices. For higher V_d, IGBTs can be used. Again, if the primary supply is ac, it can be rectified and filtered to generate the dc supply. With ac supply, the regenerated power can be dissipated in a dynamic brake or can be pumped up to the ac side through a PWM rectifier.

FIGURE 8.38   Closed-loop current control of a BLDM in the PWM/FW mode.

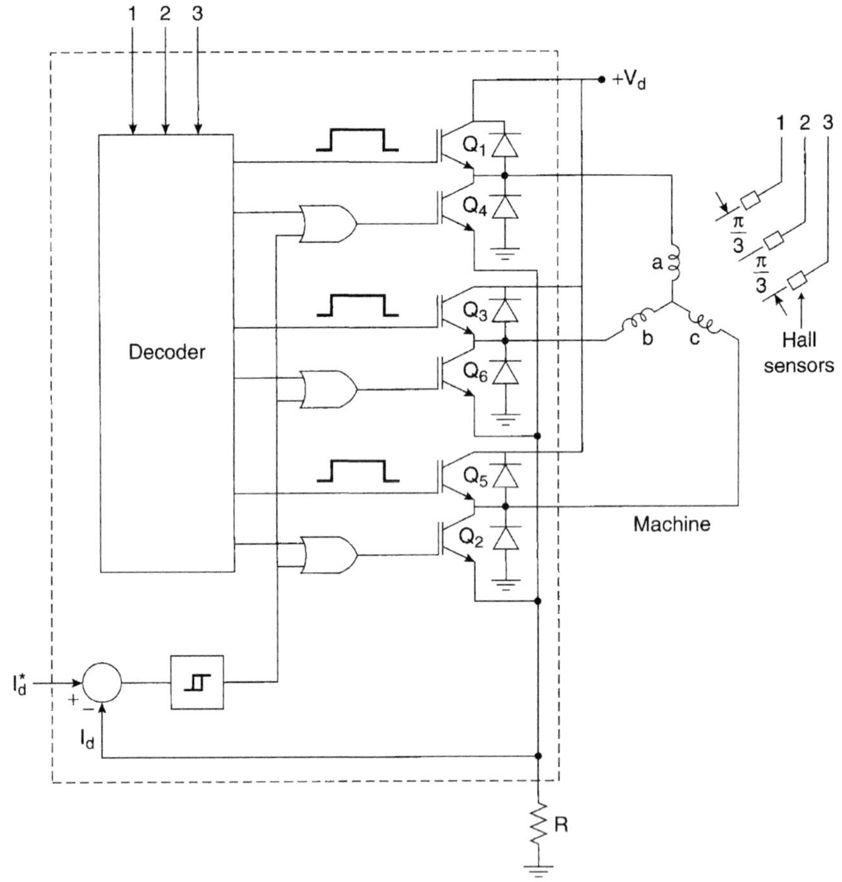

A much more economical BLDM drive, in which the inverter operates in freewheeling (FW) mode, is shown in this figure. Here, the IGBT inverter has a slightly different configuration. The BLDM is shown with torque or control of I_d current. The Hall position sensors and decoder are the same as in the previous figure. However, in this case, the upper devices of the inverter (Q_1, Q_3, and Q_5) are turned on sequentially for 120° in the middle of the respective positive voltage half cycles, whereas the lower devices (Q_4, Q_6, and Q_2) are pulse width modulated for 120° angles in the respective phase voltage negative half cycles. The current regulator operates in the HB-PWM mode with feedback current I_d and generates the switching pulses for the lower devices as shown. If, for example, Q_6 is turned on with Q_1 on, $+i_a$ and $-i_b$ will build up as usual. When Q_6 is turned off, the phase currents will freewheel through Q_1 and D_3 (bypass diode of Q_3). A single common resistance R will provide the feedback current I_d. The dashed box indicates that the whole unit is available in the form of a power integrated circuit (PIC).

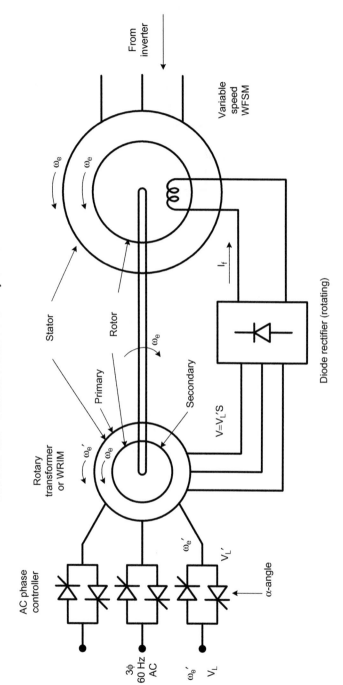

FIGURE 8.39 Brushless excitation of a WF synchronous machine.

The dc field current I_f of WFSMs is controllable, which allows for a programmable power factor (leading, lagging, or unity) at the machine stator terminal. The traditional method of excitation is rectification of ac to dc by a thyristor rectifier and then feeding the dc to the rotor field winding through slip rings and brushes. The alternate method is brushless excitation, as shown in this figure. In this method, a rotary transformer (which is the same as a wound-rotor induction motor) is mounted on the same shaft and the slip voltage induced in the secondary winding is rectified to dc by a diode rectifier and fed to the field winding as shown. There are no slip rings and brushes in WFSMs. The stator of a WRIM is supplied from 60-Hz ac through a thyristor ac voltage controller as shown. The diode rectifier input voltage varies with WRIM slip speed and ac controller firing angle α. If, for example, the WFSM speed ω_e (controlled by its inverter frequency) decreases, the slip S will increase, which will increase field current I_f. However, for constant I_f, the α angle is increased to decrease voltage V. The transient response of brushless excitation is somewhat slower than that of brush excitation, but has the advantages of elimination of brushes and slip rings. The ac phase controller can be replaced by a PWM-type converter to improve line power quality and DPF.

FIGURE 8.40   Self-control of current-fed inverter for WFSM drive.

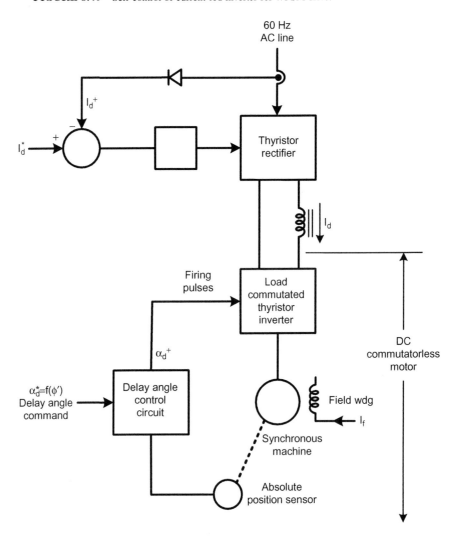

Because wound field synchronous machines can be operated at leading power factor by adjusting the field excitation current I_f, it is possible to use simple load commutation of a thyristor current-fed inverter with it. In Chapter 5, load-commutated inverters (LCI) were discussed extensively. This figure shows the basic elements of an LCI-WFSM drive system. The thyristor inverter is invariably self-controlled, i.e., its firing pulses are derived from the absolute position sensor as shown through a delay angle control circuit, which will be explained later. With self-control, the machine-inverter group behaves like a dc commutatorless machine, as explained early in this chapter. The dc current I_d at the inverter input is generated by a thyristor rectifier using closed-loop current control as shown. The rectifier and inverter circuits are essentially the same, which permits easy motoring or generating operation of the system. The LCI-WFSM drives are extremely popular in high-power, multi-megawatt drives for applications such as pumps and compressor drives and modern ship propulsion. Note that a load-commutated converter system can also be used as a variable-frequency starter for a synchronous motor. When the machine develops the rated speed and rated CEMF at the rated frequency (60 Hz), the motor is transferred to the 60-Hz line to operate at constant speed. The variable-frequency starter operation will be illustrated for ship propulsion in Figure 8.49.

FIGURE 8.41   Phasor diagrams of WFSM with load-commutated inverter for (a) motoring and (b) regeneration.

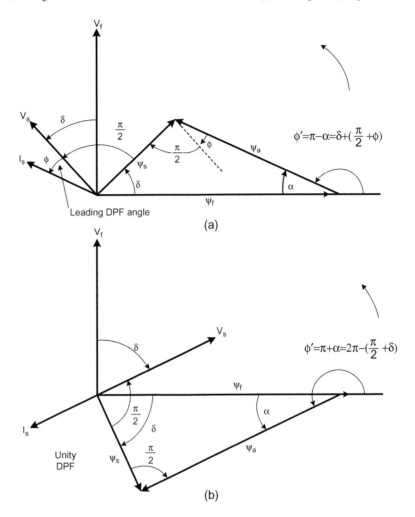

(a)

(b)

The figure shows the fundamental frequency phasor diagrams of a WFSM in motoring and regeneration modes with a load-commutated inverter (LCI). Machine saliency and stator resistance are neglected for simplicity. The figure is somewhat similar to Figure 8.5. In motoring mode, stator current I_s leads stator voltage V_s by DPF angle φ. The position of field flux ψ_f is related with that of the position sensor, and it is taken as the reference phasor. The magnitude of I_d determines stator current I_s and it controls the magnitude of armature reaction flux $\psi_a = I_s L_s$. This phasor can be positioned at a desirable delay angle α_d^* indicated in Fig. 8.40. In (a), the ψ_a phasor leads ψ_f by angle $\varphi' = \pi - \alpha = \delta + (\pi/2 + \varphi)$. In the regenerating mode, the rectifier and inverter reverse their roles, when V_s and I_s are in opposite directions in the inverter.

FIGURE 8.42 Starting of WFSMs by discrete current pulsing method.

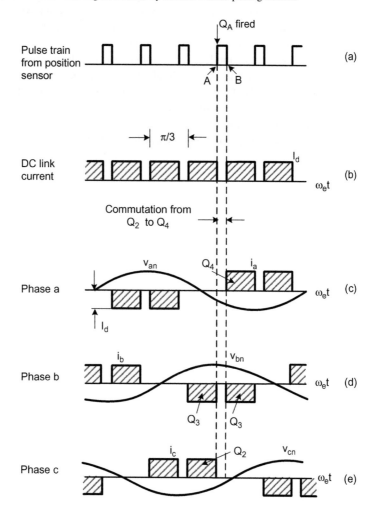

The load commutation of a synchronous machine requires the machine to run at a certain speed to develop the CEMF that causes inverter commutation. This means that if the speed is below a critical limit (typically 5%), some type of forced commutation of the inverter is needed. A type of forced commutation that is very popular is known as the pulsed or dc link current interruption method, which is described in this figure by the waveforms. The inverter is self-controlled by the position sensor as usual. However, dc link current I_d is reduced to zero every 60° for commutation of inverter thyristors. Consider Figure 5.2 and assume that the inverter uses thyristors instead of GTOs. Again, assume that in

Figure 5.2, Q_3 is conducting in the upper side, whereas Q_2 is conducting in the lower side. At the 60° angle, commutation from Q_2 to Q_4 is desired. At corresponding point A in this figure, the rectifier thyristors are blocked and the auxiliary bypass thyristor Q_A across L_d is fired. The rectifier is dragged into inverting mode when I_d current in inductor L_d goes into freewheeling mode, but the stored energy in the machine phases is pumped to the line. Thus, the current in Q_3 and Q_2 falls to zero, turning them off. Then, at point B, Q_3 and Q_4 are fired together to resume the phase currents. Thus, commutation is achieved successfully and the current is transferred from phase c to phase a. The start-up is more convenient if load current (i.e., torque) is kept at a low level. The developed torque develops the speed, and at the appropriate point the machine is transitioned to load commutation.

FIGURE 8.43 Control block diagram of LCI-WFSM drive.

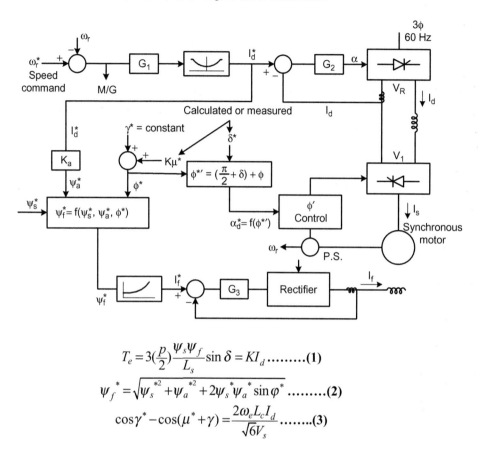

$$T_e = 3(\frac{p}{2})\frac{\psi_s \psi_f}{L_s} \sin \delta = KI_d \ldots \ldots .(1)$$

$$\psi_f^* = \sqrt{\psi_s^{*2} + \psi_a^{*2} + 2\psi_s^* \psi_a^* \sin \varphi^*} \ldots \ldots .(2)$$

$$\cos\gamma^* - \cos(\mu^* + \gamma) = \frac{2\omega_e L_c I_d}{\sqrt{6}V_s} \ldots \ldots .(3)$$

Once the motor is started and minimum speed is developed, the machine CEMF will permit load commutation of the inverter. This figure shows the control diagram of the load-commutated inverter drive with minimum turn-off angle γ^*. Because the motor is operated at near unity power factor (slightly leading), the developed torque is proportional to current I_d as shown in Eq. (1). Therefore, the speed control loop controls I_d by controlling the firing angle α of the rectifier. The two additional control variables of the system are field current I_f and phase angle φ'. These are controlled to satisfy the phasor diagrams in Figure 8.41. Equation (2) generates the flux command ψ_f^* from the flux diagram, where $\psi_s^* =$ constant, $\psi_a^* = L_s I_s = k_a I_d$ and $\varphi^* = \gamma^* + k\mu^*$ (which is less than β angle). The ψ_f signal generates the command current I_f through a function generator, where the saturation effect is taken into consideration. Then, I_f is controlled by a closed loop as shown. Input signal ψ_a is easily generated from current command I_d^*. Angle μ is a function of I_d, which can be solved by Eq. (3), where $V_s = \omega_e \psi_s^*$, but ψ_s is a constant.

The actual μ angle can also be obtained by processing the voltage signal across inverter thyristor. Torque angle δ can be estimated by solving Eq. (1), or measured from the actual waveforms. The generation of phase angle $\varphi^{*\prime}$ is shown in the figure. Corresponding delay angle α_d^* generation and control of the $\varphi^\prime$ angle are discussed in the next figure. In motoring mode, as the speed increases, dc link voltage V_R increases, ultimately saturating the rectifier and thus losing the current control. Higher range speed control is possible by weakening ψ_s inversely with speed so as to make the current control effective. The motoring/generating (M/G) mode signal is derived from the polarity of the speed loop error. At generating mode, power factor angle $\varphi = 180°$, and angle $\varphi^\prime$ is altered to put the flux triangle in the second quadrant as shown in Figure 8.41. Because estimated signals may not be very accurate, an extra margin of safety should be incorporated to avoid commutation failure in the inverter. Note that with load commutation, the control response is inherently sluggish. A sudden increase of load is likely to cause commutation failure.

FIGURE 8.44   (a) Control diagram of angle α_d and (b) motoring mode α_d control waves.

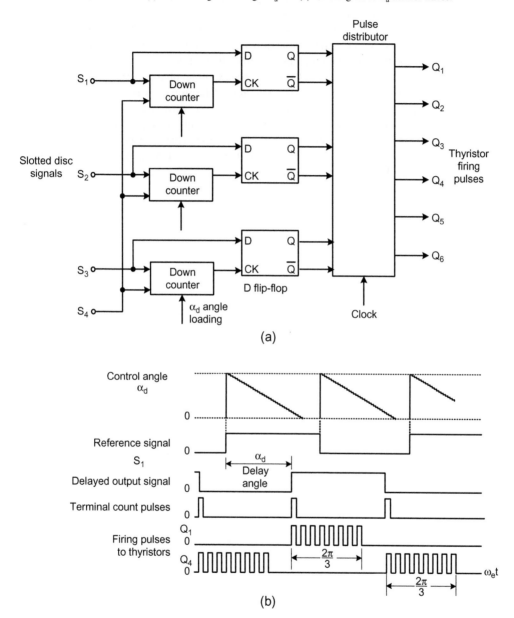

The principle of delay angle (α_d) control and its corresponding waveforms are explained in this figure. The slotted disk absolute encoder signals (S_1–S_4) from Figure 8.9 are received at the input of (a) as shown. Each channel of the respective phase delay control consists of a programmable down-counter followed by a D flip-flop. The outputs of the flip-flops are combined in the pulse distributor to generate a 120° wide pulse train for the inverter thyristors. Consider, for example, phase a firing pulse generation from the sensor S_1 signal. At the leading edge of the square wave, angle α_d is loaded to the down-counter and the corresponding sawtooth wave is generated in (b) by the clock pulses from S_4. At the terminal count, a clock pulse is applied to the D flip-flop, which generates an α_d angle phase-shifted square wave to the pulse distributor. All phase-shifted square waves for the three phases are then combined in the decoder, and the 120° wide pulse train is generated for each phase. The lower figure shows the firing pulses for Q_1 and Q_4 for the phase a group which are at a 180° phase difference.

FIGURE 8.45 (a) Phasor diagram explaining reference signal from S_1 and (b) waveforms showing relation between α_d and φ'.

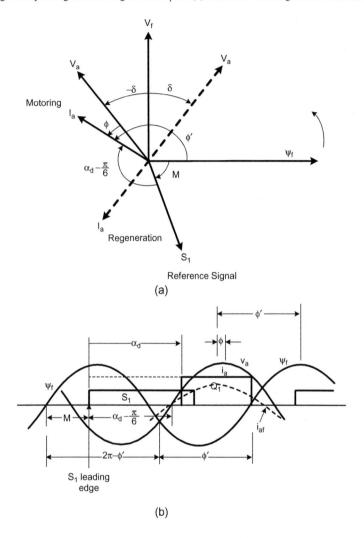

(a)

(b)

Sensor S_1 for the phase a reference signal is positioned at the lagging angle with respect to the ψ_f wave, and S_2 and S_3 are located at a 120° phase difference (not shown). Thyristor Q_1 is fired at angle α_d to generate a 120° current pulse for phase a as shown in the lower figure. This pulse leads the phase a voltage wave v_a by angle φ. Therefore, the total lag angle of i_a with respect to ψ_f is angle φ' as shown in the figure. Evidently then, the i_a wave lags the S_1 leading edge by $\alpha_d - 30°$ angle as shown. Therefore, the relation between α_d and φ' can be given from the relation $M + (\alpha_d - 60°) = 360° - \varphi'$, i.e., $\alpha_d = K - \varphi'$, where $K = 360° + 30° - M$. This expression indicates that φ' varies linearly with α_d. The waveforms show motoring mode operation only. In the regenerating mode, the current pulse is placed opposite the v_a wave, as shown in the phasor diagram.

FIGURE 8.46 (a) Machine terminal voltage processing with zero-crossing detector (ZCD) and (b) ZCD principle.

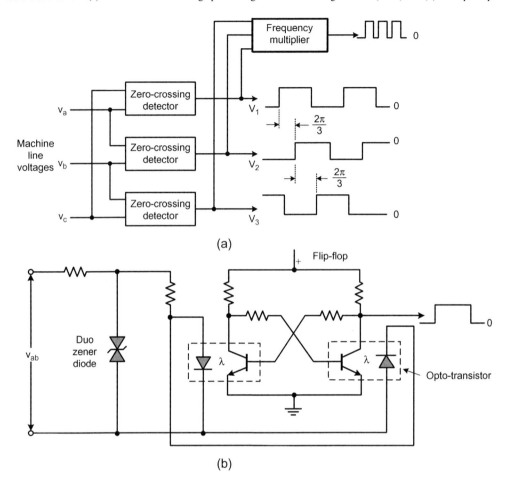

(a)

(b)

If zero speed starting, as described in Figure 8.42, is not needed and machine speed control at very low speed is not essential, the absolute shaft position encoder can be eliminated. Alternatively, the machine can be started at open loop by variable frequency. Then at a finite speed the terminal voltage waves can be sensed [10] and used for generation of self-control signals. However, the machine voltage waves contain large commutation spikes that require cleanup. Part (a) shows sensing of line voltages and converting them into 120° phase-shifted square waves through zero-crossing detectors. The details of the ZCD through an optotransistor flip-flop to provide isolation are shown in the lower figure. The flip-flop is not sensitive to commutation spikes and gives square-wave output in phase with line voltage v_{ab}. There will be a small phase error at the output (which can be compensated) because of the threshold voltage required for the optocoupler. The frequency multiplier is explained in the next figure.

FIGURE 8.47   (a) Frequency multiplier principle and (b) waveforms for phase a firing pulse generation.

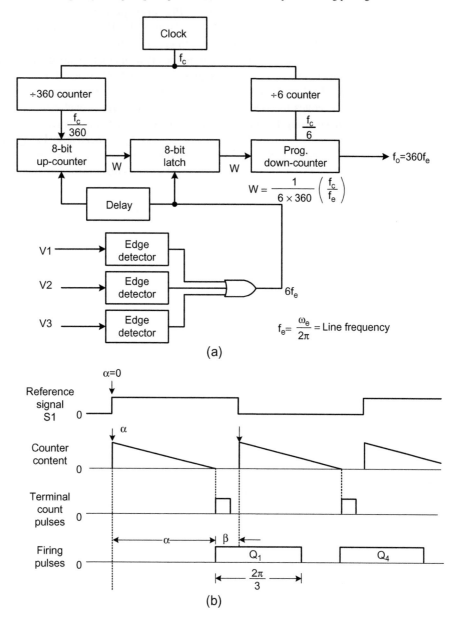

The frequency multiplier generates output frequency $f_0 = 360f_e$, i.e., at $1.0°$ interval of the line frequency f_e. The conventional PLL principle is not satisfactory because of the large frequency range involved. All square waves (V_1, V_2, and V_3) from the previous figure are combined through edge detectors to generate the $6f_e$ signal as shown. This signal loads the content of an 8-bit up-counter to the latch and then clears the up-counter with a delay. The up-counter receives the frequency signal $f_c/360$, where f_c is generated from a crystal clock. The digital count $W = (1/6fe)(fe/360)$ is loaded into a programmable down-counter that generates the output frequency as shown. The waveforms in the lower figure are generated using the principle from Figure 8.44. Square waves V_1, V_2, V_3, and clock f_0 represent the signals from S_1, S_2, S_3, and S_4, respectively. Firing angle α [or α_d in Figure 8.44(b)] in this case is generated from the relation $\alpha^* = 180° - \beta = 180° - (\gamma^* + \mu)$, where $\gamma^* = $ constant turn-off angle and $\mu = $ calculated overlap angle.

FIGURE 8.48 Phase-locked loop (PLL) control at γ^* angle of load-commutated inverter drive.

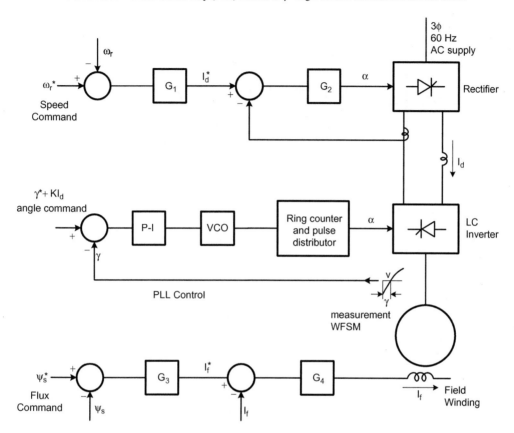

A simpler control method for a LCI-WFSM drive that uses the phase-locked loop (PLL) principle is shown in this figure. Again, it is assumed that the machine has sufficient CEMF, and it is started by some auxiliary method. The speed control loop independently controls the torque by controlling dc current I_d as before. Constant stator flux ψ_s is controlled by controlling the machine field current. Turn-off angle γ^* is controlled independently by the PLL principle. However, a parameter kI_d is added to γ^* to keep a margin of safety against commutation failure. Feedback angle γ is obtained from the six thyristors at 60° phase difference in the worst case basis. If actual γ tends to decrease due to loading, the VCO input will increase and the firing angle α will advance appropriately. For the regeneration mode detected by the error polarity of the speed loop, angle γ is advanced to near 180°. If field-weakening control is desired, ψ_s can be decreased inversely with speed. One disadvantage of PLL control is that the transient response is much slower because of the sluggish PLL control. Note that as the frequency of the drive is lowered, the turn-off time margin is increased because $\gamma = t_{off}\,\omega_e$.

FIGURE 8.49 *Queen Elizabeth 2 (QE2) cruise ship diesel-electric propulsion system.*

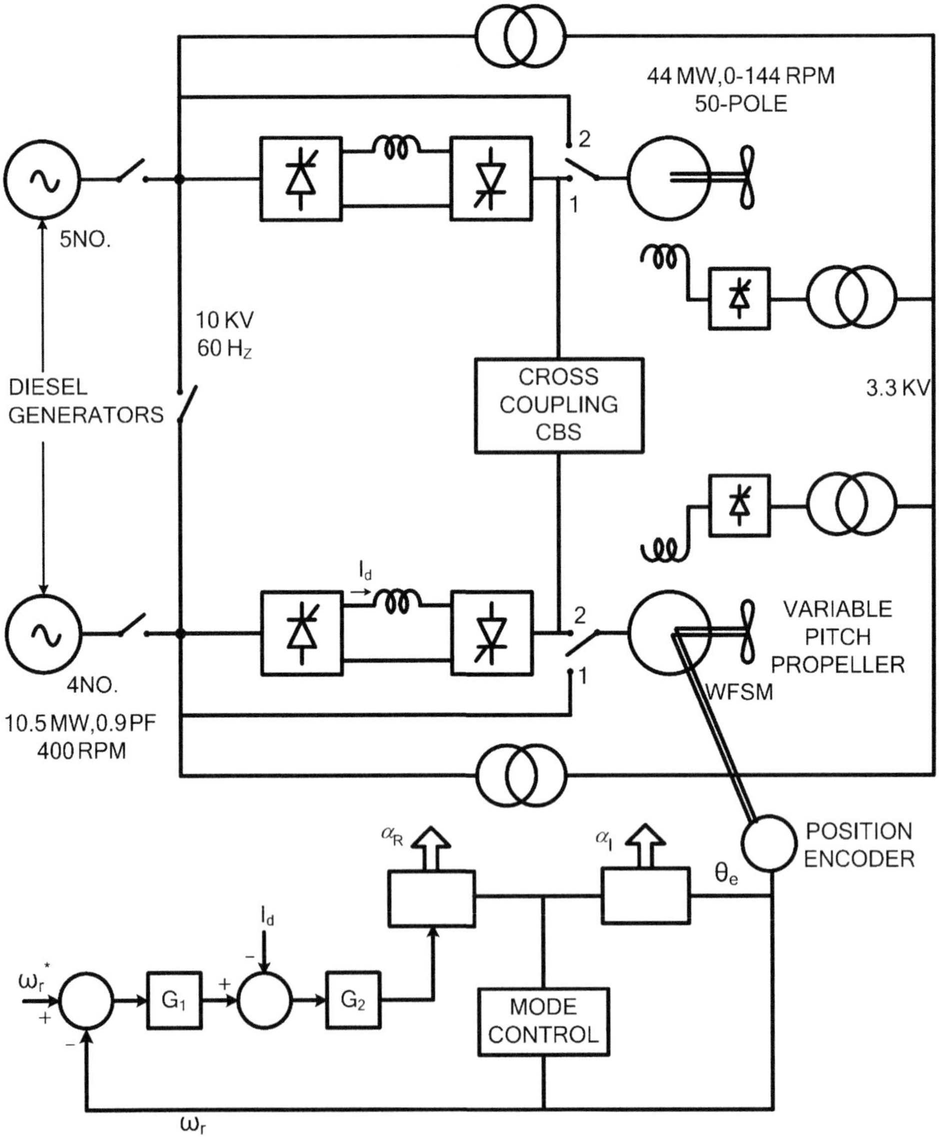

We now look at some sample applications for large drives. In recent years, variable-frequency drives have become popular in ship propulsion. Large cruise ships, such as *Carnival, Crystal,* and *Queen Elizabeth 2* (Cunard), use load-commutated inverter (LCI), variable-frequency synchronous motor drives for propulsion. The motivation for using variable-frequency drives is mainly fuel savings and flexibility of equipment installation. A simplified block diagram of the *QE2*'s propulsion system [11] is shown in this figure. Diesel engine-generators run at constant 400 RPM (i.e., 60 Hz) such that their efficiency is high. The 60-Hz bus power is converted to variable frequency at variable current to control the speed of the propulsion motor. However, when the inverter frequency rises to 60 Hz, the machine is switched directly to a 60-Hz line to improve system efficiency. Thus, the converter system is basically used as a soft-starter. The cross-connection shown permits a machine to operate from either converter system in case of a converter fault. The turbine propeller pitch angle is controlled to control load torque. The motor is started with the inverter up to 50% rated speed at no load, and then the load torque is established by controlling the propeller pitch angle. The machine excitation is controlled to maintain near unity power factor (slightly leading) angle. The drive has speed control in the outer loop with developed torque control ($T_e \propto I_d$) in the inner loop. The drives are regenerative, and their speed is reversible. The mode controller shown permits the drive to operate in four modes, i.e., harbor, ready-to-sail, combinatory (converter-driven), and free-sailing (direct bus operation). At any motor speed, the propeller pitch angle adjusts the motor load and permits the ship's speed to be varied. (*QE2* has recently been added with the world's largest cruise ship *Queen Mary 2* (*QM2*).)

FIGURE 8.50   Features of *QE2* propulsion system.

- NINE DIESEL GENERATOR UNITS – 10.5 MW, 0.9 PF, 10 kV, 60 Hz, 400 RPM (EACH)
- TWO WFSMs WITH EXTERNAL DC BRUSH EXCITATION – 44 MW, 0–144 RPM, 50-POLE, UNITY PF (EACH)
- SIX-PULSE RECTIFIERS AND SIX-PULSE LOAD-COMMUTATED INVERTER SYSTEMS
- MOTOR START-UP WITH CONVERTER, BUT SWITCH OVER TO 60-Hz LINE SUPPLY AT FULL MOTOR SPEED (144 RPM)
- CONVERTER DC CURRENT INTERRUPTION MODE AT START-UP (<10% SPEED), BUT CEMF LOAD COMMUTATION AT HIGHER SPEED
- VARIABLE-PITCH PROPELLER TO CONTROL LOAD TORQUE
- PROPULSION SPEED RANGE BY CONVERTER: 72–144 RPM
- REVERSIBLE SPEED WITH REGENERATION
- SPEED CONTROL WITH INNER LOOP I_d CURRENT CONTROL
- FULL LOAD EFFICIENCY: GENERATOR, 97.3%; MOTOR, 98%

The salient features of the *QE2* propulsion system are summarized here. The system uses a six-pulse current-fed converter system instead of a 12-pulse scheme for such a large drive in order to save space from large phase-splitting transformers. Of course, the penalty is larger harmonic loading of the generators and motors. In spite of harmonic loading and lagging PF of the rectifier, the minimum generator power factor is 0.9. The sixth harmonic pulsating torque effect on the motors is minimal because of the turbine load damping effect. The LCI drive (over the cycloconverter drive) has the advantage that it permits the output frequency to match with the line frequency (60 Hz) for successful transition of the drive to the line. The converter system is basically used as a soft-starter for the drive. Of course, an LCI-WFSM drive is the most economical and efficient for such a large drive. The motor power factor is slightly leading in the converter mode (72–144 rpm), but is unity after switchover to the line. The propeller pitch angle control has the advantage that at any speed the converter loading can be manipulated. Besides, it permits motor speed to be decoupled from ship speed. The motor develops the rated 44 MW of power when it is in "free-sailing" mode in the ocean at the highest ship speed with a motor speed of 144 rpm. The generators, motors, transformers, and converters have elaborate protection systems, which are not shown.

FIGURE 8.51 An 8800-hp cycloconverter-WFSM cement mill drive control.

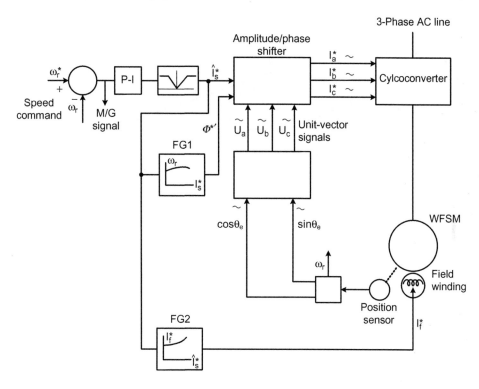

Phase-controlled cycloconverter drive with WFSM is very popular in large power cement mills, mine hoists, rolling mills, ship propulsion, etc. The machine is operated at unity PF and the cycloconverter is commutated from the line. Traditionally, cement mills had been driven by constant-speed WFSM through a large reduction gear that provided slow rotation. With a cycloconverter, the frequency can be controlled to a low value and the reduction gear can be eliminated (gearless drive). This figure shows the simplified position sensor-based self-controlled 8800-hp cycloconverter drive manufactured by ABB [12] for a cement mill drive. The speed-control loop generates the stator current command, which is proportional to torque because machine PF is unity. From this command, the field current I_f^* and MMF phase angle φ'^* signals are derived through function generators as shown. These will be further explained in the next figure. The absolute position sensor generates the feedback speed signal and unit vector angles $\cos\theta_e$ and $\sin\theta_e$ for self-control, where $\cos\theta_e$ is aligned with the ψ_f phasor. Then, three-phase unit vector signals are generated by the relations $U_a = \cos\theta_e$, $U_b = \cos(\theta_e - 120°)$, and $U_c = \cos(\theta_e + 120°)$. These signals are then used to generate the phase current commands as $i_a^* = I_s^* U_a < \varphi'^*$, $i_b^* = I_s^* U_b < \varphi'^*$, and $i_c^* = I_s^* U_c < \varphi'^*$ through the amplitude and phase shifter as explained in the next figure.

FIGURE 8.52   (a) Motoring mode phasor diagram with function generators and (b) generating mode phasor diagram.

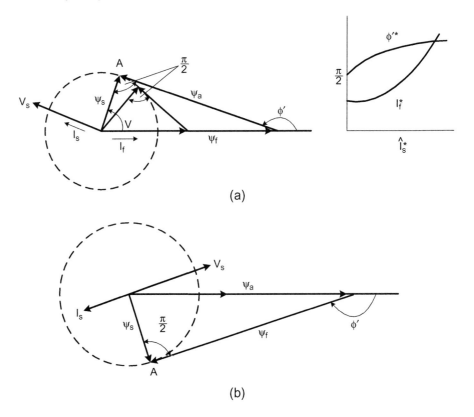

(a)

(b)

The phasor diagrams explain the operation of the cycloconverter drive from the previous figure. These are essentially the same as Figure 8.41, except the power factor angle $\varphi = 0$. Because stator flux ψ_s is constant irrespective to stator current, the locus of point A is a circle that moves in the counterclockwise direction with higher current. If $I_s = 0$, $\psi_s = \psi_f$ and are co-phasal, and the angle $\varphi'^* = 90°$. As the stator current increases, both I_f and φ'^* increase which can be calculated and plotted graphically on the right side of (a). Any saturation of ψ_f can be incorporated in the I_f function generator. The function generators used in Figure 8.51 satisfy the phasor diagrams at any load condition. The phasor diagram in the regenerative mode is similar except the flux triangle is in the second quadrant because I_s is in opposite phase with V_s. The polarity of φ' is controlled by the M/G signal of the speed control loop. Note that the drive control is scalar because the decoupling between I_s and ψ_s is not valid in the transient condition because of the sluggishness of the I_f response.

FIGURE 8.53   Features of 8800-hp cycloconverter-WFSM cement mill drive.

- GEARLESS CEMENT MILL DRIVE
- 8800-HP, 40-POLE, 0–15 RPM (0–5 Hz), UNITY PF WYE-CONNECTED WOUND-FIELD SYNCHRONOUS MACHINE
- DC BRUSH EXCITATION WITH THREE-PHASE BRIDGE RECTIFIER
- 36-THYRISTOR DUAL-BRIDGE CYCLOCONVERTER
- DELTA–WYE–WYE SINGLE-CORE TRANSFORMER SUPPLIES CYCLOCONVERTER
- BLOCKING MODE OPERATION
- SINUSOIDAL OUTPUT PHASE VOLTAGE, BUT SATURATES TO TRAPEZOIDAL WAVE AT RATED VOLTAGE AND FREQUENCY
- SCALAR SELF-CONTROL WITH VOLTS/Hz
- FOUR-QUADRANT CONTROL CAPABILITY

The figure summarizes the basic features of the commercial cycloconverter-fed synchronous machine drive for the cement mill. In a cement mill, a large drum containing steel balls rotates slowly and crushes the cement clinker. The starting torque demand of the mill is typically 60% higher than the rated torque. The speed of rotation is very slow and varies from standstill 15 rpm. The cycloconverter-WFSM drive adapts very well to such an application. Unlike an LCI drive, the machine can always be operated at unity DPF from zero speed so that there is no reactive current loading on the cycloconverter. The cyclo-converter is the standard six-pulse type (36-thyristor) with transformer isolation at the input side so that the wye-connected machine can be connected directly to the wye-connected cycloconverter. Because the output frequency is very low with the supply frequency of 50 Hz, high-quality sinusoidal voltage waves could be fabricated. At low speed, the system PF is low. However, with the linear volts/Hz relation, at the rated frequency (5 Hz), the output voltage wave is made trapezoidal in order to get high input PF (>0.86). The resulting harmonic penalty of the machine is low because of its high impedance. The machine uses brush excitation through slip rings to get a high response from the field current. The regenerative torque helps to stop the machine without wasting any power. The speed reversal is hardly used. The machine is self-controlled and, therefore, has hardly any stability or hunting problems.

FIGURE 8.54   Vector control of cycloconverter-WFSM drive with the phasor diagram.

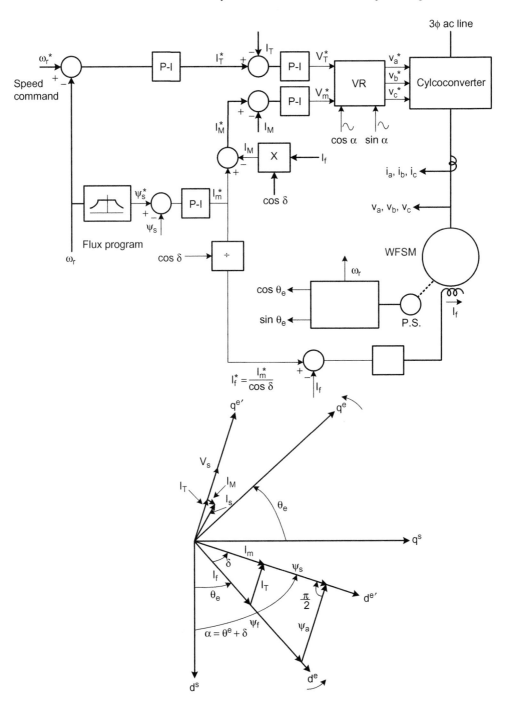

The transient response of a cycloconverter-WFSM drive can be considerably enhanced by vector control as shown in this figure. Because the response of I_f is sluggish, the decoupling phasor diagram in steady state is not satisfied during a transient. The shortage of I_f can be compensated by temporarily injecting magnetizing current from the stator side. In the figure, the speed loop increases the torque component of current I_T^*, which is aligned in the direction of the $q^{e\prime}$ axis of the phasor diagram. Stator flux ψ_s should remain constant in the constant-torque region, but decreases with speed during field-weakening vector control. The ψ_s loop controls I_m in the $d^{e\prime}$ axis to keep the flux constant. Because $I_f^* = I_m/\cos \delta$ in steady state, a sudden increase in torque angle δ will increase I_f^*. But the response delay of I_f will cause a finite I_M^* trying to maintain constant ψ_s. As I_f builds up, I_M decreases until it vanishes completely to satisfy the phasor diagram at unity DPF. The I_T^* and I_M^* control loops generate the corresponding voltage signals by synchronous current control. The unit vector aligns v_T^* and v_M^* in the directions of the $q^{e\prime}$ and $d^{e\prime}$ axes, respectively. Obviously, vector control of WFSM is not as fast as an induction motor drive. The complex signal processing required for vector control is described in the next two figures.

FIGURE 8.55 Feedback signal processing block diagram for vector control.

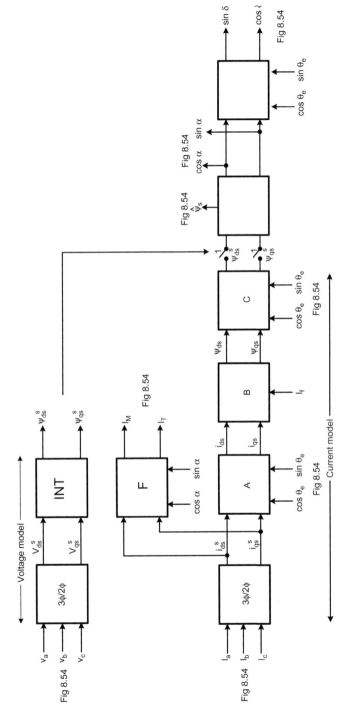

The figure gives the detailed block diagrams for computation of the feedback signals required for vector control. The equations for blocks A, B, C, D, E, and F are given in the next figure. Obviously, a powerful DSP is needed to implement complete control. It is assumed that the drive operates from zero speed up to the field-weakening region. Therefore, both the voltage model and current model are used to compute stationary frame stator fluxes ψ_{ds}^s and ψ_{qs}^s as indicated in the figure. The current model is highly parameter sensitive, but is essential in the low-speed range. As the speed develops and crosses a critical value, the flux computation is transitioned to the voltage model. The voltage model requires ideal integration but the parameter variation effect is negligible. Of course, the voltage model neglects stator resistance (R_s), which is very small in a large machine.

FIGURE 8.56 Computation equations for the various blocks in Figure 8.55.

A:
$$i_{ds} = i_{ds}{}^s \cos \theta_e + i_{qs}{}^s \sin \theta_e \ldots\ldots\ldots (1)$$
$$i_{qs} = -i_{ds}{}^s \sin \theta_e + i_{qs}{}^s \cos \theta_e \ldots\ldots\ldots (2)$$

B:
$$\psi_{ds} = i_{ds}L_{ls} + i_{dm}{}'L_{dm} = i_{ds}L_{ls} + L_{dm}(I_f + i_{ds})\frac{(R_{dr} + SL_{ldr})}{R_{dr} + S(L_{ldr} + L_{dm})}$$
$$\ldots (3)$$
$$\psi_{qs} = i_{qs}L_{ls} + i_{qm}L_{qm} = i_{qs}L_{ls} + L_{qm}i_{qs}\frac{(R_{qr} + SL_{lqr})}{R_{qr} + S(L_{lqr} + L_{qm})}$$
$$\ldots(4)$$

C:
$$\psi_{ds}{}^s = \psi_{ds} \cos \theta_e - \psi_{qs} \sin \theta_e \ldots\ldots\ldots (5)$$
$$\psi_{qs}{}^s = \psi_{ds} \sin \theta_e + \psi_{qs} \cos \theta_e \ldots\ldots\ldots (6)$$

D:
$$\hat{\psi}_s = \sqrt{\psi_{ds}{}^{s2} + \psi_{qs}{}^{s2}}, \ldots\ldots\ldots (7)$$
$$\sin \alpha = \psi_{qs}{}^s / \hat{\psi}_s, \ldots\ldots\ldots (8)$$
$$\cos \alpha = \psi_{ds}{}^s / \hat{\psi}_s, \ldots\ldots\ldots (9)$$

E:
$$\sin \delta = \sin(\alpha - \theta_e) = \sin \alpha \cos \theta_e - \cos \alpha \sin \theta_e \ldots\ldots (10)$$
$$\cos \delta = \cos(\alpha - \theta_e) = \cos \alpha \cos \theta_e - \sin \alpha \sin \theta_e \ldots\ldots (11)$$

F:
$$I_M = i_{ds}{}^s \cos \alpha + i_{qs}{}^s \sin \alpha \ldots\ldots (12)$$
$$I_T = -i_{ds}{}^s \sin \alpha + i_{qs}{}^s \cos \alpha \ldots (13)$$

The computation equations for vector control in the six blocks of Figure 8.55 are shown. Block A converts the stationary frame stator currents, which are aligned with the d^s-q^s axes (see Figure 8.54) to rotating frame currents aligned with the d^e-q^e axes. These are essentially VR^{-1} relations. Block B uses the transient d^e-q^e equivalent circuits (Figure 6.42) of a synchronous machine and calculates the rotating frame stator flux components using i_{ds}, i_{qs} currents and field current I_f. It is assumed that the machine has a damper winding. Block C converts the fluxes back to the stationary frame. Block D calculates the flux magnitude and unit vectors for alignment to the $d^{e'}$-$q^{e'}$ rotating frame axes. Block E calculates $\cos \delta$ from the unit vector signals. Finally, block F calculates I_T and I_M, which are aligned with the $q^{e'}$-$d^{e'}$ axes. All of these computed signals are used in Figure 8.54.

FIGURE 8.57 Icebreaker diesel-electric ship propulsion with cycloconverter-WFSM drive.

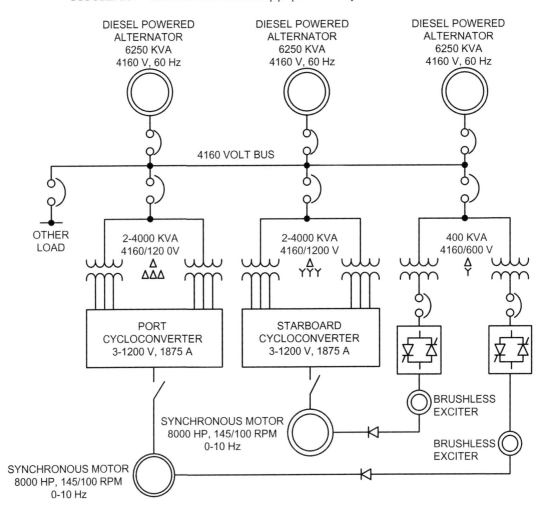

The figure shows the diesel-electric ship propulsion system manufactured by Canadian GE and installed on an icebreaker ship on the St. Lawrence River in Canada [15]. The St. Lawrence River freezes in winter, and the function of the ship is to keep it navigable throughout the year that uses the propellers as ice cutters. The propulsion system basically consists of cycloconverter-fed variable frequency synchronous motor drives that use 4160-V, 60-Hz bus power generated by diesel-powered alternators, as shown in the figure. The engine efficiency is high at constant speed, which permits a significant amount of fuel savings. The phase-controlled 36-thyristor cycloconverters operate in the 6-pulse mode. However, the input phase-splitting transformers permit 12-pulse operation at the input to reduce generator harmonic loading. The motor field current is supplied by a brushless exciter. The machine is controlled to operate at unity DPF so that there is no reactive loading on the machine and cycloconverter. In the constant-torque region, the machines are operated with direct stator flux-oriented vector control, but in the field-weakening mode, the control is scalar when the voltage waves are trapezoidal. At low speeds, the stator flux vectors are calculated from the current model, whereas at higher speed they transition to the voltage model. The instantaneous phase currents are controlled by feedback with calculated injection of feedforward CEMF to enhance the response.

FIGURE 8.58   Features of the icebreaker's propulsion cycloconverter-WFSM drive.

- INSTALLED BY CANADIAN GE FOR ICE BREAKING IN ST. LAWRENCE RIVER
- CONSTANT BUS VOLTAGE AT FIXED SPEED DIESEL ENGINE (4160 V, 60 Hz) TO IMPROVE ENGINE FUEL CONSUMPTION
- 36-THYRISTOR, 6-PULSE BLOCKING-MODE CYCLOCONVERTER
- SELF-CONTROLLED WFSM DRIVE WITH POSITION SENSOR (8000 HP, 12-POLE, 0–180 RPM, 0–18 Hz)
    - BRUSHLESS EXCITATION
    - SPEED REVERSAL BUT NO REGENERATION
    - UNITY MACHINE DPF
    - DIRECT VECTOR CONTROL WITH STATOR FLUX ORIENTATION
    - CURRENT MODEL FLUX VECTOR ESTIMATION AT LOW SPEED, BUT VOLTAGE MODEL ESTIMATION AT HIGH SPEED
    - INSTANTANEOUS PHASE CURRENT CONTROL WITH ESTIMATED FEEDFORWARD CEMF INJECTION
- SCALAR CONTROL IN FIELD-WEAKENING MODE WITH TRAPEZOIDAL VOLTAGE WAVE

The salient features of the icebreaker propulsion system are summarized here. The traditional 6-pulse, 36-thyristor isolated supply cycloconverter is used, and the 30° phase shift between the two transformer banks permits low 12-pulse harmonic loading of the generator. The generator is oversized due to harmonic loading as well as a low DPF load for the cycloconverter. The motor provides rated torque up to a speed of 145 rpm (base speed), and then it runs in the field-weakening mode up to 180 rpm. The low-output frequency ratio of the cycloconverter permits blocking mode operation and sinusoidal output voltage for vector control. Drive regeneration is possible, but is not used because the diesel-power alternator cannot absorb the power. Closed-loop speed control is used with stator flux-oriented vector control, where the flux vector is computed from the current model at low speed, but because of parameter variation problems, the voltage model is used at higher speed. The principle was described before. The absolute position sensor on the shaft helps vector rotation (VR) for the current model and supplies the feedback speed signal. In recent years, cycloconverters have been replaced by multilevel PWM converters.

FIGURE 8.59 12-MW dual cycloconverter-WFSM drive for mining ore-crushing mill.

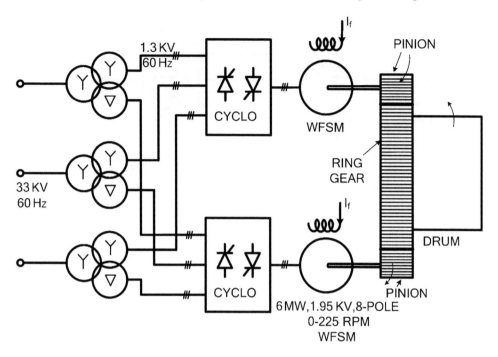

The figure shows a high-capacity (12-MW) dual cycloconverter-WFSM drive [16] for a mining ore-crushing mill. The two identical drives transmit torque to the respective pinions, which are coupled to the central ring gear with the drum. Each thyristor phase-controlled cycloconverter generates sinusoidal voltage at variable frequency (0–15 Hz) for the machine, which operates in the speed range of 0–225 rpm for four quadrants (reversible speed with regeneration) operation. The 6-pulse or 36-thyristor cycloconverter operates in blocking mode. There are three phase-shifting 60-Hz, 33-kV/1.3-kV transformers on the line side, where the upper cycloconverter is supplied from secondary wye windings and the lower unit is supplied from the delta windings. Therefore, the line harmonics correspond to 12-pulse operation. Thus, the transformers provide line isolation, voltage stepdown, and line pulse multiplication. By excitation control, the machines always operate at unity power factor so that the cycloconverter and machine loadings are minimum. However, the line DPF is low with a high degree of harmonics. A static VAR compensator and hybrid active filter are installed at the line (not shown). The drive system is fuseless and there is no circuit breaker. However, there is a high-speed breaker on the line (not shown), which is tripped in case of a line voltage dip along with cycloconverter gate blocking protection. The speeds of the machines are closely tracked with stator flux-oriented vector control in the inner loops.

FIGURE 8.60   10-MVA three-level converter-WFSM drive for rolling mill.

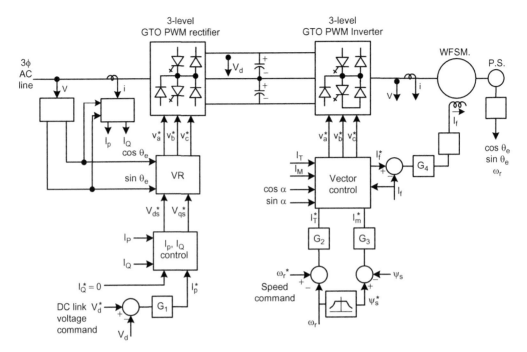

The figure shows a practical two-sided PWM voltage-fed converter-WFSM drive system for rolling mill [17]. Vector control is applied on both sides, and the principle of inverter vector control is similar to that described earlier for a cycloconverter drive. The control principles are applicable to two-level as well as multilevel converter drives. Recently, cycloconverter being replaced by PWM converter systems to solve the harmonics and power factor problems. The inverter-machine section uses closed-loop speed control with stator flux control in the constant-torque and field-weakening regions of vector control as described before. The details of feedback signal processing are same as in Figure 8.55 and are not shown here. The speed loop generates the torque component of current I_T^*, and the flux loop generates the flux component of current I_m^*. The remaining control similar to Figure 8.54 is lumped in the vector control block. The absolute position sensor mounted on the shaft generates the unit vector signals $\cos \theta_e$ and $\sin \theta_e$ as shown. The line-side PWM rectifier permits the line current to be sinusoidal at programmable power factor that can be leading, lagging, or unity. It maintains the dc link voltage V_d constant in the outer loop, which generates the active current command I_P^*. The reactive current command I_Q^* is shown as zero. Both I_P and I_Q are controlled by synchronous current control. The VR block receives the unit vector signals by processing the line voltages as explained before. The neutral point voltage of the diode-clamped converters is controlled by balancing the neutral currents. Both the converters use the space vector PWM technique. Note that a line voltage transient-induced commutation failure in a cycloconverter is not possible in this configuration.

FIGURE 8.61 Features of PWM converter synchronous motor drive for steel rolling mill.

- PWM THREE-LEVEL CONVERTER SYSTEM WITH HIGHEST GTO RATINGS (6000 V, 6000 A) – BY MITSUBISHI
- SOLVES LOW-POWER FACTOR AND HARMONICS PROBLEMS OF CYCLOCONVERTER
- DC LINK VOLTAGE: 6000 V
- REGENERATIVE SNUBBER WITH DC-DC CONVERTER GIVES 97% CONVERTER EFFICIENCY
- SPACE VECTOR PWM WITH MINIMUM PW CONTROL
- SUPPRESSED NEUTRAL VOLTAGE FLUCTUATION
- FOUR-QUADRANT OPERATION: 0–60 Hz, 0–3600 V OUTPUT
- FIELD-WEAKENING RANGE: 2.25:1
- PEAK OUTPUT – 15 MVA FOR 1.0 MINUTE
- DIRECT VECTOR CONTROL ON BOTH CONVERTERS
- MORE RELIABLE THAN CYCLOCONVERTER

Mitsubishi recently developed the rolling mill drive (Figure 8.60) to replace the traditionally used cycloconverter-WFSM drive. The principal advantages in the voltage-fed converter system are unity line power factor, sinusoidal line current, boost rectifier operation with line voltage sag, and more reliable operation. The scheme is based on NPC or three-level diode-clamped converters using the highest power GTOs manufactured by the company. To reduce the large switching losses of GTOs, energy recovery from the capacitor-diode snubbers is made through dc-dc converters. This makes the efficiency of the converter system high (97%). The drive system has the usual four-quadrant control capability required by the rolling mill and the system transient response is high because of vector control on both sides. Because the machine is operated at unity DPF, the system can deliver 10 MW power on a steady-state basis and 15 MW (for 1.0 minute) on a transient basis. The drive operates smoothly in the vector control mode in both the constant-torque and constant-power regions.

FIGURE 8.62   25-MW superconducting synchronous motor ship propulsion system.

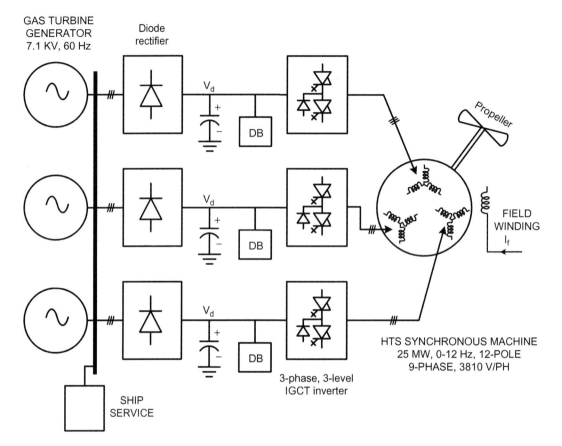

The figure shows the simplified block diagram of a naval ship propulsion system using a high-temperature superconductivity (HTS with liquid nitrogen) wound-field synchronous machine. Compared to a standard WFSM, an HTS machine is smaller, lighter, and has higher efficiency [18, 19]. The machine has ironless construction, which eliminates the acoustic noise signature for the enemy. The harmonic torque on the machine and the corresponding vibration and mechanical resonance possibilities are eliminated by its high-quality power supply. The main ratings for the machine are indicated on the figure. Field current I_f can be controlled slowly (to avoid skin effect heating) to maintain unity power factor on the machine terminal. There are nine isolated phase windings with 3 three-phase groups at 40° angular orientation as shown. Each three-phase group of the machine is fed with a variable-frequency, variable-voltage power supply from an IGCT-based three-level diode-clamped inverter. Multiplicity of three-phase groups provides redundancy in case of inverter failure. In the front end, 60-Hz gas turbine-generator power is rectified to dc by a three-phase diode rectifier, filtered by a C filter, and then fed to the inverter. All three converter channels are identical. The machine has reversible speed control, but there is no regeneration. Although inherent braking torque of the turbine is provided by the load, a dynamic brake (DB) is provided on the dc link. The converter system is also ironless to avoid acoustic noise. The drive operates in the vector control mode within the speed control loop covering the speed range of 0–120 rpm in the constant-torque region. The ship service power at 60 Hz is derived from the main supply bus.

FIGURE 8.63   Features of superconducting magnet ship propulsion system.

- SYNCHRONOUS MACHINE:
  - LIQUID NITROGEN COOLED (HTS) FIELD WINDING
  - IRONLESS CONSTRUCTION
  - RATED POWER: 25 MW
  - NUMBER OF PHASES: 9
  - PHASE VOLTAGE: 3810 V
  - NUMBER OF POLES: 12
  - FREQUENCY RANGE: 0–12 Hz
  - SPEED RANGE: 0–120 RPM
  - POWER FACTOR: 1.0
  - EFFICIENCY: 96%
- SUPPLY BUS: 7100 V, 60 Hz
- DIODE-CLAMPED NPC VOLTAGE-FED CONVERTER:
  - 6.5-kV, 3000-A (peak) IGCT WITH INTEGRATED DIODE
  - 1.0-kHz SWITCHING FREQUENCY
  - SPACE VECTOR PWM
  - HARD-SWITCHED WITH REGENERATIVE SNUBBER
  - DC LINK VOLTAGE: 10,000 V
  - C FILTER: C(SPLIT) = 5000 µF
  - NEUTRAL POINT VOLTAGE BALANCING
  - EFFICIENCY: 97%
- DIODE BRIDGE RECTIFIER:
  - 6000-V, 1000-A DIODE (TWO IN SERIES)
  - R AND RCD SNUBBER
  - EFFICIENCY: 98%
- DIRECT VECTOR CONTROL IN CONSTANT TORQUE
- SPEED CONTROL WITH FLUX CONTROL

The detailed features of the superconducting machine and converter system are listed in this figure. This machine has the advantages that it is lighter, smaller, ironless, and much more efficient than a conventional machine. Because the machine is ironless, the voltage rating of the stator winding is high and synchronous reactance is very low. The short time torque rating is typically five times higher than the rated torque because the transient stator current can be high. The multilevel inverter with SVM gives excellent harmonic quality to the voltage waves, which produce near sinusoidal current in spite of low X_s. Because no transformer is permitted, multipulsing of the diode rectifier is not possible to reduce the generator harmonic loading. The drive operates in the constant-torque region with direct vector control with speed control in the outer loop.

FIGURE 8.64   8/6-Pole SRM structure with typical phase a control current waves.

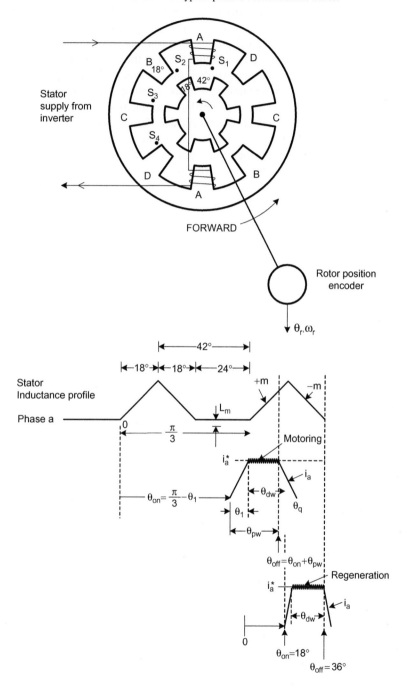

The SRM is not a synchronous machine. However, an SRM drive is included in this chapter to complete our discussion on ac drives. A typical 8/6-pole (8 stator poles and 6 rotor poles) SRM structure and its operating principle was discussed in Figure 6.47. The same machine and typical phase a motoring and regeneration control currents are described here for the drive (see Figure 8.65) in the constant-torque region. For the machine, the period of inductance profile is $60°$, of which $18°$ has a positive slope, $18°$ has a negative slope, and the remaining $24°$ is the dwell period with minimum inductance (L_m) as shown in the figure. The SRM developed torque is given by $T_e = 0.5mi^2$, where m = inductance slope and i = instantaneous current. For motoring, current i_a is turned on at the advance angle $\theta_l = I^*L_m\omega_r/V_d$, where I^* = command current, ω_r = motor speed, and V_d = converter dc link voltage. Therefore, turn-on angle $\theta_{on} = 60° - \theta_l$ as shown in the figure, and the turn-off angle $\theta_{off} = \theta_{on} + \theta_l + \theta_{dw} = \theta_{on} + \theta_{pw}$ as also shown. The current falls to zero at θ_q angle. Because of the encroachment into the $-m$ region, a pulse of negative torque will be contributed at the trailing end. For regeneration or negative torque, the conduction pulse is in the $-m$ slope with $\theta_{on} = 18°$ and $\theta_{off} = 36°$ as shown. The falling current edge does not contribute any torque.

FIGURE 8.65   Four-phase converter for SRM and simplified control diagram of SRM drive.

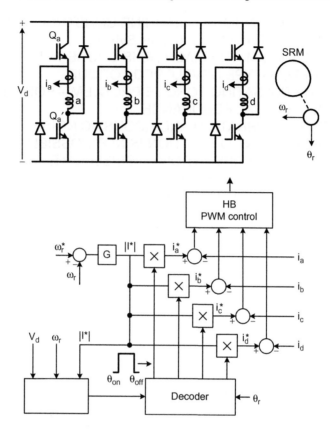

There has been a tremendous growth in SRM drive literature in recent years. An SRM is basically an electronic motor and, therefore, a converter in the front end, as shown here, is mandatory. All four stator pole pairs, A-A, B-B, C-C, and D-D (Figure 8.64) are sup-plied, respectively, by the inverter phases a, b, c, and d. For example, phase a current pulse is supplied by turning on Q_a and Q_a' at angle θ_{on} in synchronism with the rotor posi-tion and then turning them off at θ_{off}. At turn-off, the stored energy in the inductance is fed to the source by the feedback diodes. Each phase contributes a unidirectional current pulse every 60° and all the phases conduct in sequential manner. In the speed control system, the speed loop generates the absolute I^* signal, which is related to developed torque. In the con-stant-torque region, the current magnitude is controlled by the HB-PWM technique. A particular phase is enabled by the θ_{on}-θ_{off} pulse, which is generated by the decoder shown in the figure. The decoder receives the absolute position signal θ_r, dc voltage V_d, speed ω_r, and absolute I^* signal. The motoring or regeneration mode is determined by the polarity of the speed loop error. At higher speed, HB current control is lost due to high counter emf. In that case, the phase current is controlled by θ_{on} and θ_{off} only (see Figure 6.47), resulting in higher loss of torque at higher speed (field-weakening region).

FIGURE 8.66   Features of switched reluctance machine drive.

- SIMPLE AND ECONOMICAL MACHINE CONSTRUCTION
- HIGH SPEED CAPABILITY, LIKE AN SyRM
- UNLIKE IM AND SM, THE MACHINE IS CRUDE WITH HAMMER-BLOW-TYPE DEVELOPED TORQUE
   - SEVERE PULSATING TORQUE PROBLEM – POSSIBILITY OF MECHANICAL RESONANCE
- PROBLEMS OF VIBRATION AND ACOUSTIC NOISE
- PARALLEL MACHINE OPERATION IS NOT POSSIBLE
- NO BYPASS MODE OPERATION POSSIBLE LIKE IM
- NEEDS ABSOLUTE SHAFT POSITION SENSOR
- INFERIOR TRANSIENT RESPONSE – PARTICULARLY IN FIELD-WEAKENING MODE
- OPEN-LOOP OR VECTOR CONTROL IS NOT POSSIBLE
- NO FAULT TOLERANCE – EXTREME VIBRATION AT CONVERTER FAULT
- POOR MACHINE AND CONVERTER UTILIZATION DUE TO DISCONTINUOUS PHASE EXCITATION WITH OVERLAP BRAKING TORQUE
- HIGH ROTOR IRON LOSS
- FEEDBACK SIGNAL ESTIMATION IS MORE DIFFICULT THAN FOR SINUSOIDAL MACHINE

The salient features of SRM drives compared to induction and PMSM drives are summarized here. Although the machine is simple and economical, it has many inferior features, as listed in the figure. Although not mentioned, the converter dc-side capacitor is somewhat larger because of discontinuous current pulses, and there are skin effect and cooling difficulties with the stator concentrated windings. Considering so many inferior features, the SRM is not expected to compete with induction and PMSM drives for industrial applications, except in a few specialized cases.

FIGURE 8.67 Sensorless control of trapezoidal SPM machine drive.

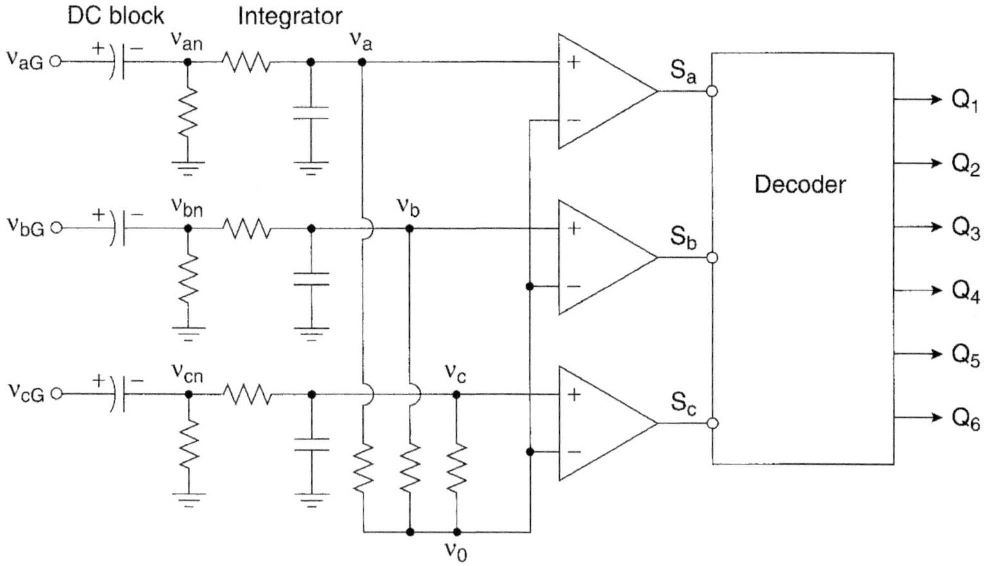

Speed sensorless control of induction motor drives was discussed in Chapter 7. It has been mentioned that an absolute position sensor is mandatory for self-controlled synchronous motor drives. However, the position sensorless control is also possible for synchronous motor drives. In fact, load-commutated inverter drives were discussed with derivation of self-control signals from machine counter emfs. We will now briefly discuss sensorless control of trapezoidal and sinusoidal PMSM drives. Position sensorless control of trapezoidal machines, based on terminal voltage sensing, is somewhat simple as described in this figure and explained with waveforms in the next figure. Consider that the inverter phase legs center-point voltages with respect to ground (v_{aG}, v_{bG}, and v_{cG}) in Figure 8.33 are tapped and connected at the input of this figure. The average voltage of these waves is $0.5V_d$, which is blocked in series capacitors to derive the trapezoidal voltage waves v_{an}, v_{bn}, and v_{cn} shown in Figure 8.34. These voltage waves are integrated, compared with the common mode voltage v_0, through op amps, and then inverter gate drive signals are generated through a decoder as shown in the figure. Note that the scheme will be ineffective at low speed (including the start-up condition) since it depends on the machine counter emf signals.

FIGURE 8.68   Waveform processing for derivation of inverter drive signals in Figure 8.67.

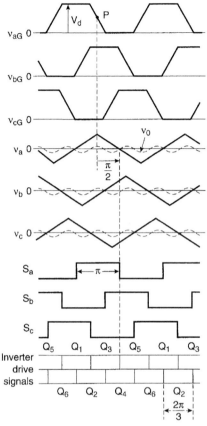

The figure shows the explanatory waveforms for Figure 8.67. The voltage waves v_{aG}, v_{bG}, and v_{cG} are trapezoidal pulses with a peak value of V_d (dc link voltage) and P is the mid-point magnitude. Evidently, the mean value of these waves is $0.5V_d$, which can be blocked to convert them into trapezoidal counter emf waves (v_{an}, v_{bn}, and v_{cn}), as shown in Figure 8.34 (but not shown in this figure). Note that on the positive side of the inverter, $Q_3 \rightarrow Q_5$ commutation occurs at 90° delay angle with respect to point P. In the same way, it can be observed in Figure 8.34 that negative-side device commutation occurs at a 90° delay angle with respect to negative $\rightarrow$ positive transition of counter emf waves. The integration of trapezoidal waves give near triangular waves (v_a, v_b, and v_c), where the v_a peak corresponds to the point P, as indicated in the figure. The common mode voltage wave v_0 has triplen harmonics and is shown as a dashed wave in correct phase relation. Therefore, the comparator outputs (S_a, S_b, and S_c) are square waves with 120° phase shifts as shown in the figure. The decoder combines these signals and generates 120°-wide logic turn-on pulses for inverter devices in correct sequence and phase position, which can be verified from Figure 8.34. Once the logic signals are established, machine torque can be controlled by varying current I_d by means of the PWM technique.

FIGURE 8.69 Trapezoidal SPM machine drive with open-loop starting and transition to terminal voltage-based drive signals.

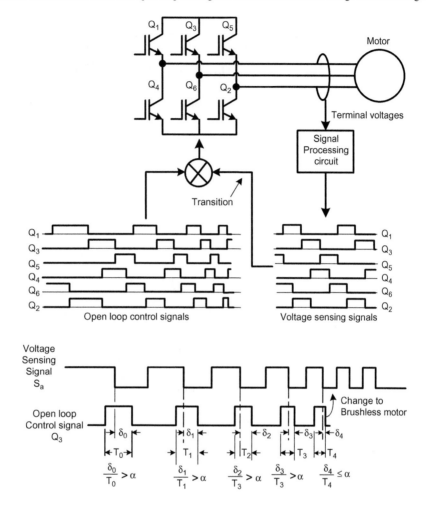

How then can the machine be started at zero speed successfully and then transitioned to counter emf-based sensorless control? This figure shows an open-loop starting scheme [21] and then transitions to counter emf-based control. The scheme was developed by Hitachi and used in residential variable-speed air conditioners. The open-loop control signals for the inverter devices are fabricated in correct sequence from a frequency signal as shown at the left. The frequency is gradually ramped up to increase the machine speed (somewhat similar to open-loop volts/Hz control). At a critical speed, the control is transitioned to voltage-sensing signals as shown at the right. However, proper synchronization is necessary at transition to prevent heavy jerking or an out-of-step condition. This is explained in the lower part of the figure. The open-loop control signal for Q_3 and sensor signal S_a (see Figure 8.68) are considered for synchronization. The phase deviation δT at the trailing edge as shown is measured. As its value falls below a critical value (α), the transition is activated. Evidently, $\alpha = 0$ at ideal synchronization.

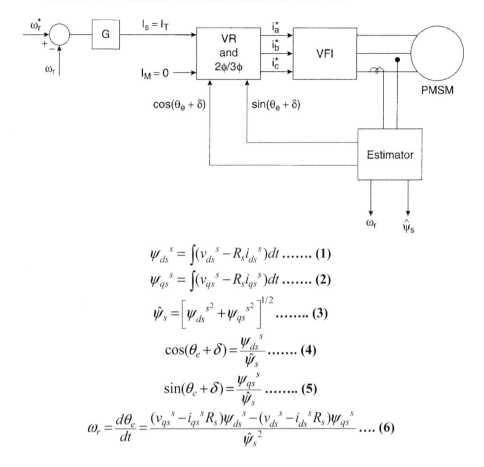

FIGURE 8.70   Sensorless vector control of sinusoidal PMSM drive.

$$\psi_{ds}{}^{s} = \int (v_{ds}{}^{s} - R_s i_{ds}{}^{s}) dt \text{ .... (1)}$$

$$\psi_{qs}{}^{s} = \int (v_{qs}{}^{s} - R_s i_{qs}{}^{s}) dt \text{ .... (2)}$$

$$\hat{\psi}_s = \left[\psi_{ds}{}^{s2} + \psi_{qs}{}^{s2} \right]^{1/2} \text{ .... (3)}$$

$$\cos(\theta_e + \delta) = \frac{\psi_{ds}{}^{s}}{\hat{\psi}_s} \text{ .... (4)}$$

$$\sin(\theta_e + \delta) = \frac{\psi_{qs}{}^{s}}{\hat{\psi}_s} \text{ .... (5)}$$

$$\omega_r = \frac{d\theta_e}{dt} = \frac{(v_{qs}{}^{s} - i_{qs}{}^{s} R_s)\psi_{ds}{}^{s} - (v_{ds}{}^{s} - i_{ds}{}^{s} R_s)\psi_{qs}{}^{s}}{\hat{\psi}_s{}^{2}} \text{ .... (6)}$$

The figure describes position sensorless direct vector control of a sinusoidal SPM machine drive with stator flux orientation. The scheme is based on machine terminal voltage sensing assuming that the machine is running above a critical speed. The machine voltages and currents are sensed and are processed through the estimator to calculate speed (ω_r), stator flux ($\hat{\psi}_s$), and unit vectors [$\cos(\theta_e + \delta)$, $\sin(\theta_e + \delta)$], and the corresponding equations are shown below the figure. Note that in stator flux-oriented vector control, magnetizing current I_M is oriented with $\hat{\psi}_s$ and torque current I_T is oriented with the V_s axis. As described earlier, the machine can be started in simple open-loop volts/Hz control, and when a critical speed is developed, can be transitioned to counter emf-based sensorless control. Zero speed (or zero frequency) sensorless vector control is very difficult, and has been attempted in the recent literature by external signal injection [24].

FIGURE 8.71 Advances and trends in synchronous motor drives.

- SYNCHRONOUS MOTORS HAVE HIGHER EFFICIENCY, BUT ARE MORE EXPENSIVE THAN INDUCTION MOTORS, i.e., LIFE-CYCLE COST IS LOWER AT HIGHER ENERGY COST
- WFSM DRIVES ARE POPULAR IN HIGHEST POWER RANGE BECAUSE OF IMPROVED EFFICIENCY AND ECONOMICAL CONVERTER SYSTEM DUE TO UNITY OR NEAR-UNITY LEADING POWER FACTOR
- DECLINING COST OF NdFeB PERMANENT MAGNET WILL MAKE PMSM DRIVES MORE POPULAR IN FUTURE (EVENTUALLY MAY SURPASS INDUCTION MOTOR DRIVES)
- ABSOLUTE POSITION SENSOR IS MANDATORY IN SELF-CONTROLLED SYNCHRONOUS MOTOR DRIVES
- SENSORLESS SELF-CONTROL IS EXTREMELY DIFFICULT AT LOW SPEED (INCLUDING ZERO FREQUENCY)
- SPM MACHINE DRIVES ARE USED IN CONSTANT-TORQUE REGION WHEREAS IPM MACHINE DRIVES CAN BE USED UP TO FIELD-WEAKENING EXTENDED SPEED OPERATION
- TRAPEZOIDAL SPM MACHINE DRIVE IS TRULY ANALOGOUS TO DC DRIVE (BLDM OR BLDC)
- MANY ADVANCED CONTROL AND ESTIMATION TECHNIQUES FOR INDUCTION MOTORS ARE ALSO APPLICABLE FOR SYNCHRONOUS MOTORS
- SWITCHED RELUCTANCE DRIVES HAVE QUESTIONABLE FUTURE EXCEPT IN SPECIALIZED APPLICATIONS

A few features of synchronous motor drives and the corresponding advances and trends are summarized in this figure. The decreasing cost of high-energy magnets will make PM machine drives more popular in the future even with nominal energy cost. With converter fault, the machine CEMF can feed fault and develop high pulsating torque.

Summary

This chapter encompasses an extensive review of synchronous motor drives that includes different converter-machine configurations, control topologies, feedback signal estimation, and discussion on sensorless control. Different types of synchronous machines were described in Chapter 6. Broadly, the machines are classified into wound-field, permanent magnet, and reluctance types. The switched reluctance machine (SRM) does not fall into this category, but is covered because of its similarity to the synchronous machine, wide literature coverage, and completeness of description in ac drives. The stepper motor drive is not covered (see Chapter 6) because it is not well-adapted for variable-speed applications. Note that line-start PM machines are not suitable for variable-frequency drives because of excessive copper loss in cage winding due to inverter harmonics. Currently, synchronous motor drives are close competitors to induction motor drives in many industrial applications, and their applications are continuously growing. Certain applications, as described in this chapter, are uniquely suited for synchronous motor drives. These are generally more expensive than induction motor drives, but the advantage is that the efficiency is higher, which tends to lower the life-cycle costs. Note that PM drives are more popular in Japan because of high energy costs. Wound-field machines, which can be brush or brushless type, have the advantage that field current or flux ψ_f is controllable unlike PM machines. High-power multi-megawatt machines are normally used at leading power factor (near unity) for load-commutated thyristor converter drives and unity power factor for cycloconverter and voltage-fed PWM converter drives. The largest power drives in the industry are of these classes. A number of practical applications of these drives have been described that include icebreaker ship propulsion with cycloconverter drive, dual mining ore-crushing drive with cycloconverter, gearless cement mill drive with cycloconverter, cruise ship propulsion with load-commutated current-fed converters, rolling mill drive with voltage-fed converters, and future ship propulsion with high-temperature superconducting (HTS) motors. One interesting application is the use of the converter system as a variable-frequency starter for the machine (see Figure 8.49). The PM machines are generally classified into radial flux (or drum) and axial flux (or disk) types. Although the latter type has a higher power density, lower inertia, and smooth operation, the radial type is more commonly used because of the ease of manufacture. The gradually declining cost of high-energy magnets (NdFeB) is promoting applications of PM machines, and eventually with this trend, their volume of applications may exceed that of induction motors. The PM machines are again classified into surface magnet and interior magnet machines, or sinusoidal and trapezoidal machines. The interior PM (IPM) machines are preferred for the extended speed field-weakening range, such as for EV/HV drives. Trapezoidal SPM machines have simple construction, are easy to manufacture, and have a somewhat higher power density than sinusoidal SPM machines. They are widely used as "brushless dc drives." The wound-field and PM machine drives have one disadvantage: in the case of a converter fault, the machine counter emf feeds the fault and causes dangerous torque pulsation (which is a problem with PM machines because field flux is not controllable). The synchronous reluctance and switched reluctance motors are somewhat comparable because both of them are magnetless, somewhat simpler in construction, and can operate at high speed. However, the latter has the disadvantage of pulsating torque that causes vibration and acoustic noise. Both of these machines are bulkier than PM machines, particularly with NdFeB magnet. The control of PM and synchronous reluctance machines can be generally classified into open-loop volts/Hz and self-control types, and the latter type is almost universally used. However, volts/Hz control is very simple, and is essential for close tracking of speed control in parallel machines, such as textile machine drives. With self-control, there is no command frequency, but the drive control is robust like dc machine drives. Although any ac machine (induction or synchronous) with self-control has dc drive analogy, a trapezoidal SPM drive is the closest, and is normally defined as a BLDM or BLDC (brushless dc). An absolute shaft position encoder is mandatory for self-control. Both scalar and vector control techniques are applicable to synchronous machine drives. Stator flux-oriented direct vector control is normally used. Like induction motor drives, various other control methods, such as sliding mode control, DTC control, MRAC control, and light-load flux-weakening efficiency optimization control, are also applicable for synchronous motor drives, but these are not separately covered in this chapter. The feedback signal estimation of sinusoidal machines is somewhat similar to that of induction machines. A position sensorless drive is possible using the machine counter emf signal as long as the machine does not operate below a critical speed. At zero frequency (i.e., zero speed), position estimation is extremely difficult, as is also true of an induction motor (indeed more difficult here than induction matter because the latter may have slip frequency at zero speed). Currently, an external signal injection technique is being proposed in the literature [24], but the success of this method remains questionable. Of course, the machine can be started with open-loop volts/Hz control, and then at a critical speed, it can be transitioned to counter emf-based self-control.

References

[1] P. Pillay (Ed.), *Performance and Design of Permanent Magnet AC Motor Drives*, IEEE-IA Society Tutorial Course, 1989.

[2] G. K. Dubey, *Power Semiconductor Controlled Drives*, Prentice Hall, Upper Saddle River, NJ, 1989.

[3] B. K. Bose, *Modern Power Electronics and AC Drives*, Prentice Hall, Upper Saddle River, NJ, 2002.

[4] P. Vas, *Sensorless Vector and Direct Torque Control*, Oxford, England, 1998.

[5] A. Chiba and T. Fukao, "A close loop control of super high speed reluctance motor for quick torque response," *IEEE IAS Annu. Meet. Conf. Rec.*, pp. 289–294, 1987.

[6] T. M. Jahns, G. R. Kliman, and T. W. Neumann, "Interior permanent magnet synchronous motors for adjustable speed drives," *IEEE Trans. Ind. Appl.*, vol. 22, pp. 738–747, July/August 1986.

[7] T. M. Jahns, "Flux-weakening regime operation of an interior permanent magnet synchronous motor drive," *IEEE Trans. Ind. Appl.*, vol. 23, pp. 681–689, July/August 1987.

[8] B. K. Bose and P. M. Szczesny, "A microcomputer based control and simulation of an advanced IPM synchronous machine drive system for electric vehicle propulsion," *IEEE Trans. Ind. Elec.*, vol. 35, pp. 547–559, November 1988.

[9] B. K. Bose, "A high performance inverter-fed drive system of an interior permanent magnet synchronous machine," *IEEE Trans. Ind. Appl.*, vol. 24, pp. 987–997, November/December 1988.

[10] H. Le-Huy, A. Jakubowicz, and P. Perret, "A self-controlled synchronous motor drive using terminal voltage sensing," *Conf. Rec. IEEE/IAS Annu. Meet.*, pp. 562–569, 1980.

[11] J. B. Borman, "The electrical propulsion system of the QE 2: Some aspects of design and development," *IMAS* 88, pp. 181–190, May 1988.

[12] H. Stemmler, "Drive system and electronic control equipment of the gearless tube mill," *Brown Boveri Review*, pp. 120–128, March 1970.

[13] R. A. Errath, "15000-hp gearless ball mill drive in cement—why not!," *IEEE Trans. Ind. Appl.*, vol. 32, pp. 663–669, May/June 1996.

[14] J. A. Allan, W. A. Wyeth, G. W. Herzog, and J. A. Young, "Electrical aspects of the 8750 hp gearless ball-mill drive at St. Lawrence cement company," *IEEE Trans. Ind. Appl.*, vol. 11, pp. 681–687. November/December 1975.

[15] W. A. Hill, R. A. Turton, R. J. Dugan, and C. L. Schwalm, "Vector controlled cycloconverter drive for an icebreaker," *IEEE IAS Annu. Meet. Conf. Rec.*, pp. 309–313, 1986.

[16] H. Stemmler, "High power industrial drives," Chapter 7, pp. 332–397, in *Power Electronics and Variable Frequency Drives*, edited by B. K. Bose, IEEE Press, New York, 1997.

[17] H. Okayama, M. Kayo, S. Tammy, T. Fujii, R. Ached, S. Mizoguchi, H. Ogawa, and Y. Shimomura, "Large capacity high performance 3-level GTO inverter systems for steel main rolling mill drives," *IEEE IAS Annu. Meet. Conf. Rec.*, pp. 174–179, 1996.

[18] S. Kalsi, B. Gample, and D. Bushko, "HTS synchronous motors for Navy ship propulsion," *1998 Naval Symp. Electric Machines*, pp. 139–146, 1998.

[19] M. Benatmane et al., "Electric propulsion full scale development on U.S. Navy surface ships," *1998 Naval Symp. Electric Machines*, pp. 125–134, 1998.

[20] B. K. Bose, T. J. E. Miller, P. M. Szczesny, and W. H. Bicknell, "Microcomputer control of switched reluctance motor," *IEEE Trans. Ind. Appl.*, vol. 22, pp. 708–715, July/August 1985.

[21] K. Lizuka, H. Uzuhashi, M. Kano, T. Endo, and K. Mohri, "Microcomputer control for sensorless brushless motor," *IEEE Trans. Ind. Appl.*, vol. 21, pp. 595–601, May/June 1985.

[22] T. M. Jahns, "Variable frequency permanent magnet AC machine drives," Chapter 6, pp. 277–325, in *Power Electronics and Variable Frequency Drives*, edited by B. K. Bose, IEEE Press, New York, 1997.

[23] K. Rajashekara, A. Kawqmura, and K. Matsuse (Eds.), *Sensorless Control of AC Drives*, IEEE Press, Piscataway, NJ, 1996.

[24] H. Kim and R. D. Lorenz, "Carrier signal injection based sensorless control methods for IPM synchronous machine drive," *IEEE IAS Annu. Meet. Conf. Rec.*, 2004.

CHAPTER 9

Computer Simulation and Digital Control

Simulation

Digital Control

FIGURE 9.1 What is simulation? Why simulation?

- VIRTUAL OR SOFTWARE REPRESENTATION OF PHYSICAL CIRCUIT OR SYSTEM
- NEEDS MODEL DESCRIPTION BY MATHEMATICAL EQUATIONS OR CIRCUIT TOPOLOGY
- COMPLEX CONVERTER AND CONTROL STRATEGY REQUIRE SIMULATION STUDY PRIOR TO BREADBOARD OR PROTOTYPE DEVELOPMENT
- VIRTUAL PERFORMANCE TEST SAVES TIME AND MONEY FOR NEWLY DEVELOPED PRODUCT
- SIMULATION STUDY IS HIGHLY EDUCATIONAL
- MODERN SIMULATION PROGRAMS ON PC ARE VERY EFFICIENT AND USER-FRIENDLY
- SIMULATION RESULTS ARE ONLY AS GOOD AS THE MODEL DESCRIPTION
- REAL-TIME CONTROL AND DIAGNOSTIC PROGRAMS CAN OFTEN BE GENERATED DIRECTLY FROM SIMULATION PROGRAM
- SIMULATION PROGRAM CAN BE ITERATED FOR OPTIMUM CIRCUIT AND SYSTEM DESIGN
- NO FEAR OF DAMAGE DUE TO FAULT OR ABNORMAL OPERATION

A newly developed converter or control system should be simulated on a computer prior to breadboard or prototype development, particularly if it is complex. The advantages of simulation studies are summarized here. A simulation study can provide the steady-state, transient, and fault performance of the system and also can help the design of system and its protection. The software emulation and virtual performance tests give the developer a lot of confidence in the product development. The FFT analysis of waveforms can aid in line power quality studies and design. Simulation results are highly educational. If a simulated system behaves abnormally, there is no fear of any damage.

FIGURE 9.2 Flowchart for prototype ac drive control development with a simulation study.

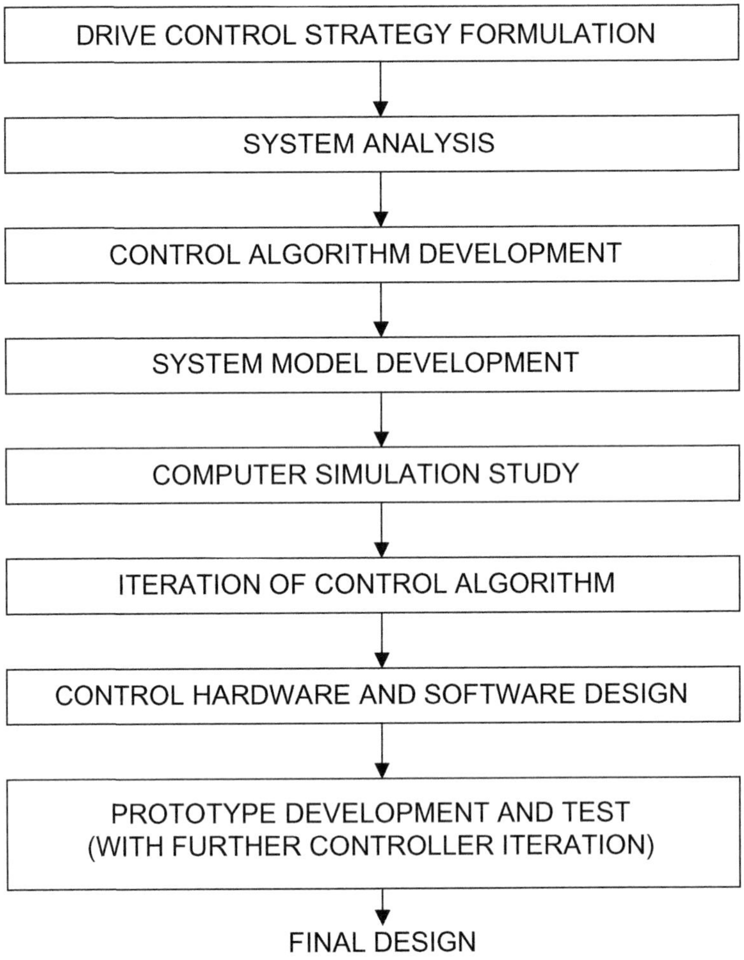

```
┌─────────────────────────────────────────────┐
│      DRIVE CONTROL STRATEGY FORMULATION       │
└─────────────────────────────────────────────┘
                      │
                      ▼
┌─────────────────────────────────────────────┐
│                SYSTEM ANALYSIS                │
└─────────────────────────────────────────────┘
                      │
                      ▼
┌─────────────────────────────────────────────┐
│         CONTROL ALGORITHM DEVELOPMENT         │
└─────────────────────────────────────────────┘
                      │
                      ▼
┌─────────────────────────────────────────────┐
│           SYSTEM MODEL DEVELOPMENT            │
└─────────────────────────────────────────────┘
                      │
                      ▼
┌─────────────────────────────────────────────┐
│           COMPUTER SIMULATION STUDY           │
└─────────────────────────────────────────────┘
                      │
                      ▼
┌─────────────────────────────────────────────┐
│         ITERATION OF CONTROL ALGORITHM        │
└─────────────────────────────────────────────┘
                      │
                      ▼
┌─────────────────────────────────────────────┐
│     CONTROL HARDWARE AND SOFTWARE DESIGN      │
└─────────────────────────────────────────────┘
                      │
                      ▼
┌─────────────────────────────────────────────┐
│        PROTOTYPE DEVELOPMENT AND TEST         │
│        (WITH FURTHER CONTROLLER ITERATION)     │
└─────────────────────────────────────────────┘
                      │
                      ▼
                 FINAL DESIGN
```

A flowchart for the steps of advanced drive control development is given in this figure. It is assumed that the machine with the converter and local controller design already exists. In the beginning, the drive control strategy is formulated based on the performance specs of the system. The drive system is then analyzed and control algorithms are developed in detail. However, to have full confidence in the performance of a newly developed controller, it is necessary to perform simulation studies of the whole drive system and modify the control strategy and its parameters as needed. Mathematical models of the converter, machine, and controller are developed and simulation studies are conducted with these models. After the simulation performance is optimized, a prototype drive and its controller hardware and software are designed, built, and tested to finalize the design.

FIGURE 9.3 Some simulation programs.

- MATLAB/SIMULINK
- PSPICE
- PSIM
- EMTP
- ACSL
- $MATRIX_x$
- SIMNON
- SABER
- C

The figure shows some common, currently used digital simulation programs [1]. Most of these programs are available for personal computers. Note that, historically, analog, digital, and hybrid computers have been widely used for simulation. However, digital simulations are universally popular now. Some of the programs listed here are good for circuit simulation, and others are good for system simulation. Of course, in principle, any program can be used for both circuit and system simulation. For example, MATLAB-based Simulink [3] is more convenient for system simulation, whereas PSPICE [4] is more convenient for circuit simulation. Recently, Simulink's capability has been advanced by addition of SimPowerSystems [10]. In this chapter, PSPICE and Simulink will be discussed (particularly the latter in detail) only, because of their popularity in power electronics. PSIM [5] provides the advantage of circuit simulation hybriding with system simulation. EMTP (Electro-Magnetic Transients Program) is circuit oriented and has traditionally been used in power systems that incorporate power electronics. Simnon is equation oriented, where each element is described by a state-space equation and then interconnected by a connection routine. ACSL (recently renamed ACSL Sim) and $MATRIX_x$ are also equation-based programs. SABER provides large and powerful sophisticated simulation of analog and digital systems that may include electric, power electronic, hydraulic, mechanical, etc., systems. The traditional C or even FORTRAN language has been widely used in simulation of power electronic systems. One advantage of C simulation is that the controller codes can be used in DSP for real-time control. One general rule in simulation programs is that the more user-friendly it is, the slower it is. Of course, with the current dramatic increases in computer speeds, this is not much of a problem.

FIGURE 9.4 Features of PSPICE.

- CIRCUIT SCHEMATIC-ORIENTED SIMULATOR
- PRINCIPAL CIRCUIT COMPONENT SYMBOLS WITH THE ATTRIBUTES AND DEVICE MODELS STORED IN THE PROGRAM LIBRARY
- INTERNAL CONVERSION OF THE SCHEMATIC TO THE NETLIST AND ANALYTICAL EQUATIONS ARE TRANSPARENT TO THE READER
- CAN PERFORM THE FOLLOWING ANALYSIS:
    TRANSIENT ANALYSIS
    FREQUENCY-DOMAIN ANALYSIS
    FOURIER ANALYSIS
    PARAMETRIC ANALYSIS
    TEMPERATURE ANALYSIS
    MONTE CARLO ANALYSIS
    SENSITIVITY/WORST-CASE ANALYSIS
    GRAPHICAL PLOTTING OF RESULTS USING "PROBE"
- DOES NOT DIRECTLY DESIGN THE CIRCUIT, BUT ITERATIVE ANALYTICAL RESULTS WITH DIFFERENT CIRCUIT COMPONENTS HELP TO OPTIMIZE THE CIRCUIT DESIGN
- CAN BE USED FOR SYSTEM SIMULATION IF DRAWN IN EQUIVALENT CIRCUIT FORM

PSPICE is circuit analysis software, one version of which was commercially introduced by MicroSim [4] for SPICE (Simulation Program with Integrated Circuit Emphasis). SPICE was originally developed by the University of California, Berkeley, for analog circuits analysis, but was later extended to power electronic circuits. The PSPICE simulation program is very popular in the power electronics community. The circuit schematic is generated by the parts stored in the program library. An extra disk can supplement this library. All component characteristics, including power semiconductor models, are stored in the library. Once the correct schematic is entered, the program captures it and converts it internally to netlist and analytical equations that are transparent to the user. The program can perform the various types of analysis indicated in the figure. The final results can be plotted in graphical form by using Probe software. If a system is represented by equivalent circuits (such as the d-q model of an induction motor), PSPICE can be used for its analysis. The analysis can be iterated with different component parameters until the user reaches the optimum circuit design. Student Version 9.1 of PSPICE and the User's Guide from MicroSim [4] can be downloaded free from the Internet.

FIGURE 9.5 PSPICE simulation flowchart.

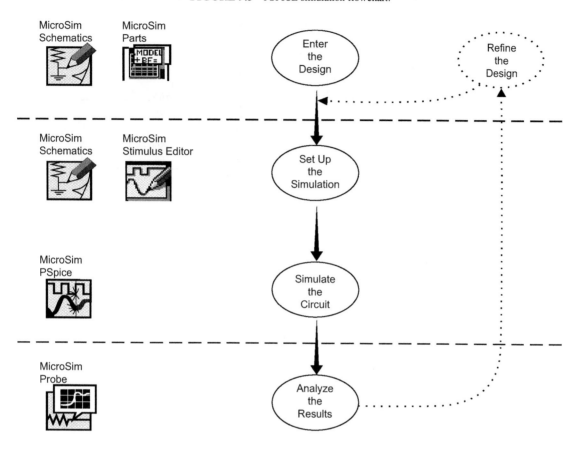

Simulation studies of power electronic circuits with PSPICE are very simple to conduct. They basically consist of three steps [4] as shown in this figure: (1) creation of circuit schematic, (2) simulation, and (3) graphical plotting of the results. The programs required in these steps are shown in the figure. The schematic is created with the help of schematics and parts programs. The parts, such as voltage and current sources, resistors, capacitors, op amps, and gates, and power devices, such as diodes, power MOSFETs, and IGBTs, are brought in from the parts library with their symbols, attributes, and models and then interconnected to generate the complete schematic. The schematic is debugged for possible errors. Then, a simulation command is used to start the simulation using PSPICE. The Stimulus Editor is a graphical input waveform editor that generates the time-domain waves to test the circuit's response during simulation. Finally, when the analysis results are generated, the Probe software plots the waveforms from the output data. The waveforms help to refine the circuit design as shown until satisfactory waveforms are obtained from the optimized circuit design.

FIGURE 9.6   PSPICE example circuit simulation of step-down dc-dc converter.

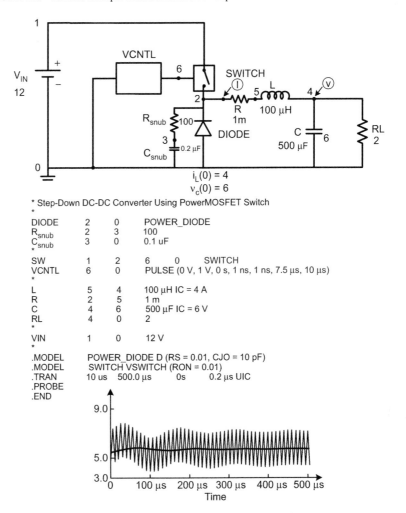

* Step-Down DC-DC Converter Using PowerMOSFET Switch
*

DIODE	2	0	POWER_DIODE	
R_{snub}	2	3	100	
C_{snub}	3	0	0.1 uF	
*				
SW	1	2	6 0	SWITCH
VCNTL	6	0	PULSE (0 V, 1 V, 0 s, 1 ns, 1 ns, 7.5 µs, 10 µs)	
*				
L	5	4	100 µH IC = 4 A	
R	2	5	1 m	
C	4	6	500 µF IC = 6 V	
RL	4	0	2	
*				
VIN	1	0	12 V	
*				
.MODEL			POWER_DIODE D (RS = 0.01, CJO = 10 pF)	
.MODEL			SWITCH VSWITCH (RON = 0.01)	
.TRAN			10 us 500.0 µs 0s 0.2 µs UIC	
.PROBE				
.END				

A very simple buck converter, as shown here, is considered for a simulation study [6]. All circuit components, along with the diode and power MOSFET models, are stored in a library. Once the circuit is designed, the schematic is drawn in a window by dragging in the elements and assigning their attributes in the dialog box. A schematic can be large and cover several pages. In this example, the diode is characterized by conduction resistance of 0.01 ohm and junction capacitance of 10 pF in open circuit. The switch conduction resistance is 0.01 ohm and open circuit resistance is infinity. The gate drive is defined by Stimulus Editor, or else the control circuit can be drawn in schematic form. In this example, a PWM wave of 1.0 V at 100 kHz with a duty ratio of 0.75 is defined. The resulting V (solid) and I (oscillatory) (as marked) waves are shown at the bottom of the figure. Instead of drawing the schematic, the user can create the file (netlist) as shown using the Text Editor. Numbering of nodes is not needed for schematic entry.

FIGURE 9.7   MATLAB toolboxes for solving power electronic problems.

- SIMULINK*
- SIGNAL PROCESSING
- CONTROL SYSTEM
- CURVE FITTING
- DATA ACQUISITION
- ROBUST CONTROL
- SYSTEM IDENTIFICATION
- OPTIMIZATION
- FUZZY LOGIC*
- NEURAL NETWORK*
- GENETIC ALGORITHM

MATLAB (The MathWorks, Inc.) [2] is an almost universally used piece of software for solving broad engineering problems that relate to various mathematical operations, data analysis, and graphical plots. These include arithmetic and logical operations; integration; differentiation; matrix manipulation; the solving of algebraic, differential, difference, and polynomial equations; Laplace and Fourier analysis and transforms; Z-transforms; regression and curve fitting; and interpolation. MATLAB supports a large number of toolboxes for problem solving. The toolboxes that are important for solving power electronics-related problems are listed in the figure. A toolbox can be defined as a subprogram in the MATLAB environment for solving specialized problems. Simulink is a simulation program with a graphical user interface (GUI). Signal Processing helps analysis of the time- and frequency-domain signals. Control System helps modeling, analysis, and design of control systems. System Identification helps generation of mathematical system models based on I/O data and provides tools for estimation and identification of signals and parameters. Fuzzy Logic, Neural Network, and Genetic Algorithm toolboxes help in the design of intelligent control and estimation of the system. Programs in different toolboxes can be interconnected to build and study a composite system. For example, a neural network designed with the Neural Network toolbox can be embedded in a Simulink-based drive simulation program, and system performance can then be studied. The toolboxes that will be discussed in the book are indicated by an asterisk (*).

FIGURE 9.8 Simulink features.

- POWERFUL MATHEMATICAL MODEL-BASED SYSTEM
  SIMULATION PROGRAM IN MATLAB ENVIRONMENT
- GRAPHICAL USER INTERFACE (GUI)
- SIMULATION OF NONLINEAR DYNAMIC SYSTEM
    - LINEAR/NONLINEAR
    - CONTINUOUS/DISCRETE TIME
    - MULTIRATE
    - HIERARCHICAL MODELS
- LIBRARY OF GRAPHICAL BUILDING BLOCKS
- USER INTERFACE FROM MATLAB, FORTRAN, OR C CODE
- CAN GENERATE C CODE FROM MODELS FOR REAL-TIME
  CONTROL
- SIMULATION CAN BE LINKED WITH OTHER TOOLBOXES
- SIMULATION RESULTS CAN BE USED FOR MATLAB
  PROCESSING
- EXTENSIVE GRAPHICS CAPABILITY

Simulink [3] is basically a digital simulation program for nonlinear dynamic systems that operates in the MATLAB environment, where MATLAB is used as its computational engine. The system can be continuous, discrete time, or multirate with different sample times. Complex models can be represented by adding a hierarchy of subsystems. The system's mathematical model is to be developed first either in the time domain in terms of state-space, integral, and algebraic equations, or with Laplace transfer functions prior to simulation study. The graphical user interface permits fast simulation setups. Simulink has a graphical block library of sources, sinks, continuous, discontinuous, discrete, look-up tables, math operations, signal attributes, signal routing, etc. The blocks can be dragged into the model window and interconnected. The parameters of the block can be defined in the dialog box. An integration method is selected before running the simulation. Often, a fixed but conservatively selected time step shortens the simulation time. The running of simulation is interactive; i.e., the parameters can be changed on the fly to observe the effect of the change. The simulation results can be observed on scope or stored in workspace or files. The system design can be optimized by means of iterative simulation studies. Once the simulation is iterated for successful design, the Real-Time Workshop program can translate the controller software into C code, which can be compiled and used in DSP for real-time control. In addition, Simulink can generate VHDL (very high speed hardware description language) software for FPGA (described later) implementation of control system. This is one of the great advantages of Simulink simulation. The simulation waves can be used, for example, to plot a harmonic spectrum by using a MATLAB-based Fourier analysis program. Several example programs will be illustrated next with Simulink.

FIGURE 9.9 Simulink simulation of a simple three-phase voltage-fed inverter. (Courtesy of Pearson.)

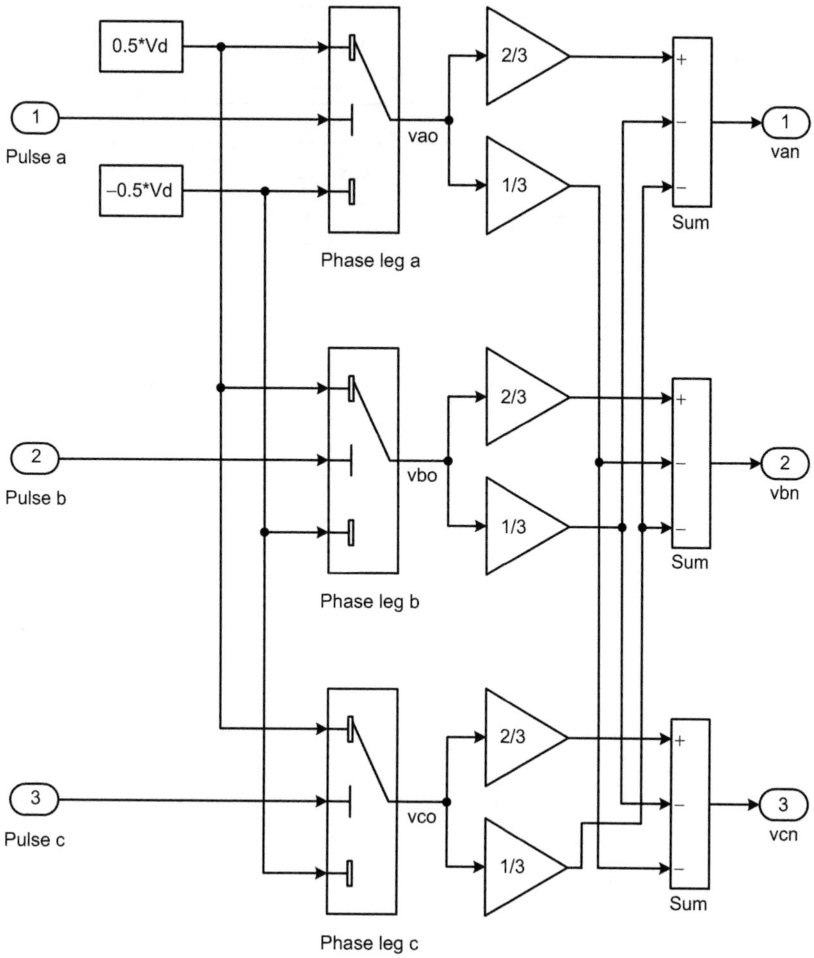

The figure shows a simple simulation block diagram for a three-phase two-level voltage-fed inverter. The power devices are considered as ideal switch, there are no snubbers, and gate drive circuits are not shown for simplicity. Each phase leg of the inverter is represented by a "switch" that has three input terminals and one output terminal. The output of a switch (v_{a0}, v_{b0}, v_{c0}) is connected to the upper input terminal $(+0.5V_d)$ if the PWM control signal (middle input) is positive. Otherwise, the output is connected to the lower input terminal $(-0.5V_d)$. Thus, the output phase voltage oscillates between $+0.5V_d$ and $-0.5V_d$, as shown, which is characteristic of the inverter. The output line-to-neutral voltages are fabricated by expressions $v_{an} = \frac{2}{3}v_{a0} - \frac{1}{3}v_{b0} - \frac{1}{3}v_{a0}$, $v_{bn} = \frac{2}{3}v_{b0} - \frac{1}{3}v_{a0} - \frac{1}{3}v_{c0}$ and

$v_{cn} = \frac{2}{3}v_{c0} - \frac{1}{3}v_{a0} - \frac{1}{3}v_{b0}$, the simulations of which are evident from the figure.

FIGURE 9.10   Development of mathematical model equations for an HF inverter.

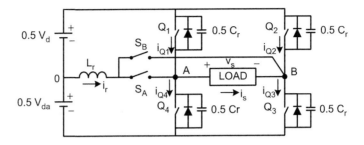

General Equations:

$$i_s = i_{Q1} + i_r - i_{Q4}, \quad i_s = i_{Q3} - i_{Q2} - i_r \tag{1}$$

$$v_{Q1} + v_{Q4} = v_d \tag{2}$$

$$v_{Ao} = v_{Q4} - 0.5V_d, \quad v_{Bo} = v_{Q3} - 0.5V_d, \quad v_{AB} = v_{Ao} - v_{Bo} \tag{3}$$

$$i_r = \frac{1}{L_r} \int (0.5V_d - v_{Q4})dt \quad (S_A \text{ is on}) \tag{4}$$

$$i_r = \frac{1}{L_r} \int (0.5V_d - v_{Q3})dt \quad (S_B \text{ is on}), \quad i_r = 0 \quad (S_A \text{ is off}, S_B \text{ is off}) \tag{5}$$

HALF-BRIDGE   Q1Q4

- MODE1:   (Q1 on, Q4 off)   • MODE2:   (Q1 off, Q4 on)
  $$v_{Q1} = 0, \quad v_{Q4} = V_d \qquad\qquad v_{Q1} = V_d, \quad v_{Q4} = 0$$
- MODE3: (Q1 off, Q4 off)

$$v_{Q1} = \frac{1}{0.5C_r} \int i_{Q1} dt \qquad v_{Q4} = V_d - v_{Q1} \qquad i_{Q1} = -i_{Q4} = \frac{1}{2}(i_s - i_r)$$

HALF-BRIDGE Q2Q3 IS SIMILAR

Next, let us attempt Simulink simulation of the complex high-frequency link converter system [7] described in Figure 4.75. Again, for simplicity, we will ignore the control circuit and model the MCT devices as simple on–off switches as before. In this figure, we will describe the model equations of the high-frequency inverter shown at the top. Because the input dc capacitors are large, the supply is represented by a split power supply. The mode of the converter changes with the state of the switches. Therefore, the model will consist of a set of equations for every switching state. The general equations of the inverter are written, where i_s = the HF transformer secondary load current reflected to the primary ($n = 1$), i_r = resonance circuit current flowing through L_r, i_{Q1}-i_{Q4} = the main device currents, v_{Q1}-v_{Q4} = main device voltages, and $v_s = v_{AB} = v_{A0} - v_{B0}$. Equations (4) and (5) for left half-bridges are valid when the auxiliary switches S_A and S_B are closed, respectively, whereas $i_r = 0$, when both of them are open. The three modes of operation of the left half-bridge are indicated in the figure. The right half-bridge that operates independently has similar three modes (not shown). The simulation diagram should satisfy the model equations in the three modes of operation. Note that instead of a mathematical model-based Simulink simulation, it is also possible—and possibly easier—to conduct a topological simulation with PSPICE. However, for a complete drive system simulation, Simulink is preferable.

FIGURE 9.11   Simulink simulation of the HF inverter.

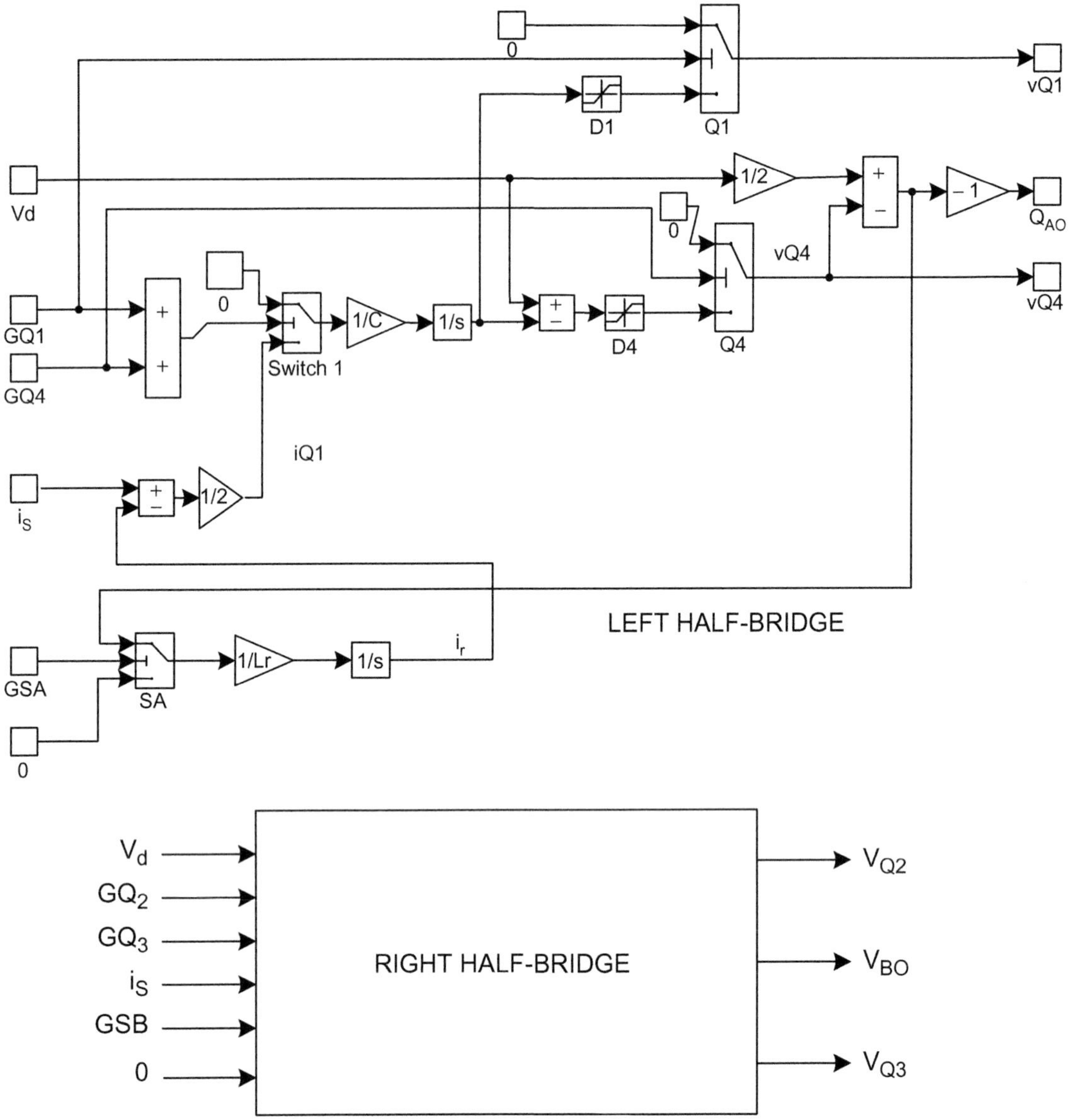

The model developed in the previous figure is used to develop the simulation diagram [7] shown in this figure. Again, for simplicity, only the simulation of the left half-bridge is shown in detail on the top. The simulations of switches Q_1, Q_4, and S_A are included in the figure. The figure receives the gate drive signals ($GQ1$, $GQ4$) for Q_1 and Q_4, gate drive signal GSA for S_A, supply voltage V_d, and load current i_s on the left, and generates v_{Q1}, v_A, and v_{Q4} signals at the output. An auxiliary fictitious switch (Switch 1) is included that receives both GQ_1 and GQ_4 drive signals. The switch is on (output = 0) if either GQ_1 or GQ_4 is high, but off and connects the output to the lower side if both are zero, i.e., Q_1 and Q_4 are off (mode 3). In modes 1 and 2, Switch 1 output is zero and, therefore, v_{Q4} and v_{Q1}, respectively, have V_d output. The clamping circuit limits the voltage between $+V_d$ and negative drop of the bypass diode (which is zero in this case). In mode 3, when both Q_1 and Q_4 are off, the Switch 1 output is i_{Q1}, which is generated by the combination of i_r and i_s, as indicated in the figure. The right half-bridge simulation is shown by a block diagram. The inverter output voltage v_s can be generated by the relation $v_{AB} = v_{A0} - v_{B0}$. The current i_s is fed back from the ac-ac converter, as discussed next. The details for the gate drive circuits and device models can be included in the model, if desired.

FIGURE 9.12   Developing model equations for ac-ac converter (IPM).

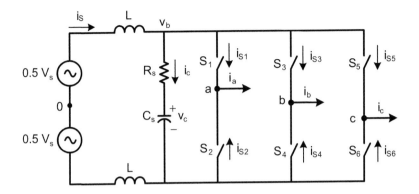

General Equations:

$$v_{ao} = 0.5v_s - L\frac{di_s}{dt} - v_{s1} \qquad \dots\dots(1)$$

$$v_s = 2L\frac{di_s}{dt} + v_b \qquad \dots\dots(2)$$

$$v_b = R_s i_c + \frac{1}{C_s}\int i_c dt \qquad \dots\dots(3)$$

$$i_s = i_c + i_{s1} + i_{s3} + i_{s5} \qquad \dots\dots(4)$$

Phase- a:
- MODE-1: (S_1 on , S_2 off)

$$i_{s1} = i_a \, , \, i_{s2} = 0 \, , \, v_{s1} = 0 \, , \, v_{s2} = v_b$$

- MODE-2: (S_1 off , S_2 on)

$$i_{s1} = 0 \, , \, i_{s2} = i_a \, , \, v_{s1} = v_b \, , \, v_{s2} = 0$$

(b and c phases are similar)

The three-phase, high-frequency, zero-voltage switched ac-ac converter described in Figure 4.75 is modeled in this figure. As usual, the devices are considered as ideal on–off switches, and the gate drive circuits are ignored for simplicity. The simplified converter circuit is shown on the top, where v_s = supply voltage generated by the inverter, i_s = supply current, $2L$ = transformer equivalent secondary leakage inductance, i_a, i_b, i_c = load phase currents, i_c = snubber current, v_b = bus voltage, and $i_{s1} - i_{s6}$ = currents carried by the respective switches (S_1–S_6). The general voltage and current equations of the circuit are given by Eqs. (1)–(4). Only the operation modes (mode 1 and mode 2) of phase a are described in the figure. There is short overlapping of the S_1 and S_2 on-switches for transfer of current i_s, but the overlap time is neglected. Phases b and c operate independently and the modes are similar; therefore, their modes are not described in the figure. The simulation of the complete converter will be described in the next figure.

FIGURE 9.13 Simulink simulation of an ac-ac converter (see Figure 4.75).

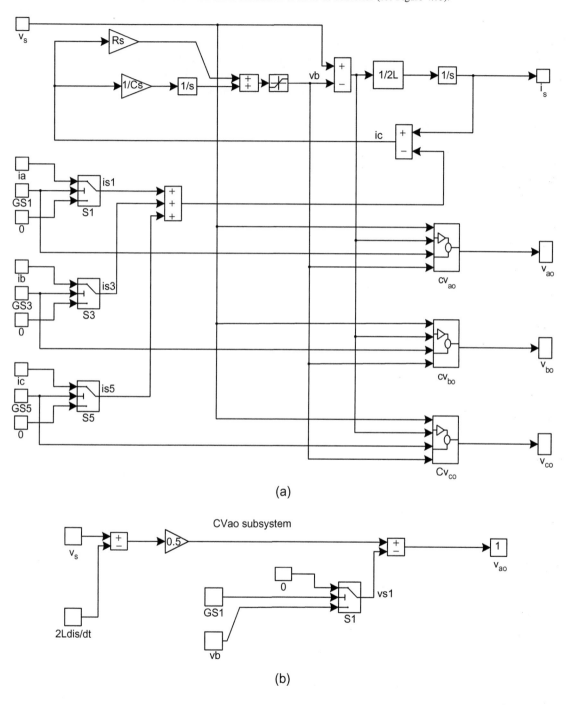

(a)

(b)

The model equations in Figure 9.12 are used to develop the simulation diagram shown in this figure [7]. Simulation of only switches S_1, S_2, and S_3 is included because operation of the lower switches (S_2, S_4, S_6) is complementary to that of the upper switches. Equation (2) is solved for i_s on the top, which receives v_b and v_s at the input, but v_b is determined from Eq. (3) with the help of Eq. (4). Note that when S_1 is closed by the GS_1 gate signal, $i_{S1} = i_a$, but otherwise it is zero. Switches S_2 and S_3 for other phases operate in a similar way. Equation (4) generates i_c by summing the output of the switches and then subtracting it from i_s. Finally, Eq. (1) is solved in the Cv_{a0} block, the details of which are shown in part (b). In the same manner, v_{b0} and v_{c0} are solved in the Cv_{b0} and Cv_{c0} blocks, respectively. Input signal v_s is obtained from Figure 9.11. The phase currents are fed back as shown from the load and the corresponding output current i_s is fed back at the input of inverter simulation. The equations may require some manipulation to avoid what is called *algebraic loops*.

FIGURE 9.14   Simulation of the d^e-q^e flux model for an induction motor.

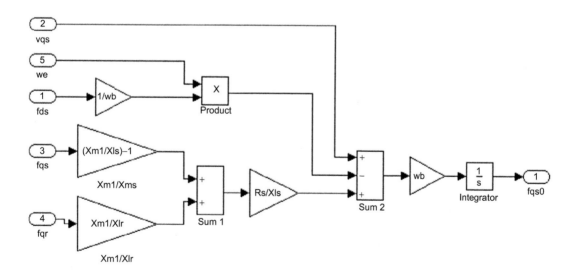

The induction motor dynamic d^e-q^e model equations in terms of flux linkages were derived in Chapter 6 and given in Figure 6.23. This figure shows the simulation diagram of this model. The model receives input voltages v_{qs} and v_{ds} and frequency ω_e at the input and solves for output currents i_{qs} and i_{ds} with the help of flux linkage equations. At the input, the "abc-syn" block converts the inverter output voltages v_{an}, v_{bn}, and v_{cn} to v_{qs} and v_{ds} by abc to d^e-q^e transformation. Similarly, output block "2_3" converts i_{ds} and i_{qs} into phase current signals i_a, i_b, and i_c by d^e-q^e to abc transformation. The torque T_e and speed ω_r calculations are done as indicated in Figure 6.23. The simulation details of subsystem block F_{qs}, for example, is shown in the lower figure. Here, F_{mq} input is derived from the $F_{qm} = (X_{ml}/X_{ls})F_{qs} + (X_{ml}/X_{lr})F_{qr}$ relation, where $X_{ml} = 1/[(1/X_m) + (1/X_{ls}) + (1/X_{lr})]$. The machine can be connected at the converter output of Figure 9.9 or 9.13. In the latter case, conversion should be made from v_{a0}-v_{b0}-v_{c0} to v_{an}-v_{bn}-v_{cn}. The ω_e signal can be independent or generated from self-control, as shown in the next figure.

FIGURE 9.15   Simulink simulation of an indirect vector-controlled induction motor drive. (Courtesy of Pearson.)

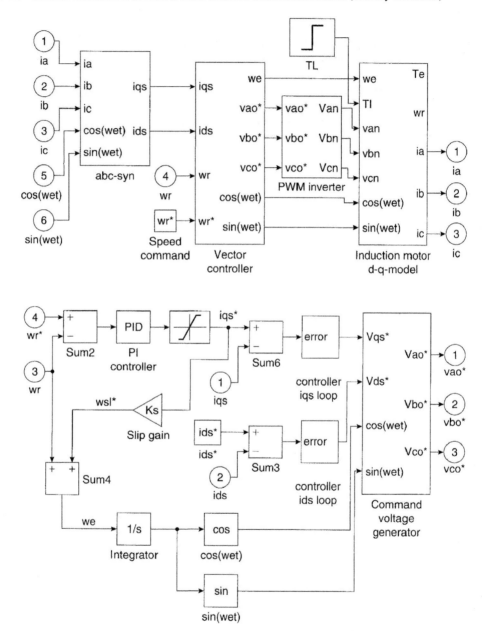

The simplified simulation block diagram for a complete indirect vector-controlled induction motor drive is shown in this figure. The upper figure consists of "abc-syn" block for signal transformation, vector controller, PWM inverter, and induction motor d-q model, and the lower figure explains the details of the vector controller. The figure should be self-explanatory. The PWM inverter receives the sinusoidal phase voltage command signals and generates the line-neutral voltages as shown (supply voltage V_d is not shown). For simplified simulation, the inverter can be represented by a unity transfer ratio block, neglecting its dynamics and harmonics generation. The control signal flow is evident in the lower figure. Slip ω_{sl}^* is generated from the command current i_{qs}^* and added with speed ω_r to generate the ω_e signal as shown in the figure. The currents i_{qs} and i_{ds} controllers indicated by the block "error" are basically P-I controllers. The function blocks "cos" and "sin" generate $\cos\theta_e$ and $\sin\theta_e$, respectively, from the ω_e signal as shown. All input currents and ω_r signals are derived from the machine simulation (details are not shown).

FIGURE 9.16 SimPowerSystems features.

- MODELING AND SIMULATION PROGRAM FOR POWER SYSTEM (INCLUDING POWER ELECTRONICS AND DRIVES)
- EXTENDED SIMULINK SIMULATION WITH GUI
- HYBRID SIMULATION CONTAINING BLOCKS OF SIMPOWERSYSTEMS AND SIMULINK
- MODELS OF POWER SYSTEM EQUIPMENT AND ELEMENTS ARE STORED IN LIBRARY
- PSPICE-LIKE SCHEMATIC-ORIENTED CIRCUIT AND SYSTEM SIMULATION
- NEEDS SIMULINK AND MATLAB ENVIRONMENT
- SIMULATION CAN BE LINKED WITH OTHER TOOLBOXES
- EXAMPLES OF POWER SYSTEM BLOCKSET (PSB) LIBRARY (powerlib) CONTENTS:
  - MODELS OF POWER ELECTRONICS COMPONENTS
  - MODELS OF DC, INDUCTION AND SYNCHRONOUS MACHINES
  - ELECTRICAL SOURCES AND CIRCUIT ELEMENTS
  - ELECTRICAL CIRCUIT MEASUREMENTS
  - CIRCUIT AND SYSTEM ANALYSIS WITH GUI

SimPowerSystems (MathWorks) [10] is basically an extension of the Simulink program that is designed for simulation of power systems that may include dc drives, ac drives, general converter systems, and power generation, transmission, and distribution systems. One important feature of this program is that mathematical models of key power system components are stored in the library where the icons can be fetched and interconnected graphically to set up the simulation. The parameters of the component can be defined in the dialog boy. Models available to the user include the electrical sources, linear and nonlinear passive elements, three-phase elements, power electronics components, electrical machines, electrical circuits measurements, electrical circuits analysis with GUI, and signal and pulse sources. For example, the power electronics library includes diodes, ideal switches, thyristors, GTOs, MOSFETs, IGBTs, two-level universal bridges, and three-level bridges. Similarly, the machine library includes dynamic models of dc machines, induction machines, synchronous machines (PM and wound-field), etc. Each machine model comes with a Mux to measure the key electromechanical response variables. The SimPowerSystems blocks can be combined with Simulink blocks to describe the schematic for a complete power system. A power electronics system is simulated in discrete time with fixed sample time. Simulation results can be displayed on scope, stored in files, or directly analyzed, for example, by FFT with a graphical interface. The simulation is schematic oriented like PSPICE, but is more powerful. Two example simulations of power electronics systems will be given for clarity.

FIGURE 9.17 SimPowerSystems simulation of chopper-fed dc motor drive. (Courtesy of MathWorks, Inc.)

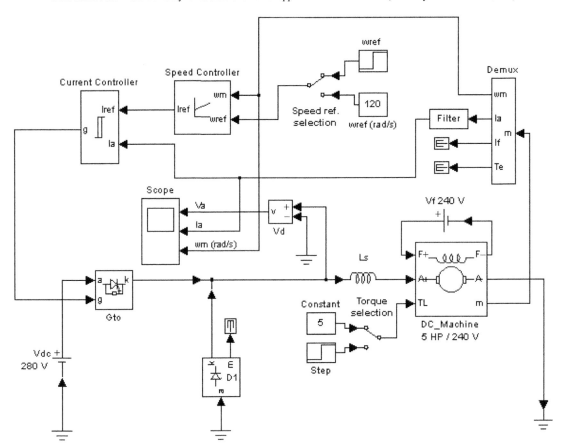

A SimPowerSystems-based simulation of a simple GTO chopper-fed separately excited dc motor drive is described in this figure [10]. The speed control loop generates the armature current command I_{ref} through the *P-I* speed controller. Then, the hysteresis-band PWM current controller generates the duty-cycle-based drive signal to the one-quadrant GTO chopper. The chopper with a bypass diode (D_1) feeds the motor armature, which has a series filter inductance (L_s). Interconnection of the system elements is evident from the simulation diagram. The DC machine block with the field winding is retrieved from the Machines group of the powerlib library, and its parameters [*Ra, La, Rf, Lf, J, Bm* (viscous friction coefficient), *Tf* (Coulomb friction coefficient), and initial speed] are selected. Load torque *TL* and machine speed *wref* can be selected either as constant or step function as shown. The machine block has an internal Mux block that permits measurement of speed (*wm*), armature current (*Ia*), field current (*If*), and torque (*Te*). The same signals can be retrieved through the Demux block available from Simulink. The GTO block with its snubber is obtained from the PowerElectronics sublibrary and its internal parameters are set. The device is modeled with *Ron, Lon*, and *Vf* in series, and with fall *(Tf)* and tail times *(Tt)*. The speed and current controllers are masked blocks implemented by Simulink. The elements *Ls, D1, Vdc, Vf,* and *Vd* and groundings are retrieved from powerlib with appropriate parameter settings. The Manual Switch blocks permit selection of Speed ref and load torque as shown. The system is discretized with the help of the powergui block with a fixed sample time.

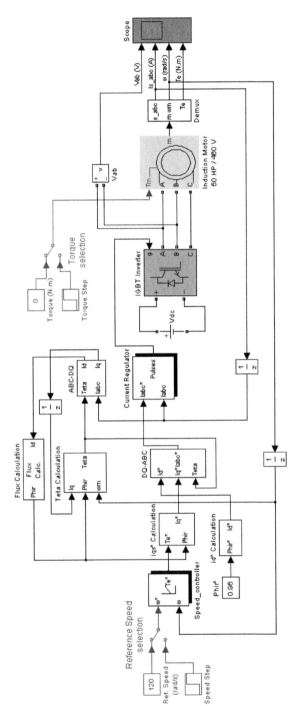

FIGURE 9.18 SimPowerSystems simulation of a vector-controlled induction motor drive. (Courtesy of MathWorks, Inc.)

The figure shows the simulation of a speed-controlled indirect vector drive for an induction motor [10]. The control strategy should be evident from the simulation diagram. The SimPowerSystems and Simulink blocks have been hybridized in the system as usual. The machine dynamic d^e-q^e model based on voltages and currents is retrieved from powerlib and all the electromechanical parameter values are entered in the dialog box. The model based on d^s-q^s or d^r-q^r (rotor reference frame) can also be selected, if desired. The external Demux permits us to read the stator currents (is_abc), speed (wm), and developed torque (Te) for the Scope display and feedback. The three-phase voltage-fed IGBT inverter is retrieved from PowerElectronics group of powerlib as a Universal Bridge block with RC snubber, where the device model and snubber parameters are entered in the dialog box. For snubberless operation, the snubber resistance is set to infinity. The inverter is controlled by a simple hysteresis-band PWM Current Regulator. The transformation blocks DQ-ABC and ABC-DQ are obtained from powerlib. However, the Current Regulator, Speed_controller, Id* Calculation ($i_{ds}^* = \psi_r^*/L_m$) from command rotor flux Phir*, $iqs*$ Calculation ($i_{qs}^* = (2/3)(2/p)(L_r/L_m)(T_e^*/\psi_r)$, Teta Calculation ($\theta_e = \int(\omega_m + \omega_{sl})dt$, $\omega_{sl} = (L_m/\psi_r)(R_r/L_r)i_{qs}^*$), and Flux Calculation ($\psi_r = L_m i_{ds}/(1 + \tau_s)$) are all synthesized by Simulink blocks. Note that the feedback signal filters ($1/z$) are introduced in appropriate places.

FIGURE 9.19 Digital signal processing features.

- HARDWARE SIMPLIFICATION
- FLEXIBLE ALGORITHM
- NO DRIFT OR PARAMETER VARIATION PROBLEMS
- NO EMI OR NOISE PROBLEMS
- COMPATIBILITY WITH HIGHER SYSTEM LEVEL DIGITAL COMMUNICATION
- POSSIBILITY OF UNIVERSAL HARDWARE
- COMPLEX CONTROL, ESTIMATION, AND DECISION MAKING
- MONITORING, WARNING, AND DATA ACQUISITION FOR STORAGE AND POSTPROCESSING
- POWERFUL DIAGNOSTICS

Digital control normally refers to control that uses a microcomputer or DSP (digital signal processor) software, dedicated digital hardware, or a combination of both, although the term is commonly used for microcomputer control. In the pre-microcomputer era, power electronic systems were controlled by hardwired analog devices and digital circuits. However, the majority of systems today is controlled digitally because digital control has a number of advantages. The simplification of control hardware and the corresponding reductions in cost are the principal advantages of microcomputer control. Microcomputer/ DSP technology is advancing very fast with higher processing speeds and more functional integration. Modern VLSI ASIC chips with integration of total control hardware for a specific application in large-volume production can be very economical. Miniature size, reliability, and reduced power consumption can be added advantages. With the same hardware, the software-based control algorithm can be flexible, i.e., it can be easily modified or additional functions can be added. In fact, universal hardware can be designed for a wide range of products, where the algorithms may be different. Digital processing eliminates the signal drift and parameter variation effects that are prevalent in analog systems. Digital computation is 100% accurate with control of signal overflow and underflow by appropriate scaling. Large hardware integration permits decoupling of large voltage and current transients in power electronic systems by nominal shielding. Conduction of EMI can be avoided by adequate filtering of the power supply and I/O signals. Microcomputer control of local power electronic subsystems permits easy compatibility with the higher level host computer in the large integrated industrial environment, where frequent information exchange may be needed. Complex control functions, feedback signal estimations, and decision making in modern power electronic systems are impossible without the help of software. It is interesting to note that although modern vector control was invented in the early 1970s using analog/digital hardware, it could not be commercialized until the 1980s because of the nonavailability of microcomputers. Digital control permits easy signal monitoring, warning, and data acquisition with storage for postprocessing. Powerful off-line or on-line diagnostic software can be developed that can be used for fault identification and fault-tolerant control. A few disadvantages of digital control are sluggish sequential computation, particularly with multitasking, quantization error due to peripheral A/D and D/A converters, lack of easy access of software signals, and, of course, software development may be time consuming.

FIGURE 9.20   Digital control classification tree.

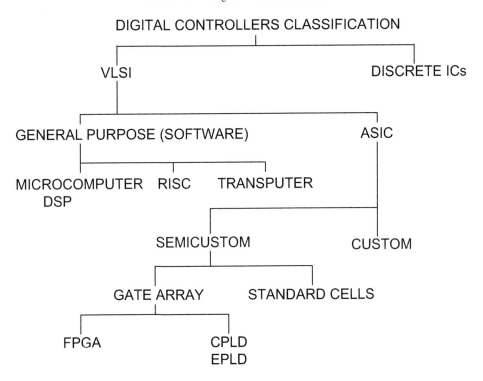

The figure shows the classifications for different types of digital controllers. Broadly, digital control of power electronic systems is possible by means of VLSI (very large scale integration) or discrete IC (integrated circuit) based hardwired controllers. The latter may combine digital and analog ICs. In a VLSI, a very large number of devices are integrated monolithically in a chip to provide great simplification of hardware. A VLSI chip may use digital, analog, or mixed signals. Current submicron (0.10 μm or smaller) CMOS and BiCMOS technologies permit very large scale integration (several million equivalent gates). Again, a VLSI chip can be either general purpose for software programs, or application specific IC (ASIC), that is, designed for a particular application. Examples of ASIC chips are the AD2S100/AD2S110 vector controller (Analog Devices) and NLX230 fuzzy controller (American Neurologix). General-purpose VLSI chips can be subdivided into microcomputer/DSPs, RISCs (reduced instruction set computers), and transputers. In the microcomputer category, the instruction set is complex, which results in a complex architecture and sluggish computation. With RISCs, the simple architecture allows for high speeds. A transputer is specially designed for high-speed parallel processing using several processors. The ASIC chip can be classified as custom or semicustom, where the latter is again subclassified into gate array and standard cells.

A custom ASIC chip (not programmable) is very economical and is designed for specific applications in large volume, such as laundry washer control. An ASIC chip can be designed with digital, analog, memory elements (ROM, RAM), and RISC or DSP core to satisfy full control function. The gate array consists of a matrix of simple logic gates that are interconnected by the user in a field programmable GA (FPGA) or programmable logic device (PLD), which can be programmed electrically (EPLD) or by mask (CPLD) similar to EPROM or PROM, respectively. Standard cells are interconnected functional circuits (such as D flip-flop, counters, etc.) to provide more efficient performance and effective chip real estate utilization than gate arrays. Any of these chips can use a combination of analog, digital, and memory devices to design a complete system, as mentioned before.

FIGURE 9.21   Basic elements of a microcomputer.

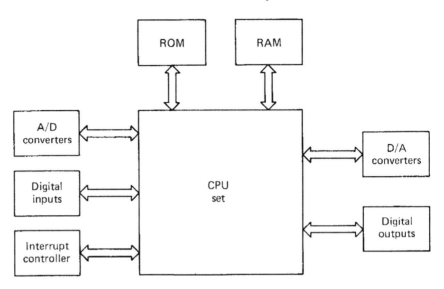

The basic elements of a microcomputer [11–13] are shown in this figure. It consists of the central processing unit (CPU) set, read-only memory (ROM), random access memory (RAM), analog and digital inputs and outputs, and an interrupt controller. The CPU set generally consists of a basic microprocessor, clock generator, bus controller, and bus drivers. It is the heart of a microcomputer and is responsible for unified operation with all elements. It is linked to all elements by the data bus, address bus, and control bus. The data bus width determines the bit size of a computer. The software program consists of a set of instructions for the CPU and is stored in the ROM. The instructions are fetched through the data bus, decoded, and then executed. The program counter in CPU sets up the sequential addresses of the program, which are sent through the address bus. The data to be manipulated by the CPU are stored in RAM or accessed through analog/digital inputs. The processed data are then stored in RAM or sent out through digital/analog outputs. The orderly sequence of program execution can be broken by interrupt signals, which can be generated by hardware or software. A ROM can be mask programmable (PROM) or electrically programmable (EPROM) and ultraviolet light erasable, or electrically programmable and electrically erasable (EEPROM). The read/write memory (RAM) can be static (SRAM) like a flip-flop, or dynamic (DRAM), where the bit information is stored as a charged capacitor that requires periodic refreshing. A flash memory is similar to EEPROM. Dynamic RAM cell size is very small, consumes low power, and is available in large capacity in a chip. The analog signals in the power electronics system are acquired by ADC in a single channel or multichannel (i.e., in a multiplexed manner) with a certain bit resolution and sampling time. Similarly, analog output signals are delivered by DAC in a demultiplexed manner with a certain bit resolution and sample time. Sometimes, ADCs, DACs, and other dedicated hardware (counters, barrel shifters, etc.) are integrated in the computer. Each computer has a characteristic instruction set in assembly language (AL) mnemonics. The source program in high-level language is compiled into object code for real-time operation. The digital program can be executed quickly by selecting a high-speed computer, by using a more efficient and lower level programming language, by using preprocessing and a look-up table method of execution, or by taking advantage of more dedicated hardware or parallel processing with multiple computers. A few commercial microcomputers/DSPs will be reviewed in the next few figures.

FIGURE 9.22   Intel 87C51GB microcontroller features (MCS-51 family) [14].

- 8-BIT FIXED POINT
- HIGHLY IMPROVED VERSION OF THE ORIGINAL MEMBER 8751/8051 IN MCS-51 FAMILY
- 4K × 8 EPROM, 128 × 8 RAM, 16-MHz, 0.75-μs INSTRUCTION CYCLE TIME
- SIX 8-BIT MULTIFUNCTION BIDIRECTIONAL PARALLEL PORTS
- THREE 16-BIT PROGRAMMABLE DEDICATED TIMERS/COUNTERS (ONE WITH UP/DOWN MODE)
- TWO 16-BIT PROGRAMMABLE COUNTER ARRAYS WITH:
    COMPARE/CAPTURE, SOFTWARE TIMER, HIGH-SPEED OUTPUT, PWM MODULATOR, AND WATCHDOG TIMER FUNCTIONS
- 8-CHANNEL ANALOG-DIGITAL CONVERTER WITH 8-BIT RESOLUTION AND COMPARE MODE
- FULL-DUPLEX PROGRAMMABLE SERIAL PORT WITH SERIAL EXPANSION PORT
- HARDWARE WATCHDOG TIMER
- IDLE AND POWER DOWN MODES
- MCS-51 FAMILY INSTRUCTION SET
- BOOLEAN PROCESSING, MULTIPLY, DIVIDE INSTRUCTIONS

Microcomputers/DSPs of various types are available from a number of manufacturers. In this chapter, we briefly review a few sample types manufactured by Intel and Texas Instruments. This figure summarizes the main features of an Intel 87C51-8-bit fixed-point microcomputer [14], which is an enhanced version of the older MCS-51 family. The architecture of the device is shown in the next figure. Recently, Intel has added the MCS-251 family (8-bit) as further upgrading of the MCS-51 family. In addition, the 16-bit MCS 96/296 family is also part of the existing product lines. The 8-bit computer can be used for small projects and when high performance is not required. The computer can be operated at either 12 or 16 MHz. The on-chip EPROM and RAM capacity is indicated. For large programs and large amounts of data, external memory can be added. One special feature of the computer is that it has an 8-channel, 8-bit A/D converter that operates in the multiplexed mode or individual selection mode. The computer can process Boolean variables. It has three 16-bit counters and two additional 16-bit programmable counter arrays that can generate PWM signals for the converter along with additional timing functions. The six bidirectional parallel I/O ports can be programmed for external hardware interrupts, serial ports, timer clock input, read/write signals for external memory, etc. The watchdog timer invokes a reset signal (unless a periodic hold-off signal prevents this reset). The timer helps detect hardware and software malfunctions. There is a set of 8-bit special function registers (SFRs), some of which are bit and byte addressable. A common instruction set is applicable to all the members of the MCS-51 family.

FIGURE 9.23 Intel 87C51GB architecture. (Courtesy of Intel Corp.)

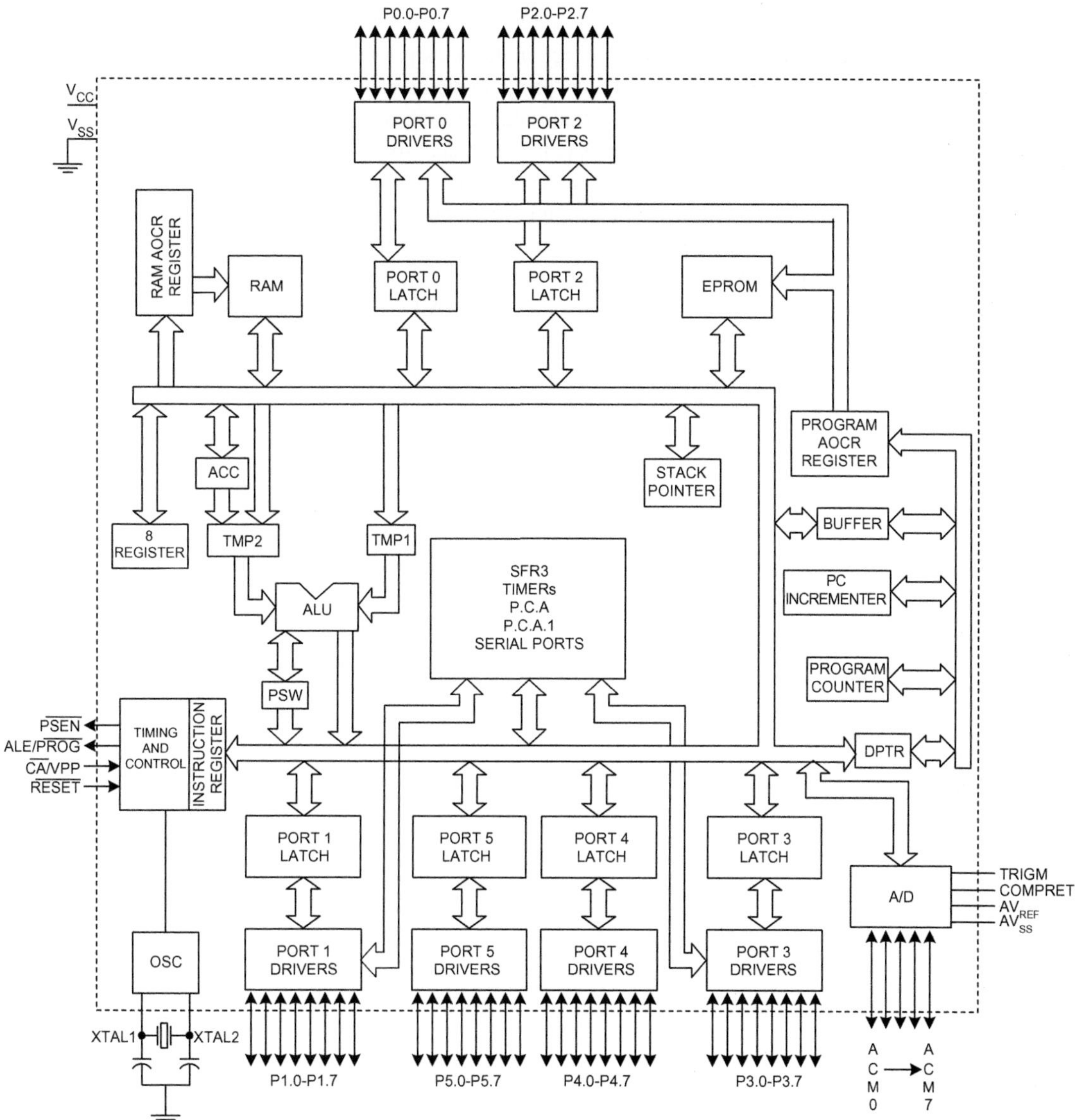

A simplified version of the Intel 87C51GB single-chip microcontroller architecture [14] is shown in this figure. All program memory address buses are 16 bits wide, but the internal RAM and SFR can be accessed by 8-bit addresses. Sixteen addresses in SFR are both byte and bit addressable. The upper two I/O ports (0 and 2) generate a 16-bit address, where port 0 is multiplexed as an 8-bit data bus for instruction fetch. The program counter generates the programs addresses sequentially, and the 8-bit instruction words are fetched to the instruction register and then executed by the CPU. For an interrupt or jump instruction, the next address is stored in the stack pointer and the PC is vectored to the new address. The eight registers, PSW (program status word), stack pointer, DPTR (data page pointer to generate 16-bit external data address), and port latches shown are all part of the SFR (special function register). The successive approximation type A/D converter can be triggered externally or internally. AV_{ss}, AV_{REF}, and COMREF (for comparison with sampled signal) are external analog voltages. The sampled signals are stored in the SFR. The ALU (arithmetic logic unit) processes the arithmetic and logic signals with the help of two temporary registers, and an accumulator (ACC) is the common destination for most arithmetic and logical operations. All five counters on the chip can be programmed for a variety of functions. The oscillator clock on the chip requires external crystals as shown. EPROM is good for initial program development, but PROM (mask programmable ROM) is economical for the final product.

FIGURE 9.24 MCS-51 assembly language instruction examples.

1. ARITHMETIC OPERATIONS

```
ADD    A, Rn
ADDC   A, direct
SUBB   A, Rn
INC    A
DEC    A
MUL    AB
DIV    AB
```

2. LOGICAL OPERATIONS

```
ANL    A, Rn
ANL    A #data
ORL    A, Rn
XRL    A, Rn
CPL    A
RL     A
RLC    A
RRC    A
SWAP   A
```

4. BOOLEAN VARIABLE MANIPULATION

```
CLR    C
ANL    C, /bit
ORL    C, bit
MOV    C, bit
JC     rel
JBC    bit, rel
```

5. PROGRAM BRANCHING

```
LCALL addr16
RET
SJMP ret
JZ
DJNZ
CJNE
NOP
```

3. DATA TRANSFER

```
MOV    A, direct
MOV    Rn, A
MOV    @Ri, A
PUSH   direct
POP    direct
```

The Intel MCS-51 family has 111 assembly language instructions that are divided into the five categories shown above, but only a few sample instructions are shown here. The software source code for the program can be developed in C, PL/M, or assembly language. Source code in C or PL/M is compiled into assembly code and then converted (by ASM-51) to object code for program memory. Each instruction (8-bit wide) consists of an operation mnemonic followed by 1 to 4 operands. A simple instruction takes one cycle (0.75 μs) for execution. For example, ADD A, R0 means the data in the accumulator (A) is added with that of register R0 and then stored in A. MUL AB multiplies the 8-bit integer in A and register B, and the result is stored in A (lower byte) and B (higher byte). MOV C, bit transfers the bit (1 or 0) in the specified bit address to carry C location in PSW.

FIGURE 9.25 Texas Instruments TMS320 family evolution [15].

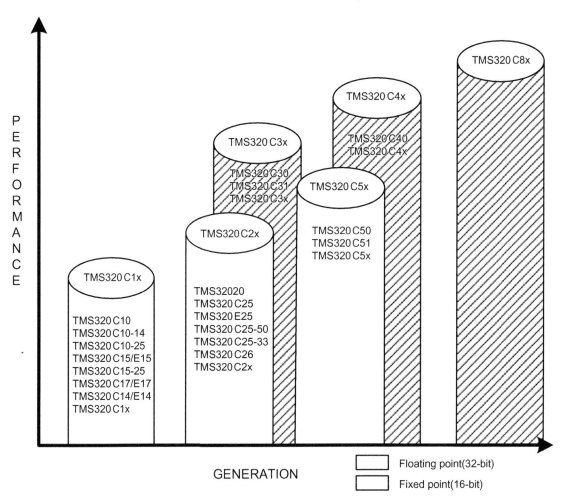

Texas Instruments (TI) has established itself as the leader in modern digital signal processors (DSPs) development, and their TMS320 family of processors is routinely used in power electronic systems. There are essentially six generations of evolution of the old 320 family as shown in this figure, and performance is indicated in the vertical direction. Recently, the 320 family has advanced even further, as discussed later. The modern DSP age practically started when TI introduced the 32010 in 1982. In the 320 family, C1x, C2x, and C5x are 16-bit fixed-point DSPs, whereas C3x, C4x, and C8x are 32-bit floating-point DSPs. The subclassification in each group is shown in the figure. C1x has a typical instruction cycle time of 160 ns, which reduces to 80 and 35 ns, respectively, for C2x and C5x. In the floating-point DSPs, MFLOPS (million floating point operations per sec) are typically 33, 60, and 100, respectively, for C3x, C4x, and C80x. Both C4x and C8x are designed for multiprocessor applications to get high execution speed (C8x is often defined as an RISC processor). The fixed-point DSPs use a Harvard-type architecture that permits overlap of instruction fetch and execution of consecutive instruction for faster program execution, whereas 32-bit DSPs use a register-based CPU architecture with multiple bus sets to get the same objective. C30, C40, and C50 DSPs are discussed later. Note that none of these DSPs has an A/D converter on the chip.

FIGURE 9.26   TMS320C30 32-bit floating-point DSP features [16].

- COMPUTATION RATE: 33 MFLOPS (60-ns CYCLE TIME)
- 2K 32-BIT RAM
- 4K 32-BIT ROM
- 64 × 32-BIT CACHE MEMORY
- 32-BIT INSTRUCTION AND DATA WORDS, 24-BIT ADDRESSES
- 40/32-BIT FLOATING-POINT/INTEGER MULTIPLIER
- 32-BIT BARREL SHIFTER
- TWO 32-BIT TIMERS
- DIRECT MEMORY ACCESS (DMA) CAPABILITY
- INTEGER, FLOATING-POINT, AND LOGICAL OPERATIONS
- TWO BIDIRECTIONAL SERIAL PORTS
- FOUR EXTERNAL INTERRUPTS
- ASSEMBLY AND C-LANGUAGE SUPPORT
- 1.0-μM CMOS VLSI CHIP, $+5-V$ SUPPLY

The TI C30 DSP has been widely used in power electronic systems. One advantage of large word-size (32-bit) floating-point computation is that the signal step is very small, i.e., the resolution is high, and the large dynamic range does not require any scaling to prevent signal overflow or underflow. Of course, the DSP permits data conversion between the floating-point and integer formats and computation in either format. The computation speed is very fast as indicated. Each RAM and ROM block is capable of supporting two CPU accesses in a single cycle. The cache memory permits fast access of the often-repeated part of the program. The barrel shifter is used to shift up to 32 bits left or right in a single cycle. The DMA access permits fast data input/output operation without interfering with CPU operation. Both the hardware counters are of the incrementing type and can be programmed in various ways. Two serial ports are independent, and can be programmed as transmitter or receiver supporting 8/16/24/32-bit data transfers. Off-chip memory and I/O devices can be accessed by compatible access time devices. The slower devices need WAIT states for data transfer.

FIGURE 9.27 TMS320C30 architectural block diagram. (Courtesy of T1, Inc.)

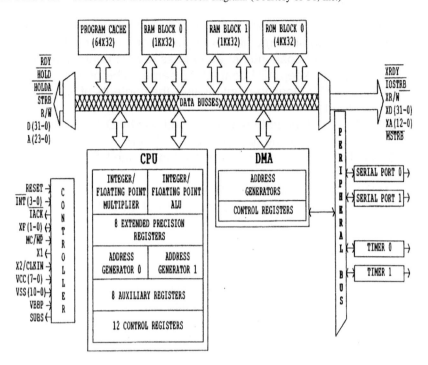

The TI C30 hardware architecture [16] is very complex and, therefore, a simplified block diagram is given here. One reason why DSP execution is very fast is because it has a pipeline architecture. This means that the processes of fetching an instruction from memory, decoding the instruction, reading the operand from memory, and executing the instruction are overlapped through separate program, data, and DMA buses to economize execution time. In a typical microcomputer, these operations are done sequentially in one instruction cycle time. The DSP can be used in microprocessor (MC/ $\overline{\text{MP}}$ = 1) mode ignoring the on-chip ROM or in the microcomputer mode. It has a total memory space of 16 million 32-bit words that can store program, data, and I/O spaces. Of the total memory space, the lower spaces (0h–7FFFFFh) are accessed by a primary bus interface (upper left), whereas the upper 800000h–0FFFFFFh spaces are mapped to the expansion bus (upper right). Both the data buses are 32 bits wide, but the expansion address bus is only 13 bits wide. The CPU block contains multiplier, barrel shifter, ALU, two ARAUs (auxiliary register arithmetic units), eight 40-bit FP/32-bit integer registers (R0–R7), eight 32-bit auxiliary registers (AR0–AR7), 12 control registers [program counter (PC), status register (ST), data page pointer (DP), stack pointer (SP), repeat counter (RC), two index registers (IRs), interrupt enable register (IE), interrupt flag register (IF), I/O flag register (IOF), etc.]. The control signals (reset, four external interrupts, two flag signals, etc.) are shown on the lower left. All peripherals (lower right) are controlled through memory-mapped registers on a dedicated peripheral bus. Note that there are no A/D converters on the chip. It is a 1.0-micron VLSI chip.

FIGURE 9.28 TMS320C30 assembly language instruction examples.

ADDF *AR4++(IR1), R5: Floating-point add of R5 content with AR4 address and store
            in R5, then modify AR4 by addition of IR1 content(1)
LDI *−AR1(IR0), R0: Load integer at address of AR1 content subtracted by IR0 content
            to R0, keep AR1 content the same as before(1)
MPYF3 R2, R7, R1: Floating-point multiply the contents of R2 and R7 and store in
            R1(1)
NOT @982Ch, R4: Generate data page address by DP and @982Ch, bitwise logically
            complement the data and load in R4(1)
RORC R4: The content of R4 is right-rotated by one bit through status register's carry
            bit(1)
ASH3 *AR3−(1), R5, R0: Arithmetic shift the content of R5 by the count located in the
            address of AR3 and store in R0. Left-shift for +count but right-
            shift for −count. Post decrement AR3 by 1(1)
NEGF *++AR3(2), R1: Add 2 to the address of AR3, negate the floating-point number
            at this address and load in R1(1)
  STF R4, *AR3−
!! STF R3, *++AR5: Parallel store the floating-point number in R4 at address of
            AR3−then decrement its address
             and floating-point number in R3 to the address at
            AR5 incremented by 1(1)
FIX R1, R2: Convert the floating-point number in R1 to nearest integer and load in
            R2(1)
FLOAT R2, R1: Convert the integer number in R2 to floating-point number and load in
            R1(1)
CALL 12345h: Jump to a new address 12345h of the program, and store next PC
            address in SP(4)

The assembly language instruction set for the C30 contains 113 instructions organized in six functional groups: load-and-store, 2-operand arithmetic/logical, 3-operand arithmetic/logical, program control, interlocked operations, and parallel operations. Most of the instructions take only one cycle time for execution. The status register stores seven condition flags (LUF, LV, UF, N, Z, V, C) as a result of arithmetic and logical instructions that are not discussed here. A few sample representative instructions are given for familiarity of the reader, and the number of cycles for execution is given at the end of description. The detailed explanation of these instructions can be found in the User's Guide [16]. Computations can be done in either floating point or in integers. Note that the addresses can be generated directly or indirectly (symbol*) by auxiliary registers (ARs) and index registers (IRs). The AR content can have predisplacement add (+X) or subtract (−X), predisplacement add modify (++X) or subtract modify (−−X), postdisplacement add modify (X++) or subtract modify (X−−). Instead of displacement, the IR content can be added or subtracted with that of AR to generate the indirect address.

FIGURE 9.29   TMS320C40 32-bit floating-point DSP features.

- COMPUTATION RATE: 60 MFLOPS /30 MIPS
- 2K 32-BIT RAM
- 128 × 32-BIT CACHE MEMORY
- 32-BIT INSTRUCTION AND DATA WORDS, 32-BIT ADDRESSES
- 40/32-BIT FLOATING POINT/INTEGER MULTIPLIER
- 32-BIT BARREL SHIFTER
- TWO 32-BIT TIMERS
- DMA COPROCESSOR WITH 6 DMA CHANNELS
- SIX HIGH-SPEED (20-MB/s) COMMUNICATION PORTS FOR MULTIPROCESSING
- FLOATING-POINT, INTEGER, AND LOGICAL OPERATIONS
- FOUR EXTERNAL INTERRUPTS
- ASSEMBLY AND C LANGUAGE SUPPORT

The features of the C40 DSP [17] are summarized here, and its simplified architecture is shown in the next figure. The DSP is very similar to that of the C30 except that its design is optimized for high-speed parallel processing with multiple DSPs. The computation rate is up to 60 MFLOPS/30 MIPS, which is somewhat faster than the C30. The external memory space is very large (4 G words) because of its wide external buses. The special features of the DSP are that it has a DMA coprocessor with six DMA channels, and six high-speed communication ports for data exchange with other C40 units connected to it. The DMA coprocessor can read from or write to any location in the memory map without interfering with CPU operation. It can also be used for two-way memory-to-communication port transfers for multiprocessor operations. Communication port data transfer is also possible by CPU. These ports provide rapid processor-to-processor communication through each port's dedicated communication interfaces. The total system tasks can be distributed among several processors in a parallel processor system to attain fast and high performance of the system. The DSP has 145 instructions and most of the instructions are executed in one cycle. Its assembly language source codes are compatible with those of the C30. The DSP supports divide, square root instructions, and supports IEEE standard 754 format conversion so that IEEE codes can be executed.

FIGURE 9.30   Simplified architecture of the TMS320C40 DSP. (Courtesy of T1, Inc.)

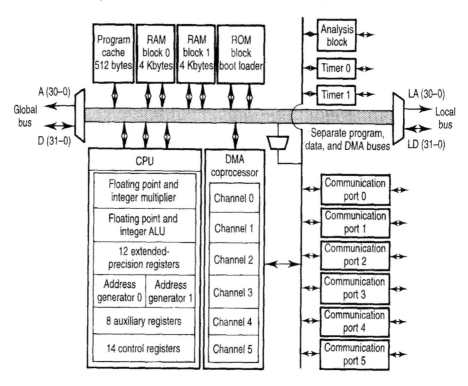

A simplified architecture for the C40 DSP is shown in the figure. The detailed architecture and description are given in Ref. [17]. Again, the fast instruction cycle time is possible because of the pipeline architecture (i.e., overlap of program fetch, decode, read, and execute operations). The DSP can be used in either microprocessor or microcomputer mode. In the latter mode, the ROM block boot loader becomes active. This on-chip code transfers program from an external memory or from a communication port to its RAM at power-up reset. The 512-byte (128-word) on-chip instruction cache is provided to store often-repeated sections of code, thus reducing the number of slower off-chip accesses. The 31-bit address bus and 32-bit data bus for local and global buses support large external memory space and help in parallel processing. The "analysis block" shown permits hardware and software debugging for a parallel processor system. Six identical communication ports permit bidirectional 8X32-bit FIFO (first-in-first-out) mode data transfers either under the control of the CPU or DMA coprocessor. The DSP can support a wide variety of multiprocessor architectures (such as rings, trees, hypercubes, bidirectional pipelines, two-dimensional Euclidean grids, hexagonal grids, and three-dimensional grids) to speed up complex system operations. The DMA coprocessor that supports six DMA data transfer channels (32-bit wide) operates in two basic modes: unified mode (memory-to-memory transfer) and split mode (two-way memory-to-communication port transfers). The DSP does not have any serial ports.

FIGURE 9.31   TMS320C50 DSP features.

- 16-BIT FIXED POINT
- COMPUTATION RATE: 33 MIPS (30-ns CYCLE TIME)
- 9K WORDS PROGRAM/DATA RAM (SINGLE-CYCLE)
- 1K WORD DATA RAM (DUAL-ACCESS)
- 2K WORDS BOOT ROM
- ONE 16-BIT TIMER
- TWO SERIAL PORTS (ONE FULL DUPLEX)
- 192K WORDS ADDRESSABLE MEMORY
- 16×16 -BIT MULTIPLIER (32-BIT PRODUCT)
- $64K \times 16$ PARALLEL PORTS
- SOURCE CODE COMPATIBLE WITH C1X/C2X FAMILIES
- 16-BIT PARALLEL LOGIC UNIT (PLU)
- 16 S/W PROGRAMMABLE WAIT STATE GENERATORS
- ASSEMBLY AND C LANGUAGE SUPPORT
- FULLY STATIC 0.72-µM CMOS FOR LOW POWER CONSUMPTION

The TI C50 DSP is a 16-bit fixed-point device that is suitable for use in most power electronic systems. However, the data step size is larger (as inherent in a fixed-point DSP), and, therefore, its dynamic range is smaller compared to similar floating-point DSPs. Therefore, to prevent data overflow/underflow, suitable scaling may be needed. One special feature of the device is that it uses a fully static CMOS device that contributes to low power consumption and, therefore, a battery power supply can be used. The high speed of computation is due to its Harvard architecture (in which two separate bus sets are used to transport instructions and date independently). The addition of a PLU provides the power to manipulate bits in data memory without using accumulator and ALU. The programmable wait states permit the DSP to interface with slower devices. The DSP supports a large number of parallel ports which may be convenient in some applications.

FIGURE 9.32 Texas Instruments advanced TMS320 DSP families.

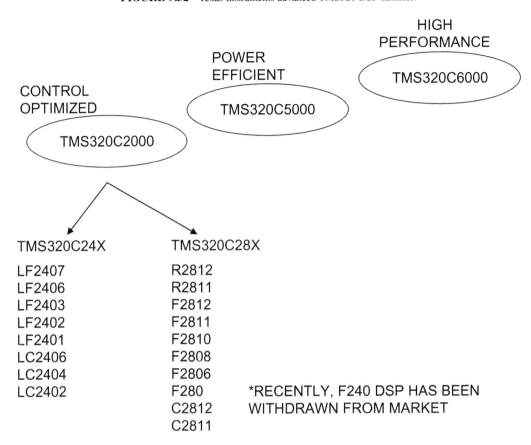

The DSP technology has recently advanced tremendously because of advancements in submicron VLSI circuits. The TMS320 family evolution was discussed in Figure 9.25. This figure shows the advanced TMS320 families developed in recent years [15]. These new DSP platforms include the C6000, C5000, and C2000 families as indicated in the figure. The C6000 DSPs are the industry's fastest devices and are available in fixed-point (C64X and C62X) and floating-point (C67X) types. The former types have a speed range of 1200–8000 MIPS, whereas the latter is available in the range of 600–1350 MFLOPS. These DSPs are used in products such as broadband infrastructure and audio and imaging applications. The C5000 family has the combination of high performance, peripheral options, small packaging, and best power efficiency performance in the industry. Power consumption is typically 0.45 mA/MHz and performance is up to 600 MIPS. These DSPs are used typically in portable applications, such as digital music players, GPS (global positioning system) receivers, medical equipment, and cell phones.

The C2000 family design is optimized for control applications, which combine high performance and code-efficient control with the on-chip multichannel A/D converter, PWM outputs, etc., that make them especially important for power electronics systems. This family is further subdivided into C24X and C28X classes as indicated in the figure. In later pages, we will discuss the details of the LF2407 and F2812 sample types. Note that the popularly used DSP F240 in the C24X group has recently been discontinued. DSPs are evolving at such a rapid rate that it is difficult to keep track of them.

FIGURE 9.33 TMS320LF2407 DSP features.

- 16-BIT FIXED POINT
- 40-MHz FREQUENCY
- COMPUTATION RATE: 40 MIPS (25-ns CYCLE TIME)
- 2.5K WORDS RAM
- 32K WORDS FLASH MEMORY
- BOOT LOADER ROM
- 10-BIT, 16-CHANNEL A/D CONVERTER – A/D CONVERSION
  TIME of 0.5 µs
- 16 PWM CHANNELS
- WATCHDOG TIMER
- FOUR 16-BIT TIMERS
- 16 × 16-BIT MULTIPLIER
- 16-BIT BARREL SHIFTER
- FIVE EXTERNAL INTERRUPTS
- FOUR POWER-DOWN MODES
- C AND ASSEMBLY LANGUAGE SUPPORT

As mentioned before, the LF2407 is a control-optimized DSP that is very well suited for power electronics and motor drive systems. Basically, it is an upgraded version of the F240, which was very popular in power electronics applications. The DSP is a 16-bit, fixed-point type with a pipelined (Harvard) architecture and an instruction cycle time of 25 ns. It has an on-chip multichannel (16 number) A/D converter with 10-bit resolution and conversion time of 500 ns, in addition to 16 PWM channels. The device has 18K words of data SRAM, 32K words of flash memory (similar to EEPROM) for program storage, and boot loader ROM for program booting. The hardware multiplier, timers, and barrel shifter help for real-time application. The power-down modes help conservation of supply power. The ANSI C compiler, assembler/linker, C-source debugger, and emulator are available (similar to other TI DSPs) for software design.

FIGURE 9.34   Simplified architecture of the TMS320LF2407 DSP. (Courtesy of T1, Inc.)

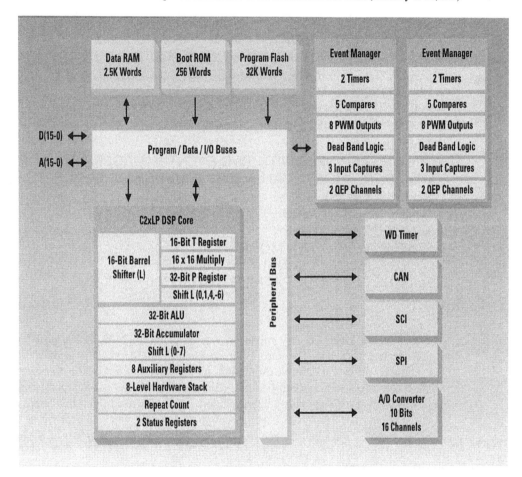

FIGURE 9.34   Simplified architecture of the TMS320LF2407 DSP. (Courtesy of T1, Inc.)

A simplified architecture for the LF2407 is shown in this figure. The DSP uses submicron static CMOS technology. This, along with four power-down modes (idle, standby, halt, and disable peripheral clocks), conserve the power supply. The 16-bit address and data buses permit the total memory space of 224K words. However, the on-chip RAM and flash memory should be adequate for most applications. Two Event Managers, as shown, permit up to 16 compare/PWM channels with continuously running UP/DOWN counting of the hardware counters. Some compare units have programmable deadband logic. There are six input capture units [four with QEP (quadrature encoder-pulse interface) capability]. The Event Managers generate PWM waves with variable carrier frequency and duty cycles for the converters. Other peripheral devices include a WD (watchdog) timer, CAN (control area network), SCI (serial communication interface), and SPI (serial peripheral interface). The WD timer monitors DSP hardware/software healthiness, and the CAN module permits control of multiple motors with one DSP. The core of the DSP is indicated in the figure.

FIGURE 9.35   TMS320F2812 DSP features.

- 32-BIT FIXED POINT
- 150-MHz FREQUENCY
- STATIC CMOS WITH HARVARD ARCHITECTURE
- COMPUTATION RATE: 150 MIPS (6.67-ns CYCLE TIME)
- 18K WORDS RAM AND 128K WORDS FLASH MEMORY
- BOOT LOADER ROM
- 12-BIT, 16-CHANNEL A/D CONVERTER – A/D CONVERSION
  TIME of 80 ns
- 16 PWM CHANNELS WITH TWO EVENT MANAGERS
- WD TIMER, SCI, SPI, CAN, McBSP (MULTICHANNEL BUFFERED
  SERIAL PORTS)
- THREE 32-BIT TIMERS
- 32 × 32-BIT MULTIPLIER
- FOUR POWER-DOWN MODES
- C AND ASSEMBLY LANGUAGE SUPPORT
- SOURCE CODE COMPATIBLE WITH C2000 FAMILY

The TI F2812 DSP is a highly enhanced version of the LF2407, and is possibly the world's highest performance control-optimized and code-efficient DSP available today. For example, all of the control, estimation, monitoring, and diagnostic software of a modern sensorless vector-controlled induction motor drive can be incorporated into a DSP with high speed and high signal resolution. The 32-bit fixed-point DSP has a computation speed of 150 MIPS (6.7-ns instruction cycle time), which is nearly four times faster than that of the LF2407. The on-chip memory includes 18K words of RAM and 128K words of flash memory—considerably higher than LF2407. The A/D conversion has better resolution and faster conversion time (more than six times) as indicated. Many features of the LF2407 and F2812 are similar except for the performance enhancement of the latter and, therefore, its architecture is not described separately.

FIGURE 9.36 Microcomputer/DSP-based control flowchart.

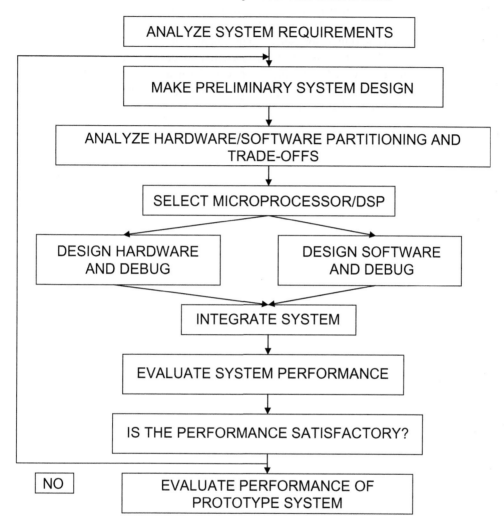

Once the decision is made that the system should have microcomputer-based control, the project normally proceeds according to this flowchart. After the simulation study fine-tunes the control strategy, further analysis and preliminary design are required from the viewpoint of software control. These include identification of all control and estimation tasks, monitoring, warning, sequencing, and diagnostic needs, their execution time, responses, and accuracy needs, multitasking requirements, the corresponding sampling times, etc. If necessary, flowcharts can be drawn in detail for all tasks to be executed in sequence. Some of these tasks, such as PWM algorithm and feedback signal processing, can be implemented either by software or by dedicated hardware. Software implementation may be slow but economical, whereas the hardware implementation may be fast but expensive and needs additional real estate. This partitioning and trade-off should be considered carefully before selecting the microcomputer. Hardware and software design can proceed in parallel to save time. Once they are designed and debugged, the performance should be tested after integration. If performance specs are not met, further iteration may be necessary. Finally, the total prototype system should be integrated and tested thoroughly before acceptance.

FIGURE 9.37 General software development flow diagram.

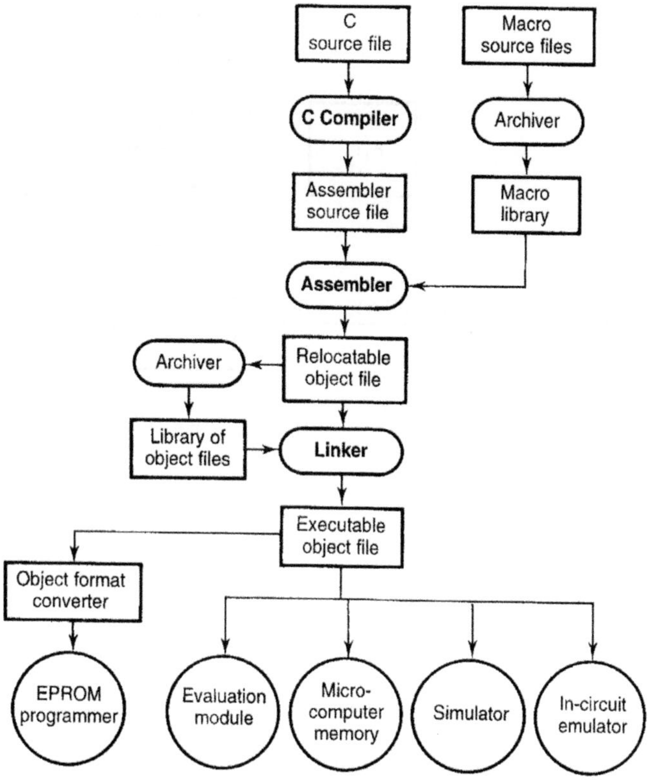

Software development is generally a large and time-consuming task for a microcomputer-based control project. Most microcomputers support a high-level language, such as C or C++, that makes the software development process user-friendly and permits its portability. However, Assembly language program is most time-critical and code-size efficient, although programming in this language is very tedious. A practical program may use a strategic mix of C and Assembly language source programs (not shown). For example, time-critical protection software should be in Assembly language. The C compiler converts the C source code into Assembly language code. Then, the Assembler translates it to a machine relocatable object file, which is then converted to an executable object file by Linker after assigning absolute memory addresses. This file is then downloaded to the microcomputer ROM or EPROM through the Object format converter, which alters the file into specific format. The Archiver collects a group of files into a single archive file. The various debugging tools shown are the software-based Simulator (simulates the program in host PC), hardware-based In-circuit emulator (connected to target system for full-speed emulation), and the Evaluation module (developmental board for full speed emulation and hardware debugging).

FIGURE 9.38   Digital implementation of *P-I* compensator.

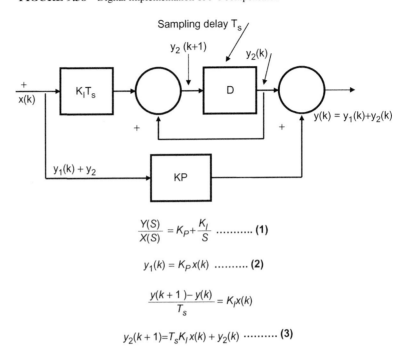

$$\frac{Y(S)}{X(S)} = K_P + \frac{K_I}{S} \quad \ldots\ldots\ldots \text{ (1)}$$

$$y_1(k) = K_P x(k) \quad \ldots\ldots\ldots \text{ (2)}$$

$$\frac{y(k+1) - y(k)}{T_s} = K_I x(k)$$

$$y_2(k+1) = T_s K_I x(k) + y_2(k) \quad \ldots\ldots\ldots \text{ (3)}$$

The software implementation of a *P-I* (proportional-integral) compensator given by Eq. (1) (in Laplace transfer function) in a feedback system is shown in the figure. Because the microcomputer operates in the time domain and in sequential form, it is necessary to convert the equation in time-domain finite difference form. Note that there are two parallel signal paths: One is proportional given by Eq. (2) and the other is integral. The integral part is converted to Eq. (3), where k and $k+1$ are consecutive sampling instants at intervals of T_s. The implementation of Eqs. (2) and (3) is shown in the figure, where the output $y(k) = y_1(k) + y_2(k)$. The sampling delay T_s essentially performs the integration, which can be easily verified. For a given input $x(k)$, the output $y(k)$ is computed every sampling time T_s by a sequence as follows:

$$x(k).K_I T_s + y_2(k) = y_2(k+1)$$

$$y_2(k+1) \rightarrow y_2(k)$$

$$y_2(k) + y_1(k) = y(k)$$

The other types of compensators, such as *P-I-D*, lead-lag, etc., can be implemented in a similar manner.

FIGURE 9.39 Typical system sequencing diagram.

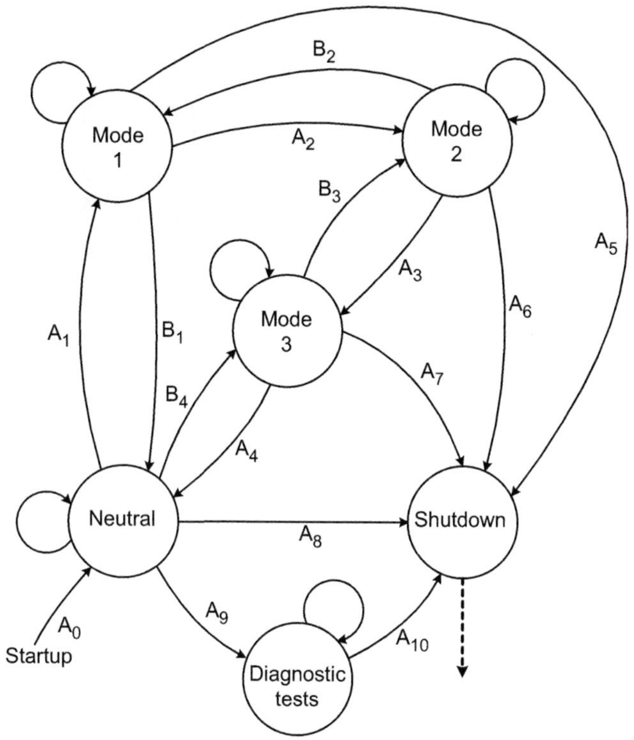

The sequencing control in finite state form, as shown here, is an important function in a complex control system. For example, a drive system can have different operating modes, such as start-neutral, constant-torque motoring, constant-torque regeneration, field-weakening motoring, and field-weakening regeneration. In the figure, the circles represent the modes or states, and the arrows indicate permissible transition paths between the states. A transition from one state to another is dictated by a set of conditionals in the form of Boolean variables (A_1, B_1, etc.). When the conditionals are satisfied, the transition is initiated smoothly by execution of an action routine. The sequencing software is executed periodically with a sampling time. If the transition conditionals are not satisfied, the state falls back to itself repetitively. If a fault develops in any state, the drive shuts down. If, for example, the operator presses the Start button in the beginning, the drive acquires the Neutral state after power-up reset and hardware–software initialization. The diagnostic tests can be initiated before starting the drive. If the Forward button is on and the speed is below the base speed, Mode 1 (forward constant-torque motoring mode) can be initiated from the Neutral mode. In a complex system, there may be many more states, and careful design will prevent a bumpy transition. Complex sequencing software is normally designed as a finite state machine (FSM).

FIGURE 9.40 Simplified structure of software.

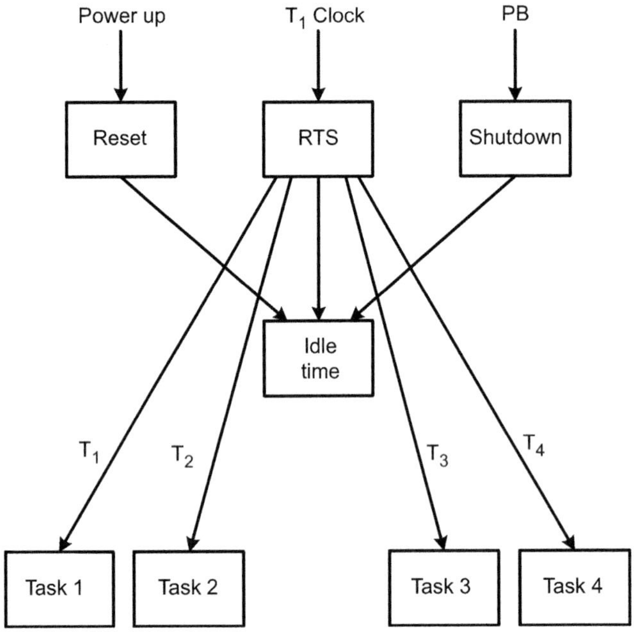

The figure shows a simple and typical software structure for a microcomputer. As mentioned earlier, the functions or tasks (such as control, sequencing, feedback estimation, monitoring/warning, diagnostics) to be performed by the microcomputer are identified in the beginning. All tasks are performed in a sequential manner, and some tasks are executed at a faster rate (lower sampling time) than others. Therefore, different tasks and their sampling times of execution are to be designed for satisfactory system operation. This figure shows that the execution of the four tasks at the corresponding sampling rate is under the control of executive software RTS (real-time scheduler). The RTS gets the hardware interrupt clock of period T_1 at the input, which then generates the other sampling times by software down-counters as shown. The RTS can be custom designed or a commercial RTOS (real-time operating system) can be used. When all four tasks are successfully performed, the microcomputer goes into idle mode as shown. At power-up reset or initialization of the microcomputer, it vectors into the idle time mode and then waits for task execution to begin by interrupts. If the operator pushes a button (PB in the figure) to shut down the system, the idle mode is invoked, and from there the shutdown is activated. There are, of course, a number of methods for structuring the software. Example software structures of some practical projects will be given later.

FIGURE 9.41 Task timing diagram.

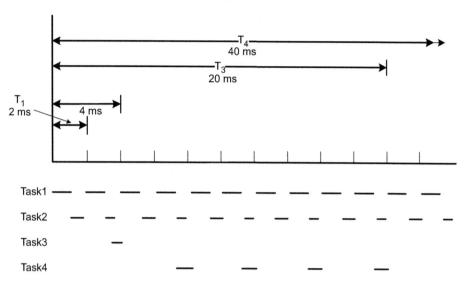

Task1(T_1 period): Current loop control and diagnostics-1
Task2(T_2 period): Flux and speed loop control
Task3(T_3 period): Sequencing and diagnostics-2
Task4(T_4 period): Background functions

The figure shows the task timing diagram, which lists the typical four classes of drive system tasks according to priority. The background functions, for example, may involve filling a memory buffer with the software variables that are needed for diagnostic purposes. Time interval T_1 is due to real-time external interrupt, and the other times are generated by scaled-down software timers. Current loop control along with fast diagnostics for protection requires the fastest response (highest bandwidth) and, therefore, sample time T_1 is the fastest. With a modern DSP, the time can be much shorter than the 2 ms shown in the figure. Task 1 is always completed within every sampling interval T_1 before any low-priority task is executed. If a low-priority task is being executed, it is suspended until completion of task 1. In this figure, task 2 is suspended once (shown by gap), and task 4 is suspended three times as indicated. After completion of task 4, if any spare time is left, the computer will idle until the T_1 interrupt is exerted. The loading factor (LF) can be defined as the fraction of time the computer is busy during the longest sampling time T_4, i.e., LF = Σ(computing time)/T_4. It is an extremely important design parameter and is discussed further in Chapter 12. During software development, the execution time of each task is carefully monitored to prevent any possible overflow.

FIGURE 9.42   Simplified flowchart of real-time scheduler (RTS).

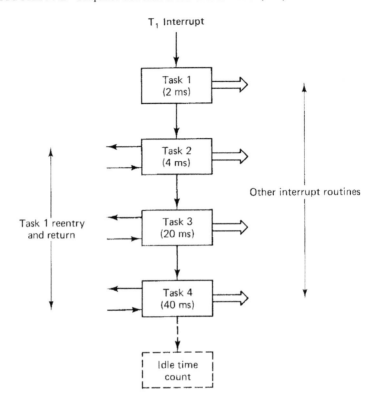

The figure shows the flow diagram of RTS driven tasks in sequence according to their priority levels. The reentry of task 1 during execution of low-priority tasks 2, 3, and 4 is shown by interrupts. As soon as a T_1 interrupt comes in from the hardware clock, the next PC address is pushed in the stack, the task 1 routine address is loaded in the PC, and execution of the task begins. After execution of the routine, the stack address is popped back to PC and the usual course of execution resumes. Of course, any higher priority interrupt (such as from protection system) can enter any time and request service, suspending all the other activities. The software counting of idle time gives an estimate of spare time that can be utilized gainfully.

FIGURE 9.43 Thyristor gate firing control of three-phase bridge converter.

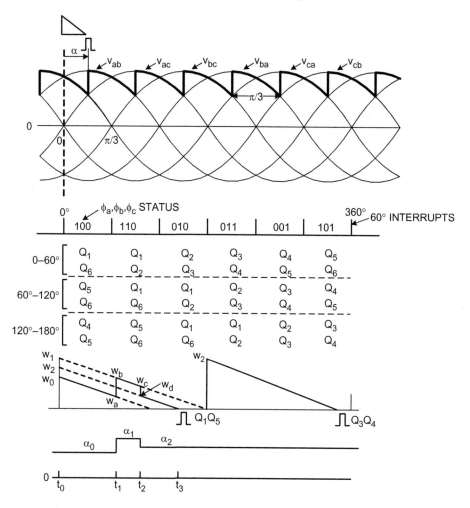

One simple application of a microcomputer is for gate firing control [12] of a three-phase thyristor bridge converter (see Figure 3.13). For simplicity, it is assumed that the converter always operates in continuous conduction mode in two quadrants, and the desired α angle resolution is 0.5°. In the upper part of the figure, the line voltage waves (v_{ab}, v_{bc}, and v_{ca}) are shown. From these waves, hardware interrupts at each 60° crossover points and polarity information (φ_a, φ_b, and φ_c) of the respective line voltage waves in these intervals are generated. There is also a phase-locked loop (PLL) that generates a clock frequency $f_{ck} = 720 f_s$, where f_s is the supply frequency [60 Hz]. The PLL can be synthesized by hardware or software. The 60° interrupts, polarity signals (φ_a, φ_b, and φ_c), and PLL output are received by the computer, which then generates the firing pulses for the six thyristors. Internally, the computer has a look-up table (shown in the middle of the figure) of thyristors to be fired in each 60° segment. For a command dc voltage output (V_d*), the computer calculates α by the relation $\alpha = \cos^{-1}(V_d*)/1.35 V_L$. The α is then scaled so that 180° corresponds to a 360 count with 0.5° resolution. The count identifies one of the three angular segments (0–60°, 60–120°, and 120–180°). The digital α count is loaded to a down-counter (hardware or software) in the computer and clocked by f_{ck}. When the counter is cleared, an interrupt is generated. The interrupt identifies the status of φ_a, φ_b, and φ_c, and also determines the thyristor pair to be fired from the look-up table. On the fly, the counter can be updated to make angle corrections as shown in the figure.

FIGURE 9.44 Three-phase thyristor bridge speed control of dc motor with nonlinearity ($\Delta\alpha$) compensation due to discontinuous conduction.

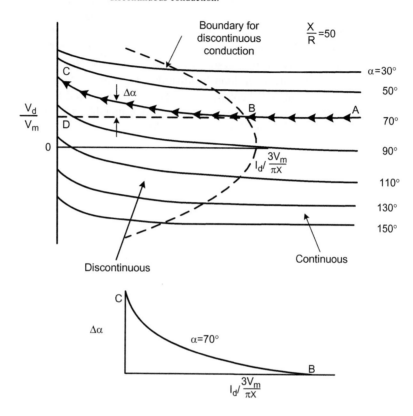

A practical thyristor converter may fall into discontinuous conduction mode at light load, particularly with a dc motor-type counter emf, as shown in this figure [9]. At discontinuous conduction, the converter transfer characteristics become nonlinear, giving variable gain at different operating points that can result in a stability problem. The figure (similar to Figure 3.18) has been plotted for a particular drive with $X/R = 50$ in the armature circuit. Note that dc voltage $V_d \propto$ speed and dc current $I_d \propto T_e$ in Figure 3.18. If, for example, $\alpha = 70°$ and the current I_d is decreased from point A, conduction is continuous with practically horizontal trajectory. In segment BC, the conduction is discontinuous, where V_d increases with decreasing I_d. The converter transfer characteristics will be linearized if voltage V_d is maintained constant to the level of continuous conduction by adding a compensating angle $\Delta\alpha$ so that the BC curve stays on line BD, as shown in the figure. This $\Delta\alpha$–I_d compensation curve for $\alpha = 70°$ is shown separately in the lower part of the figure. A two-dimensional look-up table with $\Delta\alpha = f(I_d)$ for increments of the α angle can be precomputed for a particular drive and stored in microcomputer memory. Of course, instead of a look-up table, polynomial equations can also be described and solved in real time to derive the $\Delta\alpha$ angle. The block diagram of the complete drive system with $\Delta\alpha$ compensation is shown in Figure 3.19. Because the current loop has higher bandwidth than speed loop, it should be sampled at a faster rate than the speed loop. The current is sensed and averaged (filtered) digitally in a 60° period. Similarly, the speed signal is sensed by a resolver or optical encoder. Note that delays in feedback signals seriously affect the loop bandwidths. The current loop compensator $G(S)$ may be a simple gain or *P-I* compensator like that of the speed loop discussed earlier. The $\cos^{-1}$ function is usually implemented by a look-up table. The other elements in the controller can be implemented without difficulty.

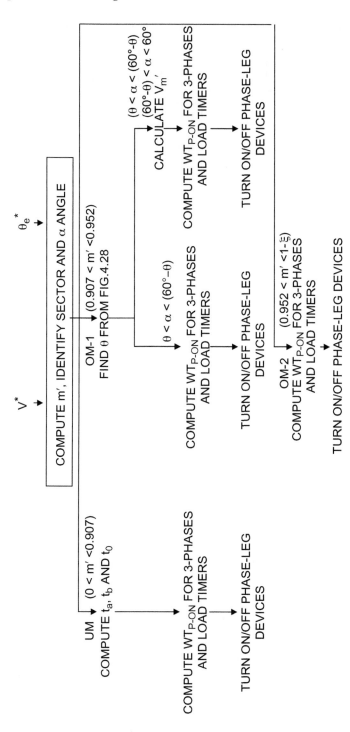

FIGURE 9.45 Flow diagram for a microcomputer-based space vector PWM implementation for an IGBT bridge converter.

The figure shows the computation flow diagram of space vector PWM for a three-phase, two-level bridge converter, and includes both undermodulation and overmodulation modes. Note the analogy with Figure 4.38, which describes a similar flow diagram for a three-level converter. The computer receives the command voltage (V^*) and angle (θ_e^*) signals at the input and generates the IGBT (or other device) drive signals for the three phases at the output. For a given V^*, modulation factor m' is calculated, and corresponding undermodulation (UM) and overmodulation modes (OM-1 and OM-2) are identified. Then the computation flow proceeds according to theory [9]. The precomputed plots of turn-on time (T_{P-ON}) similar to Figure 4.37 are needed as indicated. The internal timers of the computer can be used to convert the digital signals into timing waves.

FIGURE 9.46   Digital controller block diagram for an IPM synchronous motor drive for EV propulsion.

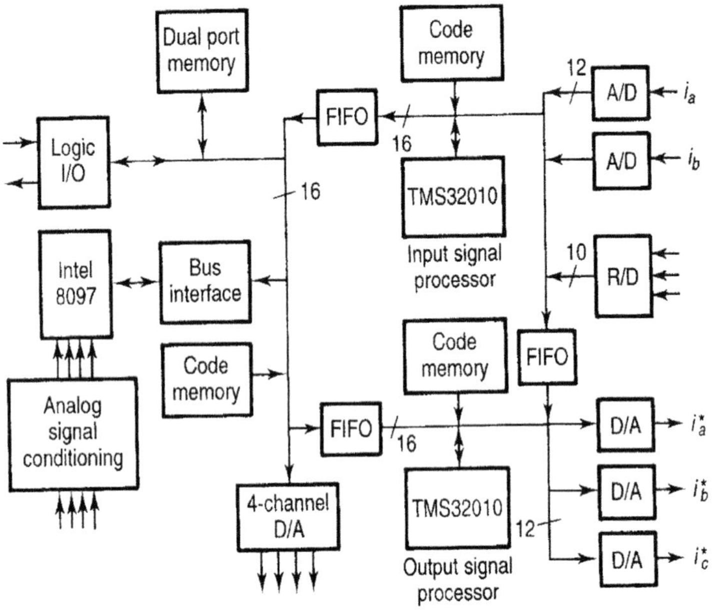

The figure shows the simplified hardware block diagram for multiprocessor-based control [18] of the IPM synchronous motor drive described in Figures 8.22–8.32. Several important features of microcomputer control are illustrated in the figure. The system uses one Intel 8097 (MCS-96 family) microcomputer and two TI TMS32010 DSPs [one as an input signal processor (ISP), the other as an output signal processor (OSP)] for sharing of data acquisition, control, feedback signal processing, and diagnostic functions. Both are 16-bit, fixed-point computers with hardware multipliers, but the 8097 has an integrated 10-bit, 8-channel unipolar A/D converter with the analog multiplexer. The ISP is mainly responsible for acquiring and processing the motor currents and speed encoder signals through A/D and R/D (resolver/digital converter with analog resolver) converters, respectively, whereas the OSP processes the output signals and delivers the output PWM current commands through D/A converters as shown. The 8097 essentially handles the feedback and signal estimation functions. All interprocessor communications are accomplished with 16-bit, 16-deep FIFO registers as shown. All the signals are scaled appropriately to prevent any overflow and underflow. This problem normally does not exist in floating-point processors. Because the drive is designed for EV propulsion, data exchanges occur through the vehicle computer system. The parallel processing system adds the speed of computation and strategic sharing of complex computation and diagnostic functions.

FIGURE 9.47 Simplified software structure of Intel 8097 for IPM motor EV drive.

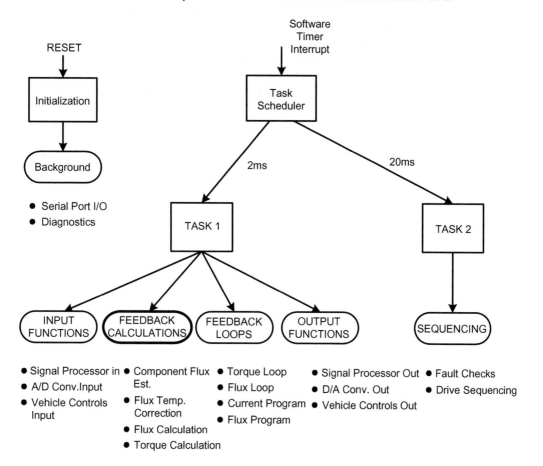

The figure shows the simplified software structure of an Intel 8097 [18] in the EV propulsion system described in the previous figure. Essentially, two task groups are handled by the computer. Task 1 is computed at the sampling rate of 2 ms and consists of input functions, feedback calculations, feedback control loops, and output functions, the details of which are listed in the figure. Task 2 consists of drive sequencing and fault checks and, being of lower level, it is executed every 20 ms. The sequencing logic serves to select the appropriate control algorithm for either the PWM or square-wave mode, and open-loop test modes are incorporated for debugging. The software also permits various feedback loop configurations for orderly debugging of the system starting with the core drive elements. These configurations include (1) open all loops with $\delta = 0$, (2) open all loops with released δ, (3) close overlay current loops and initialize torque loop integrator, (4) close torque loop using the current program, (5) initialize flux loop integrator, (6) close all PWM loops,

(7) open vector rotator square-wave mode loop, (8) close vector rotator square-wave mode, (9) conduct PWM to square-wave mode transition, (10) conduct square-wave to PWM transition, etc. The interrupt for task 1 is generated by a software timer, whereas the 20-ms interval is generated by another software timer that is triggered every 2 ms by the 2-ms timer. At the power-up reset of the computer, hardware and software initialization occurs, and then it enters into background mode and waits for 2-ms interrupts. Some diagnostics and serial port I/O data are processed in the background.

FIGURE 9.48 Advances and trends in simulation and digital control.

- ANALOG AND HYBRID SIMULATION, ONCE SO POPULAR IN POWER ELECTRONICS, IS NOW OBSOLETE
- MODERN SIMULATION LANGUAGES ARE VERY USER-FRIENDLY BUT SLUGGISH – HIGHER COMPUTER SPEED WILL COMPENSATE IT
- SIMULINK WITH SIMPOWERSYSTEMS CAN BE UNIVERSAL FOR POWER ELECTRONICS AND MOTOR DRIVE SYSTEMS
- GROWING TREND TOWARDS USING SIMULATION MODULES IN REAL TIME
- POWER OF REAL-TIME SIMULATION
- DIGITAL CONTROL IS PRACTICALLY UNIVERSAL
- MODERN SUBMICRON VLSI ASIC CHIPS AND DSPs ARE BECOMING EXTREMELY POWERFUL
- A SINGLE VLSI CHIP IN FUTURE WILL BE ADEQUATE FOR COMPLEX SENSORLESS VECTOR DRIVE WITH ON-LINE DIAGNOSTICS AND FAULT-TOLERANT CONTROL
- INTELLIGENT POWER CONVERTER IC WITH EMBEDDED CONTROLLER ASIC CHIP IN FUTURE

Simulation and digital control technologies are going through a revolution mainly because of the availability of submicron and super high speed VLSI chips. In the first generation, analog computer simulation and then hybrid computer simulation were very popular in the power electronics area, but they essentially became obsolete in the 1980s. Simulation is now easy because of the availability of so many powerful and user-friendly digital simulation languages. More user-friendly languages tend to be slower, but higher computer speeds tend to compensate for that disadvantage. Real-time simulation of a complete system in a high-speed computer is extremely useful for sophisticated control systems, where estimation variables become available in real time. However, real-time simulation of a complex system with varying parameters remains a challenge. Simulink with SimPowerSystems has the potential to be a universal simulation language in power electronic systems. However, currently, it is sluggish and power system block sets require more accurate modeling without sacrificing simulation speeds. Progressively, the trend will be to use simulation software in real time. Both ASIC chips and DSPs are becoming very powerful. We can visualize a single ASIC chip with control and diagnostics being embedded in a power IC in the future.

Glossary for Microcomputers/DSPs

Access time   The delay between the time when a memory receives an address and the time the data from that address are available at the outputs

Accumulator   A register that is the source of one operand and the destination of the result for most arithmetic and logical operations

Address   The location of program code or data in memory

ALU   The part of the CPU (central processor unit) that performs arithmetic and logic operations

Architecture   Hardware structure of a microcomputer or DSP

Assembly language   A programming language in which the programmer can use mnemonic instruction codes, labels, and names to refer directly to their binary equivalents

Bit, byte, word   A binary digit, a group of eight bits in length, and a group of bits that a computer can manipulate

Boot loader   An on-chip code that transfers code from an external memory or from a communication port to the DSP RAM at power-up

Cache memory   An on-chip memory that stores often-repeated instructions. This permits the bulk of the code to be stored off-chip in slower and lower cost memories

Compiler/assembler   A compiler translates source code in high-level language program into Assembly language. An assembler translates Assembly language code into object code

Clock   A periodic timing signal generated by the oscillator that makes possible synchronized operation of all computer operations

Direct addressing   Generation of memory address directly by concatenation of the instruction word and data page pointer (DPTR)

Direct memory access (DMA)   A fast method of transferring contents of memory locations without processor help

DSP   Digital signal processor; basically the same as a microcomputer (and often called a computer)

Dual access memory   A memory that can be accessed twice in a single clock cycle

Dynamic/static memory   A memory that loses its contents gradually unless periodically refreshed by a clock. A static memory's contents remain unchanged without refreshing. Usually applies to RAM memory (DRAM or SRAM)

EEPROM (or E²PROM)   A programmable read-only memory (PROM) that can be written and erased by electrical signals

EPROM   A PROM that can be completely erased by exposure to ultraviolet light and reprogrammed

FIFO register   A set of registers that permit memory transfer by a first-in/first-out order, like a conveyer belt

Fixed point (or integer)   A number that is directly represented by digital word integers (XXXXs)

Flash memory   A memory that can be quickly electrically programmed and electrically erased. It is similar to an EEPROM

Floating point A number that is represented by a mantissa and exponent of 2 ($XXXXX.2^{YYYYY}$)

In-circuit emulator (ICE) A special tool connected to the target system microprocessor socket through a probe to control and debug the target system

Indirect addressing Generating a memory address through the contents of an auxiliary register

Instruction A group of bits that defines a computer operation

Instruction cycle time Time needed to fetch, decode, and execute an instruction

Instruction set A set of instructions available to a particular microcomputer/DSP

Interrupt A signal that interrupts microcomputer operation and forces it to execute an important routine first. The signal can be generated externally or internally

MFLOPS Millions of floating-point instructions per second

MIPS Million instructions per second

Mnemonics Symbolic names or abbreviations for instructions, registers, memory locations, etc., which indicate their actual functions

Multitasking system System operation that involves multiple tasks being executed by one or multiple sample times

Object code Code that can be executed by a microcomputer/DSP

Polling Determining the state of peripherals or other devices by checking

Program counter A register that specifies the address of the next instruction to be fetched from program memory

RAM/SRAM Random access memory can be read or written (often defined as scratch pad or R/W memory). Static RAM does not require any refreshing

ROM/PROM Read-only memory/programmable read-only memory. A PROM is programmed permanently by using a mask

RISC Reduced instruction set computer, in which higher execution speed is possible because the architecture has been simplified

Simulator A software program running on a host computer simulates microcomputer/DSP program operation

Source code Original code generated by the programmer

Stack A block of memory reserved for storing and retrieving data on a first-in/last-out basis. Normally used for storing return addresses for the program counter

Transputer A microcomputer that is specially designed for high-speed parallel processing by reducing the time required for interprocessor communications

Wait state Period of time that the CPU must wait to access external memory

Watchdog timer Timer operation that indicates the healthy state of hardware and software operation

Summary

The discussion in this chapter is subdivided into two topics: simulation and digital control. Both are extremely important in the development and design of power electronics and drive systems. A newly developed converter or control system, particularly if it is complex, should be simulated with a digital computer considering the many user-friendly simulation languages currently available. Two simulation languages are emphasized that have recently become very popular: circuit-oriented PSPICE simulation and system-oriented Simulink simulation. Recently, the Simulink simulation has been enhanced by the addition of SimPowerSystems, which integrates both PSPICE and Simulink features. A number of practical examples are given including a PSPICE-based buck converter, Simulink-based voltage-fed PWM inverter, high-frequency converters, induction motor and vector-controlled induction motor drive, and SimPowerSystems-based dc and vector-controlled induction motor drives simulation. Digital control normally means micro-computer (or DSP)-based control, although it may also be discrete digital/analog IC or ASIC chip-based (which may include hardware and software) control. Recently, ASIC chips with a DSP core have been very powerful, but no discussion is included in this area. Developing microcomputer control is complex and time consuming, but it is now almost universally used in power electronics and drive systems considering its many advantages. Commercial microcomputers/DSPs of various types are available from a large number of vendors, and the technology is advancing at a fast rate. A few sample and popular types from Intel and Texas Instruments (TI) are reviewed that are important in power electronic systems: the Intel 8751 8-bit, fixed-point microcontroller and the TI C50 and LF2407 16-bit fixed-point, F2812 32-bit fixed-point, and C30 and C40 32-bit floating-point DSPs. Note that recently the popular F240 DSP was withdrawn from the market. The F2812 is possibly the ultimate DSP for speed and performance. Microcomputer application methodology is discussed, and a few example applications are illustrated, including *P-I* control, thyristor gate-firing control, converter linearization, space vector PWM, and IPM machine drives for electric vehicles. Finally, a comprehensive glossary has been added for digital control that will be useful for newcomers in this area.

References

[1] N. Mohan, W. P. Robbins, T. M. Undeland, R. Nilssen, and O. Mo, "Simulation of power electronic and motion control systems," *Proc. IEEE,* vol. 82, pp. 1287–1302, August 1994.

[2] *Using MATLAB, Version* 6, MathWorks, Natick, MA, www.mathworks.com, August 2002.

[3] *Using Simulink, Version 5,* MathWorks, Natick, MA, www.mathworks.com, April 2003.

[4] *MicroSim PSpice & Basics User's Guide,* MicroSim, Irvine, CA, www.microsim.com.

[5] *PSIM Simulation,* www.powersimtech.com.

[6] N. Mohan, T. M. Undeland, and W. P. Robbins, *Power Electronics,* 2nd ed., Wiley, New York, 1995.

[7] L. Hui, B. Ozpineci, and B. K. Bose, "A soft-switched high frequency non-resonant link integral pulse modulated dc-ac converter for ac motor drive," *IEEE IECON Conf. Rec.,* pp. 726–732, 1998.

[8] C. M. Ong, *Electric Machinery,* Prentice Hall, Upper Saddle River, NJ, 1998.

[9] B. K. Bose, *Modern Power Electronics and AC Drives,* Prentice Hall, Upper Saddle River, NJ, 2002.

[10] *SimPowerSystems User's Guide, Version 3,* MathWorks, Natick, MA, www.mathworks.com, February 2003.

[11] B. K. Bose (Ed.), *Microcomputer Control of Power Electronics and Drives,* IEEE Press, New York, 1987.

[12] B. K. Bose, *Power Electronics and AC Drives,* Prentice Hall, Upper Saddle River, NJ, 1986.

[13] H. Le-Huy, "Microprocessors and digital ICs for control of power electronics and drives," Chapter 10 in *Power Electronics and Variable Frequency Drives,* edited by B. K. Bose, IEEE Press, New York, 1997.

[14] *Intel MCS 51 Microcontroller Family User's Manual,* www.intel.com, 1994.

[15] *Texas Instruments DSP Platforms,* http://dspvillage.ti.com/docs/catalog/dspplatform, November 2004.

[16] *TMS320C3X User's Guide,* Texas Instruments, 1991.

[17] *TMS320C4X User's Guide,* Texas Instruments, 1996.

[18] B. K. Bose and P. M. Szczesny, "A microcomputer-based control and simulation of an advanced IPM synchronous machine drive system for electric vehicle propulsion," *IEEE Trans. Ind. Electron.,* vol. 35, pp. 547–559, November 1988.

CHAPTER 10

Fuzzy Logic and Applications

FIGURE 10.1   What is artificial intelligence (AI)?

- HUMAN BRAIN WITH BIOLOGICAL NEURAL NETWORK HAS
  NATURAL INTELLIGENCE – ABILITY
  TO LEARN, REASON, AND COMPREHEND
- GOAL OF AI – PLANTING HUMAN INTELLIGENCE IN COMPUTER
  SO THAT A COMPUTER CAN THINK INTELLIGENTLY LIKE A
  HUMAN BEING
- CAN A COMPUTER REALLY THINK AND MAKE INTELLIGENT
  DECISIONS?
- COMPUTER INTELLIGENCE – FAR INFERIOR TO NATURAL
  INTELLIGENCE. HOWEVER, IT HELPS TO SOLVE COMPLEX
  PROBLEMS
- AI TECHNIQUES ARE USED EXTENSIVELY IN
  - INDUSTRIAL PROCESS CONTROL
  - GEOLOGY
  - MEDICINE
  - INFORMATION MANAGEMENT
  - MILITARY SYSTEM
  - SPACE TECHNOLOGY, ETC.

Fuzzy logic is a part of artificial intelligence (AI), which is an important branch of computer science or computer engineering. Recently, AI techniques are making a serious impact in electrical engineering, particularly in the area of power electronics and motor drives. AI is basically computer emulation of human thinking (called *computational intelligence*). The human brain is the most complex machine on earth. Neurobiologists have taken a bottom-up approach to understand the brain structure and its functioning, and the psychologists and psychiatrists have taken the top-down approach to understanding the human thinking process. However, our understanding of the brain and its behavior has been extremely inadequate. The goal of AI is to mimic human intelligence so that a computer can think like a human being. However complex the human thought process is, there is no denying the fact that computers have adequate intelligence to help solve problems that are difficult to solve by traditional methods. Today, the AI techniques are used in many areas that include power electronics and motor drives.

FIGURE 10.2   Artificial intelligence (AI) classifications.

- EXPERT SYSTEM (ES)
- FUZZY LOGIC (FL)
- ARTIFICIAL NEURAL NETWORK (ANN)
  OR NEURAL NETWORK (NNW)
- GENETIC ALGORITHMS (GA)

AI techniques are principally classified into four different areas, as shown in this figure. The ES and GA will be briefly reviewed here, whereas FL and NNW will be discussed in this and the following chapter, respectively. The ES is basically an "intelligent" computer program based on Boolean logic that is designed to implant the expertise of a human being in a certain domain so that it can solve a problem, thus replacing the human expert. In 1854, George Boole published an article "Investigations on the Laws of Thought," where he propounded that human thinking is basically based on yes–no principles or 1–0 logic. As a result, Boolean algebra was born, and gradually modern-day powerful digital computers were ushered in. However, considering the limitation of solving algorithmic-type problems only, the ES techniques based on the principle of "IF .... THEN ..." rules were developed in the 1970s and applied extensively in the 1980s. Lotfy Zadeh defines ES as "hard" or precise computing, whereas FL, NNW, and GA are defined as "soft" or approximate computing. The software for "knowledge" or rule-base is organized such that it has easy learning, altering, and updating capabilities. A knowledge engineer acquires the knowledge from a technical expert and plants it in a knowledge base, and an inference engine executes the ES with the user interface. The ES technique has been applied in the power electronics area for *P-I* tuning of control, fault diagnostics, automated drive test and performance evaluation, selection of drive products, design and simulation of converter-fed drives, etc. The GA theory of AI is more modern and was proposed by John Holland in 1974. The GA technique, also known as a part of evolutionary computation, uses a probabilistic method of solving optimization problems. It follows the principles of biological genetics, or Darwin's survival of the fittest theory of evolution. Biological terms such as *population, offspring, chromosomes, reproduction, generation, crossover* (or *mating*), *mutation, objective* (or *fitness*) *function,* and *convergence* are commonly used in GA. In GA solutions, an initial population (a series of binary strings) is assumed and then the optimization is reached over several generations that involve the steps of reproduction, crossover, and mutation. GA theory has been applied in the power electronics area to the solution of nonlinear transcendental equations, optimization of FL controller (membership functions and rule base), NNW training, optimization of *P-I-D* parameters in control systems, etc. Note that the MATLAB environment provides a Toolbox for GA-based design that can be merged with MATLAB/Simulink simulations of the total system. Application of ES and GA in the power electronics area is somewhat limited compared to that of FL and ANN techniques.

FIGURE 10.3   What is intelligent control?

- CONTROL AND ESTIMATION BASED ON
  ARTIFICIAL INTELLIGENCE
- OFTEN DEFINED AS:
     - LEARNING CONTROL
     - SELF-ORGANIZING CONTROL
     - SELF-ADAPTIVE OR ADAPTIVE CONTROL
- NEEDS POWERFUL DSP

Control that is based on AI techniques is often defined as intelligent control. Intelligent estimation also is defined in a similar way. Traditional control techniques are based on mathematical models of the plants. For example, the control parameters of *P-I* or *P-I-D* control for a linear system can be determined by Nyquist or Bode analysis. Intelligent control, on the other hand, may not need any mathematical model. Many processes, such as nuclear reactor control, combustion in a boiler, chemical fermentation, etc., do not have mathematical models, or models may be ill defined. Even a well-defined plant such as a *d-q* dynamic model of an induction motor can have a parameter variation problem. Intelligent control may be a good candidate for such plants. The steam generation plant mentioned in the next figure uses intelligent control based on FL. As a simple example, the proportional (K_P) and integral gain (K_I) parameters of a system can be tuned on-line for an experimental system based on its actual performance. Starting with a set of initial K_P and K_I so that the plant is well within stability and knowing its *a priori* behavior, these parameters can be tuned on-line for a desired response by observing the transient behavior of the plant. The system under control can also be nonlinear and of higher order. Such control can be defined as learning control or self-adaptive or adaptive control. An adaptive *P-I* tuned control can be designed by using ES, FL, NNW, or GA. Of course, for computer simulation study of the system, a mathematical model of the plant is essential. Often the experimental performance of a plant can be captured, and this can be the basis for developing intelligent control. The implementation of intelligent control and estimation needs the help of powerful DSPs.

FIGURE 10.4 Features of fuzzy logic (FL).

- BOOLEAN OR CRISP LOGIC: 1 (Yes), 0 (No)
  FUZZY LOGIC: MULTIVALUED (0 TO 1)
- EMULATION OF FUZZY HUMAN THINKING
  FL EXAMPLE:
    IF: SPEED OF THE DRIVE MOTOR IS *LOW*
      THEN: CURRENT SHOULD BE *LARGE*
  ES EXAMPLE:
    IF: SPEED < 1000 RPM
      THEN: CURRENT > 50 A
- FUZZY VARIABLES (SPEED, CURRENT) AND LINGUISTIC
  FUZZY SETS (*LOW, LARGE*) ARE REPRESENTED BY
  MEMBERSHIP FUNCTIONS
- INVENTED BY LOTFY ZADEH OF UNIVERSITY OF
  CALIFORNIA, BERKELEY, IN 1965
- FIRST INDUSTRIAL APPLICATION (1975) – STEAM
  GENERATION PLANT CONTROL BY MAMDANI AND
  ASSILIAN IN LONDON QUEEN MARY COLLEGE

FL is another class of AI, but its history and applications are more recent than ES. In 1965, Lotfy Zadeh, a computer scientist, propounded the theory of FL. He argued that human thinking is often fuzzy, vague, or imprecise in nature and, therefore, cannot be represented by yes (1) or no (0) type precision logic used in ES. While both ES and FL are rule based, the latter is basically based on multivalued logic (between 0 and 1). A simple example illustrates a fuzzy rule and compares FL to the corresponding rule in ES. Classical set theory is based on Boolean logic, where a particular object or variable is either a member of a given set (logic 1), or it is not (logic 0). On the other hand, in fuzzy set theory based on FL, a particular object has a degree of membership in a given set that may be anywhere in the range of 0 (completely not in the set) to 1 (completely in the set). The illustrated fuzzy rule contains the fuzzy variables (speed, current), which are defined, respectively, by a number of fuzzy sets. Each fuzzy set is defined by a linguistic variable (low, large), which is again defined by a multivalued membership function (MF). Note that although FL deals with imprecise IF . . . THEN . . . rules, the information is processed in sound mathematical theory, which has seen advances in recent years. It is interesting to note that the first application of fuzzy control was proposed for a steam generation plant after a time lag of 10 years. FL applications in industrial products, such as washing machines, autofocus cameras, camcorders, and air conditioners, were pioneered by Japanese engineers. Gradually, the interest in applications spread in the United States and other countries.

FIGURE 10.5   Concept of membership function between "tall" and "not tall" persons. (Courtesy of Mathworks, Inc.)

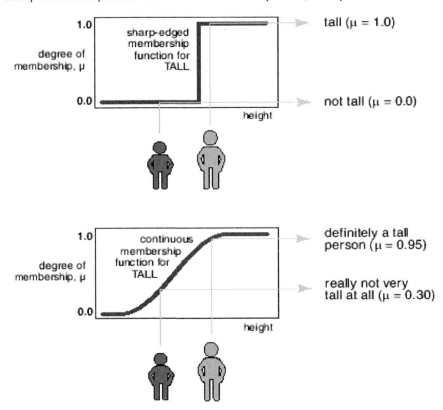

As mentioned earlier, FL uses multivalued logic (between 0 and 1) instead of Boolean logic (0 and 1). This figure explains the concept of membership function (MF) with multivalued logic as used by FL by means of a simple example of the definition of a tall person. We can define a person as *tall* if his height is 6.0 ft or above, and *not tall* if his height is below 6.0 ft. Such a definition follows the Boolean logic shown in the upper curve. The curve is defined as MF and the vertical axis is defined as degree of membership (μ) or membership value. The MF of the curve varies abruptly at 6.0 ft. The value of $\mu = 0$ below 6.0 ft and 1 at and above 6.0 ft. However, such a definition of a tall person is unfair. A more practical definition is shown in the lower curve, where the degree of membership gradually varies between 0 and 1. For example, if the person is 4.0 ft tall or less, we can define him as *not tall* ($\mu = 0$). If the height is 4.5 ft with $\mu = 0.3$, the person can be defined as really not very tall at all. For the height of 5.5 ft ($\mu = 0.95$), he can be defined as almost definitely a tall person, whereas for 6.0 ft and above ($\mu = 1$), he is a tall person. The continuously varying MF in the lower figure where the μ varies between 0 and 1 (multivalued) is appropriate in FL. In this case, *height* is the fuzzy variable, its value is defined by the fuzzy set or linguistic expression *tall*, and different values of height constitute the *universe of discourse*.

FIGURE 10.6 Representation of "temperature" using (a) fuzzy sets and (b) crisp sets.

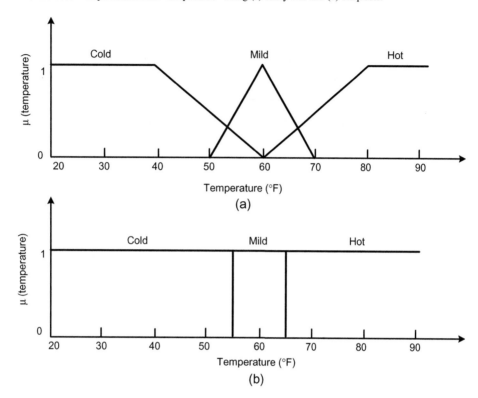

In this figure, the temperature of a motor as shown in the upper figure is the fuzzy variable, and its values are defined by the qualifying linguistic variables (fuzzy sets or subsets) Cold, Mild, or Hot, where each set is represented by a triangular or straight-line segment MF. The fuzzy sets can have additional subdivisions such as Zero, Very Cold, Medium Cold, Medium Hot, and Very Hot for more precise descriptions of temperature. In (a), if the temperature is below 40°F, it belongs completely to the set Cold, i.e., MF = 1, whereas at 55°F, it is in the set Cold by 30% (MF = 0.3) and in the set Mild by 50% (MF = 0.5). At temperature 60°F, it belongs completely to the set Mild (MF = 1) and not in the sets Cold and Hot (MF = 0). If the temperature is above 80°F, it belongs completely to the set Hot (MF = 1), where MF = 0 for Cold and Mild. In (b), the corresponding crisp or Boolean classification of the variable is given for comparison. Note the abrupt variation of μ between 0 and 1. For if the temperature is below 55°F, it belongs to the set Cold (MF = 1); between 55°F and 65°F, it belongs to the set Mild (MF = 1), and above 65°F, it belongs to the set Hot (MF = 1) only. The value of MF or degree of membership μ beyond the defined ranges is zero. The universe of discourse for temperature is from 20°F to 90°F as shown.

FIGURE 10.7   Different types of membership functions.

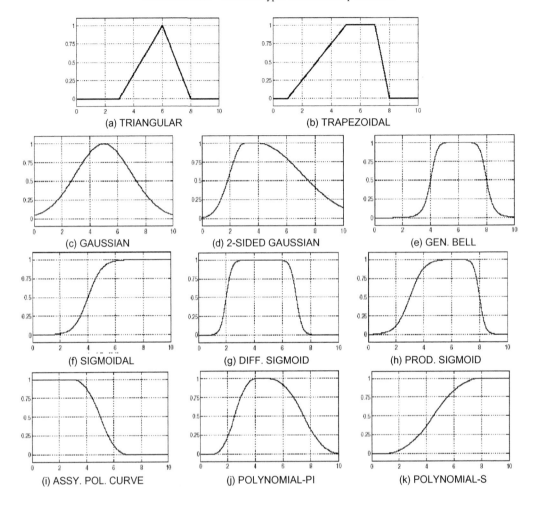

An MF varying between 0 and 1 can have various shapes as indicated in this figure. If X is the universe of discourse and its elements are denoted by x, then a fuzzy set A in X is defined as a set of ordered pairs: $A = \{s,\ \mu_A(x) \mid x \in X\}$, where $\mu_A(x) = $ MF of x in A. The MATLAB-based Fuzzy Logic Toolbox [6] (discussed later) for solving FL problems has all of the above MFs built in. These MFs are defined, respectively, as triangular (triamf), trapezoidal (trapmf), Gaussian (gaussmf), two-sided Gaussian (gauss2mf), generalized bell (gbellmf), sigmoidal-right (sigmf), difference sigmoid (dsigmf), product-sigmoid (psigmf), asymmetrical polynomial curve (zmf), polynomial-PI (pimf), and polynomial-S (smf). The Toolbox permits creation of new MFs, if desirable. A singleton is a special type of MF (not shown) that has a value of 1 at one point on the universe of discourse and zero elsewhere (a vertical spike). The MFs can be represented in a computer by look-up tables, mathematical functions, or straight-line segments. The triangular MFs are most common in solving fuzzy logic problems.

FIGURE 10.8   Logical operation on crisp sets and fuzzy sets.

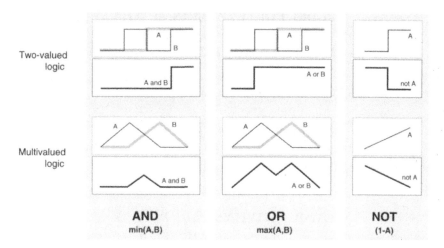

AND Operation: $\mu_{A \cap B}(x) = min\,[\mu_A(x),\, \mu_B(x)] = \mu_A(x) \wedge \mu_B(x)$. . . . .**(1)**

OR Operation: $\mu_{A \cup B}(x) = max\,[\mu_A(x),\, \mu_B(x)] = \mu_A(x) \vee \mu_B(x)$. . . . .**(2)**

NOT Operation: $\mu_A(x) = 1 - \mu_A(x)$ . . . . **(3)**

The logical operations on fuzzy sets and crisp sets are similar in nature, as explained in this figure. The basic AND, OR, and NOT operations are shown assuming fuzzy MFs have triangular shape. The AND or Intersection operation on the left selects the min(A,B) envelope and is given by mathematical expression (1), where $\mu_A(X)$, $\mu_B(X)$ = respective degree of membership, $\cap$ = intersection operator, and $\wedge$ = the minimum operator. The OR or Union operation, shown in the middle, is given by Eq. (2), where $\cup$ = union operator and $\vee$ = the maximum operator. The NOT or Complement operation (also called Negation) is given in Eq. (3). The respective Boolean operation on two-valued functions is shown in the second line.

FIGURE 10.9   Additional properties of fuzzy sets.

PRODUCT OF FUZZY SETS: $\mu_{A.B}(x) = \mu_A(x).\mu_B(x)$

MULTIPLICATION BY A CRISP NUMBER: $\mu_{kA}(x) = k.\mu_A(x)$

POWER: $\mu_{A^m}(x) = \left[\mu_A(x)\right]^m$

DOUBLE NEGATION: $(\bar{\bar{A}}) = A$

IDEMPOTENCY: $A \cup A = A$
$\qquad\qquad\qquad A \cap A = A$

COMMUTATIVITY: $A \cap B = B \cap A$
$\qquad\qquad\qquad\quad A \cup B = B \cup A$

ASSOCIATIVE PROPERTY: $(A \cup B) \cup C = A \cup (B \cup C)$
$\qquad\qquad\qquad\qquad\quad (A \cap B) \cap C = A \cap (B \cap C)$

DISTRIBUTIVE PROPERTY: $A \cup (B \cap C) = (A \cup B) \cap (A \cup)C$
$\qquad\qquad\qquad\qquad\quad A \cap (B \cup C) = (A \cap B) \cup (A \cap)C$

ABSORPTION: $A \cap (A \cap B) = A$
$\qquad\qquad\qquad A \cup (A \cap B) = A$

De MORGAN'S THEOREM: $(A \cup B) = \bar{A} \cap \bar{B}$
$\qquad\qquad\qquad\qquad\quad (A \cap B) = \bar{A} \cup \bar{B}$

The additional properties of fuzzy sets of which many are similar to Boolean algebra are indicated here, where A, B, and C are the fuzzy sets defined over a common universe of discourse X and μ is the degree of membership. The symbols $\cup$ and $\cap$ are union and intersection operators, respectively, and the bar indicates the complementation or NOT process. These properties permit additional operations with fuzzy sets when solving FL-based problems.

FIGURE 10.10 Interpretation of fuzzy rule. (Courtesy of MathWorks, Inc.)

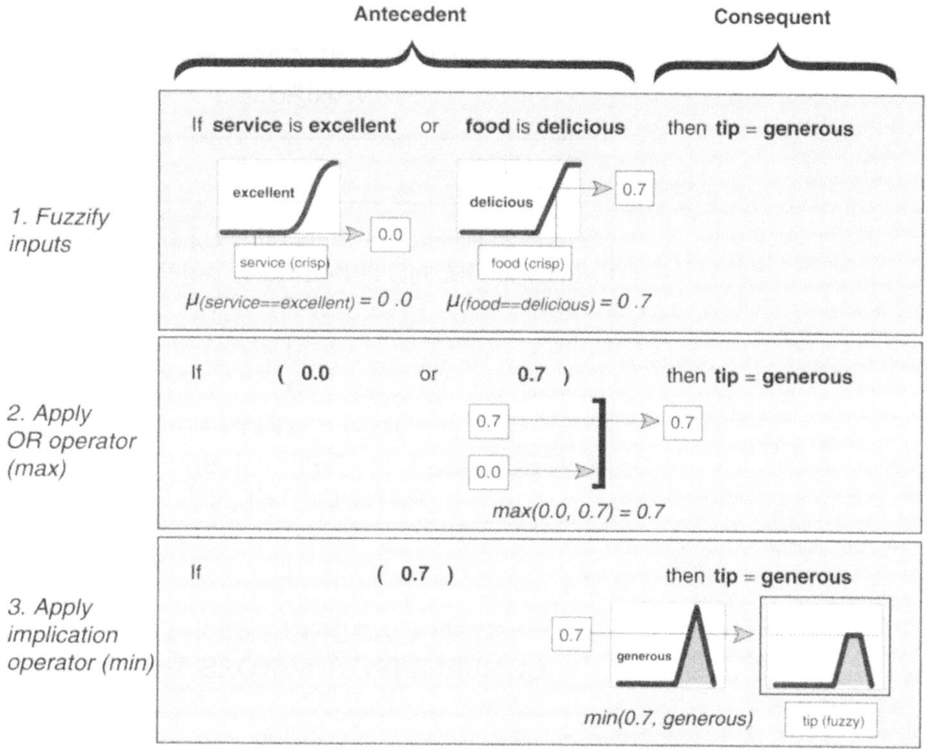

Let us take a simple fuzzy rule related to restaurant tipping and interpret it. The IF part of the rule is defined as Antecedent and THEN part is defined as Consequent. There are three fuzzy variables (Service, Food, and Tip) in the rule of which two are input (Service and Food) variables and one is an output (Tip) variable. The corresponding fuzzy set of each is defined by linguistic variable Excellent, Delicious, and Generous, respectively. In this case, Excellent, Delicious, and Generous sets are respectively described by sigmoidal, straight line, and triangular MFs as shown in the figure. Assume that in the range of 0–10, the score on Service is 3 and the score on Food is 7. These scores are crisp values. These values are fuzzified by the respective MFs, and the corresponding fuzzy outputs are 0.0 and 0.7. Then OR logical operation is applied and the output of the antecedent part is 0.7 (middle row). Then, finally in the consequent part (last row), the process of implication generates the fuzzy output of Tip by truncating the triangle as shown. The middle point of the base of the triangle can be considered as crisp output. The last step is called defuzzification.

FIGURE 10.11   Basic steps in fuzzy inference system (FIS).

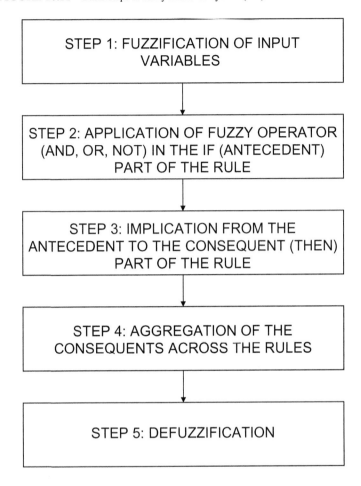

A fuzzy system (FS), also called a fuzzy inference system (FIS), may have a large number of rules of which only a few are validated at one time. The validation of one simple rule was illustrated in the previous figure. The valid cluster of rules is solved in FIS by the five steps shown in this figure. If FIS is considered as a black box, note that both input and output values of fuzzy variables are crisp. The steps in the figure will be made clear via examples given later.

FIGURE 10.12 Fuzzy inference system illustrated for restaurant tipping. (Courtesy of MathWorks, Inc.)

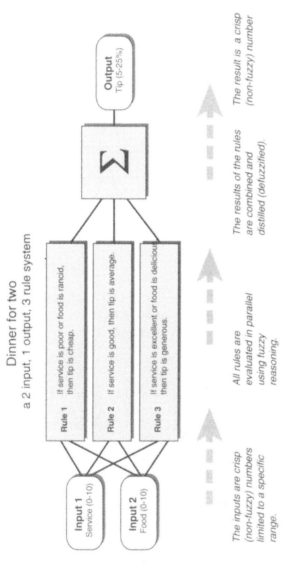

Dinner for two
a 2 input, 1 output, 3 rule system

Input 1
Service (0–10)

Input 2
Food (0–10)

Rule 1 If service is poor or food is rancid, then tip is cheap.

Rule 2 If service is good, then tip is average.

Rule 3 If service is excellent or food is delicious, then tip is generous.

Output
Tip (5–25%)

The inputs are crisp (non-fuzzy) numbers limited to a specific range.

All rules are evaluated in parallel using fuzzy reasoning.

The results of the rules are combined and distilled (defuzzified).

The result is a crisp (non-fuzzy) number

Let us consider a fuzzy system for restaurant tipping that has three rules as shown in the figure. Normally, the tipping amount is determined in our mind by fuzzy human thinking. In this case, fuzzy theory has been applied to determine the tipping amount. The Service and Food are the input variables (0–10 range), and Tip is the output variable (5%–25%). The output value is arrived at by parallel evaluation of the three rules explained in the next figure.

FIGURE 10.13   Information processing of fuzzy inference system for restaurant tipping.

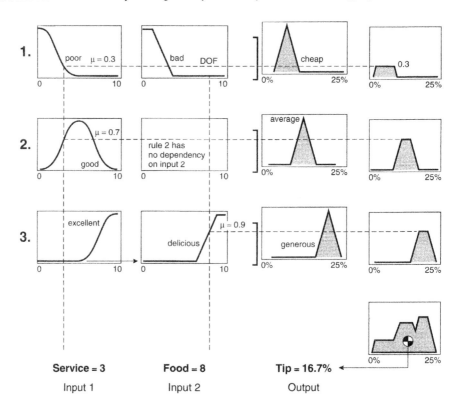

In this example, the fuzzy variable Service is represented by three fuzzy sets (Poor, Good, and Excellent) and the corresponding MFs are defined by Assy. Polynomial Curve, Gaussian, and Sigmoidal, respectively (first column). Similarly, Food is represented by two sets (Bad, or Rancid, and Delicious), which correspond to straight-line MFs (second column), and the output Tip is represented by the sets Cheap, Average, and Generous, which correspond to triangular MFs as shown (third column). The universe of discourse for the input variables is 0–10%, whereas for the output variable it is 0–25%. Processing of the three rules in the horizontal direction is shown. Consider, for example, a score for Service of 3 and 8 for Food. For rule 1, the corresponding fuzzy outputs are $\mu = 0.3$ in MF Poor and $\mu = 0$ in MF Bad, respectively, as a result of fuzzification (Step 1). In the rule, OR or max operation is specified, and therefore, $\mu = 0.3$ is selected (Step 2). This is also defined as *degree of fulfillment* (DOF) of a rule. The implication step (Step 3) of this rule truncates the output MF Cheap at 0.3 to give the fuzzy output as shown on the right. All three rules are evaluated in this manner to generate the respective output indicated in the figure. These results are combined or aggregated in a cumulative manner to result in the final fuzzy output (Step 4) shown at the bottom right. Finally, the fuzzy output given by the area is converted to crisp output by defuzzification (Step 5). The example in restaurant tipping is also applicable for FL control, which will be discussed later.

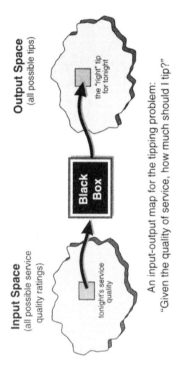

FIGURE 10.14 Input–output mapping problem. (Courtesy of MathWorks, Inc.)

Input Space
(all possible service quality ratings)

tonight's service quality

Black Box

Output Space
(all possible tips)

the "right" tip for tonight

An input-output map for the tipping problem:
"Given the quality of service, how much should I tip?"

The fuzzy system for restaurant tipping discussed in the previous figures can be defined as a general input–output mapping problem, as illustrated in this figure. Of course, only Service is shown here as the input variable. The input information is defined in the input space, and it is processed in the black box, and the solution appears in the output space. The mapping is static and nonlinear because there are no dynamics in the system. The solution can be given by a look-up table, which is nothing but input–output mapping. For example, in Figure 10.13, the solution can be given by a three-dimensional nonlinear look-up table, where Service = 3 and Food = 8 gives the output Tip = 16.7%. In general, the input–output mapping can also be interpreted as a pattern matching or pattern recognition problem. When we see an apple, the input–output mapping can also be interpreted as a pattern matching or pattern recognition problem. When we see an apple characterized by its color and shape, we recognize that it is an apple. Similarly, when we see a person's face, we remember his name. These are examples of pattern recognition by the associative memory property of the human brain, which is represented by the black box in this figure. The mapping can be static as well as dynamic, and the mapping characteristics are determined by the black box's characteristics. The black box can represent not only a fuzzy system, but also an ES, NNW, or general mathematical system where input excitation creates the output response. The concept of input–output mapping properties is extremely important in fuzzy logic and neural networks and will be further discussed later.

FIGURE 10.15   Fuzzy inference system using Mamdani method.

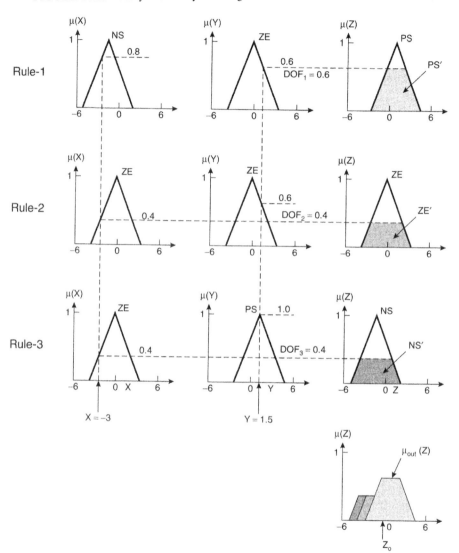

Rule 1: IF: X IS NEGATIVE SMALL (*NS*) AND Y IS ZERO (*ZE*)
         THEN: Z IS POSITIVE SMALL (*PS*)
Rule 2: IF: X IS ZERO (*ZE*) AND Y IS ZERO (*ZE*)
         THEN: Z IS ZERO (*ZE*)
Rule 3: IF: X IS ZERO (*ZE*) AND Y IS POSITIVE SMALL (*PS*)
         THEN: Z IS NEGATIVE SMALL (*NS*)

We will now discuss several implication methods (Step 3 in Figure 10.11) using only triangular MFs. This figure explains the most commonly used Mamdani method for evaluating the three example rules shown in the figure. In fact, we have already used the Mamdani method in the tipping examples. In this case, X and Y are the input variables, Z is the output variable, and $NS, ZE,$ and PS are the fuzzy sets described by triangular MFs. Assuming $X = -3$ and $Y = 1.5$, the evaluation of the three rules is shown in the figure. Because all the rules have an AND logical operator, the DOF (degree of fulfillment) of each rule selects only the minimum value. Therefore, $DOF_1 = 0.6$, $DOF_2 = 0.4$, and $DOF_3 = 0.4$ as shown in the figure. The fuzzy output of each rule is given by the shaded area of the output sets $PS, ZE,$ and NS, respectively. The total fuzzy output is the superposition of these areas. Finally, defuzzification will give the crisp output Z_0 as indicated.

FIGURE 10.16   Fuzzy inference system using Lusing Larson method.

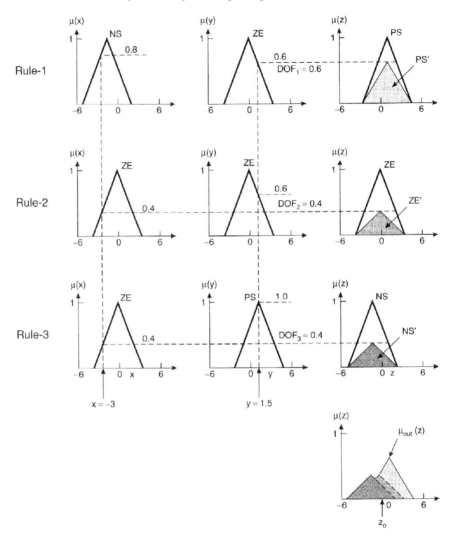

The implication method discussed in this figure is the Lusing Larson method, in which the output MF for each rule is scaled down instead of using truncation as in the Mamdani method. The processing of the antecedent part of each rule remains the same as before. The same three rules in Figure 10.15 with same values of X (= −3) and Y (= 1.5) are considered here. Therefore, $DOF_1 = 0.6$, $DOF_2 = 0.4$, and $DOF_3 = 0.4$ as before. The output MF *PS* of Rule 1 is scaled down so that the fuzzy output is triangle *PS* with the peak value of 0.6. Similarly, Rules 2 and 3 give fuzzy outputs *ZE′* and *NS′* as shown. The total fuzzy output is given by the superposition of these output scaled triangles, which is then defuzzified to generate the crisp output Z_0.

FIGURE 10.17 Fuzzy inference system using Sugeno (zero-order) method.

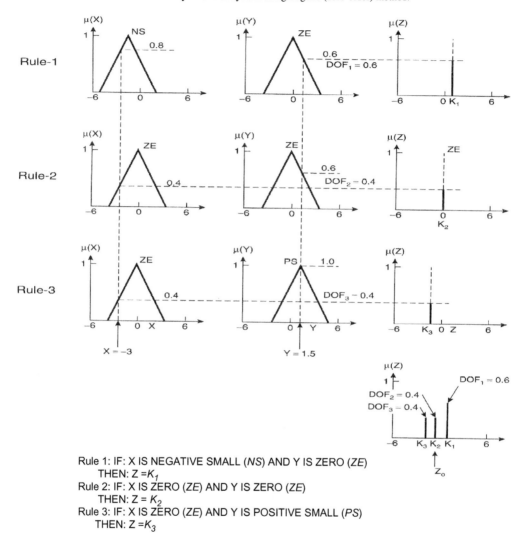

Rule 1: IF: X IS NEGATIVE SMALL (*NS*) AND Y IS ZERO (*ZE*)
   THEN: Z =K_1
Rule 2: IF: X IS ZERO (*ZE*) AND Y IS ZERO (*ZE*)
   THEN: Z = K_2
Rule 3: IF: X IS ZERO (*ZE*) AND Y IS POSITIVE SMALL (*PS*)
   THEN: Z =K_3

The figure illustrates the zero-order Sugeno method of implication. The difference in this case compared to the Mamdani and Lusing Larson methods is that the output fuzzy sets are defined by singleton MFs at constant values of Z (K_1, K_2, and K_3, respectively) in the universe of discourse. The three rules are modified for the zero-order Sugeno method and shown in the lower part of the figure. The *DOF* of each rule truncates the singleton to the respective amplitude. These truncated vertical segments are then aggregated to constitute the total fuzzy output as shown in the figure.

FIGURE 10.18   Fuzzy inference system using Sugeno (first-order) method.

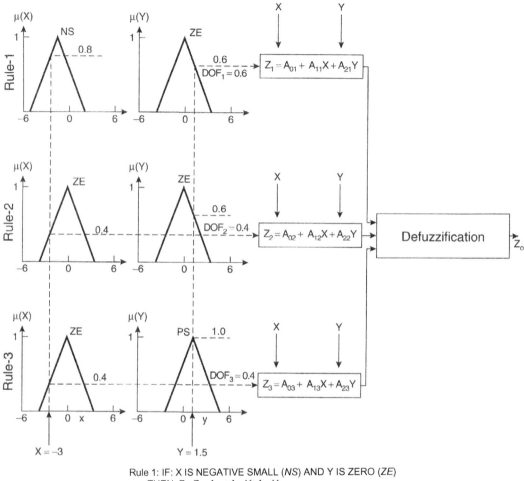

Rule 1: IF: X IS NEGATIVE SMALL (*NS*) AND Y IS ZERO (*ZE*)
   THEN: Z =Z_1=A_{01}+A_{11}X+A_{21}Y
Rule 2: IF: X IS ZERO (*ZE*) AND Y IS ZERO (*ZE*)
   THEN: Z = Z_2 = A_{02} + A_{12}X + A_{22}Y
Rule 3: IF: X IS ZERO (*ZE*) AND Y IS POSITIVE SMALL (*PS*)
   THEN: Z =Z_3=A_{03}+ A_{13}X+A_{23}Y

A more general first-order Sugeno method of implication is shown here, and the corresponding rules are given in the lower part of the figure. In the zero-order method, the output singletons were located at constant values (*K*'s) of Z. In this case, K_1, K_2, and K_3 are replaced, respectively, by Z_1, Z_2, and Z_3, which are linear modifications of input signals X and Y with the help of constants (*A*'s). Therefore, the first-order system can be visualized as moving singletons that are modified linearly in the universe of discourse depending on the magnitude of input signals. Fortunately, the application of different implication methods in a fuzzy system will make little difference in the output. We will discuss a demo application example later using Mamdani, Lusing Larson, and zero-order Sugeno methods.

FIGURE 10.19 Defuzzification methods.

- **CENTER OF AREA :** $Z_0 = \dfrac{\int Z.\mu(Z)dZ}{\int \mu(Z).dZ} \cong \dfrac{\sum\limits_{i=1}^{n} Z_i \mu(Z_i)}{\sum\limits_{i=1}^{n} \mu(Z_i)}$ .... (1)

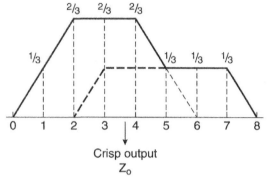

Crisp output
Z_0

FUZZY OUTPUT FOR TWO-RULE SYSTEM

$$Z_0 = \frac{1.1/3 + 2.2/3 + 3.2/3 + 4.2/3 + 5.1/3 + 7.1/3}{1/3 + 2/3 + 2/3 + 2/3 + 1/3 + 1/3 + 1/3} = 3.7 \ \ldots\ldots\ldots (2)$$

- **HEIGHT:** $Z_0 = \dfrac{3.2/3 + 5.1/3}{2/3 + 1/3} = 3.67$ .... (3)

- **SUGENO (ZERO-ORDER):** $Z_0 = \dfrac{K_1.DOF_1 + K_2.DOF_2 + K_3.DOF_3}{DOF_1 + DOF_2 + DOF_3}$ ..... (4)

- **SUGENO (FIRST-ORDER):** $Z_0 = \dfrac{Z_1.DOF_1 + Z_2.DOF_2 + Z_3.DOF_3}{DOF_1 + DOF_2 + DOF_3}$ ..... (5)

The conversion of fuzzy output in a fuzzy inference system into crisp output is defined as *defuzzification*. The fuzzy output is constructed by superimposing the outputs of individual rules. For triangular and singleton output MFs, the construction of fuzzy output is shown in Figures 10.15 to 10.18. Similar principles can be applied for other types of MFs. For non-singleton output, there are generally two methods of defuzzification: center-of-area method and height method. The center-of-area method (also called centroid or center-of-gravity method) involves integration covering the outer envelope shown by Eq. (1) and is valid for arbitrary MF shapes. It is often simplified into discrete form as indicated. A simple numerical example is given for the output of a two-rule system with triangular MFs. The discrete form of calculation used in Eq. (2) gives output of 3.7. The height method shown by Eq. (3) is simpler and considers the heights of the trapezoids and their mean base values. Sugeno's methods are similar to the height method as shown by Eqs. (4) and (5) for Figures 10.17 and 10.18, respectively.

FIGURE 10.20   Typical FL applications in power electronics.

- ROBUST CONTROL OF DRIVE WITH PARAMETER VARIATION AND LOAD TORQUE DISTURBANCE*
- SINGLE- OR MULTIDIMENSIONAL NONLINEARITY COMPENSATION IN CONVERTERS AND CONTROL*
- TUNING OF CONTROL PARAMETERS IN FEEDBACK SYSTEM
- OPTIMIZATION PROBLEM BASED ON ON-LINE SEARCH*
- DESIGN AID IN POWER ELECTRONICS SYSTEM
- ESTIMATION FOR DISTORTED WAVES*
- SLIP GAIN TUNING OF VECTOR DRIVE WITH INDUCTION MOTOR*
- ESTIMATION OF MACHINE PARAMETERS
- FUZZY BEHAVIORAL MODEL OF NONLINEAR COMPLEX POWER ELECTRONICS PLANT

Some typical applications of FL are given in this figure. There are many more applications, and new applications will be proposed in the future. The applications discussed in this chapter are indicated by an asterisk (*). One very important FL application is adaptive and robust control for a drive system with parameter variation and load disturbance (moment of inertia or load torque). Fuzzy control is basically nonlinear adaptive control and can give robust deadbeat-type performance in both linear and nonlinear systems. It has been argued that fuzzy control is possibly the best robust control. Because of input–output nonlinear mapping properties, FL is good for multidimensional nonlinear look-up table generation. This principle can be applied to inverse mapping of the nonlinearity to be compensated in converter and control. For a *P-I* or *P-I-D* controlled feedback system, transient performance can be characterized by fuzzy IF... THEN... statements, which can be used to tune proportional, integral, and derivative control parameters. In general, system performance such as efficiency and power output can be optimized by on-line search without depending on the plant model. A fuzzy ES (expert system) has been proposed to design a converter that includes its snubber. For estimation of distorted waves in power electronics, such as derivation of DPF, PF, etc. has been found helpful. Machine parameters such as stator and rotor resistances can be approximated by FL. In general, for complex power electronics plants such as arc furnaces and electrochemical plants a fuzzy model (in terms of IF... THEN... statements) can be defined and used for fuzzy control.

FIGURE 10.21 Fuzzy speed controller for a vector-controlled induction motor drive in a variable inertia system.

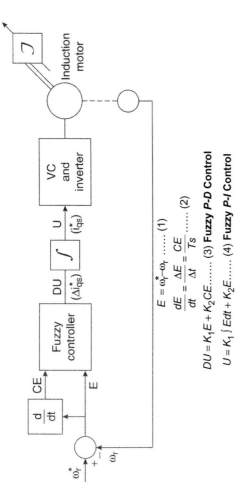

$$E = \omega_r^* - \omega_r \ \dots\dots \ (1)$$

$$\frac{dE}{dt} = \frac{\Delta E}{\Delta t} = \frac{CE}{T_s} \ \dots\dots \ (2)$$

$$DU = K_1 E + K_2 CE \dots\dots (3) \ \textbf{Fuzzy P-D Control}$$

$$U = K_1 \int E dt + K_2 E \dots\dots (4) \ \textbf{Fuzzy P-I Control}$$

Let us now apply fuzzy control in a feedback vector drive for an induction motor as shown in the figure. The fuzzy controller receives two input signals: the speed loop error E and its derivative dE/dt. As shown in Eq. (2), the change in error CE is proportional to dE/dt for constant sampling time T_s of the DSP. These are the two input fuzzy variables. The output variable DU or Δi_{qs}^* is integrated to generate i_{qs}^*, which is the torque component of stator current in the vector drive. Imagine that the fuzzy controller consists of a human being who observes the E and CE signal magnitudes and their polarity, and controls the DU signal based on his judgment (IF... THEN... fuzzy rules). Because a fuzzy controller basically uses input–output nonlinear mapping, we can write Eq. (3), where K_1 and K_2 are nonlinear gains. Equation (3) can be used directly without an integrator. This is then basically a nonlinear P-D (proportional-derivative) controller. With the integration of output shown in Eq. (4), we have nonlinear P-I (proportional-integral) control (also known as fuzzy P-I control). Similarly, fuzzy P-I-D control is also possible.

FIGURE 10.22   One-rule fuzzy speed control concept.

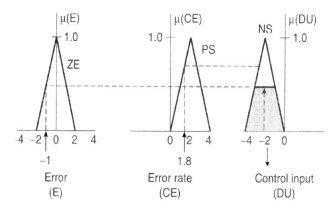

RULE: IF: SPEED LOOP ERROR (E) IS ZERO (*ZE*) AND
ERROR RATE (*CE*) IS POSITIVE SMALL (*PS*)
THEN: CONTROL OUTPUT (*DU* OR $\Delta I_{qs}{}^*$) IS NEGATIVE
SMALL (*NS*)

Let us assume that the fuzzy controller in Figure 10.21 is based on only one rule, the one given at the bottom of this figure. The rule is interpreted here with the Mamdani method using triangular MFs because they are the most commonly used MFs [6]. The rule should be framed to be consistent with the physical behavior of the system. The physical basis of the rule should be checked by the reader. Formulation of such a rule (or rules) requires that the designer should have intuitive experience with the system's operation. As before, E and CE are the input variables and DU or $\Delta I_{qs}{}^*$ is the output variable. In this example, the crisp values of E and CE are –1 and 1.8, respectively. The corresponding fuzzification outputs are $\mu(ZE) = 0.5$ and $\mu(PS) = 0.8$. These component outputs are combined (by AND) to give the resulting output membership value (or degree of membership) $\mu(NS) = 0.5$. Therefore, the fuzzy output is given by the shaded area of *NS*, and the resulting defuzzified crisp (center of gravity) output can be calculated as –2, as indicated in the figure. The torque component of current should be decremented by 2 amps so that actual speed ω_r approaches the command speed $\omega_r{}^*$.

FIGURE 10.23 Two-rule fuzzy speed control concept.

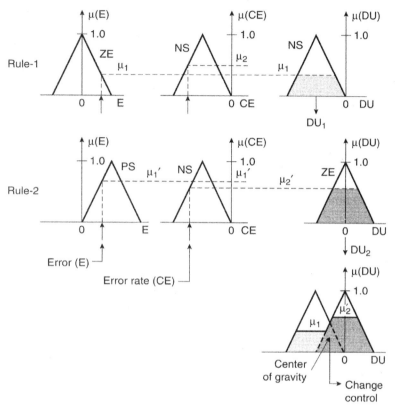

RULE 1 – IF: ERROR (E) IS ZERO (ZE) AND CHANGE IN ERROR
(CE) IS NEGATIVE SMALL (NS)
THEN: OUTPUT (DU_1 or $\Delta i_{qs1}{}^*$) IS NEGATIVE SMALL (NS)

RULE 2 – IF: ERROR (E) IS POSITIVE SMALL (PS) AND CHANGE IN
ERROR (CE) IS NEGATIVE SMALL (NS)
THEN: OUTPUT (DU_2 or $\Delta i_{qs}{}^*$) IS NEGATIVE SMALL (NS)

Let us now consider fuzzy speed control based on two rules as given in this figure. For a practical system, usually more than one rule is validated as shown by the tipping example in Figure 10.12. Each rule generates a meaningful control action depending on the input values of the variables. The corresponding fuzzy outputs of the NS and ZE MFs are combined to generate the resultant output area as shown. Then it is defuzzified to generate the crisp control output. For fuzzy P-D control, this output is used directly, whereas for fuzzy P-I control, the DU signal has to be integrated.

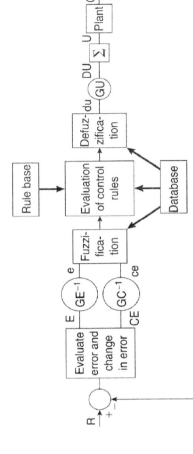

FIGURE 10.24   Structure of fuzzy controller in a feedback system. (Courtesy of Pearson)

The practical structure of a fuzzy *P-I* controller in a feedback system is shown in this figure. The reference (*R*) and output (*C*) signals of the plant are compared and the *E* and *CE* signals are calculated. These signals are then converted into per-unit signals by dividing with the respective scale factors, i.e., *e* = *E*/*GE* and *ce* = *CE*/*GC*. The fuzzification, evaluation of control rules, and defuzzification blocks receive input from the database. Both database and rule base constitute the knowledge base of the system. The processed output signal *du* appears in per unit of *DU* signal which is descaled by multiplying with the factor *GU* so that actual output *DU* = *du·GU*. These input and output scale factors can be either constants or function of some variables. The advantage of fuzzy control design in terms of *pu* values is that the same algorithm becomes valid for all the plants in the same family. Besides, the gains can be made adaptive for performance optimization. The database supplies the MFs needed for fuzzification, evaluation of control rules, and defuzzification processes. The control is based on a cluster of rules fed to the system as shown.

FIGURE 10.25 Membership functions of fuzzy speed controller variables.

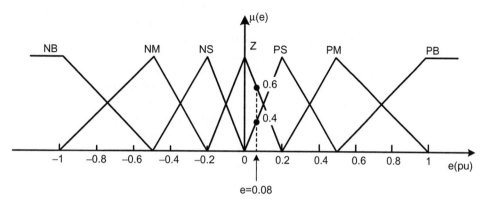

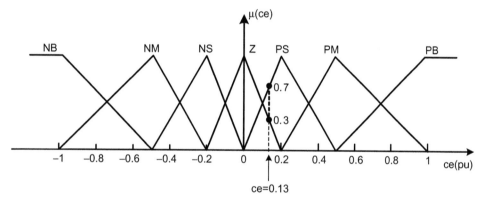

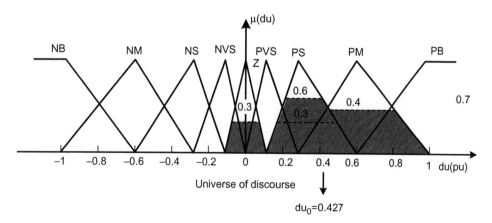

Let us now consider practical example applications of FL in power electronics and drives areas. In the first example, the fuzzy control of the vector drive discussed thus far will be illustrated. There are two basic design elements in fuzzy control, i.e., description of the MFs of the fuzzy variables as shown in this figure, and the rule matrix shown in the next figure. Asymmetrical triangular MFs have been selected in this design which causes crowding near the origin (steady state) and, therefore, gives more precision. The universe of discourse for all the fuzzy variables *e(pu)*, *ce(pu)*, and *du(pu)* spreads in the region −1 to +1, and the MFs are symmetrical on both positive and negative sides. The input variables have seven MFs, whereas the output variable has nine MFs. The selection of MFs and their description are again based on experience, although sophisticated GA and NNW-based design techniques do exist. Sometimes, large numbers of iterations are needed to reach an optimal design.

FIGURE 10.26   Rule table for fuzzy speed controller.

ce(pu) \ e(pu)	NB	NM	NS	Z	PS	PM	PB
NB	NVB	NVB	NVB	NB	NM	NS	Z
NM	NVB	NVB	NB	NM	NS	Z	PS
NS	NVB	NB	NM	NS	Z	PS	PM
Z	NB	NM	NS	Z	PS	PM	PB
PS	NM	NS	Z	PS	PM	PB	PVB
PM	NS	Z	PS	PM	PB	PVB	PVB
PB	Z	PS	PM	PB	PVB	PVB	PVB

The figure describes the rule table for a fuzzy speed controller in a vector drive. The top row and left column describe the sets for the variables *e(pu)* and *ce(pu)*, respectively, whereas the body of the table in the third dimension describes the sets of the output variable *du(pu)*. Because there are 7 sets for each input variable, there are altogether $7 \times 7 = 49$ rules in the table. A typical rule in the table can be read as:

IF: *e(pu)* = PS and *ce(pu)* = Z, THEN: *du(pu)* = PS

This table is designed from the experience about the system, as mentioned before. Some of the outer boxes can remain empty to simplify the design. If the typical values are $e = 0.08$ and $ce = 0.13$ (see Figure 10.25), the vertical lines for both intersect Z and PS, and corresponding numerical values for μ are shown. This indicates validation of the four rules shown by the shaded area at the center of this figure. The corresponding fuzzy outputs are shown by the areas in Figure 10.25 from which the defuzzified output can be calculated as $du_0 = 0.427$. Fortunately, computer programs, such as the MATLAB-based Fuzzy Logic Toolbox [6], are available for easy and fast design of fuzzy controllers.

FIGURE 10.27   Fuzzy controller response in a vector drive with inertia (J) change and load torque (T_L) disturbance: (a) $J = J_1$ and (b) $J = 4J_1$.

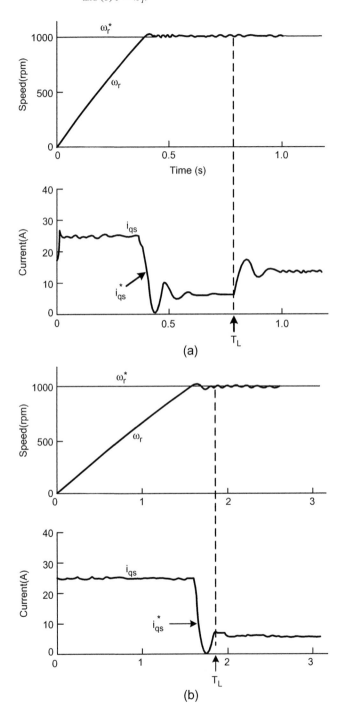

The figure shows the response of a fuzzy-controlled vector drive with a step speed command. Both speed and current loop responses with nominal shaft inertia (J_1) are shown in (a), whereas those with four times the nominal inertia ($4J_1$) are shown in (b). The speed loop response is almost deadbeat, but in the latter case, it takes longer time to settle due to the same limit of torque component of current (i_{qs}), i.e., one-fourth of the acceleration. A step load torque (T_L) disturbance is applied at steady state in both cases, but the resulting speed deviation is practically negligible. The current response in the former case is more oscillatory. The performances are definitely superior to conventional *P-I* control. The robustness in the response in fuzzy control is evident in the figures.

FIGURE 10.28 Difficulties in fuzzy system design.

- THERE IS NO SYSTEMATIC DESIGN APPROACH
- NO DEFINITE CRITERIA FOR SELECTION OF:
    - FUZZY SETS OF VARIABLES
    - SHAPE AND SLOPE OF MFs
    - DEGREE OF OVERLAP OF MFs
    - RULE TABLE
- TUNING OF MFs AND RULE TABLE FOR OPTIMUM RESPONSE MAY BE TIME CONSUMING IF ITERATION IS BASED ON EXPERIMENTAL RESULTS
- SYSTEM ANALYSIS AND PERFORMANCE PREDICTION ARE DIFFICULT
- USE ITERATION BY SOFTWARE PROGRAM ON SIMULATED SYSTEM IF MATHEMATICAL PLANT MODEL IS AVAILABLE
- COMPUTATION RESULTS ARE ALWAYS APPROXIMATE (SOFT COMPUTATION)

The application example given before indicates that although the results of fuzzy systems are excellent, the controller does not have any systematic design approach. We do not know precisely how to select the distribution of fuzzy sets for each variable, the shape and slope of MFs, their degree of asymmetry if needed, and the overlap. Similar problem exists for design of the rule table. The whole design approach is approximate and depends on the designer's or operator's experience with the behavior of the plant. Therefore, fuzzy system design may sometimes be tedious and time consuming. With such a design approach, performance prediction is difficult. If there is no model of the system, performance of the fuzzy system is to be iterated based on experimental results. Fortunately, user-friendly software programs (such as the MATLAB-based Fuzzy Logic Toolbox [6]) exist for easy synthesis of MFs and the rule table. If a mathematical model of the plant exists (such as the vector drive example given before), the fuzzy system can be designed quickly with the software program and performance can be verified on system simulation. With such an approach, faster system design is possible. Recently, NNW and GA techniques have been used for tuning fuzzy systems. Several more application examples will be included in this chapter before discussing the FL Toolbox and its demo examples.

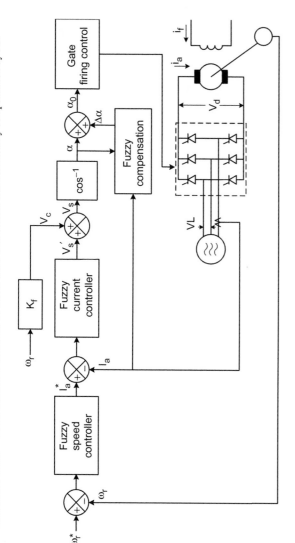

FIGURE 10.29 Phase-controlled converter dc motor drive with control and nonlinearity compensation by FL.

The figure shows an FL-controlled separately excited dc motor drive [8], where FL is applied not only in the speed and current control loops, but also for nonlinear gain compensation of the converter. The methodology for speed and current control remains the same as for a vector-controlled induction motor drive and will not be discussed further. The gain or transfer characteristic of the thyristor converter becomes nonlinear at discontinuous conduction at high speed due to higher counter emf, which was discussed in Chapter 3. In the figure, the speed loop has the usual inner armature current control loop that generates the armature voltage signal V_s'. This is then added with the feedforward counter emf signal V_c as a function of speed. The firing angle α of the converter is then generated from it by the $\cos^{-1}$ function. This angle goes to the gate firing control directly in continuous conduction mode, but adds with the auxiliary angle $\Delta\alpha$ in discontinuous conduction as shown. The objective of fuzzy compensation is to generate the $\Delta\alpha$ angle as a function of armature current I_a and firing angle α such that the converter behaves like that in continuous conduction.

FIGURE 10.30 $V_d(pu)$ vs. $I_a(pu)$ characteristics at discontinuous conduction: (a) without compensation and (b) with compensation.

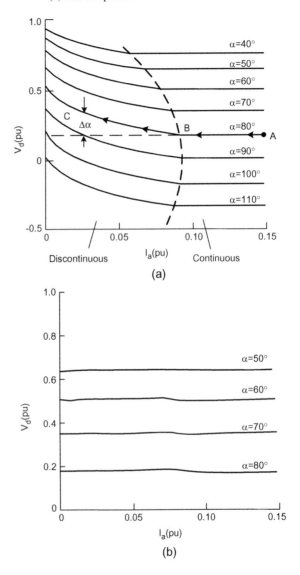

The motor $V_d(pu) - I_a(pu)$ characteristics at different firing angles α for continuous and discontinuous conduction modes shown in (a) of this figure follow the characteristics of the speed-torque curves shown in Figure 3.18, since $V_d(pu)$ is proportional to speed and $I_a(pu)$ is proportional to torque. The dashed curve shows the boundary between these modes. The gain of the converter is constant at continuous conduction, but varies nonlinearly at different points of discontinuous conduction. The objective of the fuzzy compensation is to generate the compensating angle $\Delta\alpha$ in order to linearize the converter transfer characteristics shown in part (b).

FIGURE 10.31   Membership functions for $\Delta\alpha$ compensation.

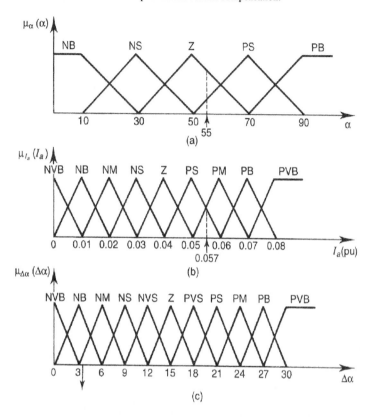

In Figure 10.29, α angle and $I_a(pu)$ current are the input fuzzy variables and $\Delta\alpha$ angle is the output variable. This figure shows the MF plots of the variables covering the region of discontinuous conduction only. Symmetric triangular MFs have been selected for all the variables. Note that two variables are in actual values, whereas one variable is expressed in per-unit values for convenience. Although all of the signals are unipolar, the linguistic terms used for the sets are for convenience only and must be interpreted in a "context-free" grammar. The sensitivity of a variable determines its number of sets. In this case, α is described by five sets, whereas $I_a(pu)$ and $\Delta\alpha$ are described by 9 and 11 sets, respectively. The MFs are shown with 50% overlap. At typical values of $\alpha = 55°$ and $I_a(pu) = 0.057$ shown in the figure, four rules will be fired (or activated).

Figure 10.32   Rule matrix for fuzzy $\Delta\alpha$ compensation.

I_a \ α	NB	NS	Z	PS	PB
NVB	NVB	PB	PB	PB	PB
NB	NVB	Z	Z	Z	Z
NM	NVB	NS	NVS	NVS	NVS
NS	NVB	NM	NS	NS	NS
Z	NVB	NB	NM	NM	NS
PS	NVB	NVB	NB	NM	NM
PM	NVB	NVB	NB	NB	NB
PB	NVB	NVB	NVB	NB	NB
PVB	NVB	NVB	NVB	NVB	NB

EXAMPLE RULES:

   Rule 1: IF α = Z and I_a(pu) = PS, THEN $\Delta\alpha$ = NB
   Rule 2: IF α = Z and I_a(pu) = PM, THEN $\Delta\alpha$ = NB
   Rule 3: IF α = PS and I_a(pu) = PS, THEN $\Delta\alpha$ = NM
   Rule 4: IF α = PS and I_a(pu) = PM, THEN $\Delta\alpha$ = NB

The rule base for a dc drive $\Delta\alpha$ compensation is shown in this figure. The five sets of α and nine sets of $I_a(pu)$ give $5 \times 9 = 45$ rules. The four rules corresponding to the given values of α and $I_a(pu)$ in Figure 10.31 are shown at the bottom of the table. This final rule base was arrived at after a number of iterations on the simulation-based drive system.

Figure 10.33 FL-based current loop response in dc drive: (a) with fuzzy $\Delta\alpha$ compensation and (b) without $\Delta\alpha$ compensation.

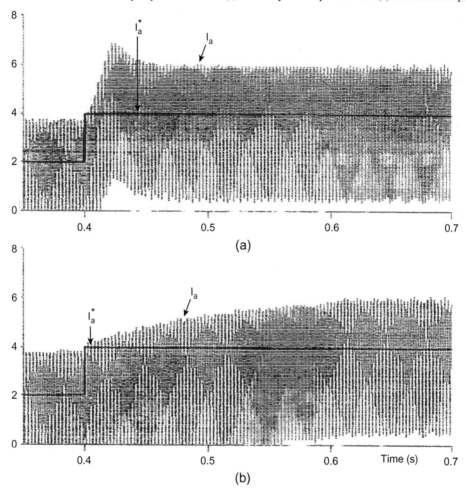

Once the fuzzy speed, current controllers, and firing angle compensator in Figure 10.29 had been designed successfully after a few iterations, system performance was tested in sequence starting with the inner loop. This figure shows the typical current loop response with step command in discontinuous conduction mode with both a fuzzy current controller and fuzzy compensator in (a), and compares it with the response without the compensator in (b). Evidently, the compensator helps the loop response to be very fast. The speed response of the dc drive is similar to the response of the vector drive (see Figure 10.27) and will not be given separately. The fuzzy compensator in this case acts as a three-dimensional nonlinear look-up table.

FIGURE 10.34 Efficiency optimization of induction motor vector drive with on-line search-based flux (ψ_r) programming.

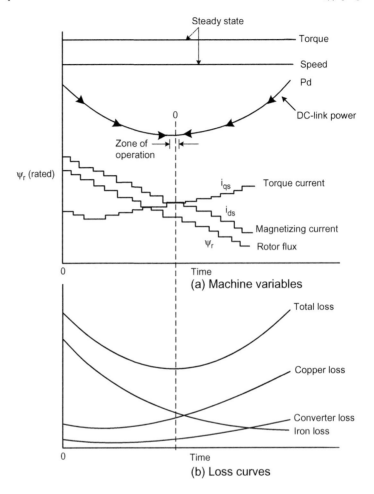

The efficiency improvement of an induction motor drive by flux programming was discussed in Chapter 7. This figure explains the on-line search based method of efficiency
optimization by flux programming in an indirect vector drive. Consider that the
machine operates initially at the rated flux in steady state with the constant load torque
and speed as indicated. The rotor flux is decremented in steps by reducing the magnetizing component (i_{ds}) of stator current. This causes an increase in i_{qs} (torque component
of current) normally by closed-loop speed control so that the developed torque remains
the same. As the core loss decreases with the decrease in flux, the copper loss increases,
but the system (converter and machine) loss decreases, thus improving the total efficiency.
This is reflected in the decrease of dc link power P_d as shown. The search continues
until the system settles at the minimum input power point O. In practical operation, the
operation oscillates about point O. The on-line search algorithm is universal and does
not depend on system parameters.

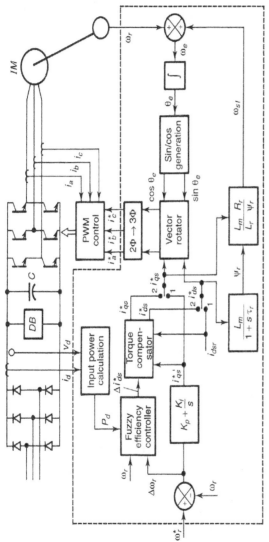

FIGURE 10.35 Vector-controlled induction motor drive with fuzzy flux programming efficiency optimizer.

The figure shows an indirect vector drive for an induction motor that incorporates a fuzzy efficiency optimizer [9]. The FL method of flux programming has the advantage of adaptive decrementing of i_{ds} current so that fast convergence is attained. In addition, it operates well in a noisy signal environment. In the figure, the speed control loop generates the $i_{qs}*$ current as shown. The vector rotator receives $i_{qs}*$ and $i_{ds}*$ current commands, respectively, from two positions of SPDT switches, i.e., transient position (1), where $i_{ds}* = i_{dsr}$ is the rated excitation current, and the $i_{qs}*'$ current from the speed loop, and steady-state position (2), where $i_{ds}*$ and $i_{qs}*$ are generated by the fuzzy efficiency controller and feedforward torque compensator, which will be explained later. The fuzzy controller becomes effective at steady-state condition, i.e., when $\Delta \omega_r$ approaches 0. Note that minimization of dc link power P_d by the fuzzy controller also minimizes the input power, i.e., the system operates at optimum efficiency for the same load torque and speed condition.

FIGURE 10.36   (a) Fuzzy efficiency optimizer block diagram and (b) MFs for $L\Delta I_{ds}^*$.

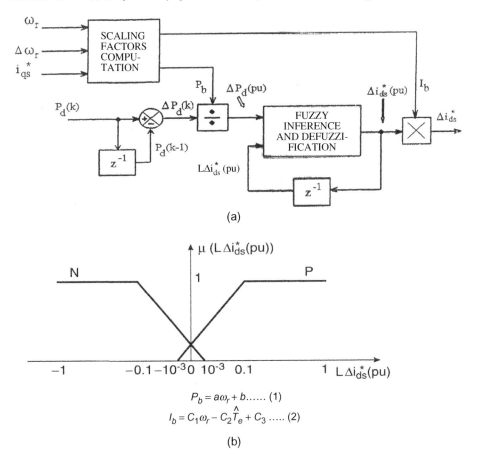

(a)

(b)

The block diagram for a fuzzy efficiency optimizer is shown in the upper part of the figure. The mean dc link power P_d is sampled and compared with the previous value to determine the increment $\Delta P_d(k)$. This input and the polarity of the last excitation current $L\Delta i_{ds}^*(pu)$ are reviewed in the fuzzy system, which generates the correct magnitude and polarity of the $\Delta i_{ds}^*(pu)$ signal to satisfy Figure 10.34. The P_b is the scale factor for $\Delta P_d(k)$, and I_b is the descaling factor for $\Delta i_{ds}^*(pu)$. Both of these factors are programmable and are shown by Eqs. (1) and (2), respectively, where $\hat{T}_e$ is the estimated torque. Variables $\Delta P_d(pu)$ and $\Delta i_{ds}^*(pu)$ are each described by seven asymmetrical triangular MFs (similar to Figure 10.25) and are not shown in the figure. But the polarities MFs of $L\Delta i_{ds}^*(pu)$ are included in the figure. A small overlap avoids the zero value of μ.

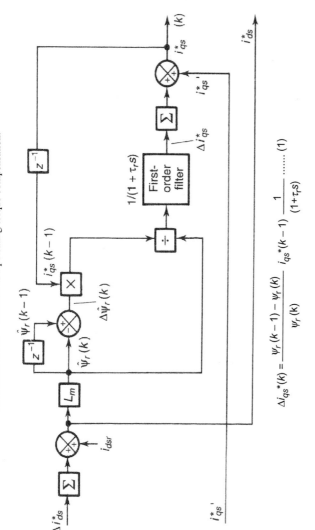

FIGURE 10.37 Feedforward pulsating torque compensation.

$$\Delta i_{qs}^{*}(k) = \frac{\psi_r(k-1) - \psi_r(k)}{\psi_r(k)} \quad i_{qs}^{*}(k-1) \quad \frac{1}{(1+\tau_r s)} \quad \ldots\ldots (1)$$

The feedforward pulsating torque compensator functions to compensate the temporary loss of torque due to decrementing of i_{ds} by the efficiency optimizer. Normally, the speed control loop pumps up the equivalent Δi_{qs} so that the developed torque and speed remain unaltered. However, the speed loop is slow, and will induce a pulsating torque at low frequency (sampling frequency of fuzzy optimizer) that might induce harmful mechanical resonance. In the figure, Δi_{ds}^{*} subtracts from the rated i_{dsr} to constitute the actual i_{ds}^{*}. The compensating current Δi_{qs} is calculated by Eq. (1), which is summed and adds with the current i_{qs}^{*}' to generate the actual $i_{qs}^{*}(k)$ as shown in the figure. The response of compensating current is instantaneous, but Δi_{ds} responds with the delay of rotor time constant (τ_r). Therefore, the feedforward compensating current is processed through the first-order filter with machine rotor time constant so that the original torque remains unaltered.

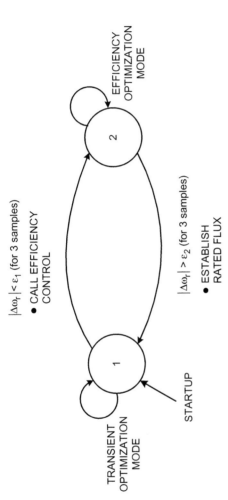

FIGURE 10.38   Transition between efficiency optimization mode and transient optimization mode.

The figure shows the transition of the drive in Figure 10.35 between efficiency optimization mode (mode 2) and transient response optimization mode (mode 1). The drive starts up in mode 1 with the rated flux and then at steady state ($|\Delta\omega_r| < \varepsilon_1$) it falls to mode 2 by changing the switch position from 1 to 2. During mode 2, if a transient condition arises due to a change in speed or load torque, $\Delta\omega_r$ increases ($|\Delta\omega_r| > \varepsilon_2$). Mode 2 is then abandoned, the rated ψ_r is established by altering the switch in Figure 10.35 to position 1, and the drive falls back to mode 1. A hysteresis band is provided so that no chattering occurs between the two modes.

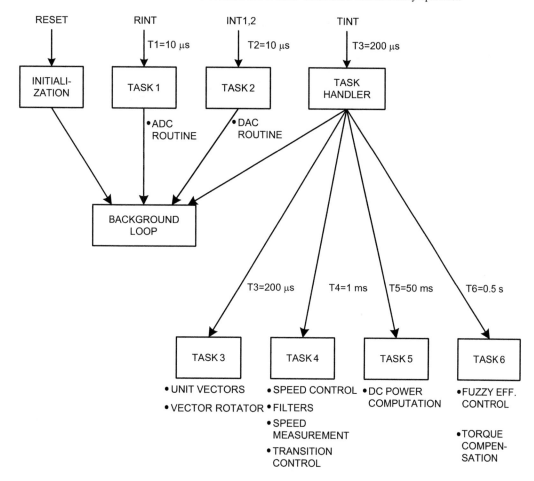

FIGURE 10.39 Software structure of FL-based vector drive with efficiency optimizer.

The software structure of the complete drive system in Figure 10.35, including the efficiency optimizer and pulsating torque compensator, is given in this figure. The control is based on a Texas Instruments DSP type TMS320C25 (16 bit, fixed point with 100-ns instruction cycle time). The six tasks and the subroutines performed in each task are summarized in the figure. Tasks 1 and 2 (ADC and DAC routines) are interrupt driven and have a sampling frequency of 50 kHz. The hardware timer interrupt (TINT) was set at 200 µs and was the sampling interval for vector control routines. The remaining tasks were executed at larger sampling times as shown, which were generated by software down counters. The control dwells in the background loop after execution of tasks 3 to 6. The power-up RESET initializes the system hardware and software and waits in the background loop for the external interrupts. Note that fuzzy efficiency optimizer is executed with a long sampling time (0.5 sec), which can induce mechanical resonance without feedforward pulsating torque compensation.

FIGURE 10.40   Efficiency optimization performance with pulsating torque compensation.

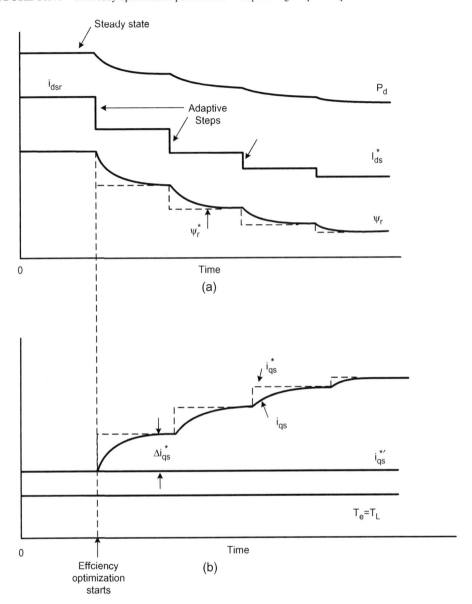

The figure shows the time-domain response of the fuzzy efficiency optimizer along with the pulsating torque compensator. Initially, at steady-state condition, the flux is rated with the rated i_{ds}^* current. As the optimizer works, i_{ds}^* decrements with adaptive step size as shown for fast convergence. However, ψ_r decreases smoothly due to rotor time constant effect. The improvement in efficiency by a gradual decrease in P_d is shown in the figure. The current i_{qs}^* increases gradually with the inverse profile of ψ_r so that T_e remains the same as T_L.

FIGURE 10.41   Family of torque-speed curves for a wind turbine at different wind velocities.

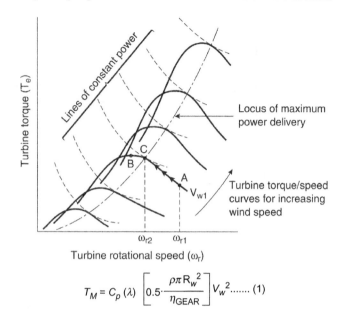

$$T_M = C_p\,(\lambda) \left[0.5 \cdot \frac{\rho \pi R_w^2}{\eta_{GEAR}} \right] V_w^2 \dots\dots\ (1)$$

Let us now apply fuzzy logic in the control of the wind generation system discussed in Chapter 7 (see Figures 7.52 and 7.53). This figure shows the torque-speed characteristics of a vertical Darrieus (egg-beater) type wind turbine at variable wind velocities (V_w) but with a constant pitch angle of the blades. A vertical turbine (compared to a horizontal turbine) has the advantage that it can be installed on the ground and can accept wind from any direction. However, the disadvantages are that the turbine is not self-starting and pulsating torque is developed that depends on wind velocity and turbine speed. The developed torque (T_M) as a function of the power coefficient (C_p), turbine radius (R_w), gear ratio (η_{GEAR}), and wind velocity (V_w) is given in Eq. (1), where C_p is a function of tip-speed ratio λ. At a particular wind velocity (V_{w1}), if the turbine speed decreases from ω_{r1}, the developed torque increases, reaches peak point B, and then decreases again. The locus of constant power curve indicates that maximum power is developed at point C where the speed is ω_{r2}. The locus of maximum power delivery curve is shown for different wind velocities. This means that as the wind speed varies, the turbine speed is to be varied to track with it so that the maximum power can be captured.

FIGURE 10.42 Fuzzy logic-based control of a wind generation system.

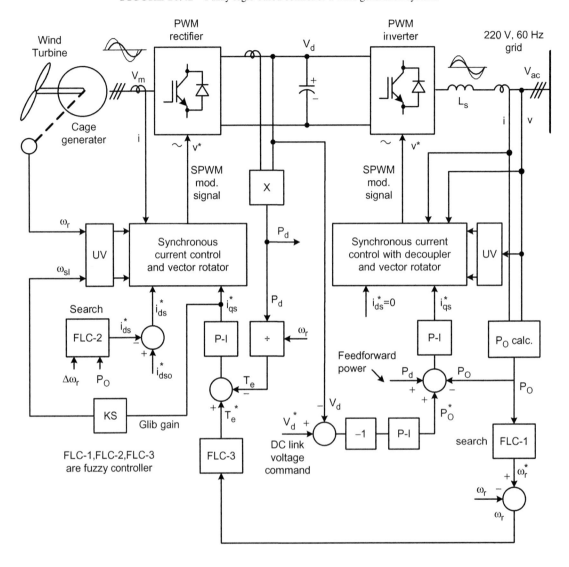

The figure shows a wind generation control system block diagram incorporating the fuzzy control elements (FLC-1, FLC-2, and FLC-3) [10]. The variable speed turbine on the left is coupled to the cage-type induction generator through a speed-up gear (not shown). The variable-voltage, variable-frequency power from the generator is rectified to dc by the PWM rectifier, which also supplies the machine excitation. The dc link voltage (V_d) is then inverted to 220-V, 60-Hz sinusoidal ac and fed to the utility system at unity DPF. Generator speed (ω_r) is controlled by indirect vector control to track the variable wind velocity. Machine flux ψ_r is programmed by the current i_{ds}^* in an open-loop manner. The line-side converter is direct vector controlled with voltage (V_d) control in the outer loop and power (P_0) control in the inner loop. Both the converters use synchronous current control as indicated. The control system uses three fuzzy controllers as shown: (1) FLC-1, for on-line search of generator speed to maximize the output power (P_0) according to Figure 10.41; (2) FLC-2, for on-line search of machine excitation current to optimize generator efficiency, i.e., maximize P_0 at light load (see Figure 10.34); and (3) FLC-3, for robust control of generator speed (see Figure 10.21). The power flow to the left permits start-up of the generator and building up of the dc voltage (V_d). At abnormal wind velocity, the pitch angle is adjusted to release all the torque.

FIGURE 10.43   Why fuzzy control in a wind generation system?

1. FEATURES OF FUZZY OUTPUT POWER MAXIMIZER (FLC-1)
- ON-LINE SEARCH OF SPEED (ω_r) FOR MAXIMUM P_0
- NO NEED FOR WIND VELOCITY INFORMATION
- UNIVERSAL PARAMETER-INSENSITIVE ALGORITHM
- ADAPTIVE ω_r STEPS GIVE FAST CONVERGENCE
- APPROPRIATE WITH INACCURATE SIGNALS
- PLANT PARAMETER VARIATION DOES NOT AFFECT PERFORMANCE

2. FEATURES OF FUZZY EFFICIENCY OPTIMIZER OF GENERATOR-CONVERTER SYSTEM (FLC-2)
- ON-LINE SEARCH OF MACHINE FLUX FOR CONVERTER-MACHINE LOSS MINIMIZATION
- ADAPTIVE ΔI_{ds}^* STEPS GIVES FAST CONVERGENCE
- APPROPRIATE WITH NOISY SIGNALS

3. FEATURES OF FUZZY SPEED CONTROLLER (FLC-3)
- UNIVERSAL PARAMETER-INSENSITIVE CONTROL
- ROBUST SPEED CONTROL AGAINST WIND VORTEX AND TURBINE PULSATING TORQUE
- DEADBEAT RESPONSE WITH SPEED CHANGE
- PREVENTS TURBINE MECHANICAL RESONANCE
- APPROPRIATE WITH INACCURATE SIGNALS

The advantages of using three fuzzy controllers in the wind generation system are explained in detail in this figure. The FLC-1 increments (or decrements) the generator speed ω_r and observes the system output power P_0 by means of a real-time search so that with variable wind velocity maximum power is always extracted, as explained in the previous figure. This is basically optimization of the aerodynamic efficiency of the turbine. Note that measurement of wind velocity is not needed. The FLC-2 is the fuzzy efficiency optimizer by flux programming that was explained before. Note that this controller minimizes losses of the system, which consists of the machine and two converters. Both FLC-1 and FLC-2 operate at the steady-state condition. The wind turbine has the characteristics of a fan ($\omega_r = kT_L^2$), but it rotates in the reverse direction. Therefore, most of the time the generator operates at a light load that permits flux-programming efficiency improvement control. FL-based robust speed control (FLC-3) is essential because of oscillatory turbine torque and wind vortex condition, although stiff torque loop is provided in the inner loop.

FIGURE 10.44   Performance of fuzzy controllers (FLC-1 and FLC-2) to optimize line power output.

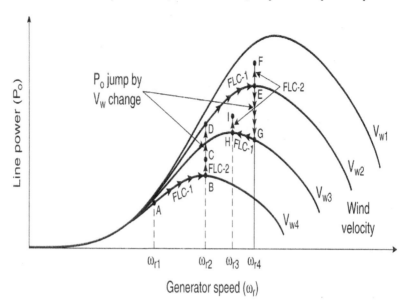

The figure explains the performance of the FLC-1 and FLC-2 controllers of Figure 10.42. The torque-speed curves in Figure 10.41 at different wind velocities can be converted to the corresponding power-speed curves shown in this figure. For simplicity, all system losses are neglected so that the turbine output power is the same as line power (P_0). Assume that initially the wind velocity is steady at V_{w4} and generator speed is ω_{r1} so that output power is at point A. The activation of FLC-1 control alters the generator speed to ω_{r2}, where P_0 is maximum at B. At the termination of FLC-1 control, FLC-2 is activated, which increases P_0 further to point C. If wind velocity increases to V_{w2}, the power jumps to point D. Then, sequential control of FLC-1 and FLC-2 raises the power to F. If at this point wind velocity decreases to V_{w3}, the locus of operation is indicated on the figure. Evidently, during an optimization search, if wind velocity changes, the search is abandoned.

FIGURE 10.45   Fuzzy estimation of three-phase diode rectifier line current.

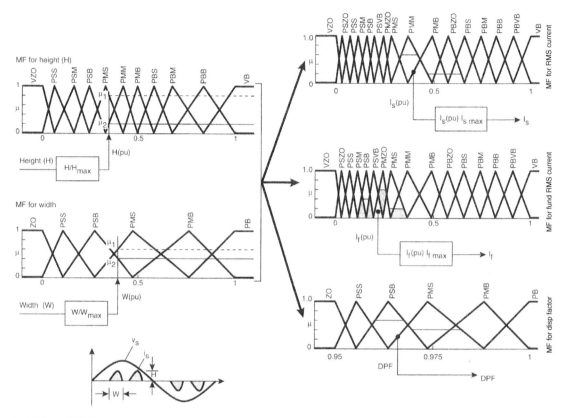

EXAMPLE RULE:
    IF: $H(pu)$ = PMM and $W(pu)$ = PMS
    THEN: $I_s(pu)$ = PBZO , $I_f(pu)$ =PMS and DPF = PMS

The distorted line current wave of a three-phase diode rectifier for a variable-frequency inverter drive is shown in the lower left-hand corner of the figure. Can FL be applied for reasonably accurate estimation of the parameters, such as rms line current (I_s), fundamental rms current (I_f), displacement factor (DPF), and power factor (PF) from this wave? As mentioned before, a fuzzy system is characterized by input–output nonlinear mapping. In this case, the wave pattern is characterized by its width (W) and height (H) as shown. All estimated parameters are essentially nonlinear mapped outputs of the input wave pattern. Such an estimation is fast, easy, and does not involve complex mathematical computation. The Mamdani method of fuzzy estimation is shown in the figure [11] with multiple outputs, where all the variables are expressed in *pu* values. Because the current varies with loading condition, parameters $H(pu)$ and $W(pu)$ are defined by

11 and 6 MFs, respectively. The number of MFs for output variables $I_s(pu)$ and $I_f(pu)$ is 16 because of accuracy requirements, but their distribution is different. The DPF has only 6 MFs. There are altogether $11 \times 6 = 66$ rules (not given). An example rule is shown below the figure. The actual values of I_s and I_f are calculated by multiplying their respective maximum values. Note that PF can be calculated by the simple relation $PF = DPF\Sigma I_f / I_s$. The tuning of MFs and rule matrix is based on simulation results. For different load conditions, the input wave is analyzed and the corresponding I_s, I_f, and DPF are obtained. This analytical result helps fine-tune the fuzzy estimator (mapping) in order to get the desired accuracy.

FIGURE 10.46   Induction motor indirect vector drive showing a fuzzy slip slip gain (K_s) tuning controller.

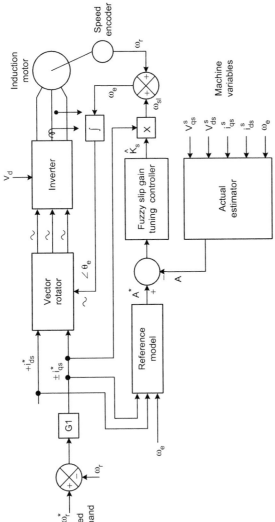

Next, let us discuss the application of FL in slip gain (K_s) tuning of an indirect vector drive system by means of the MRAC (model referencing adaptive control) method [12, 13], as explained in this figure. A detuned K_s, caused by variation of machine parameters, particularly the rotor resistance (R_r), gives undesirable transfer characteristics for torque and flux, as discussed in Chapter 7. The basic principle here is that reference model output A^*, which is valid in the tuned condition of the drive, is compared with actual estimator output A in a detuned condition and the resulting error modifies the K_s by a fuzzy tuning controller so that A approaches A^*. This will be further explained in the next figure.

FIGURE 10.47 Fuzzy logic-based MRAC slip gain tuning control block diagram.

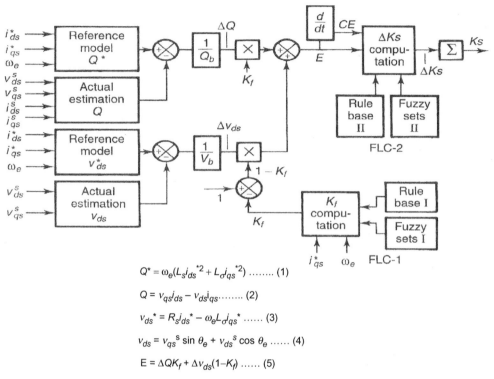

$$Q^* = \omega_e(L_s i_{ds}^{*2} + L_\sigma i_{qs}^{*2}) \ \text{........ (1)}$$

$$Q = v_{qs} i_{ds} - v_{ds} i_{qs} \text{........ (2)}$$

$$v_{ds}^* = R_s i_{ds}^* - \omega_e L_\sigma i_{qs}^* \ \text{...... (3)}$$

$$v_{ds} = v_{qs}^s \sin \theta_e + v_{ds}^s \cos \theta_e \ \text{...... (4)}$$

$$E = \Delta Q K_f + \Delta v_{ds}(1 - K_f) \ \text{...... (5)}$$

Rule 1:IF: speed (ω_e) is high (H) and torque (I_{qs}) is low (L),
      THEN: weighting factor (K_f) is low (L).

The figure shows the details of an MRAC-based fuzzy K_s tuning controller. The algorithm depends on tuning the reference models (which are valid in tuned condition) of reactive power (Q^*) and d^e-axis voltage (v_{ds}^*) shown in Eqs. (1) and (3) with the actual estimate of the corresponding values given in Eqs. (2) and (4), respectively. The loop errors are converted to *pu* values by dividing with the corresponding base values to generate the signals ΔQ and Δv_{ds}, respectively. These signals are multiplied by the respective gain factors K_f and $(1 - K_f)$ and added to generate the total E signal as shown in Eq. (5). FLC-1 generates the K_f parameter, which permits appropriate distribution of Q and v_{ds} tuning on the torque-speed ($i_{qs}^*-\omega_e$) plane. The objective here is to assign Q-based tuning dominantly in the low-speed, high-torque region, whereas v_{ds}-based tuning is used in the high-speed, low-torque region. A typical rule for FLC-1 is shown above. FLC-2 generates the corrective incremental slip gain (ΔK_s) based on the combined detuning error (E) and its derivative (CE), similar to Figure 10.21. Basically, it is an adaptive feedback controller for fast convergence at any operating point irrespective of E and CE signals. At tuned condition, $\Delta Q = 0$, $\Delta v_d = 0$. Therefore, $E = 0$. Note that there is some parameter dependence in the reference models of Eqs. (1) and (3) that will make the tuning imperfect.

FIGURE 10.48 Normalized control loop errors vs. normalized slip gain [$\omega_r(pu) = 0.5$].

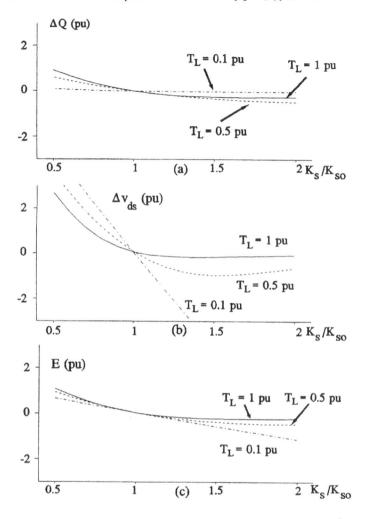

To design fuzzy controller FLC-1 of the previous figure, it is necessary to study the ΔQ and Δv_{ds} sensitivity characteristics on simulation at different speeds and load torques by detuning slip gain K_s. Figures 10.48(a) and (b) show the detuned characteristics of ΔQ (pu) and $\Delta v_{ds}(pu)$, respectively, at constant speed by varying the K_s value, but disabling the fuzzy controller ($\Delta K_s = 0$). As indicated in the figure, ΔQ (pu) = $\Delta v_{ds}(pu)$ = 0 at tuned condition, i.e., $K_s = K_{s0}$. The effect of load torque (T_L) variation is shown on the figures. If $K_s > K_{s0}$ and T_L increases, $\Delta Q(pu)$ becomes slightly negative, but for $K_s < K_{s0}$, the error becomes positive. However, the sensitivity on the $\Delta v_{ds}(pu)$ variation is very large, as shown in the figure. These characteristics indicate that gain factor K_f should be large in the high-torque region so that $\Delta v_{ds}(pu)$ compensation becomes more effective. Figure 10.48(c) shows the combined error $E(pu)$ as a function of K_s/K_{s0} with optimum K_f distribution on the torque-speed plane by FLC-1.

FIGURE 10.49 Fuzzy slip gain tuning performance: (a) detuned slip gain and (b) tuned slip gain.

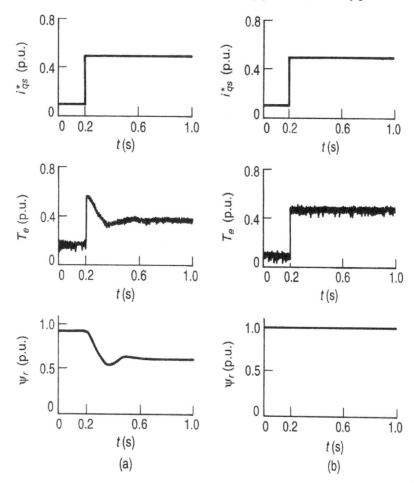

(a)

(b)

Once the fuzzy tuning controller was designed successfully after several iterations, the torque or i_{qs} control loops were tested. The effect of detuning control on torque and flux is shown in part (a). In this case, the slip gain was set at twice the correct value (K_s/K_{s0}), resulting in higher order dynamics for torque and flux transients. Also, the underfluxing condition is evident. The correct tuning, on the other hand, results in ideal response of torque and flux, as shown in (b).

FIGURE 10.50   Quasi-fuzzy estimation block diagram for the stator resistance (R_s) of an induction motor.

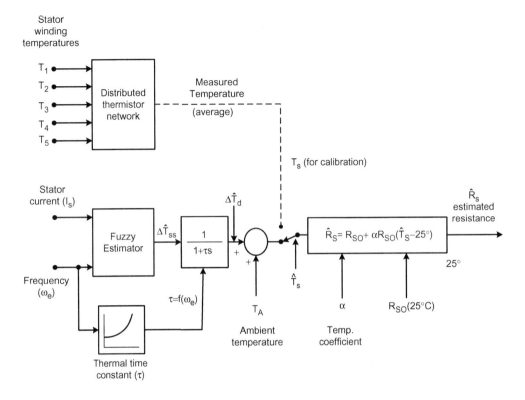

The stator resistance (R_s) of an induction motor can be estimated by using FL, as explained in this figure [14]. The R_s parameter varies essentially with stator temperature (T_s), and it was mentioned in Chapter 7 that R_s variation affects the estimation accuracy of flux, torque, speed, etc. As shown in the figure, the fuzzy estimator receives the stator current (I_s) and frequency (ω_e) or speed signals as input and estimates the steady-state temperature rise $\Delta \hat{T}_{ss}$ at the output. The reason is that copper loss and core loss essentially raise the stator temperature. The signal $\Delta \hat{T}_{ss}$ is converted to time-delayed dynamic temperature rise ($\Delta \hat{T}_d$) with the help of a machine first-order thermal time constant τ as shown. The τ is generated as a nonlinear function of frequency. For a particular machine, $\tau = f(\omega_e)$ can be generated as follows: The machine is run on a dynamometer with an adjustable speed setting. At each speed or ω_e, steps of stator current (torque component in a vector drive) were injected and the corresponding time constant τ of temperature rise was determined by the thermistor network shown in the upper part of the figure. In fact, the stator has six distributed thermistors, and the average temperature (T_s) was taken into consideration. The $\Delta \hat{T}_d$ signal is added with T_a to determine the estimated $\hat{T}_s$, and this signal is used to estimate $\hat{R}_s$, as shown in the figure. The measured T_s was used initially to calibrate (i.e., to tune its MFs and rule base) the fuzzy estimator. After calibration, the thermistor network was removed. Because the estimation is not entirely fuzzy, it is defined as a quasi-fuzzy estimator. Because the algorithm is based on measurements on a particular machine, it is not valid for another machine in a different class.

FIGURE 10.51   Steps in quasi-fuzzy estimation of R_s.

- GET STEADY-STATE STATOR TEMPERATURE RISE (ΔT_{ss}) CURVES WITH STATOR CURRENT (I_s) AT DIFFERENT FREQUENCY (ω_e) OR SPEED

$$\Delta T_{ss} = f(I_s, \omega_e)$$

- GET MACHINE THERMAL TIME CONSTANT CURVE WITH FREQUENCY OR SPEED

$$\tau = f(\omega_e)$$

- DEVELOP FUZZY ESTIMATOR MFs AND RULE TABLE
- ESTIMATE $\Delta \hat{T}_{ss}$ BY FUZZY ESTIMATOR FROM I_s AND ω_e VALUES
- CALCULATE DYNAMIC TEMPERATURE RISE $\Delta \hat{T}_d$ DUE TO THERMAL TIME CONSTANT

$$\Delta \hat{T}_d = \Delta \hat{T}_{ss} \cdot \frac{1}{1 + \tau S}$$

- CALCULATE $\hat{T}_s$

$$\hat{T}_s = T_A + \Delta \hat{T}_d$$

- CALIBRATE $\hat{T}_s$ WITH MEASURED T_s FOR FUZZY ESTIMATOR TUNING (ONLY INITIAL STAGE)
- ESTIMATE STATOR RESISTANCE $\hat{R}_s$

$$\hat{R}_s = R_{s0} + \alpha R_{s0}(\hat{T}_s - 25°)$$

The steps of R_s estimation described in Figure 10.50 are summarized here. The most crucial step is the database generation for $\Delta T_{ss} = f(I_s, \omega_e)$ from which fuzzy MFs and the rule base can be generated. This will be discussed in the next figure. The machine thermal model and corresponding determination of thermal time constant are extremely complex. However, a simplified first-order model is determined here experimentally. The steady-state temperature rise is converted to a dynamic function by the first-order thermal time constant. Note that the thermistor network is used for initial calibration of estimated stator temperature only. Once the stator temperature is estimated, the estimation of stator resistance is easy.

FIGURE 10.52 Experimental stator temperature rise curves with stator current and frequency at steady state.

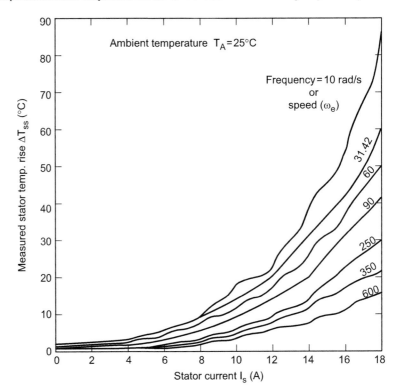

The figure shows the experimentally determined ΔT_{ss} curves as a function of stator current I_s and frequency ω_e at ambient temperature $T_A = 25°C$ for a particular machine. The machine (a 5-hp, NEMA Class B machine with shaft-mounted fan) with rated flux at vector control is operated on a dynamometer in the constant-speed mode but with different speed or frequency (slip is negligible) settings. At a particular frequency (ω_e) setting, a steady-state temperature rise ΔT_{ss} with the stator current I_s was recorded, and with the data of different speed settings this figure was plotted. Note that at higher frequency, the iron loss increases, and that will tend to give higher ΔT_{ss}, but higher machine speed with a shaft-mounted fan gives a dominant cooling effect that decreases ΔT_{ss}. These experimental curves were used to formulate the MFs and rule table for the fuzzy estimator. Basically, the estimator algorithm interpolates the ΔT_{ss} signal as a function of stator current and frequency. This is again, basically, nonlinear input–output mapping phenomena, as discussed earlier.

FIGURE 10.53   Stator resistance estimation performance (at 357 rpm).

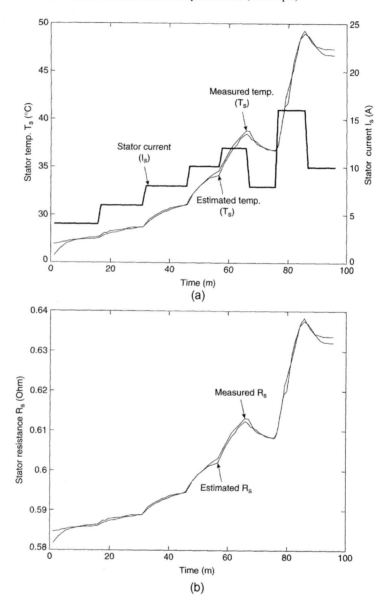

The upper part of the figure shows the estimated and measured stator temperatures at different steps of stator current but at constant speed (357 rpm), where the time delay effect on temperature rise is evident. The two curves follow closely. The lower part of the figure shows the estimated stator resistance and measured stator resistance corresponding to the above T_s curves. Similar measurements and estimation were done at different frequencies, but not included in the figure. Note that stator resistance is small in large machines and is often neglected, particularly at higher speeds, where the counter emf and stator leakage reactance drop are large compared to stator resistance drop. The effect of R_s becomes extremely important near zero-frequency operation.

FIGURE 10.54   MATLAB Fuzzy Logic Toolbox environment. (Courtesy of MathWorks, Inc.)

Once the fuzzy inference system (FIS) for a project is designed on paper, it should be fabricated, tested, and iterated with the help of software in order to optimize the design, and then used in a prototype system. Among the number of software tools available on the market, the MATLAB-based Fuzzy Logic Toolbox [6] is very user-friendly, powerful, and popularly used software. We will describe the Toolbox and then give some demo examples later. The Toolbox can be used easily with a graphical user interface (GUI) rather than command line functions. There are five primary GUI tools in the Toolbox for building, editing, and observing fuzzy systems as shown in this figure. These are FIS Editor, Membership Function Editor, Rule Editor, Rule Viewer, and Surface Viewer. These functional elements are linked dynamically and help to build and edit the fuzzy system rapidly. Once the fuzzy system has been designed, its performance can be tested by embedding it in the Simulink simulation environment of the system and fine-tuned iteratively. The Toolbox also lets the program run independently in the C program without the need for Simulink. After fine-tuning the fuzzy program, C language source program can be generated from it, compiled, and then the resulting object code can be downloaded to a DSP or RISC for real-time implementation.

FIGURE 10.55   FIS Editor of the Toolbox. (Courtesy of MathWorks, Inc.)

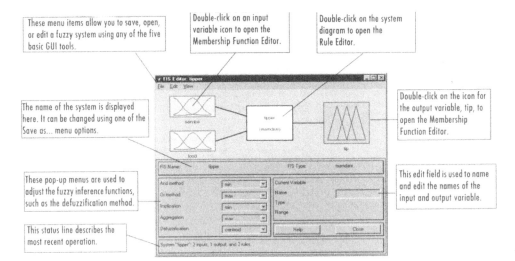

The FIS Editor of the Toolbox gives general information about the fuzzy system, and its features are pictorially displayed in this figure. At the top left, the names of the defined input fuzzy variables are indicated, and at right, the output variables are shown. The MFs shown in the boxes are simple icons and do not indicate the actual MFs used. Below this figure, the system name and inference method (Mamdani or Sugeno) are indicated. At the lower left, various steps of the inference process that are user-selectable are shown. At the lower right, the name of the input or output variable, its associated MF type, and its range are displayed. All the Editor and Viewer boxes illustrate the development of the restaurant tipping system described in earlier figures.

FIGURE 10.56   Membership Function Editor of the Toolbox. (Courtesy of MathWorks, Inc.)

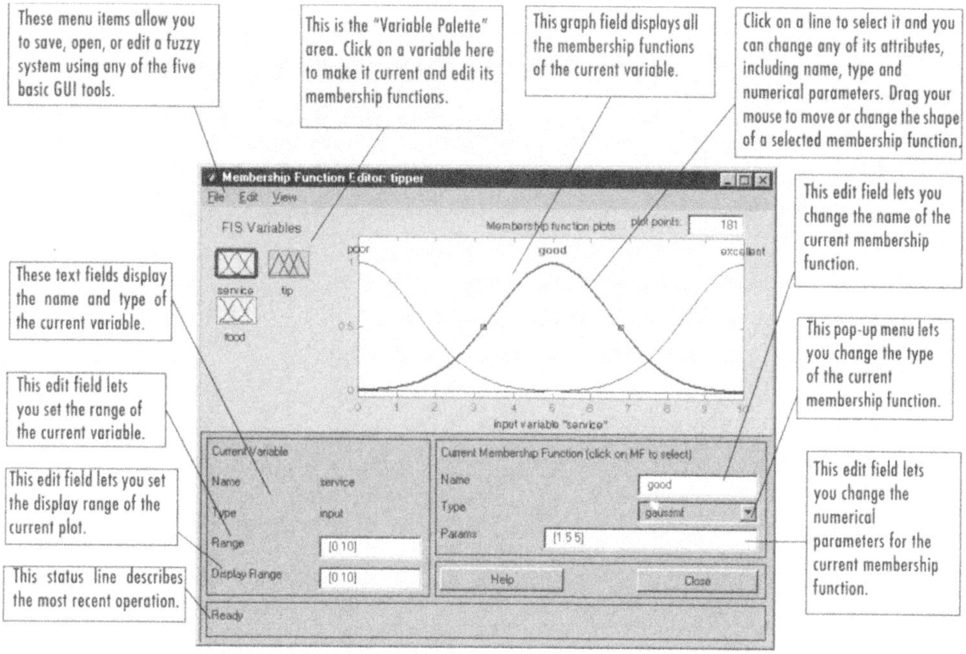

The Membership Function (MF) Editor, as shown in this figure, generates displays and permits editing of all MFs of the input and output variables. At the upper left, the variables whose MFs can be set are shown. Each setting includes a selection of the MF type and the number of MFs of each variable. At the lower right, the controls permit the name, type, and parameters (shape) of each MF to be changed once the MF has been selected. The MFs of the current variable that are being edited are shown in the graph. At lower left, the information about the current variable is given. Next to it, the range or universe of discourse and display range of the current plot of the variable under consideration can be changed.

FIGURE 10.57   Rule Editor of the Toolbox. (Courtesy of MathWorks, Inc.)

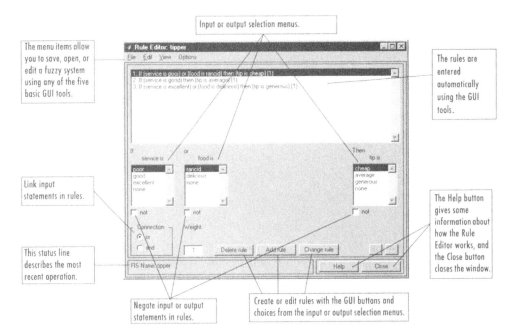

The Rule Editor, as shown in this figure, permits rules to be constructed one by one by selecting and clicking the appropriate boxes, once the rule base is designed on paper, and the input and output variables are described in the FIS Editor. The logical connectives of rules, AND, OR, and NOT, are then selected. Choosing "none" for a variable will exclude that variable from a given rule. The rules can be changed, deleted, or added by clicking the appropriate button.

FIGURE 10.58 Rule Viewer of the Toolbox. (Courtesy of MathWorks, Inc.)

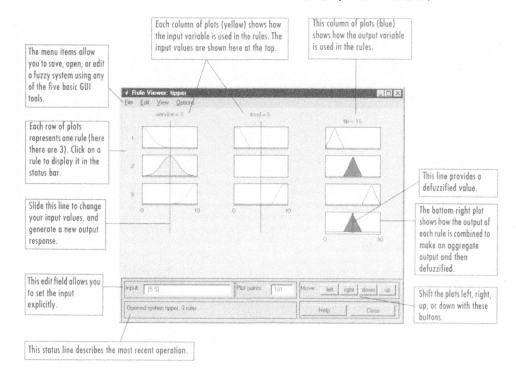

After building the fuzzy system with FIS Editor, MF Editor, and Rule Editor, the Rule Viewer permits the operation and contribution of individual rules to the output to be checked. In this figure, each rule is a row of plots, and each column is a variable. The effect of three rules in the restaurant tipping system (see Figure 10.12) is illustrated here. These rules are all connected by OR logic. In the figure, the input variables are set to arbitrary values, i.e., Service = 5 and Food = 5. The contribution of the input variable is shown by the shaded area, and the resulting contribution of the output area is indicated by shading. In the figure, rules 1 and 2 do not contribute any output, whereas rule 2 contributes maximum output. The Rule Viewer permits fine-tuning of the fuzzy system by checking the contribution of individual rules for arbitrary input values.

FIGURE 10.59   Control Surface Viewer of the Toolbox. (Courtesy of MathWorks, Inc.)

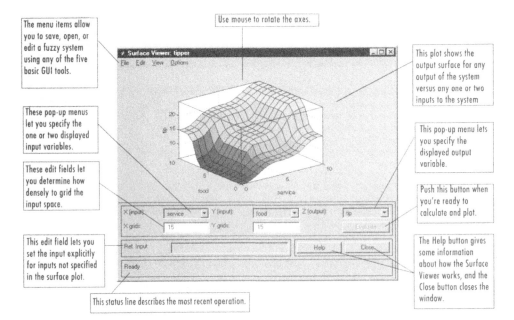

The menu items allow you to save, open, or edit a fuzzy system using any of the five basic GUI tools.

These pop-up menus let you specify the one or two displayed input variables.

These edit fields let you determine how densely to grid the input space.

This edit field lets you set the input explicitly for inputs not specified in the surface plot.

Use mouse to rotate the axes.

This plot shows the output surface for any output of the system versus any one or two inputs to the system

This pop-up menu lets you specify the displayed output variable.

Push this button when you're ready to calculate and plot.

The Help button gives some information about how the Surface Viewer works, and the Close button closes the window.

This status line describes the most recent operation.

The Surface Viewer permits us to view the mapping relations between the input and output variables after building the fuzzy system. The figure illustrates the Surface Viewer for the restaurant tipping system, where the input variables are *Food* and *Service* and the output variable is *Tip*. The three-dimensional plot gives a clear picture of contributions of inputs to the output. The plot may also be two-dimensional between a selected input and the output. For a higher dimensional system, it is possible to select three variables for viewing the mapping surface.

FIGURE 10.60 Simulink simulation of "vector controller" showing fuzzy i_{ds} and i_{qs} control (see Figure 9.15).

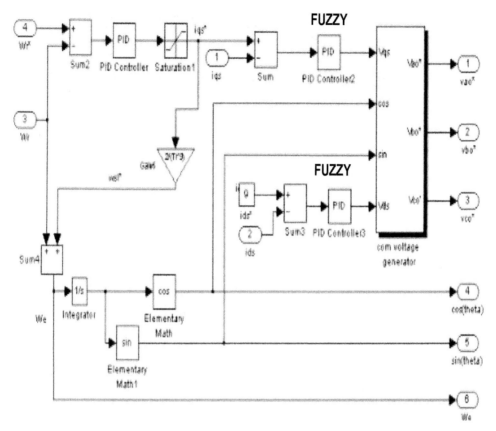

We will now demonstrate the application of Fuzzy Logic Toolbox on the indirect vector-controlled induction motor drive discussed in Figure 9.15 and compare the performance with *P-I* control. In fact, fuzzy controllers will be developed for i_{qs} and i_{ds} loops using Mamdani, Lusing Larson, and Sugeno zero-order methods to see their performance differences [15]. This figure repeats the block in Figure 9.15(b), where the i_{qs} and i_{ds} control loops are shown with fuzzy control instead of *P-I* control. However, the speed loop is left with *P-I* control. Note that open-loop flux control is used and the command current $i_{ds}*$ corresponds to the rated rotor flux. An identical fuzzy *P-I* controller was designed for both the i_{qs} and i_{ds} loops, and essentially the same were also used for other methods of fuzzy control. After design and iteration, the fuzzy controllers were embedded in a Simulink simulation of the drive system and performance was tested in each case. For simplicity, the MFs and rule table are not given.

FIGURE 10.61   Response of i_{ds} and i_{qs} with *P-I* control.

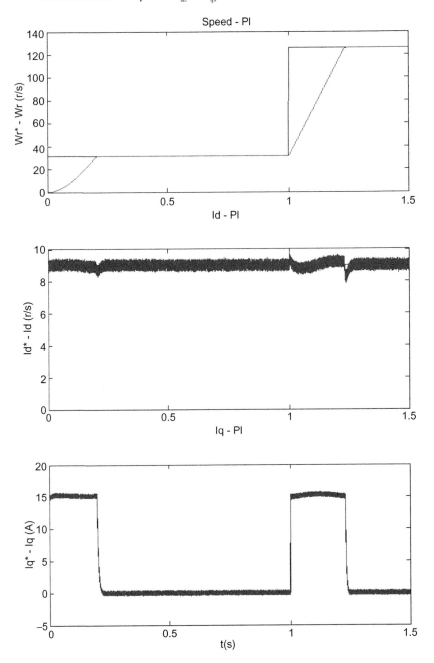

The figure shows the response of speed (ω_r), currents i_{ds} and i_{qs} (marked as i_d and i_q) with the step speed command in Figure 10.60, where the fuzzy controllers in the i_{ds} and i_{qs} loops are replaced by *P-I* control. With vector control, the current i_{qs} is almost instantaneous as usual. However, the i_{ds} response has some overshoot and undershoot as shown due to a small decoupling in vector control.

FIGURE 10.62   Response of i_{ds} and i_{qs} with fuzzy control (Mamdani method): (a) control surface and (b) responses.

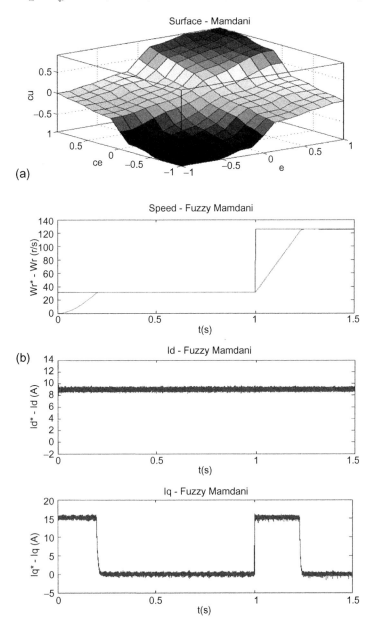

The figure shows the speed and current loop responses with the Mamdani method of fuzzy control. In this case, the fuzzy input variables are loop error e and change in error ce, and the output variable is cu (or du), where all are expressed in per-unit form. The e and ce have seven MFs in asymmetrical triangular form, and cu has 11 similar MFs so that the rule matrix consists of $7 \times 7 = 49$ rules altogether (not shown). The output cu is descaled and summed to implement fuzzy *P-I* control. With proper iteration of the fuzzy controllers, the i_{ds} response was found to be robust irrespective of torque and speed variations. Strictly speaking, fuzzy control is not needed for the i_{qs} loop, where the response remains the same as that with *P-I* control. In fact, i_{qs} current shows an increase of ripple current with fuzzy control. The control surface for the i_{ds} loop shown in (a) is a three-dimensional plot of the variables e, ce, and cu. The input–output nonlinear mapping feature of the fuzzy controller is evident from this plot.

FIGURE 10.63   Response of i_{ds} and i_{qs} with fuzzy control (Lusing Larson method): (a) control surface and (b) responses.

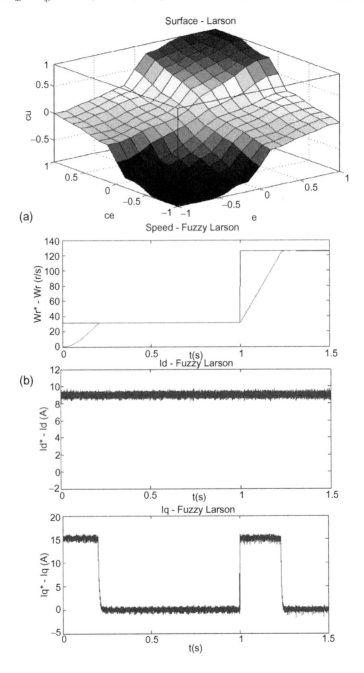

The figure shows the response of the variables when the same fuzzy controllers as in Figure 10.62 are implemented by the Lusing Larson algorithm. The MFs and the rule table remain identical to those of the Mamdani method. The response and the control surface look almost identical to those of the Mamdani method, indicating that both algorithms are equally effective.

FIGURE 10.64   Response of i_{ds} and i_{qs} with fuzzy control (zero-order Sugeno method): (a) control surface and (b) responses.

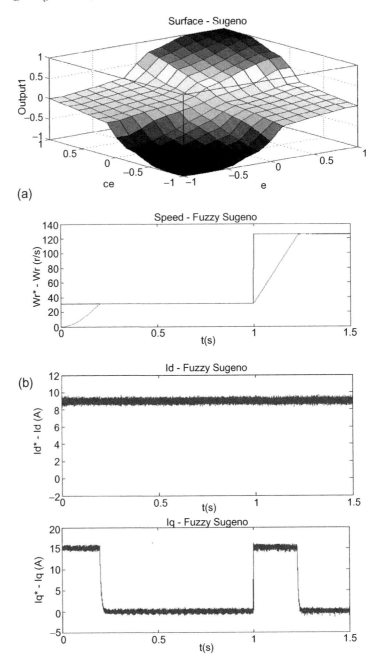

(a)

(b)

Finally, we will apply the zero-order Sugeno method to the same fuzzy controllers as in Figure 10.62 and verify their performance under identical conditions. Of course, in this case, the output MFs are given by 11 singletons (instead of triangular MFs), and defuzzification is calculated by Eq. (4) in Figure 10.19. The response in speed and current loops remains essentially the same as that for the Mamdani and Lusing Larson methods. However, a slight difference in the control surface is evident.

FIGURE 10.65   Advances and trends in FL applications.

- BROADLY, AI-BASED INTELLIGENT CONTROL AND ESTIMATION TECHNIQUES WILL PLAY SIGNIFICANT ROLE IN POWER ELECTRONICS AND MOTOR DRIVES
- FUZZY CONTROL POSSIBLY PROVIDES THE BEST ROBUST PERFORMANCE IN NONLINEAR FEEDBACK SYSTEM WITH PARAMETER VARIATION AND LOAD TORQUE DISTURBANCE
- FL WILL PLAY INCREASING ROLE IN ON-LINE FAULT DIAGNOSTICS AND FAULT-TOLERANT CONTROL
- APPLICATIONS OF FL ARE YET VERY LIMITED IN POWER ELECTRONICS AND DRIVE AREAS
- FUZZY LOGIC IS COMPETITIVE WITH NEURAL NETWORK IN INPUT–OUTPUT MAPPING (FUZZY CONTROL CAN BE IMPLEMENTED BY NEURAL NETWORK)
- FUZZY DYNAMIC MODEL DEVELOPMENT NEED OF PLANTS, SUCH AS ARC FURNACE AND STEEL MELTING CONVERTER
- FUZZY PLANT PARAMETER AND STATE` ESTIMATION
- PROBLEM OF AVAILABILITY OF LARGE FUNCTIONAL ASIC CHIP FOR FUZZY CONTROL

Although AI technology has advanced tremendously in recent years, and its potential impact on the power electronics and drive areas is great, the actual applications so far have been very limited. This has been particularly true for FL. The importance of FL has been clearly demonstrated for linear and nonlinear control systems, where robust performance is essential despite parameter variations and load disturbances. In addition, in the presence of a noisy environment, FL control performance is good. FL is potentially very important for identification of fault and fault-tolerant control in complex power electronic systems. This area remains practically unexplored. FL can be used for plant parameter and state estimation, and also for development of a fuzzy model for a plant, where mathematical model generation is difficult. A fuzzy model is basically a behavioral model, which is very useful for fuzzy control of the plant. Fuzzy control is mostly used for static mapping, but mapping in dynamic systems is also possible. (Note that a fuzzy *P-I* compensator is a dynamic system.) It will be shown in the next chapter that neural networks also provide nonlinear input–output mapping like FL, and therefore neural network technology is competitive with FL. Unfortunately, the availability of ASIC chips for FL application is very limited. Most of the FL applications are implemented by microprocessors and RISCs. Being a mapping problem, it is possible to implement a fuzzy system by means of precomputed look-up tables. It is likely that large functional ASIC chips will be available in the future, permitting inexpensive single-chip dedicated hardware control of the system.

Glossary

Adaptive neuro-fuzzy inference system (ANFIS)   A neural network technique for automatically tuning Sugeno-type fuzzy inference system based on training data (will be discussed in next chapter)

Antecedent   The initial or IF... part of a fuzzy rule

COA (center-of-area) defuzzification   A method of converting fuzzy output from the center of gravity of the output MF; also called *centroid defuzzification*

Consequent   The final or THEN... part of a fuzzy rule

Defuzzification   The process of converting a fuzzy output of a fuzzy system into a crisp output

Degree of membership   The fuzzy output of MF given by μ, where the value lies between 0 and 1; also called *membership value*

DOF (degree of fulfillment)   The degree to which the antecedent (IF...) part of a fuzzy rule is satisfied. Also called "firing strength." The value is given by μ (membership value)

Fuzzification   The process of generating membership values (μ) for a fuzzy variable using MFs

Fuzzy inference system (FIS) or fuzzy system (FS)   The overall name for a system that uses fuzzy reasoning to map an input space to an output space

Fuzzy operators   AND, OR, and NOT operators; also known as *logical connectives*

Fuzzy rule   IF... THEN... rule relating input fuzzy variables to output fuzzy variables

Fuzzy set   A set consisting of elements having degrees of membership varying between 0 and 1. It is characterized by MF and is associated with membership values

Fuzzy variable   A variable that can be defined by fuzzy sets

Height defuzzification   A method of converting fuzzy output to crisp output from a composed fuzzy value by considering the heights of each fuzzy set

Implication   The process of shaping a fuzzy set in the consequent (THEN...) based on the results of the antecedent (IF...) in a Mamdani type FIS

Linguistic variable   A fuzzy variable whose values are defined by language, such as large, medium, and small

Membership function (MF)   A function that defines a fuzzy set by associating every element in the set with a number between 0 and 1

Singleton   A fuzzy set with an MF that is unity at one particular point and zero everywhere else

Sugeno type inference   A type of fuzzy inference where the consequent of each rule is a truncated singleton (zero-order) or linear combination of the (first-order) inputs

Universe of discourse   The range of values associated with a fuzzy variable

Summary

The main theme of this chapter is the description of FL principles and its application in power electronics and motor drive systems. The FL is a discipline under AI that also includes the areas of expert systems (ESs), artificial neural networks (ANNs), or NNWs, and genetic algorithms (GAs). All of the main areas in AI, except ES, are defined as "soft computing." ESs and GAs will not be discussed further, but NNW will be discussed in detail in Chapter 11. The concept of FL and its characteristics emerge from Zadeh's theory propounded in 1965. Any FL application uses a knowledge base that consists of MFs describing the fuzzy variables, and the rule table consisting of IF... THEN... statements. The knowledge base is developed on the basis of the behavioral nature of the system. The mathematical plant model is not essential, but is convenient for conducting simulation studies that help optimize FL performance. The trial-and-error approach of FL algorithms may be time consuming, but user-friendly computer programs (such as the MATLAB-based Fuzzy Logic Toolbox) help speed the process. Of course, recently, NNW and GA-based optimization capabilities have been developed but were not discussed in the chapter. After discussing the FL principles and characteristics, a number of applications were illustrated that include speed control of induction motor vector drives, linearization of the transfer characteristics of thyristor converters at discontinuous conduction, efficiency optimization of induction motor vector drives by flux programming, wind generation systems, estimation of distorted waveforms, MRAC slip gain tuning control of vector drives, and induction motor stator resistance estimations. These examples should generate ideas for the reader about applying FL principles to solving new problems. The popularly used Fuzzy Logic Toolbox was described briefly. The Toolbox permits merging of the developed fuzzy system in the Simulink environment and simulation study of the whole system. The fuzzy system software can also be compiled and downloaded for real-time operation with the help of DSP. Finally, a demo program using this Toolbox was discussed that compares different methods of implication. A one-page glossary was provided for the convenience of newcomers to this area.

References

[1] B. K. Bose, "Expert system, fuzzy logic, and neural network applications in power electronics and motion control," *Proc. IEEE*, vol. 82, pp. 1303–1323, August 1994.

[2] B. K. Bose, *Modern Power Electronics and AC Drives*, Prentice Hall, Upper Saddle River, NJ, 2002.

[3] L. H. Tsoukalas and R. E. Uhrig, *Fuzzy and Neural Approaches in Engineering*, Wiley, New York, 1997.

[4] J. Zhao and B. K. Bose, "Evaluation of membership functions for fuzzy logic controlled induction motor drive," *IEEE IECON Conf. Rec.*, 2002.

[5] J. Zhao and B. K. Bose, "Membership function distribution effect on fuzzy logic controlled induction motor drive," *IEEE IECON Conf. Rec.*, pp. 214–219, 2003.

[6] Math Works, *Fuzzy Logic Toolbox User's Guide*, January 1998.

[7] I. Miki, N. Nagai, S. Nishigama, and T. Yamada, "Vector control of induction motor with fuzzy PI controller," *IEEE IAS NNWu. Meet. Conf. Rec.*, pp. 342–346, 1991.

[8] G. C. D. Sousa and B. K. Bose, "A fuzzy set theory based control of a phase-controlled converter dc drive," *IEEE Trans. Ind. Appl.*, vol. 30, pp. 34–44, January/February 1994.

[9] G. C. D. Sousa, B. K. Bose, and J. G. Cleland, "Fuzzy logic based on-line efficiency optimization control of an indirect vector controlled induction motor drive," *IEEE Trans. f Ind. Elec.*, vol. 42, pp. 192–198, April 1995.

[10] M. G. Simoes, B. K. Bose, and R. J. Spiegel, "Design and performance evaluation of a fuzzy logic based variable speed wind generation system," *IEEE Trans. Ind. Appl.*, vol. 33, pp. 956–965, July/August 1997.

[11] M. G. Simoes and B. K. Bose, "Application of fuzzy logic in the estimation of power electronic waveforms," *IEEE IAS NNWu. Meet. Conf. Rec.*, pp. 853–861, 1993.

[12] G. C. D. Sousa, B. K. Bose, and K. S. Kim, "Fuzzy logic based on-line tuning of slip gain for an indirect vector-controlled induction motor drive," *IEEE IECON Conf. Rec.*, pp. 1003–1008, 1993.

[13] T. M. Rowan, R. J. Kerkman, and D. Leggate, "A simple on-line adaptation for indirect field orientation of an induction machine," *IEEE Trans. Ind. Appl.*, vol. 42, pp. 129–132, March/April 1995.

[14] B. K. Bose and N. R. Patel, "Quasi-fuzzy estimation of stator resistance of induction motor," *IEEE Trans. Power Elec.*, vol. 13, pp. 401–409, May 1998.

[15] J. O. P. Pinto and L. Galotto, "Fuzzy logic demo—fuzzy controller for induction motor indirect vector control," Memo, January 2005.

CHAPTER 11

Neural Network and Applications

FIGURE 11.1   Features of neural networks.

- MOST GENERIC FORM OF AI FOR EMULATION OF HUMAN THINKING
- NEUROCIMPUTATION IS INSPIRED BY BIOLOGICAL NEURAL NETWORK OF HUMAN BEING
- BASICALLY INPUT-OUTPUT NONLINEAR MAPPING PHENOMENA LIKE FUZZY SYSTEM
- MASSIVE HIGH SPEED PARALLEL COMPUTATION WITH FAULT-TOLERANCE AND NOISE FILTERING CAPABILITY
- KNOWLEDGE IS ACQUIRED BY LEARNING (OR TRAININING) THROUGH EXAMPLES OF INPUT-OUTPUT DATA SETS
- PROPERTIES OF
  - PATTERN CLASSIFICATION AND RECOGNITION
  - FUNCTION APPROXIMATION
  - ASSOCIATIVE MEMORY
- TYPICAL APPLICATIONS:
  - CONTROL AND ESTIMATION IN POWER ELECTRONIC SYSTEMS
  - GENERAL INDUSTRIAL PROCESS CONTROL
  - ROBOT VISION
  - ON-LINE DIAGNOSTICS, ETC.

Among all the AI techniques, artificial neural network (ANN) or neural network (NNW) is the most important discipline, and its potential impact on power electronics area is tremendous. The technology has a long history, but its development was camouflaged by the glamorous evolution of modern digital computers. From the early nineties, the momentum of its R & D and applications has surged dramatically. As mentioned before, neurocomputer attempts to mimic the capability of biological nervous system, but obviously its performance is far inferior. Like fuzzy system, NNW basically performs input-output mapping which can be static or dynamic. The result is pattern recognition, pattern classification, function approximation and associative memory properties which will be discussed later. One important feature of NNW is that it normally requires supervised training (or learning) by input-output example data sets unlike conventional programming of digital computer. NNW is a vast subject [1, 2]. We will discuss its principles and applications in power electronics and motor drives in this chapter.

FIGURE 11.2   Structure of (a) a biological neuron and (b) an artificial neuron.

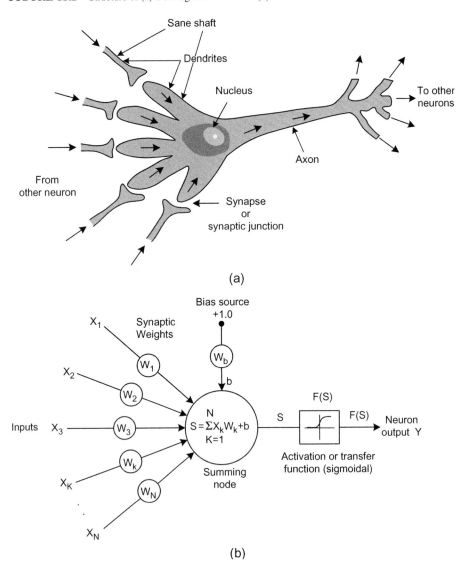

A NNW consists of a number of interconnected artificial neurons. The structure of an artificial neuron is inspired by the concept of a biological neuron as shown in this figure. A neuron is the basic processing element (PE) in the nervous system of the brain that receives and combines signals from other similar neurons through thousands of input paths called *dendrites*. Each input signal (which is electrical in nature), flowing through a dendrite, passes through a synapse or synaptic junction as shown. The junction gap is filled with neurotransmitter fluid, which either accelerates or retards the flow of the signal. These signals are then summed up in the nucleus, nonlinearly modified at the output before flowing to other neurons through the branches of an axon as shown. The synaptic junction contributes the intelligence or memory property of the cell. The model for an artificial neuron that closely matches that of a natural neuron is given by an op-amp summer-like structure. It is also called a *PE* (processing element), *neurode, node,* or *cell.* Each input signal flows through a gain or weight (called a synaptic weight) that can be positive or negative, integer or noninteger. The summing node that accumulates the weighted input signals also receive a weighted bias signal and then passes to the output through the nonlinear (or linear) transfer or activation function as shown.

FIGURE 11.3   Several activation or transfer functions of artificial neurons.

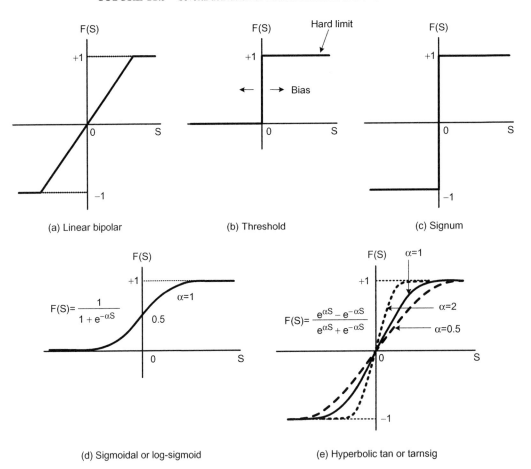

(a) Linear bipolar

(b) Threshold

(c) Signum

$$F(S)= \frac{1}{1 + e^{-\alpha S}}$$

(d) Sigmoidal or log-sigmoid

$$F(S)= \frac{e^{\alpha S} - e^{-\alpha S}}{e^{\alpha S} + e^{-\alpha S}}$$

(e) Hyperbolic tan or tarnsig

Several common-type activation or transfer (TF) functions are linear, threshold, signum, sigmoidal (or log-sigmoid), and hyperbolic-tan (or tan-sigmoid), as shown in this figure. Another type of function not included here is the Gaussian function. The magnitudes of these functions vary between 0 and 1 or −1 to +1. The linear function can be unipolar or bipolar. With a slope of infinity, it transforms to the threshold or signum function, respectively. The sigmoidal and hyperbolic tan functions are commonly used in power electronics systems. Their mathematical expressions are shown in the figure, where α is the gain or coefficient that adjusts the slope or sensitivity. The hyperbolic tan function is shown with different values of α. These two functions are differentiable, and the derivative dF/dS is maximum at $S = 0$. All of these functions are characterized as squashing functions because they squash or limit the neuron response between the asymptotic values. Note that the nonlinear activation function contributes to the nonlinear characteristics of a neuron, which permits nonlinear input–output mapping in a NNW. With a linear activation function, on the other hand, this nonlinearity is lost.

FIGURE 11.4   Simple example of NNW application: $Y = A \sin X$ for $-\pi \le X \le +\pi$.

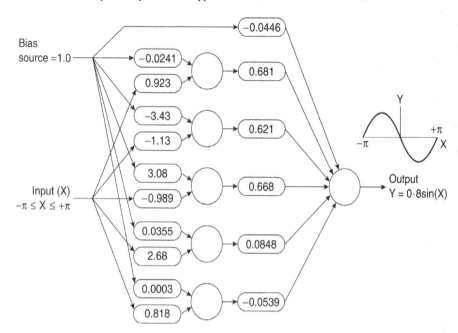

Before proceeding further, let us illustrate a simple neural network application. The problem here is to generate a function $Y = A \sin X$ for given X values in the range shown in the figure. Basically, it is a nonlinear, two-dimensional (X-Y) mapping problem, which can be easily solved with a DSP look-up table. How can we solve the same problem with NNW and what are the advantages? The figure shows an NNW with six neurons of which one is at the output and the remaining five are at the input layer. The output neuron uses a bipolar linear transfer function, whereas the input neurons have a hyperbolic tan function (TFs are not shown). The input signal X (bipolar) is connected to all the input layer neurons through a weight that is noninteger and can be positive or negative. The output of each neuron is connected to the output neuron through a similar weight. The bias source is connected to all the neurons through a weight as shown. Such a NNW is defined as a fully connected network because no connection is missing. Without the bias, if $X = 0$, $Y = 0$. The bias is not needed here, but it is connected for generality. Numerical computation with a computer program should verify the X-Y relation. Actually, the weights are determined by a computer-based training algorithm, which will be discussed later. The NNW can be implemented with an analog or digital ASIC chip. If one or two weights drift, or if a connection is severed, the distortion in the output should not be significant. This is a fault-tolerance property of NNWs. Also, if some noise is mixed with the input signal, this will not practically show up at the output. This is a noise filtering property of ANNs.

FIGURE 11.5 Some neural network models.

1. FEEDFORWARD NNW
- PERCEPTRON
- ADALINE AND MADALINE
- BACKPROPAGATION (BACKPROP) NNW
- RADIAL BASIS FUNCTION NNW (RBFN)
- GENERAL REGRESSION NNW (GRNN)
- MODULAR NEURAL NETWORK (MNN)
- LEARNING VECTOR QUANTIZATION (LVQ) NNW
- PROBABILISTIC NNW (PNN)
- FUZZY NNW (FNN)

2. RECURRENT NNW
- REAL-TIME RECURRENT NNW
- ELMAN NNW
- HOPFIELD NNW
- BOLTZMANN MACHINE
- KOHONEN'S SELF-ORGANIZING FEATURE MAP (SOFM)
- RECIRCULATION NNW
- BRAIN-STATE-IN-A-BOX (BSB)
- ADAPTIVE RESONANCE THEORY (ART) NNW
- BIDIRECTIONAL ASSOCIATIVE MEMORY (BAM) NNW

The NNWs can be classified as feedforward and feedback or recurrent networks [1], some of which are listed in this figure. In the former type, the signals flow only in the forward direction (see Figure 11.4), whereas in a recurrent network, the signals can flow forward, backward, or laterally. For static mapping, the feedforward networks are important, whereas for dynamic or temporal mapping, the recurrent networks are important. For power electronics applications, currently, the backprop and real-time recurrent networks are the most important types. Therefore, these networks and their applications will be discussed in detail. The perceptron and Adaline/Madaline will also be briefly reviewed.

FIGURE 11.6 (a) Single-layer perceptron network and (b) illustration of pattern classification boundary for upper perceptron only.

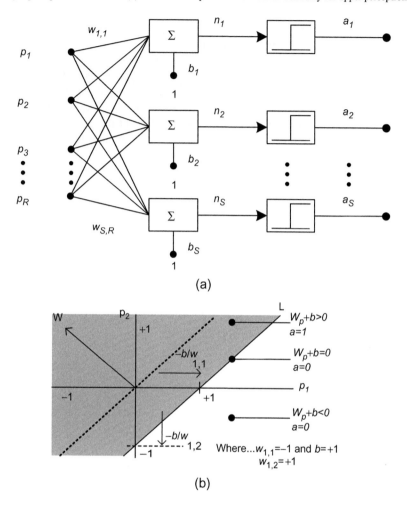

(a)

(b)

The simple perceptron NNW [5] is used for classification of input signal patterns that are linearly separable. It was invented by Rosenblatt in the 1950s. The structure of the single-layer NNW with S neurons that use a hard limit transfer function (TF) is shown in part (a). Note that each neuron (or perceptron) is basically represented by a summer and TF only. The network input signal vector p has R elements (p_1, p_2, ..., p_R) and is multiplied by the weight matrix W ($R \times S$) before summation. An element of W is given by $W_{i,j}$, where the weight connects a j input element with the ith neuron. The resulting Wp vector is added with the bias vector b (b_1, b_2, ..., b_S) to constitute the vector $n = Wp + b$. Each bias signal of a neuron is generated from a source $+1$ through a weight as shown. The output vector a is then given by $a = \text{hardlim}(Wp + b)$. The weights in NNW are normally indicated by dots, but are not shown here for simplicity. The pattern classification property for the upper neuron only is shown in (b) for two-dimensional (p_1 and p_2) input only, where $w_{1,1} = -1$ and $w_{1,2} = +1$. The shaded area shown in the figure classifies the p_1 and p_2 inputs that give +1 output, whereas the corresponding unshaded area gives 0 output. The "decision" boundary line is denoted by line L. The bias shifts the boundary line on the horizontal axis with the slope remaining the same. For more than two input elements, the classification boundary is given by a hyperplane. The boundary hyperplane for each perceptron can be designed to be different by assigning different weights and biases. A computer-based training algorithm (perceptron learning rule) can design all the weights automatically for the desired classification boundaries. A perceptron network can be used, for example, in the identification of a drive response state (steady state, rise time state, oscillation state, and disturbance state) by observing the command and response signals at the input. An example design for a five-input NAND gate will be discussed later.

FIGURE 11.7 (a) Adaline/Madaline network and (b) decision boundary of upper Adaline.

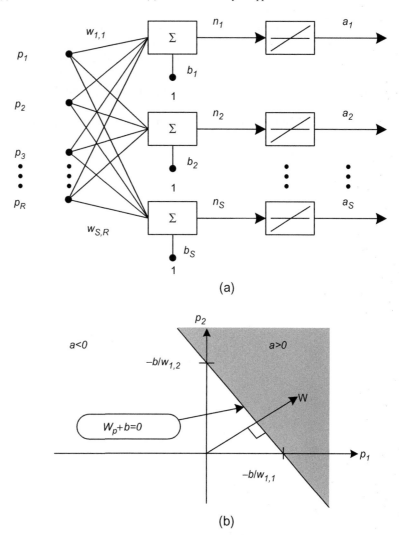

(a)

(b)

The linear NNW shown in (a) has a perceptron-like architecture, but its transfer function is linear rather than hard-limiting. A single-channel linear NNW is known as Adaline (adaptive linear network), whereas the multichannel NNW as shown is known as multiple Adalines or a Madaline [5]. The NNW output can be given as $a =$ purelin $(Wp + b)$, where $p =$ input vector, $W =$ weight matrix, $b =$ bias vector, and purelin is the bipolar linear transfer function. Therefore, the output can have any value between -1 and $+1$ instead of 0 or $+1$ in a perceptron. The linear NNW can give linear input–output mapping only. This means that the NNW is useful for linearizing nonlinear functions (linear function approximation) or pattern associations. The decision or classification boundary for the upper Adaline with p_1 and p_2 input only is shown in (b), where $a = w_{1,1}p_1 + w_{1,2}p_2 + b$. The shaded area in the lower figure shows positive output ($a < 0$), the unshaded area gives negative output ($a < 0$), and the boundary line gives zero output ($a = 0$). With more than two inputs, the boundary line will be converted to a boundary hyperplane, and different hyperplanes can be designed for different channels by adjusting the weights and biases. A linear NNW can be designed directly or trained with a computer algorithm so that for different input vectors, the error between the corresponding output and desired (or target) vectors follows the least-minimum-squares (LMS) algorithm.

FIGURE 11.8 Three-layer backpropagation network (3-5-2).

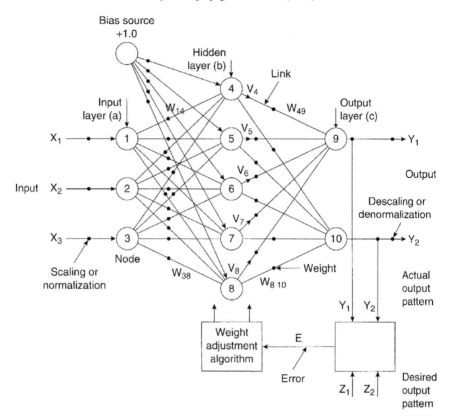

The type of NNW most commonly used in power electronics is the feedforward multi-layer backpropagation (backprop or BP) network shown in this figure. The term *back-propagation* comes from the method of supervised training used for the NNW shown by the two extra blocks in the figure. The network is commonly called a *multi-layer perceptron* (MLP), although the TF can be different from the threshold function. The NNW shown has three input signals (X_1, X_2, and X_3) and two output signals (Y_1 and Y_2). The circles represent the neurons that have associated TFs (not shown), and the weights are indicated by dots (often omitted). The network shown has three layers: an input layer, hidden layer, and output layer. With five neurons in the hidden layer as shown, the NNW is defined as a 3-5-2 network. The input layer shown is nothing but the nodes that distribute signals to the middle layer. Therefore, this topology is often called a two-layer network. If the signals are bipolar, the hidden layer neurons usually have a hyperbolic tan TF and the output layer has a bipolar linear TF. On the other hand, for unipolar signals, these TFs can be sigmoidal and unipolar linear, respectively.

Occasionally, the output layer has a nonlinear TF also. The signals within the NNW are processed in a per-unit manner. Therefore, there is input scaling or normalization and output descaling or denormalization as shown. A constant bias source normally links all of the neurons through weights, but in this figure, the bias connection is not shown for the output layer. The network output signals can be continuous or clamped to 0, 1 or −1, +1 levels. Theoretically, a 3-layer NNW is capable of approximating any function with any desired degree of accuracy (universal function approximation), but practically, often more than one hidden layer is used.

FIGURE 11.9 Features of backpropagation networks.

- LOGICAL OR CONTINUOUS INPUT AND OUTPUT SIGNALS, WHICH CAN BE UNIPOLAR OR BIPOLAR
- SYNAPTIC WEIGHTS CONSTITUTE DISTRIBUTED INTELLIGENCE – SIMILAR TO HUMAN MEMORY OR INTELLIGENCE
- NONLINEAR INPUT–OUTPUT MAPPING OR PATTERN RECOGNITION PROPERTY
- FAST PARALLEL COMPUTATION BY MEANS OF AN ASIC CHIP, INSTEAD OF SLOW SEQUENTIAL COMPUTATION BY A DSP
- FAULT-TOLERANCE PROPERTY
- NOISE IMMUNITY PROPERTY
- REQUIRES SUPERVISED TRAINING BY EXAMPLE DATA SETS
    – SIMILAR TO SUPERVISED ALPHABET TRAINING FOR A CHILD
- BACKPROPAGATION TRAINING ALGORITHM BY A COMPUTER PROGRAM

Considering the importance of backprop networks, its features (some of which were discussed before) are summarized here. Although the network handles continuous signals, the input and output signals can be continuous, logical, or discrete bidirectional. The NNW is analogous to a biological neural network, as mentioned before. Like a biological network, where the intelligence or memory is contributed in a distributed manner by the synaptic junctions of neurons, the NNW synaptic weights contribute similar distributed intelligence. This intelligence permits the basic input–output mapping or pattern recognition property of NNWs. This is also the associative memory property, which is similar to that of a biological nervous system. The NNW, when implemented with an ASIC chip, provides a massive parallel and fast computational capability that has fault tolerance and noise immunity, as mentioned before. Of course, a NNW can also be implemented in a serial manner by a DSP. One unique characteristic of NNWs is that they require supervised training with an example data table instead of the traditional programming used by a computer to solve a problem. This is analogous to the supervised alphabet training of a child with the help of a tutor. After training, the child recognizes the alphabet characters (pattern or character recognition) when exposed to them. The backpropagation training block diagram was indicated in the previous figure. The network is exposed to the input example data sets and the corresponding outputs are calculated. The deviation from the desired or target output data is calculated, and the weights are altered by a gradient descent technique so that the error converges to a minimum. Normally, a computer program (such as the MATLAB-based Neural Network Toolbox [5]) performs the training to determine the weights. The toolbox essentially helps to design and test the NNW, which will be discussed later.

FIGURE 11.10   Flowchart for training backpropagation networks.

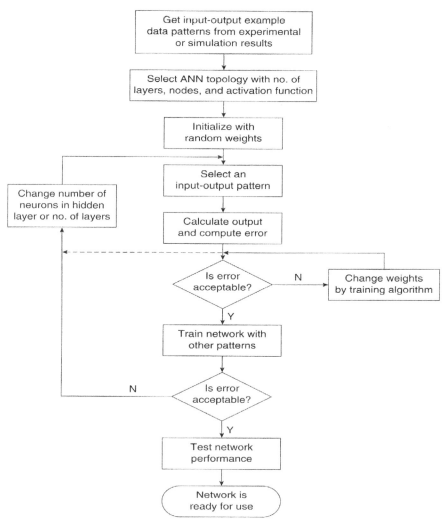

$$SSE = \sum_{p\text{-}1}^{P} \sum_{q\text{-}1}^{Q} [D(q)\text{--}a(q)]^2 \; ....(1), \; P = \text{Patterns}, \; Q = \text{output neurons}$$

$$MSE = 1/Q \; [SSE]$$

The flowchart for backpropagation training is shown in this figure. In the beginning, input–output example data patterns are obtained from the experimental results, or from simulation results if simulation is possible with the mathematical model of the plant. An initial NNW configuration is created with the desired input and output layer neurons dictated by the number of signals, a hidden layer with a few neurons, and appropriate TFs. Small random weights are selected so that neuron outputs do not get saturated. With one input pattern, the output is calculated (defined as forward pass) and compared with the desired output pattern. The weights are then altered in the backward direction by a backpropagation algorithm until the error between the calculated pattern and the desired pattern is very small and acceptable. A round-trip (forward and reverse passes) of calculations is defined as an *epoch*. Similar training is repeated with all the patterns so that the total SSE (sum of squared error) given by Eq. (1), i.e., the objective function is minimum and acceptable. Sometimes, mean square error (MSE) is taken as the objective function. If the error does not converge sufficiently, it may be necessary to increase the number of neurons in the hidden layer or to add an extra layer. However, if the hidden layer neurons are too many, it can result in overfitting of the training data leading to poor generalization. Instead of selecting one pattern at a time, a batch method of training may be used, in which all the patterns are presented to the network simultaneously. Because the training is time consuming, it is normally done off-line. A number of backpropagation training methods have been proposed, but the Levenberg-Marquardt (L-M) algorithm is frequently used because of its fast convergence. Once the NNW is trained properly, it should be adequately tested with intermediate data to verify that training is correct and complete. The entire design can be done, for example, with the help of the MATLAB-based Neural Network Toolbox, which will be described later.

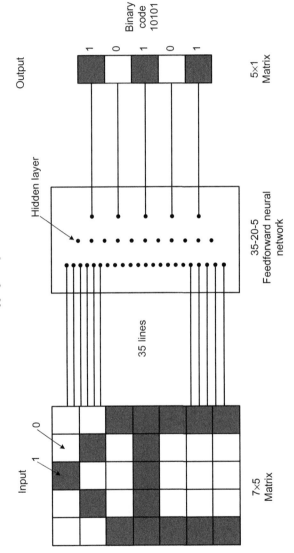

FIGURE 11.11 Mapping of input letter "A" by 5-bit code.

Before discussing any power electronics application of NNWs, let us review the popular optical character recognition (OCR) problem with a backprop network. Here the problem is to represent 32 alphabet characters with a 5-bit binary code so that each code represents a character. The figure shows that the letter A is being coded by 10101. In this case, A is represented by a 5×7 matrix of inputs consisting of logical 0s and 1s, where the shaded square corresponds to 1 and the unshaded square corresponds to 0. Thus, the input vector of 35 logical signals is connected to the input of a 35-20-5 backprop network that does not use any bias signals. The NNW uses sigmoidal TFs in all the neurons. The input–output mapping is performed by supervised learning (or tuning) of all the weights to the desired values. Altogether, the network has 800 weights, which give it 800 degrees of freedom for pattern mapping. The letter B is coded by 10001. After training with A, if B is given as input, the network output will be totally distorted. Another round of training tunes the weights so that B gives the desired code. However, if A is now impressed again, the output will be distorted. The network is trained back and forth with all 32 input vectors and the corresponding target binary codes so that each vector generates the corresponding code after successful training. Evidently, the nonlinearity of the network with so many degrees of freedom and logical clamping at the output permit such recognition of patterns.

FIGURE 11.12   (a) Inverse mapping of the letter "A" and (b) autoassociative mapping of "A."

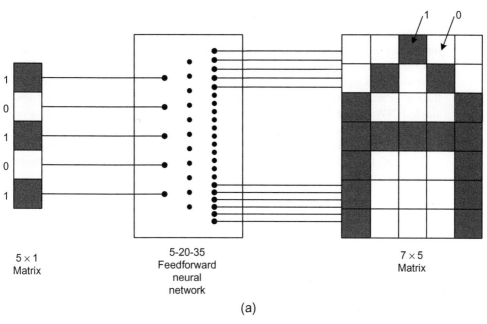

(a)

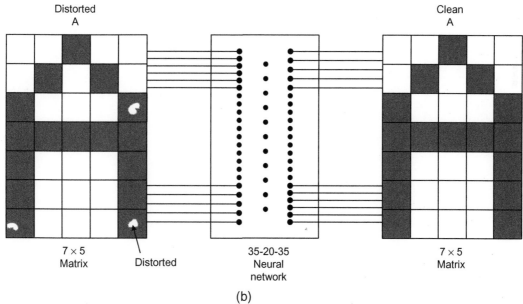

(b)

Instead of mapping the alphabet characters by a 5-bit code as discussed in the previous figure, it is possible to use inverse mapping as shown in part (a). Here, the input vector, for example, is 10101 and the output character is the letter A. In this way all of the 5-bit codes for the 32 numbers can generate the respective letters of the alphabet. The method of training for the 5-20-35 backprop network is the same as before. It is possible to cascade Figure 11.11 and part (a) of this figure so that the same letter is produced. This arrangement permits transmission of alphabet characters over a distance through a narrowband or compressed data channel. Instead of cascading as above, a single NNW with a 35-20-35 configuration as shown in (b) is possible. This type of network is called an *autoassociative network* instead of a *heteroassociative NNW,* as discussed earlier. The advantage of an autoassociative NNW is that if the input pattern is distorted, the output pattern will be clean because the network has been trained to produce the nearest crisp pattern. This is the noise filtering property of ANNs.

FIGURE 11.13   Generalized three-layer backpropagation network.

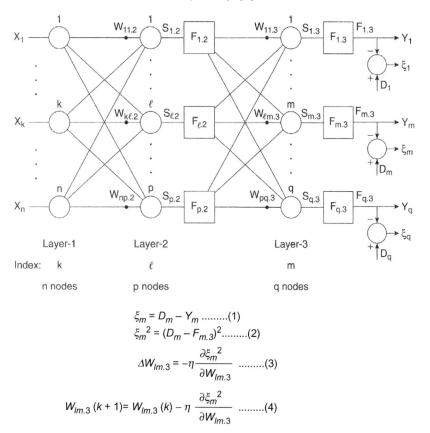

$$\xi_m = D_m - Y_m ....(1)$$

$$\xi_m{}^2 = (D_m - F_{m.3})^2 ....(2)$$

$$\Delta W_{lm.3} = -\eta \frac{\partial \xi_m{}^2}{\partial W_{lm.3}} ....(3)$$

$$W_{lm.3}(k+1) = W_{lm.3}(k) - \eta \frac{\partial \xi_m{}^2}{\partial W_{lm.3}} ....(4)$$

The principle of a backpropagation algorithm will be explained by tuning one weight in the generalized three-layer network shown in this figure. The input layer, or layer 1, has n neurons (or nodes); layer 2, or the hidden layer, has p neurons; and layer 3, or the output layer, has q neurons. The bias network is not shown. The sigmoidal type TF shown in block F is considered in the hidden and output layers, but there is no TF in layer 1. Consider the neuron m in the output layer and the problem of tuning its input weight $W_{lm.3}$, which receives signal $F_{l.2}$ from neuron l. Assume that all other input signals to m remain constant. The output of neuron m ($F_{m.3}$) is compared with desired or target value D_m to calculate the error ξ_m given by Eq. (1), where $Y_m = F_{m.3}$. The square error function to be minimized is given by Eq. (2). The output $F_{m.3}$ and the corresponding $\xi_m{}^2$ will vary with the variation of weight $W_{lm.3}$. To minimize $\xi_m{}^2$ by the gradient descent method, the change in weight $\Delta W_{lm.3}$ must satisfy Eq. (3), where η = learning rate. Therefore, the new weight expression is given by Eq. (4). These steps should be repeated until $\xi_m{}^2$ falls to minimum value (see next figure). A similar procedure is adopted to adjust all of the weights of the output layer and then the input layer (one at a time) so that the SSE falls to a minimum value.

FIGURE 11.14   Minimization of square error (ξ_m^2) by the gradient descent method.

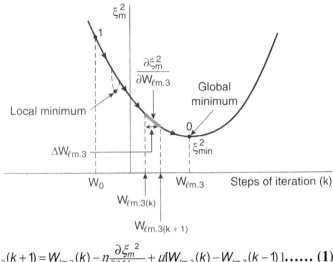

$$W_{lm.3}(k+1) = W_{lm.3}(k) - \eta \frac{\partial \xi_m^2}{\partial W_{lm.3}} + \mu[W_{lm.3}(k) - W_{lm.3}(k-1)] \dots \dots \textbf{(1)}$$

η = learning rate, μ = momentum factor

This figure illustrates the gradient descent method of minimization of ξ_m^2 by adjusting the weight $W_{lm.3}$, as discussed in Figure 11.13. Assume that the initial operating point is 1 and the corresponding weight is W_0. The weight is increased in steps until the operating point goes to 0, where ξ_m^2 is at global minimum (ξ_{min}^2). The learning rate η determines the speed of convergence. Initially, it is selected to be high and then gradually decreases to a low value (adaptive). However, a value for η that is too high can saturate the neuron outputs. The local or false minimum as shown is normally avoided by the momentum factor μ shown in Eq. (1).

FIGURE 11.15 (a) Structure of real-time recurrent network and (b) block diagram for training.

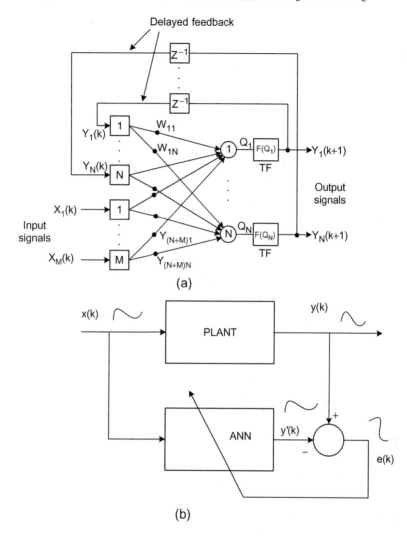

(a)

(b)

So far, we have discussed only feedforward NNWs, which give only static input–output mapping. In many applications, the NNW is required to be dynamic, i.e., it should emulate a dynamic system with temporal behavior, such as identification of a machine model or estimation of speed or flux, etc. A recurrent or feedback network with time delay (Z^{-1}) as shown in this figure can emulate a dynamic system. The output in this case not only depends on the present input, but also prior inputs, thus giving the temporal behavior of the network. If, for example, the input is a step function, the response will reverberate in the time domain until a steady-state condition is reached at the output. The network can emulate nonlinear differential equations that are characteristic of a nonlinear dynamic system. Of course, if the TFs of the neurons are linear, it will represent a linear system. Such a network can be trained by a dynamic backpropagation (real-time temporal supervised learning) algorithm, where the desired time domain output from the reference dynamic system can be used to force the NNW output to track by tuning the weights dynamically sample by sample as indicated in part (b). To be clear, consider a one-input, one-output network that is desired to emulate a series nonlinear *R-L-C* circuit (plant). Apply a step voltage signal to the NNW and the *R-L-C* circuit simultaneously. The current response in the *R-L-C* circuit is the target signal that is used to tune the NNW weights. Then the NNW will emulate the *R-L-C* circuit model. This is defined as identification of a dynamic system. This NNW has been used in power electronic systems, which will be discussed later.

FIGURE 11.16   Time–delayed NNW (TDNNW) with tapped delay line.

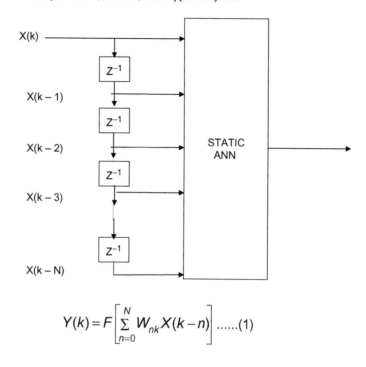

$$Y(k) = F\left[\sum_{n=0}^{N} W_{nk} X(k-n)\right] \ldots\ldots(1)$$

The dynamic NNW plays an important part in the identification and control of dynamic systems as mentioned earlier. Besides the RNN in Figure 11.15, another topology that is popular in a dynamic system is the time-delayed neural network (TDNNW) shown in this figure. In this case, the single input $X(k)$ is fed to a multiple-input static NNW through a tapped delay line. The tapped delay line generates a sequence of signals with unit time delay (Z^{-1}). The signals are multiplied by the respective weights and generate the output $Y(k)$ through the TF (assuming one output neuron only) as shown in Eq. (1). The equation represents an nth-order nonlinear dynamic system that is expressed in finite difference form. Note that there is no feedback from output to input. The network can be trained by a static backpropagation method for the desired dynamic function. Equation (1) can represent a linear dynamic system if the TF is linear. The TDNNW has also been used as a nonlinear predictor [1], adaptive transverse filter, and FFT analyzer of a wave. In the latter case, for example, sequentially sampled distorted current waves can be presented to a multilayer NNW that can be trained to generate the fundamental amplitude for each wave.

FIGURE 11.17   Dynamic plant identification by NNW with time-delayed input and output.

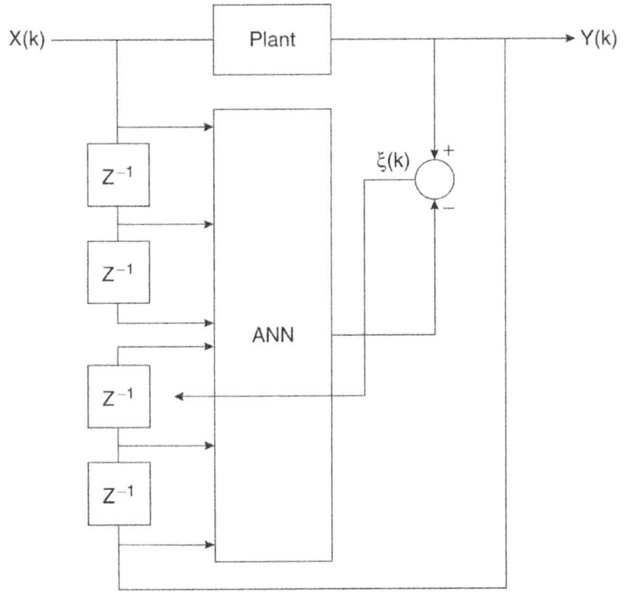

$$Y(k) = F[X(k), X(k-1), X(k-2), Y(k-1), Y(k-2)]\ldots\ldots(1)$$

In this case, the NNW is required to emulate the dynamic system given in the form of Eq. (1), where there are time-delayed inputs as well as time-delayed outputs as feedback signals. The NNW configuration is shown in the figure. This is defined as a series-parallel configuration for a NNW instead of a parallel configuration, where the feedback signal $Y(k)$ is generated directly from the NNW output. The training convergence of this topology is somewhat easier. The NNW is trained from the input–output temporal data of the plant by the dynamic backpropagation method. The training data can be generated experimentally from the plant or from simulation if a mathematical model of the plant is available. Note that this type of plant model identification in this and previous figures is done off-line. If the plant parameters vary, the model is not valid. In such a case, on-line identification or a NNW with adaptive weights (or adaptive NNW) is essential. On-line tracking may be difficult because of the time delay involved in the training.

FIGURE 11.18   (a) Training of inverse dynamic model for a plant and (b) inverse dynamic model-based adaptive control for a plant.

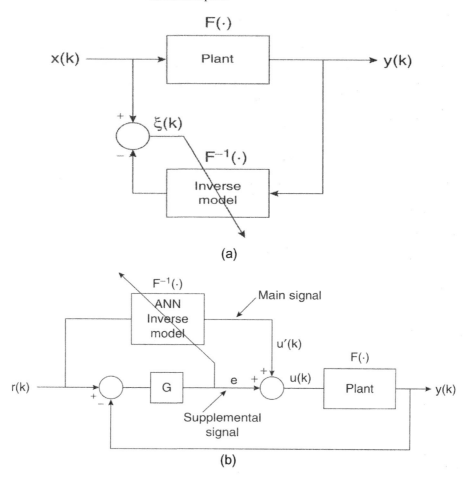

(a)

(b)

So far, we have discussed identification of a forward dynamic model for a plant. It is also possible to identify the inverse plant model (F^{-1}) by training as shown in (a). In this case, the plant response data $Y(k)$ is impressed as the input of the NNW, and its calculated output is compared with the plant input, which is the target data. The resulting error $\xi(k)$ trains the network so that $\xi(k)$ falls to the acceptable minimum value. After satisfactory training and testing, the NNW represents the inverse dynamic model of the plant. This NNW-based inverse model can be placed in series as a controller with the actual plant as shown in (b) so that the plant forward dynamics is totally eliminated, i.e., $F^{-1} \cdot F(\cdot) = 1$. Then, ideally the output signals follow the input signals and no feedback control is necessary. However, the actual output will deviate from the input because of an imperfect inverse model and/or plant parameter variation effect. The feedback control shown generates the supplemental error signal e for the control, which can also be used for on-line tuning of the NNW.

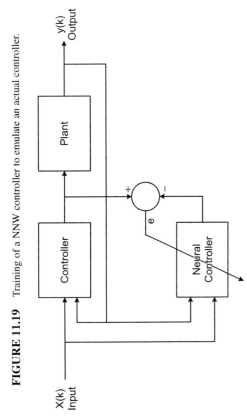

FIGURE 11.19   Training of a NNW controller to emulate an actual controller.

The controller in the feedback system shown in this figure can be simple or complex. It can be static or dynamic. Whatever it is, a neural controller can be trained to emulate the actual controller and then the NNW can substitute the controller. Once the neural controller has been trained and replaced the controller, it can be retrained on-line for plant parameter variation to make it adaptive. Besides, there is the advantage of ASIC chip-based NNW implementation of the controller instead of conventional DSP implementation . A fuzzy controller, for example, the one shown in Figure 10.42 (FLC-3), which is static and nonlinear, can be *P-I*, *P-D*, or *P-I-D* type. The controller can be easily replaced by a NNW since both FL and NNW have similar nonlinear input–output mapping characteristics.

FIGURE 11.20 Model referencing adaptive control (MRAC) via a neural network: (a) direct method and (b) indirect method.

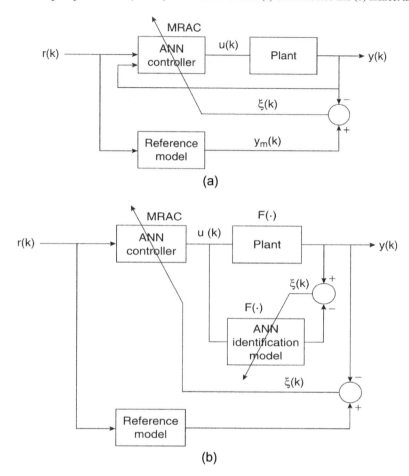

(a)

(b)

The MRAC is an adaptive control method that was discussed in Chapter 7. Figure 11.20(a) shows MRAC-based direct control, where the plant output along with the NNW controller is desired to track the dynamic response of the reference model. The reference model can be represented by a dynamic equation and solved in real time by a DSP. The error signal $\xi(k)$ between the actual plant and the reference model trains the NNW controller on-line so as to make the tracking error zero. Therefore, the plant with the NNW has a response that is identical to that of the reference model. One problem of the direct method is that the plant lies between the controller and the error, and there is no way to propagate the error backward in the controller by backpropagation training. This problem is solved in (b) by the indirect method. In this case, the NNW identification model $F(\cdot)$ is first generated to emulate the forward model of the plant. This model is then placed in series with the NNW controller to track the reference model as shown. The tuning of the NNW controller is now convenient through the NNW model.

FIGURE 11.21   Adaptive neuro-fuzzy inference system (ANFIS).

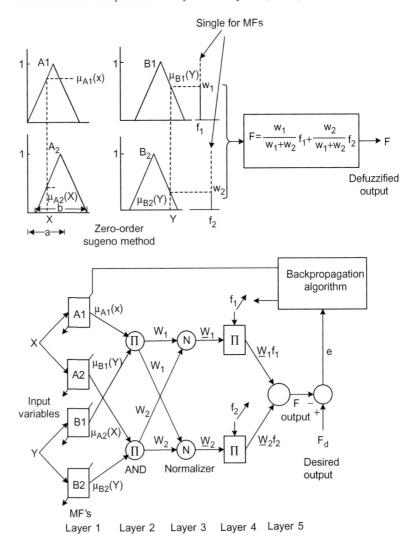

The ANFIS [6], as shown in this figure, is basically a neural network-based training or designing of a fuzzy system. This means that if the desired input–output data sets are available for a fuzzy system, the MFs and rule table for a fuzzy model can be designed using the NNW training method. The simple zero-order Sugeno system, described in Figure 10.17, is used in this case. Assume that X and Y are the input variables, A_1-A_2 and B_1-B_2 are the triangular MFs for input variables, f_1-f_2 are the singleton MFs for the output variables, and F is the defuzzified output as shown in the upper right of the figure. Each symmetric triangle is characterized by parameters a (peak) and b (support). The feedforward ANFIS structure is shown in the lower figure, where the functions A_1, A_2, B_1, B_2, f_1, and f_2 are being tuned by a backpropagation algorithm for the desired output F_d. The ANFIS has five layers of computation. Layer 1 generates the membership degrees from the triangular MFs, layer 2 generates DOFs with the AND logic of the rules, layer 3 normalizes the output (i.e., $\underline{w}=w_1/(w_1+w_2)$), layer 4 generates the consequent parameters by multiplying the input with the f function, and layer 5 sums all the component outputs. The calculated output F is compared with the desired value F_d, and the error signal is used to train the network parameters as shown. The MATLAB/Fuzzy Logic Toolbox can be used to design the ANFIS. Instead of a zero-order system, a Sugeno first-order system can also be used in ANFIS. The ANFIS can be used, for example, as a fuzzy controller in a feedback system, where $X = E$, $Y = CE$, and $F = DU$ (see Figure 10.21). The network can be trained off-line or on-line.

FIGURE 11.22   Some typical applications for neural networks in power electronics.

- PULSE WIDTH MODULATION IN INVERTER*
- MULTIDIMENSIONAL NONLINEAR LOOK-UP TABLE*
- WAVEFORM PROCESSING AND DELAYLESS FILTERING*
- ESTIMATION FOR DISTORTED WAVEFORMS*
- WAVEFORM FFT ANALYSIS
- ADVANCED ADAPTIVE CONTROL OF DRIVES*
- VECTOR CONTROL OF DRIVE AND FEEDBACK SIGNAL ESTIMATION*
- ON-LINE DIAGNOSTICS AND FAULT-TOLERANT CONTROL
- NEURO-FUZZY-BASED VECTOR AND DTC CONTROL*

Applications of neural networks in power electronics and motor drives are still in their infancy, and it appears that future growth potential is tremendous. Currently, a majority of the applications use a feedforward backpropagation type network, and only a few use dynamic NNWs. The application of other types of feedforward and recurrent networks remains practically unexplored. In this chapter, we will discuss the applications listed in this figure that are marked by an asterisk (*). Pulse width modulation, particularly space vector type (SVM), will be discussed in detail for two-level, three-level, and five-level voltage-fed converters. The nonlinear look-up table type applications are very obvious. Extending the same concept of nonlinear input–output mapping, the application of NNWs in waveform processing and delayless filtering, distorted waveform estimation, vector control of drives, and feedback signal estimation will be discussed. The real-time RNN application will be illustrated by adaptive estimation of stator flux vectors in a variable-frequency induction motor drive. Vector drive with neuro-fuzzy control and adaptive neuro-fuzzy system (ANFIS)-based DTC control will be discussed briefly at the end. Many other innovative applications of NNWs are left to the imagination of the readers.

FIGURE 11.23   Selected harmonic elimination (SHE) PWM via a neural network.

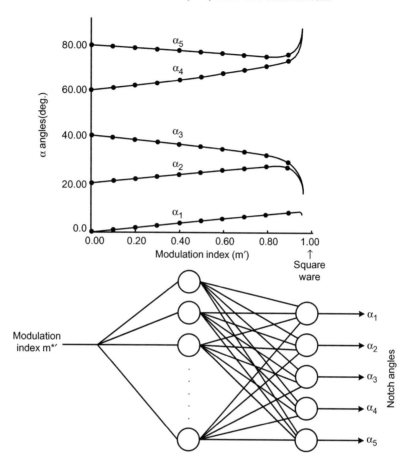

The selected harmonic elimination method of PWM was discussed in Chapter 4. In this method, the notch angles of a PWM wave are precomputed and stored in the form of a look-up table in DSP memory. The upper part of this figure shows the plot of five notch angles (α_1–α_5) as a function of modulation index m', where $m' = 1$ corresponds to a square wave. With five α angles, four lower order significant harmonics (such as 5th, 7th, 11th, 13th) can be eliminated and the fundamental voltage can be controlled. A feedforward NNW [9] can be trained as shown to generate the appropriate α angles as a function of m'. This is basically five-dimensional look-up table generation, and is one of the simplest applications of NNWs for input–output nonlinear mapping. The advantage of NNWs is that the α angles are easily interpolated by the NNW avoiding the need for a very large precision look-up table by DSP. The network (1-8-5) uses a sigmoidal TF for each neuron. The trained NNW was actually implemented in Intel's ETANN (Electrically Trainable Analog Neural Network) ASIC chip type 80170NX. The chip has 64 neurons and 10,240 EEPROM-based analog synaptic weights. However, because of a signal drift problem, the chip was recently withdrawn from the market.

FIGURE 11.24   (a) Square-wave delayless filtering and (b) NNW training MSE curve.

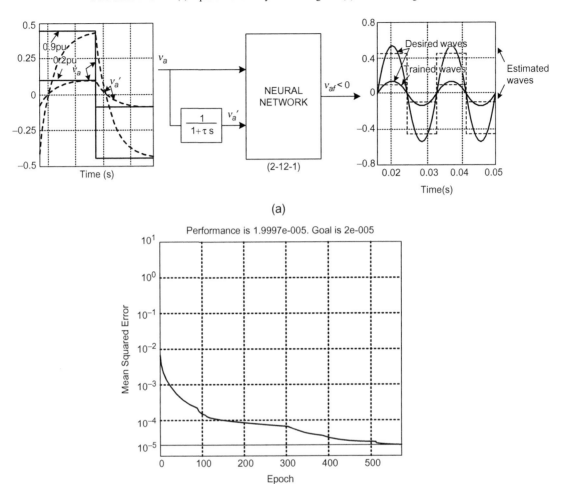

(a)

(b)

A neural network can be used for delayless filtering [10] of a harmonic-rich wave as illustrated in this case. A conventional low-pass filter (LPF) causes frequency-sensitive phase delay and attenuation at the output. A constant-frequency (60-Hz) square-wave line current which is usually encountered for a thyristor or diode rectifier is considered in this case, although other types of waves can also be considered. The wave (v_a) and its auxiliary form (v_a') through an LPF are given as input to the NNW as shown in (a). The output $(v_{af} < 0°)$ comes out as a sine wave at the same frequency and proportional to the square-wave amplitude but is locked at $0°$ phase. The auxiliary wave is needed because with constant–amplitude, square-wave input, variable-amplitude sine-wave output is not possible. The feedforward ANN was trained off-line with actual v_a and v_a' data at input and the desired output sine wave at 10% amplitude steps and 0.72° angular intervals. The desired wave was generated by prior FFT analysis of the input square wave. The MSE (mean square error) approaches the target value at the end of 600 epochs as shown in (b). The NNW (2-12-1) uses a bipolar linear TF at the input and output layers, but a tan-sig TF is used in the hidden layer, and all signals are processed in a normalized manner. Similar delayless filtering is also possible by TDNNW (Figure 11.16). The output will be distorted if the NNW is exposed to any unrecognized signals. This means that the network should be trained for all the possible wave shapes that might be encountered. Otherwise, on-line training is required. Of course, a small deviation in frequency and distortion from the trained wave will result in a tolerable error.

FIGURE 11.25   Desired and estimated output waves showing noise filtering effect: (a) 2-12-1 NNW and (b) 2-15-15-1 NNW.

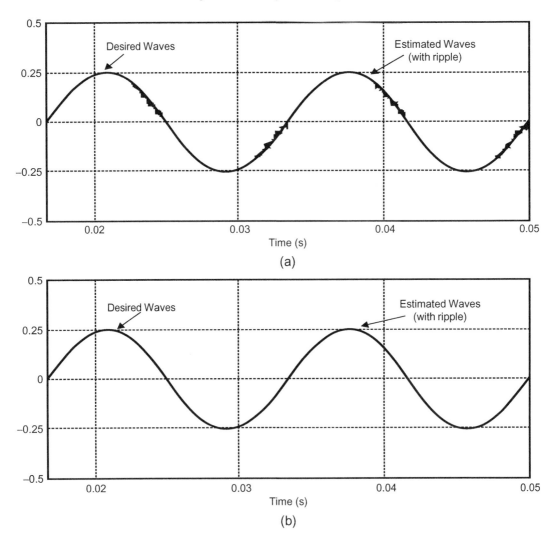

These waves demonstrate the noise filtering capability of NNWs. In Figure 11.24(a), the square-wave input signal was mixed with a ±1.0% noise signal at 2.0-kHz frequency. The top wave shows that with the same 2-12-1 network, the desired and estimated output waves nearly match, but the actual MSE was 1.9997e–005, whereas the goal was 2e–005. In (b), with a more complex four-layer topology (2-15-15-1), the MSE could be reduced to 1.1739e–005, i.e., the wave matches better with the desired wave. Note that one hidden layer is sufficient for universal function approximation. However, the training time and total number of neurons may be less in some cases with two hidden layers.

FIGURE 11.26   Single-phase, square-wave delayless filtering and multiphasing at (a) three-phase output and (b) six-phase output.

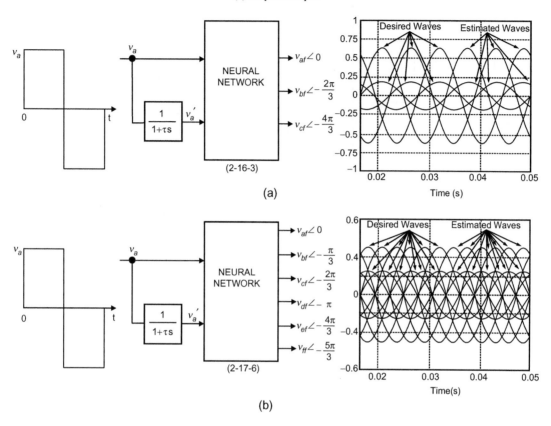

(a)

(b)

A NNW cannot only function as a delayless filter, but also as a multiphase sine-wave generator at output with symmetrical phase-shift angles. This is illustrated in this figure with single-phase square-wave input. In both (a) and (b), the output sine waves vary linearly with the square-wave amplitude, and the $v_{af}<0°$ wave remains locked in phase with the square wave. The LPF at the input in each case generates the variable-amplitude auxiliary wave (v_a') as usual. This type of waveform processing is important for three-phase 60-Hz (or any constant frequency) power supply with sinusoidal PWM inverter or cycloconverter, where the command signal waves can be generated from a simple square-wave signal. In fact, any arbitrary waveform can be transformed to an arbitrary output wave at single or multiphase with the help of a NNW, and the input–output magnitude tracking relation can be programmed to be linear or nonlinear with arbitrary functional relation. This study also suggests that NNWs can be used as zero-crossing detectors (ZCDs) for a distorted wave.

FIGURE 11.27   Two-phase square-wave delayless filtering and multiphasing by NNW at (a) three-phase output and
(b) six-phase output.

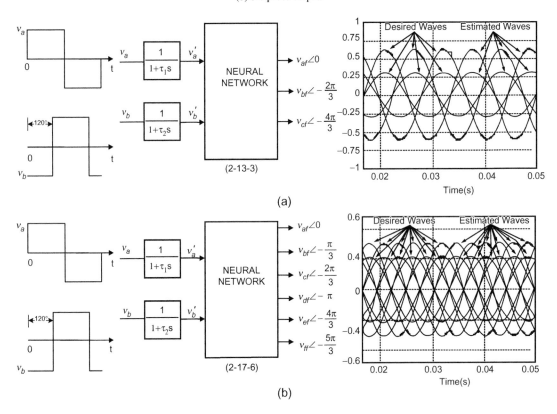

(a)

(b)

Let us consider two-phase, square-wave input at 60 Hz, where the phase difference is
120° as shown. With the same value of τ ($\tau_1 = \tau_2 = 0.002$ sec), it can be shown that the
v_a' (or v_b') fundamental component has a phase lag of 37° and a magnitude attenuation
of 0.7986. This means that the output sine waves should not only be filtered by the
NNW, but the additional phase and amplitude deviation due to the LPF are to be compensated by it. If, on the other hand, the input wave is varying continuously instead of
being a flat-top, no LPF is necessary. The NNW training was such that the output varies
linearly with the input magnitude. Signals v_a and v_b may be for a balanced three-phase
system, where $v_a + v_b + v_c = 0$. An example application for this type of waveform processing is retrieving the three-phase fundamental components from distorted two-phase
voltage or current waves of an UPS system for control and feedback signal estimation.

FIGURE 11.28   Three-phase, square-wave delayless filtering and multiphasing by NNW at (a) three-phase output and
(b) six-phase output.

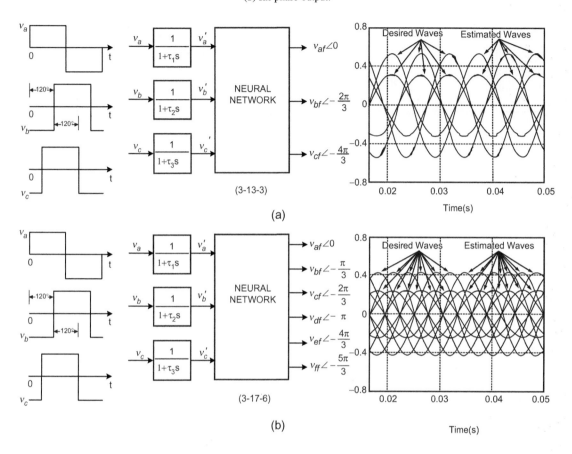

Continuing our studies further, let us consider three-phase, square-wave input to the
NNW and the corresponding generation of polyphase output with delayless filtering as
shown in this figure. Again, instead of square waves, six-step or other wave shapes are
also possible. For waveform processing and delayless filtering, it is better to consider
all three phases instead of two-phase signals. In this figure, all LPFs are identical
($\tau_1 = \tau_2 = \tau_3 = 0.002$ sec). The matching of desired and estimated waves is excellent in
the whole magnitude range (0–1.0 pu) after 1,000 epochs of training. As a conclusion
of this study, we can generalize that N-phase input signals of any wave shape can be
processed into M-phase sine waves or any arbitrary waveform at the same frequency
with desired phase angle and amplitude variation with the help of NNWs. This follows
from the generalized input–output static nonlinear mapping property of a feedforward
network. Complex polyphase waveform processing is important in multiphase rectifiers
and inverters. The fact is that NNWs must recognize the waveform. Any unrecognized
waves give erroneous output. This is also a property of biological NNWs.

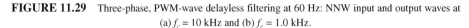

FIGURE 11.29 Three-phase, PWM-wave delayless filtering at 60 Hz: NNW input and output waves at (a) $f_c = 10$ kHz and (b) $f_c = 1.0$ kHz.

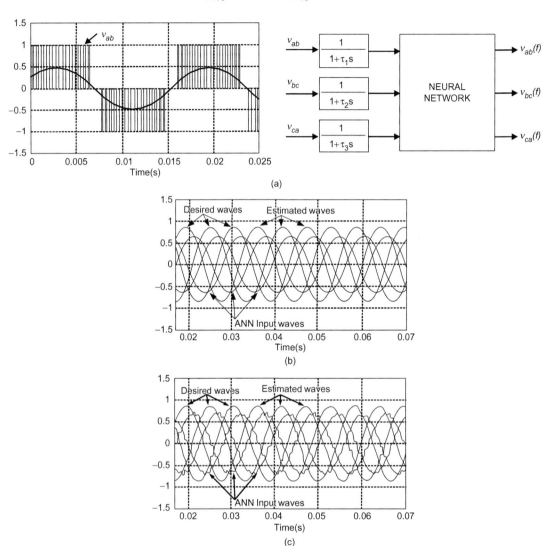

Let us consider a three-phase induction motor excited by a three-phase SPWM inverter, where the fundamental frequency is 60 Hz, but the amplitude is variable. The v_{ab}, v_{bc}, and v_{ca} are the line voltage waves (only v_{ab} is shown). It is desirable to perform delayless filtering of the PWM waves with the help of NNWs as shown in (a). The fundamental voltages at zero phase shift angle permit correct calculation of feedback signals for a vector drive. The PWM waves are initially filtered by identical LPFs as shown to convert them into continuous waves before processing through the NNW. In part (b), the NNW input waves, and the corresponding desired and estimated waves, are shown for a 10-kHz carrier frequency. The NNW performs the functions of (1) compensation of phase delay due to LPF, (2) compensation of amplitude attenuation due to LPF, and (3) additional harmonic filtering. The same waves are shown in (c) for carrier frequency of 1.0 kHz, where the additional distortion of NNW input and output waves is evident. With a more complex NNW structure, the harmonic quality in (c) can be made to match that of (b).

FIGURE 11.30   Three-phase, PWM-wave delayless filtering at variable frequency and variable voltage (f_c = 5 kHz):
(a) f = 60 Hz, (b) f = 30 Hz, (c) f = 15 Hz, and (d) f = 5 Hz.

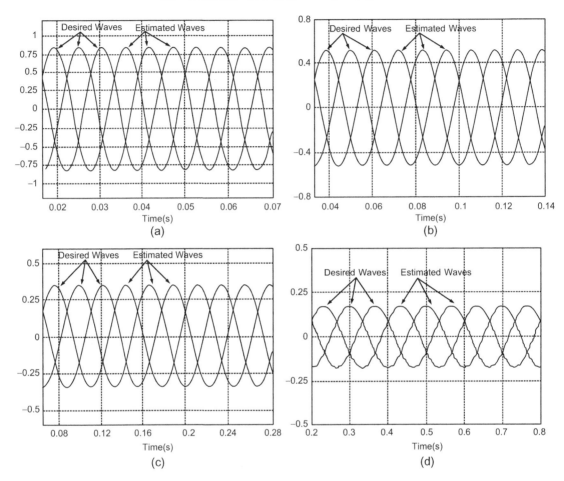

Once the performance of the NNW was found satisfactory at 60 Hz with variable magnitude, it was decided to train and test the NNW at variable-frequency, variable-magnitude SPWM waves as encountered in variable-frequency drives. A configuration similar to that in Figure 11.29(a) was used, and the frequencies under consideration for training were 60, 30, 15, and 5 Hz with the same carrier frequency of 5 kHz. For each value of f, the magnitude was varied so that the NNW training remains valid in the whole range of drive operation. After training, the NNW was tested thoroughly in these and intermediate frequencies, and test output waves at the training frequencies are shown in this figure. The harmonic quality at low f is somewhat inferior, but the overall performance was satisfactory. However, the network structure was somewhat large (3-23-23-3) and it took 1000 epochs of training. Instead of computing the fundamental voltages only as shown, additional computations (such as voltage vector generation) can be incorporated easily in the same NNW.

FIGURE 11.31 (a) Delayless harmonic filtering and processing of variable-frequency current waves ($f_c = 5$ kHz) at (b) $f = 60$ Hz, (c) $f = 20$ Hz, and (d) $f = 5$ Hz.

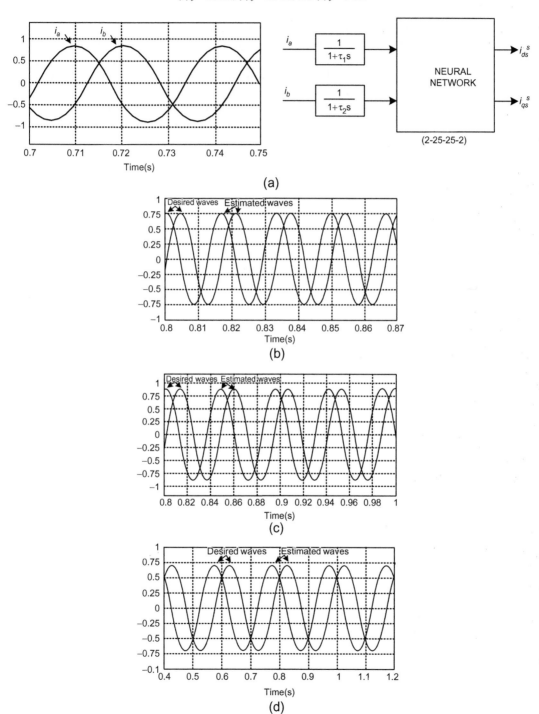

Extending the studies for variable-frequency PWM voltage waves, the NNW-based delayless filtering for variable–frequency, variable-magnitude machine current waves can also be studied as shown in this figure. Only currents i_a and i_b are considered because in a three-phase isolated neutral machine $i_a + i_b + i_c = 0$. Because the machine current waves are continuous with nearly sinusoidal shape, additional LPF as shown may not be needed. However, the NNW size was found to be somewhat less with LPFs. Instead of generating delayless sine waves at the output, it was decided to incorporate $i_{qs}{}^s$ and $i_{ds}{}^s$ computations (see Figure 6.18) in the same NNW. The estimated and desired output waves at variable frequency and variable magnitude shown in (b)–(d) were found to be very good. After satisfactory training for 600 epochs, the ANN configuration was found to be 2-25-25-2.

FIGURE 11.32 Feedback signal estimation equations for induction motor vector drives.

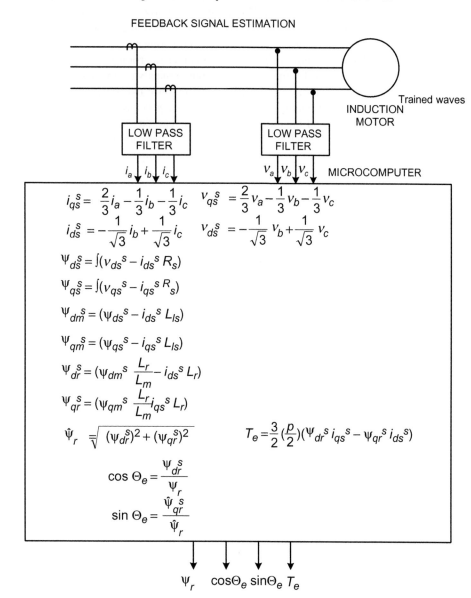

FEEDBACK SIGNAL ESTIMATION

Next, the capability of NNWs to provide feedback signal estimation for a vector drive will be demonstrated. This figure summarizes the computational equations [4] for a rotor flux–oriented, direct vector-controlled induction motor drive, where standard symbols have been used. The microcomputer/DSP receives the three-phase machine voltage and current signals and computes the rotor flux (ψ_r), unit vector ($\cos \theta_e$, $\sin \theta_e$), and developed torque (T_e). All input signals vary in magnitude and frequency. For an isolated, neutral, three-phase machine, only two phase currents are sufficient. Again, line voltage signals instead of phase voltages can be used. In this case, the effect of LPF for phase and magnitude variation is ignored. Note that a feedforward NNW cannot perform integration as shown by the two equations. Otherwise, addition, subtraction, multiplication, division, and square root computations can be easily handled by NNW.

FIGURE 11.33 Direct vector drive of an induction motor showing DSP and NNW-based feedback signal estimation.

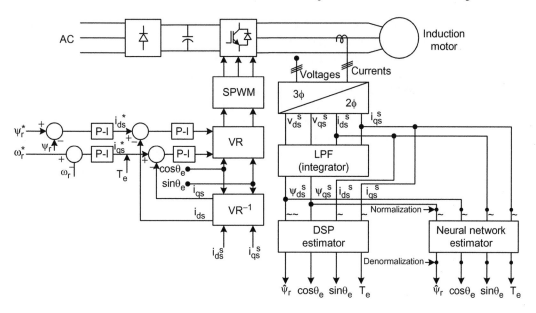

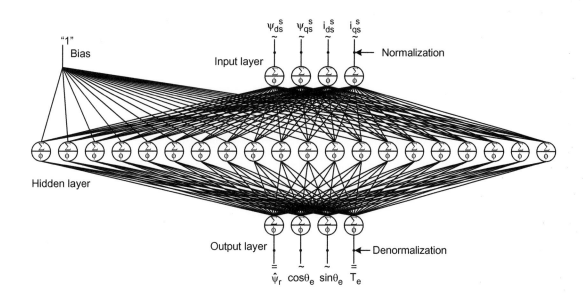

The block diagram of the direct vector-controlled drive is shown in the upper figure, which incorporates a separate low-pass filter (LPF) that performs integration to calculate the stator fluxes $\psi_{ds}{}^s$ and $\psi_{qs}{}^s$. In the figure, both the DSP estimator and neural network estimator [11] receive the $\psi_{ds}{}^s$, $\psi_{qs}{}^s$, $i_{ds}{}^s$, and $i_{qs}{}^s$ signals, which are variable-frequency, variable-magnitude waves. The DSP estimator input–output signal table is used to train the NNW. Of course, simulation-based data can also be used for training. Once the NNW is trained and tested, it replaces the DSP, thus relieving the computational burden of the system DSP controller. The NNW topology (4-20-4) with the back-propagation training blocks and the bias coupling in the hidden layer only is shown in the lower part of the figure. Because the SPWM carrier frequency is high (15 kHz), the NNW input waves have small distortion. All neurons in the output and hidden layers use hyperbolic tan TF, thus giving the NNW more computational capability.

FIGURE 11.34   NNW-based feedback signal estimator performance: (a) torque, (b) rotor flux, (c) cos θ_e, and (d) sin θ_e.

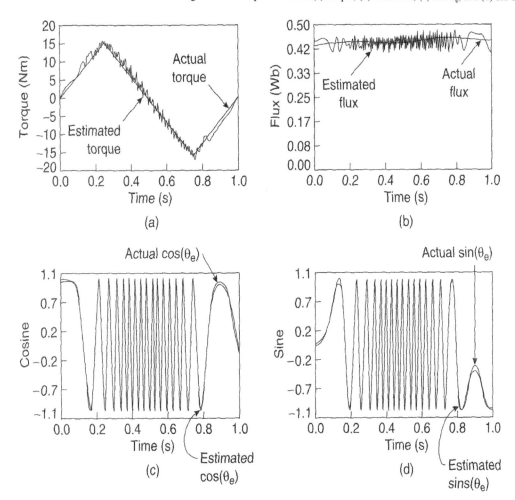

The figure shows the performance of the NNW-based feedback signal estimator and compares it to the results of a DSP-based estimator. The results match very well. The developed torque is bidirectional and describes a triangular profile. The estimated torque curve follows closely the actual curve except that the NNW-based estimator gives some additional ripple. The rated constant flux is also the same except for some additional ripple in the NNW-based estimation. Signals cos θ_e and sin θ_e are also reasonably accurate.

FIGURE 11.35   (a) Single-phase thyristor controller, (b) current waves at variable firing angle α, (c) current waves at variable impedance Z, and (d) current waves at variable impedance angle φ.

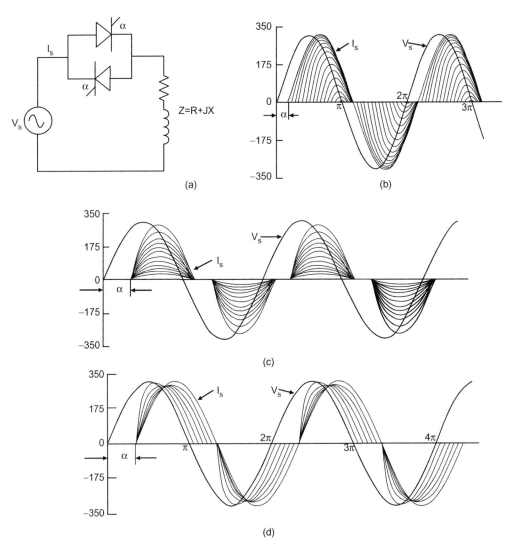

The application of NNWs for the estimation of distorted waveforms [12] will be discussed next. Let us consider a simple thyristor ac phase controller with R-L load as shown in (a) with a constant 220-V, 60-Hz power source. With constant $Z = 1.0\angle 30°$, the current waves at variable firing angle α are shown in (b) for the 30° to 180° range. With fixed $\alpha = 60°$ and impedance angle $\varphi = 30°$, the current waves at variable impedance Z_b/Z ($Z_b = 220\sqrt{2}$) are shown in (c). Again, with constant $\alpha = 60°$ and $Z = 1.0$ ohm, the current waves at variable angle φ are shown in (d). The problem here is that in each case, a NNW needs to be trained so that it gives the estimation of rms current (I_s), fundamental rms current (I_f), DPF, and PF. Analytically, it is difficult to make such estimations in real time.

FIGURE 11.36 NNW estimator performance: (a) variable α, (b) variable Z, and (c) variable φ.

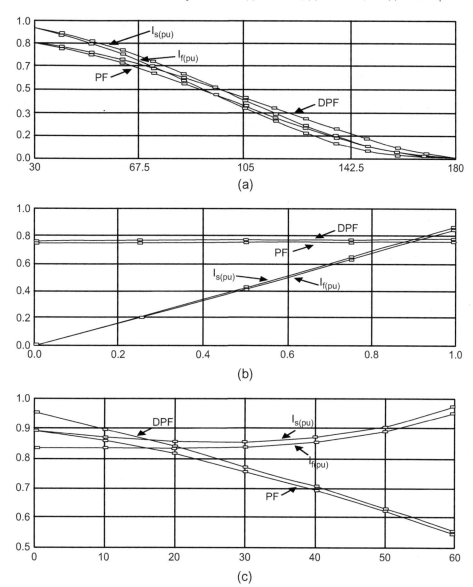

For each of the input variables (α, Z, and φ), a separate NNW with topology 1-4-4 was trained, which gives the output of $I_{s(pu)}$ (per unit rms current), $I_{f(pu)}$ (per unit fundamental rms current), DPF (displacement power factor), and PF (power factor), respectively, for the corresponding magnitude of the input variable. This merely requires the solving of an input–output mapping problem. While the DPF and PF show the actual values (not sensitive to Z amplitude variation), $I_{s(pu)}$ and $I_{f(pu)}$ values were denormalized by the scale factor Z_b/Z. The example data sets for training were created by FFT analysis of the simulated waveforms. The NNW has the inherent capability of interpolating with any intermediate value of input variable. By testing, it was found that all the estimations have accuracy within 0.1%. Note that in (b), $I_{s(pu)}$ and $I_{f(pu)}$ are practically linear with Z_b/Z variation, but DPF and PF remain nearly constant. In (c), DPF and PF increase continuously as the circuit becomes more resistive, i.e., as angle φ decreases.

FIGURE 11.37   (a) NNW topology (3-16-16-4) when α, φ, and Z_b/Z vary and (b) NNW-based $I_{s(pu)}$ estimator performance.

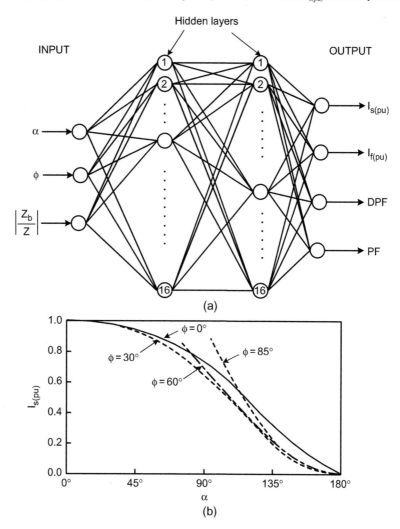

(a)

(b)

After validating the NNW-based estimation for individual α, φ, and Z inputs, the decision was made to train a composite network [12], in which all three input variables can change. Such a network (3-16-16-4) structure is shown in this figure. The training data were generated for (1) α range: $0°–180°$ in 20 steps ($10°$ intervals, plus extra steps at $5°$ and $175°$), (2) φ range: $0°–90°$ in 10 steps, and (3) Z_b/Z range: $0–1.0$ pu in four steps. The corresponding desired output data were generated by simulation. The supply was constant at 220 V, 60 Hz. If V_s varies, it should be considered as an extra input variable. Any combination of inputs that results in continuous conduction was excluded, i.e., $\alpha < \varphi$ angle. A large number of Z_b/Z values are not desirable because DPF and PF outputs will not be affected by it, and $I_{s(pu)}$ and $I_{f(pu)}$ will have only linear scaling effects by impedance variation as shown in part (b). With a large number of training steps, the error was found to converge below 0.2%. The estimator performance for $I_{s(pu)}$ is shown in (b), where $\alpha = 0$ means resistive load. Note that the estimation curve for constant φ terminates at $\alpha = \varphi$ so that conduction is always discontinuous. As usual, $I_{s(pu)}$ is denormalized by the scale factor Z_b/Z. The signal normalization for α and φ at the input is not indicated. Other performance results are shown in the next figure.

FIGURE 11.38   NNW-based estimator performance curves: (a) I_f(pu), (b) DPF, and (c) PF.

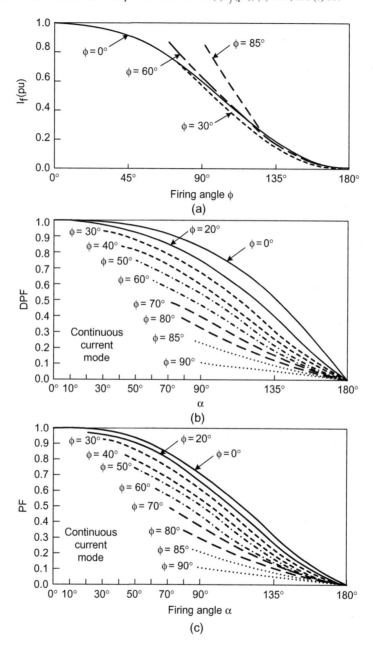

The estimation curves for $I_{f(pu)}$, DPF, and PF for the network in Figure 11.37 are extended in this figure. All conditions described before are also valid here. For simplicity, only a few φ values are shown in (a). Parts (b) and (c), which show *DPF and PF* outputs, respectively, are almost similar. With any φ value, the DPF or PF increases continuously as the α angle decreases, but each curve terminates at $\alpha = \varphi$ where continuous conduction starts. All performance results were found to be excellent.

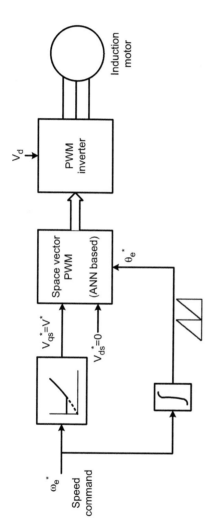

FIGURE 11.39   Volts/Hz speed control of an induction motor with NNW-based SVM modulator for two-level inverter.

The SVM (space vector PWM) principles of two-level and multilevel voltage-fed inverters were discussed in Chapter 4, and the discussion covered both undermodulation and overmodulation regions. Instead of using DSP for SVM, it is possible to implement it by feedforward NNW [13] because the SVM algorithm can be looked on basically as a nonlinear input/output mapping. This figure shows the block diagram of an open–loop, volts/Hz controlled induction motor drive with NNW-based SVM. The NNW receives the reference voltage vector magnitude V^* and the angle θ_e^* at the input, and generates the corresponding pulse width pattern for the three phases. This means that the input vector maps a certain location in the hexagon switching states of the inverter, and the output map generates the three corresponding digital words, which are then converted to pulse widths by a timer.

FIGURE 11.40   (a) Symmetrical three-phase PWM waves within sample time T_s and (b) turn-on time equations for phase A
in the undermodulation region.

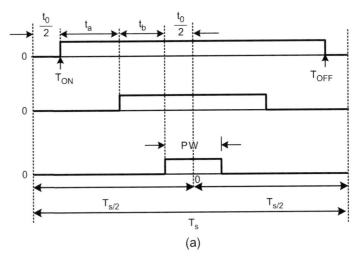

(a)

$$T_{A-ON} =$$

$$\begin{cases} \dfrac{t_0}{2} = \dfrac{T_s}{4} + K \cdot V^* \left[-\sin\left(\dfrac{\pi}{3} - \alpha^*\right) - \sin(\alpha^*) \right], S = 1,6 \\[2mm] \dfrac{t_0}{2} + t_b = \dfrac{T_s}{4} + K \cdot V^* \left[-\sin\left(\dfrac{\pi}{3} - \alpha^*\right) + \sin(\alpha^*) \right], S = 2 \\[2mm] \dfrac{t_0}{2} + t_a + t_b = \dfrac{T_s}{4} + K \cdot V^* \left[\sin\left(\dfrac{\pi}{3} - \alpha^*\right) + \sin(\alpha^*) \right], S = 3,4 \\[2mm] \dfrac{t_0}{2} + t_a = \dfrac{T_s}{4} + K \cdot V^* \left[\sin\left(\dfrac{\pi}{3} - \alpha^*\right) - \sin(\alpha^*) \right], S = 2 \end{cases}$$

$$\text{.......(1)}$$

$$= \dfrac{T_s}{4} + f(V^*) \cdot g(\alpha^*) \text{.......(2)}$$

$g(\alpha^*) = $ TURN-ON PULSE WIDTH FUNCTION AT UNIT AMPLITUDE

$f(V^*) = $ VOLTAGE AMPLITUDE SCALE FACTOR

$T_{A-OFF} = T_S - T_{A-ON}$ ....(3)

(b)

The SVM algorithm generates three-phase symmetrical PWM waves within a sampling
interval T_s as shown in (a), where the inverter switching states in the initial $T_s/2$ interval
are 000, 100, 110, and 111, respectively. In phase A (upper wave), for example, the lower
switching device is on for the $t_0/2$ interval. Then the upper switch closes at turn-on time
T_{A-ON} (or T_{ON}), and remains on for the interval $2(t_a + t_b + t_0/2)$. Then it turns off symmetrically at T_{A-OFF} (or T_{OFF}) as shown in the figure, where $t_0 + t_a + t_b = T_s/2$. Both t_a and t_b
are functions of V^* and α^*, where $\alpha^* = $ angle in a sector (see Figure 11.41). Using the
same principle, the expression for T_{A-ON} in all six sectors ($S = 1–6$) can be derived as given
in Eq. (1). Equation (1) can be written in the form of Eq. (2), where $f(V^*) = $ voltage amplitude scale factor and $g(\alpha^*) = $ turn-on pulse width function at unit amplitude ($f(V^*) = 1$).
In a similar manner, T_{B-ON} and T_{C-ON} for phases B and C, respectively, can be derived.

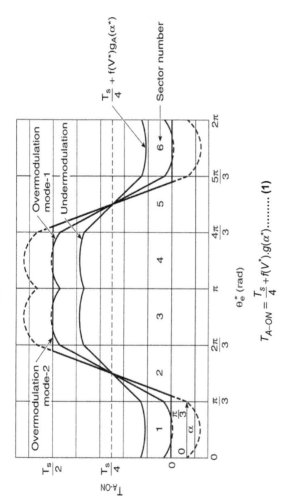

FIGURE 11.41 Turn-on time (T_{A-ON}) of phase A as a function of angle θ_e in the six sectors.

$$T_{A-ON} = \frac{T_s}{4} + f(V^*).g(\alpha^*)\ldots\ldots (1)$$

The figure shows the plot of Eq. (2) of the previous figure [repeated here as Eq. (1)], i.e., the turn-on time of phase A pulse width as a function of voltage vector angle α^* (or θ_e^*) at different magnitude of V^*. The turn-on pulse width function $g(\alpha^*)$ is a nonsinusoidal ac wave (with rich triplen harmonics), which becomes symmetrical to the $T_s/4$ axis with the bias time ($T_s/4$) added to it. In the undermodulation region, $f(V^*) = V^*$. Therefore, T_{A-ON} increases linearly with increases in V^*. If $V^* = 0$, T_{A-ON} is clamped to $T_s/4$. The undermodulation region ends at the upper value of V^* when the maximum and minimum values of T_{A-ON} are $T_s/2$ and 0, respectively. In the PWM waves (Figure 11.40), the corresponding values of $t_0/2$ are $T_s/2$ and 0, respectively. At overmodulation, T_{A-ON} saturates in the upper and lower levels because it (shown by dotted curves) tends to exceed the limit values. This means that $f(V^*)$ becomes a nonlinear function of V^*. In overmodulation modes 1 and 2, as discussed in Chapter 4, rigorous calculations can be made for $f(V^*)$ and T_{A-ON} and plotted in figures. Note that plots of T_{B-ON} and T_{C-ON} curves (not shown) for phases B and C, respectively, are identical, but mutually phase shifted by 120°.

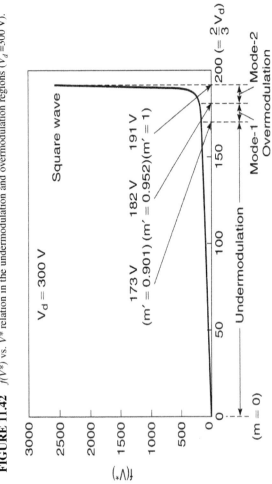

FIGURE 11.42 $f(V^*)$ vs. V^* relation in the undermodulation and overmodulation regions ($V_d = 300$ V).

The relationship between $f(V^*)$ and V^* covering both undermodulation and overmodulation regions for the dc link voltage $V_d^* = 300$ V is shown in this figure. As mentioned before, in the undermodulation region, this relation is linear, i.e., $f(V^*) = V^*$. This region ends at $V^* = 173$ V, where modulation factor $m' = 0.901$. In the overmodulation region, the question is this: For a reference voltage V^*, what should the relation between $f(V^*)$ and V^* be such that linear transfer characteristics are maintained between V^* and inverter output voltage like that in the undermodulation region? To satisfy this criterion, rigorous calculations can be made for $f(V^*)$ and $T_{A\text{-}ON}$ to establish the curves in this and the previous figure. However, it can be shown that the undermodulation curve in the previous figure can be expanded to mode 1 and mode 2 by using a nonlinear scale factor $f(V^*)$, which increases with a steep slope in overmodulation modes 1 and 2 until a square wave is reached at $V^* = \infty$. The idea is the same as that for the SPWM overmodulation principle in which the modulating voltage is magnified.

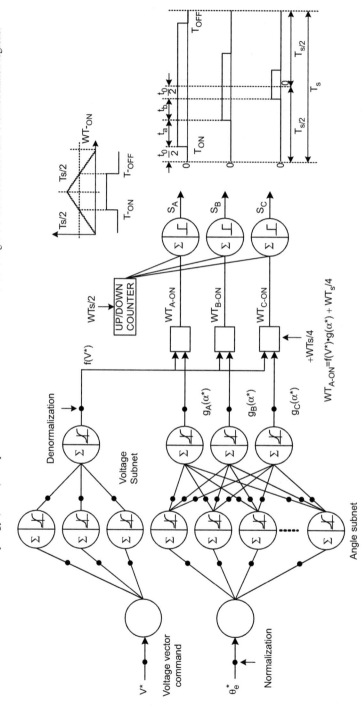

FIGURE 11.43 NNW topology (2-20-4) for space vector PWM of a two-level inverter covering the undermodulation and overmodulation regions.

The data corresponding to Figures 11.41 and 11.42 give us sufficient information to design and train a feedforward NNW [13] as shown in this figure. The NNW basically consists of two subnets. One is the angle subnet (shown in the lower part), which receives the vector angle θ_e^* at the input and solves for the phase pulse width functions $g_A(\alpha^*)$, $g_B(\alpha^*)$, and $g_C(\alpha^*)$. Note that $g_A(\alpha^*)$, $g_B(\alpha^*)$, and $g_C(\alpha^*)$ have the same wave shapes, but they are mutually phase shifted by 120°. The upper voltage subnet receives the reference voltage magnitude V^* and converts it into $f(V^*)$ as shown in the figure. Both the voltage subnet (1-3-1) and angle subnet (1-17-3) have three layers with sigmoidal TF in the middle and output layers. The angle subnet outputs are multiplied by $f(V^*)$ and added with the equivalent bias signal $WT_s/4$ to generate the digital words corresponding to the turn-on time of the three phases. Then, a single UP/DOWN counter converts these words into the corresponding symmetrical pulse widths as shown in the figure. Note that the reference voltage vector basically maps at a certain point in the hexagon, and the corresponding digital outputs ($WT_{A\text{-}ON}$, $WT_{B\text{-}ON}$, $WT_{C\text{-}ON}$) are generated by nonlinear input–output mapping.

FIGURE 11.44   Motor current waves with NNW-based SWM: (a) undermodulation (45 Hz), (b) overmodulation mode 1 (56 Hz), and (c) overmodulation mode 2 (59 Hz).

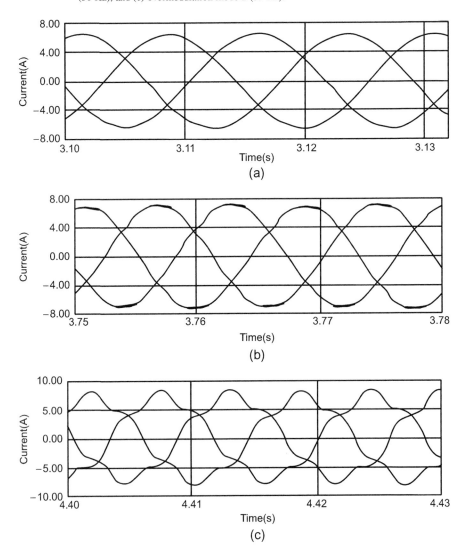

Once the NNW in Figure 11.43 was designed and trained, it was tested thoroughly to validate its performance. Then, the NNW was integrated with the volts/Hz controlled induction motor drive shown in Figure 11.39 and the drive performances were tested using Simulink. This figure shows the motor current waves at undermodulation (45 Hz), overmodulation mode 1 (56 Hz), and overmodulation mode 2 (59 Hz). The dominance of odd harmonics in (c) near the square wave (at 60 Hz) is evident. These waves and other drive performances match very well with a DSP-based simulation.

FIGURE 11.45   Three-level, diode-clamped inverter induction motor drive with NNW-based SVM.

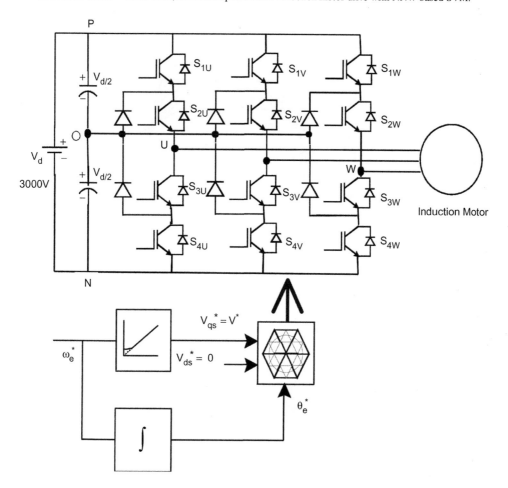

Extending the principles of NNW-based SVM for a two-level inverter, we will now introduce a NNW-based SVM for a three-level inverter [14–16]. The principles for a three-level inverter, its SVM technique in the undermodulation and overmodulation regions, and DSP-based implementation were discussed in detail in Chapter 4. This figure shows a block diagram for open-loop volts/Hz control of an induction motor drive using a NNW-based SVM. The modulator receives the reference voltage magnitude (V^*) and vector angle (θ_e^*) and generates the PWM signals for the 12 IGBTs. Because DSP-based SVM implementation is very complex, a NNW-based implementation with an ASIC chip is desirable using the principle of input–output mapping.

FIGURE 11.46 Turn-on time plots for U phase covering the undermodulation and overmodulation regions (V_d = 3000 V, T_s=1.0 ms): (a) $T_{UP\text{-}ON}$ for P state and (b) $T_{UN\text{-}ON}$ for N state.

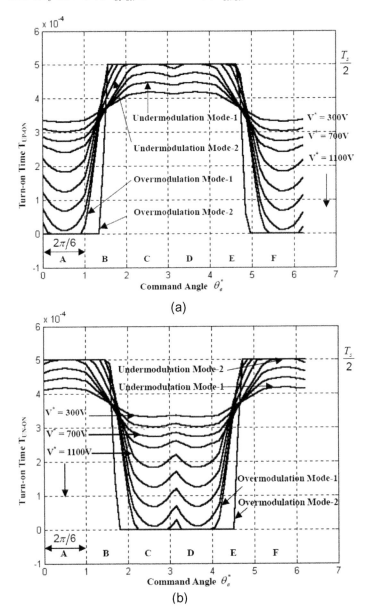

(a)

(b)

The calculated turn-on times for the P state and N state of the U phase of the inverter in Figure 11.45 have been plotted in this figure for all six sectors. The analytical expressions of the curves as a function of reference voltage magnitude (V^*) and vector angle (θ_e^*) can be derived [15, 16] from the PWM waves shown in Figures 4.35 and 4.37, respectively, for undermodulation and overmodulation regions. In undermodulation mode 1 (see Figure 4.34), the curves remain unsaturated, but reach saturation at 0.5 ms ($T_s/2$) at the end of this mode, and then mode 2 starts. In overmodulation modes 1 and 2, part of the curves reach clamping level 0 as indicated in the figure. With an increase in the modulation factor, the curves become trapezoidal in shape and ultimately approach the square wave as shown. For the V and W phases, the curves are the same but are mutually phase shifted by 120°. Referring to Figure 4.35, it can be shown that in odd sectors (A, C, E), $T_{UP\text{-}ON}$ generates a pulsed wave and $T_{UN\text{-}ON}$ generates a notched wave, whereas in even sectors (B, D, F) their roles are reversed. The mapping of turn-on times in this figure as a function of V^* and θ_e^* can generate data to train a NNW for SVM implementation.

FIGURE 11.47   NNW topologies for SVM of three-level inverter: (a) five-layer (2-9-9-9-6) topology, (b) four-layer
(2-15-15-6) topology, and (c) four-layer with two subnets [2 × (2-10-10-3)].

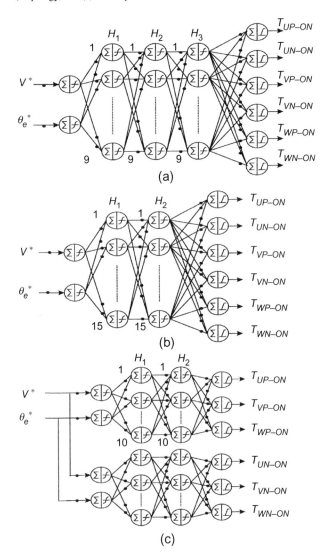

It is possible to train a number of NNW configurations for SVM, as shown in this
figure. Consider first, for example, part (a). The five-layer feedforward NNW receives
two input signals (V^* and θ_e^*) and generates the six outputs that correspond to digital
words for turn-on times of P and N states of the three phases as shown. The training
data were generated from Figure 11.46 (considering all three phases) with voltage steps

of 50 V in the V^* range of 0 to 1900 V ($V_d = 3000$ V), and angle step of 6° in the θ_e^* range of 0 to 2π. All NNW signal processing was handled in a normalized manner. The topology of the trained NNW has five layers (2-9-9-9-6), i.e., 35 neurons are used. Both the input and output layers use bipolar linear TF, whereas the hidden layers use hyperbolic tan TFs. The training took altogether 295 epochs to reach an acceptable MSE value. The testing at any intermediate V^* and θ_e^* values shows results within the same MSE. For curiosity, it was decided to train two more NNW topologies shown in the figure: one with four layers (2-15-15-6) that requires 38 neurons, and the other with two subnets [2×(2-10-10-3)] requiring 50 neurons. The number of epochs required for satisfactory training for each NNW are, respectively, 329 and 320. Therefore, the 2-9-9-9-6 topology was chosen because it uses fewer neurons and has a shorter training time.

FIGURE 11. 48 Five-layer NNW with the interface logic and up/down counter.

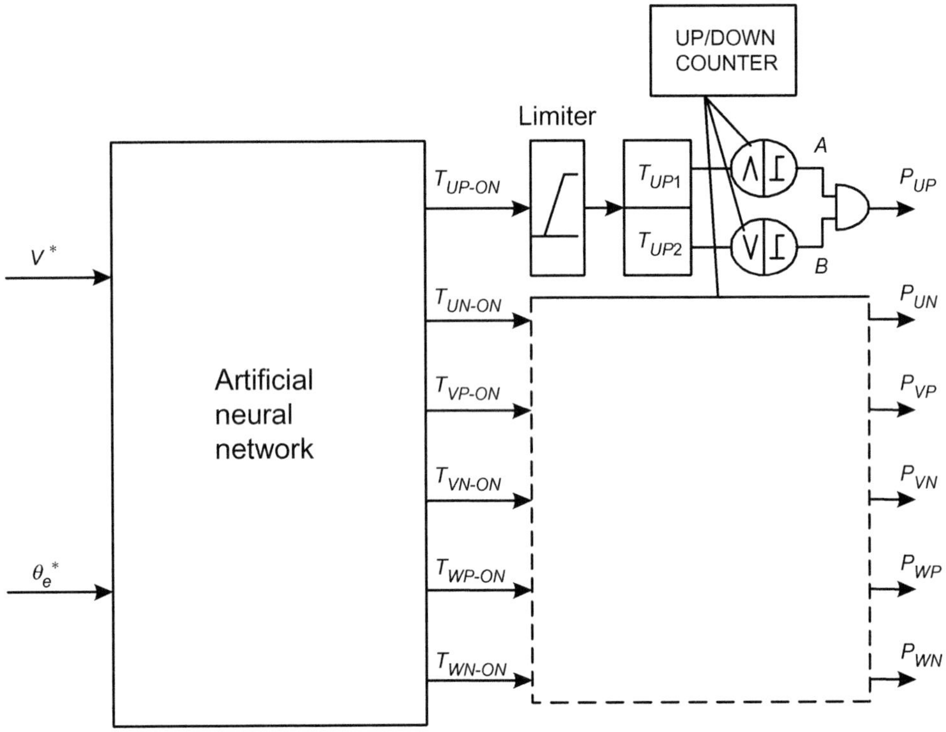

The digital words generated in the NNW of Figure 11.47(a) are converted to *P*-state and *N*-state switching logic signals for the inverter as shown by a simplified block diagram in this figure [16]. These signals are then translated into actual logic waves to drive the 12 devices. Only the processing of the $T_{UP\text{-}ON}$ signal is explained in this figure. The digital word for $T_{UP\text{-}ON}$ is segmented into two channels, T_{UP1} and T_{UP2}. The signal T_{UP1} is compared with the output of the UP/DOWN counter (with period T_s) to generate the symmetrical logic pulse *A*, whereas T_{UP2} is compared with the inverted output of the counter to generate the logic pulse *B*. These pulses are then logically ANDed to generate the pulse width for the *P* state of the *U* phase. Similar segmentation and processing are done in all six of the NNW's output channels.

FIGURE 11.49 Sector segmentation of NNW output for U-phase P-state turn-on time ($T_{UP\text{-}ON}$).

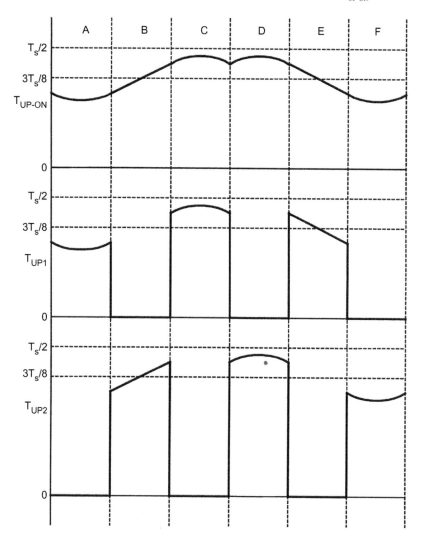

The figure illustrates sector-wise segmentation of the $T_{UP\text{-}ON}$ signal, where the T_{UP1} component is clamped to zero in even sectors (B, D, and F) and the T_{UP2} component is clamped to zero in odd sectors (A, C, and E). This segmentation complexity arises because, as mentioned before, in the odd sectors, the P states appear as pulsed waves (see Figure 4.35), whereas the N states appear as notched waves. On the other hand, in even sectors, the P states appear as notched waves and N states appear as pulsed waves. The details of segmentation are not shown here, but it can be done easily by logic waves generated through sector identification

FIGURE 11.50   Timer and logic operation to generate pulse width P_{UP}.

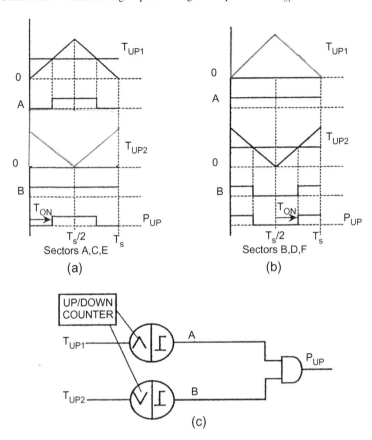

(a)   (b)

(c)

The segmented signals in the previous figure are received in the upper and lower channels as shown in part (c) of this figure. Consider, for example, sector A (odd) operation, where T_{UP1} is a valid signal but $T_{UP2} = 0$. The explanation for transporting the T_{UP1} pulse to P_{UP} in the odd sector is explained in (a) of the figure. Similarly, transporting pulse T_{UP2} in the even sector to P_{UP} is explained in (b) of the figure. Note that operation is similar in all the channels using only one counter output. Once the complete NNW-based modulator was designed and integrated with the volts/Hz controlled drive, the performances were tested thoroughly. The machine voltage and current waves are similar to those with DSP-based SVM (see Figure 4.39) and are not given separately.

FIGURE 11.51   Simplified representation of five-level inverter.

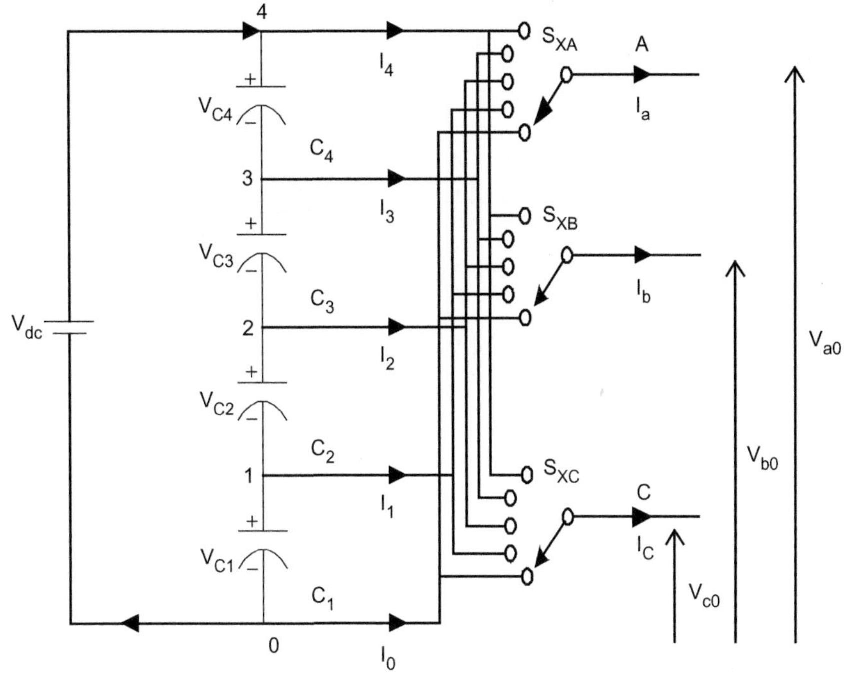

As a final phase of our discussion on NNW-based SVM, let us consider the same for a simplified five-level inverter indicated in this figure. The reason for our emphasis on NNW-SVM is that the NNW application is so well-suited to this area, whereas the DSP implementation is so complex, particularly with a higher number of levels. The topology for a five-level diode-clamped inverter was given in Figure 4.40, and this figure shows its simplified representation. A five-pole switch in each phase establishes the five levels of the output voltage with a step size of $V_d/4$. Consider, for example, the output phase voltage V_{A0}, which is established by the five-position switch S_{XA}, which selects the series capacitor taps. In the five taps shown, the levels of the output voltages are V_4, V_3, V_2, V_1, and V_0, respectively, where the lowest tap (V_0) gives the reference or zero voltage. If the phase voltage wave is established with respect to the usual center tap (tap 2) position, then the V_{A0} wave has two levels on the positive side and two levels on the negative side and one zero level.

FIGURE 11.52   (a) Switching states of five-level inverter and (b) sector A triangles showing the switching states (t_a, t_b, and t_c).

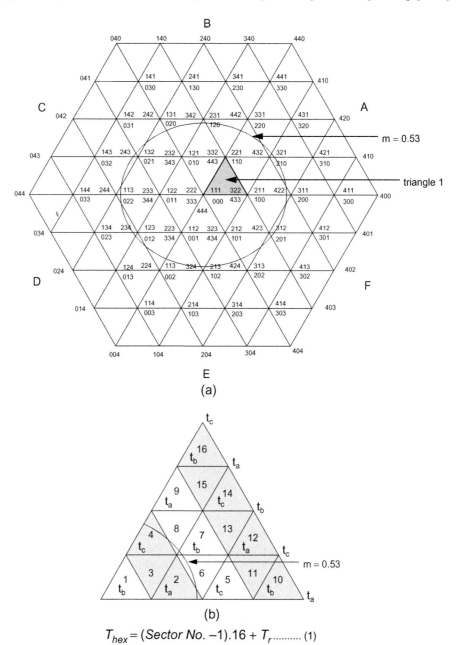

(a)

(b)

$$T_{hex} = (Sector\ No. -1).16 + T_r .... (1)$$

The hexagon switching states of the inverter that were discussed in Figure 4.42 have been repeated in (a) of the figure. There are 125 switching states, 96 triangles, 61 effective space vectors, and a multiplicity of switching states for inner triangle corners as shown. The six sectors (or sextants) of the hexagon have been numbered A to F, respectively. Sector A is drawn separately in (b) showing the 16 triangles and their switching intervals t_a, t_b, and t_c for the respective corner vectors, where $t_a + t_b + t_c = T_s$ (sampling time). The reference voltage trajectory for modulation factor $m' = 0.53$ is illustrated by the dotted circle. The location of the reference vector in a triangle (T_{hex}) of the hexagon can be determined [17] by Eq. (1) by identification of the sector number and T_r, where T_r = equivalent triangle located in sector A. In principle, the SVM switching waves can be determined by mapping of the reference vector location in any of the 96 triangles of the whole hexagon.

FIGURE 11.53 Pulse width synthesis for phase voltage A in a sampling interval T_s.

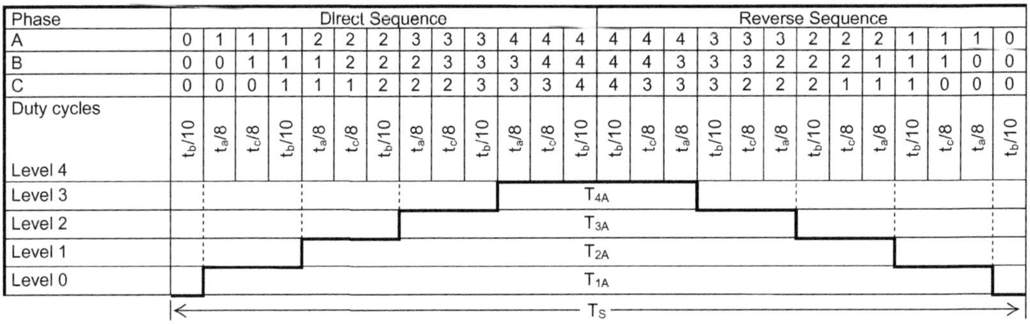

Phase	Direct Sequence													Reverse Sequence												
A	0	1	1	1	2	2	2	3	3	3	4	4	4	4	4	4	3	3	3	2	2	2	1	1	1	0
B	0	0	1	1	1	2	2	2	3	3	3	4	4	4	4	3	3	3	2	2	2	1	1	1	0	0
C	0	0	0	1	1	1	2	2	2	3	3	3	4	4	3	3	3	2	2	2	1	1	1	0	0	0
Duty cycles	$t_b/10$	$t_a/8$	$t_c/8$	$t_b/10$	$t_a/8$	$t_c/8$	$t_b/10$	$t_a/8$	$t_c/8$	$t_b/10$	$t_a/8$	$t_c/8$	$t_b/10$	$t_b/10$	$t_c/8$	$t_a/8$	$t_b/10$	$t_c/8$	$t_a/8$	$t_b/10$	$t_c/8$	$t_a/8$	$t_b/10$	$t_c/8$	$t_a/8$	$t_b/10$

Level 4
Level 3 — T_{4A}
Level 2 — T_{3A}
Level 1 — T_{2A}
Level 0 — T_{1A}
T_s

$$T_{4A} = K_{a4A}(T_{hex}).t_a + K_{b4A}(T_{hex}).t_b + K_{c4A}(T_{hex}).t_c \quad..... (1)$$

$$T_{3A} = K_{a3A}(T_{hex}).t_a + K_{b3A}(T_{hex}).t_b + K_{c3A}(T_{hex}).t_c \quad..... (2)$$

$$T_{2A} = K_{a2A}(T_{hex}).t_a + K_{b2A}(T_{hex}).t_b + K_{c2A}(T_{hex}).t_c \quad..... (3)$$

$$T_{1A} = K_{a1A}(T_{hex}).t_a + K_{b1A}(T_{hex}).t_b + K_{c1}A(T_{hex}).t_c \quad..... (4)$$

The figure shows, for example, the typical pulse width synthesis in proper sequence for the reference vector location in triangle 1 shown in Figure 11.52(a). The five levels of the output voltage (levels 0 to 4) with V_0 as the reference voltage , the corresponding pulse widths in the four levels (T_{1A} to T_{4A}), the switching states (0 to 4), and the corresponding duty cycles of t_a, t_b, and t_c in direct and reverse sequence for symmetrical pulse widths are indicated in the figure. For example, the first duty cycle of $t_b/10$ for 000 states establishes zero voltage in all phases. Next $t_a/8$ for 100 states establishes level 1 in phase A as shown but zero voltage in phases B and C. The states of 444, for example, establish level 4 in all phases. The selection of the states and the corresponding duty cycles can be calculated based on reference voltage magnitude, neutral points voltage balancing, and symmetry of pulse widths. The pulse width intervals in the figure can be expressed by Eqs. (1)–(4), where all constants shown are a function of the T_{hex} location. In this case, $T_{hex} = 1$, and the corresponding coefficients of T_{4a}, for example, are $K_{a4A} = 1/8 + 1/8 = 0.25$, $K_{b4A} = 1/10 + 1/10 = 0.2$ and $K_{c4A} = 1/8 + 1/8 = 0.25$. In the same way, phases B and C waves can be plotted and their corresponding coefficients can be determined.

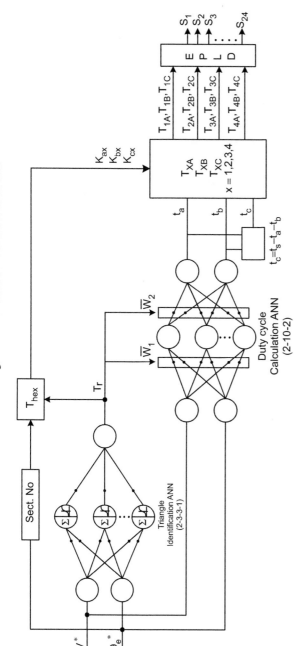

FIGURE 11.54 Block diagram of NNW-based SVM of five-level inverter.

The figure shows the NNW-based SVM implementation [17] of the complete inverter in the undermodulation range. Instead of using one large composite NNW, it is segmented into two parts: the first one is the Triangle Identification ANN(2-3-3-1) and the second one is the Duty Cycle Calculation ANN(2-10-2), as shown in the figure. The first NNW receives the reference voltage magnitude (V^*) and the vector angle (θ_e^*), and identifies the vector location in one of the corresponding 16 equivalent triangles (T_r). This information about T_r is also used to calculate T_{hex} by Eq. (1) in Figure 11.52 by knowing the sector location of the reference vector. Each of the 16 numbers of T_r generates a pair of weight vectors, $\overline{W}_1$ and $\overline{W}_2$ (altogether 16 sets of $\overline{W}_1$, $\overline{W}_2$), by look-up table and loads the second NNW as shown. The Duty Cycle ANN estimates the duty cycles t_a, t_b, and t_c for location of reference voltage anywhere in the hexagon. Again, each of the 96 T_{hex} locations stores a set of 12 coefficients as shown. These coefficients are used to calculate the four levels of pulse widths for the three phases (T_{1A}- T_{4A}, T_{1B}-T_{4B}, and T_{1C}-T_{4C}) as indicated in the figure. Finally, an EPLD (electrically programmable logic device) generates the switching waves of 24 devices (S_1-S_{24}). Each of the component NNWs was trained individually from the simulation data, and then verified by test. The extreme complexity of SVM implementation is apparent in the figure.

FIGURE 11.55 System performance at modulation factor $m' = 0.53$ (31.8 Hz): (a) line voltage (V_{AB}), (b) line current (I_A), and (c) line voltage spectrum.

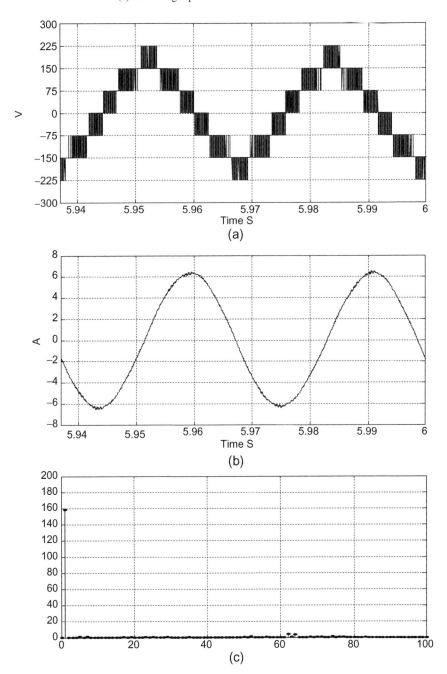

Once the NNW system in the previous figure was properly trained and tested, it was integrated with the volts/Hz controlled induction motor drive for simulation study. The performance of the NNW-based drive was found to be excellent. This figure shows typical simulation waves for line voltage, line current, and the corresponding line voltage spectrum at modulation factor $m' = 0.53$.

FIGURE 11.56   Recurrent neural network-based speed estimation for an induction motor by means of the MRAC technique.

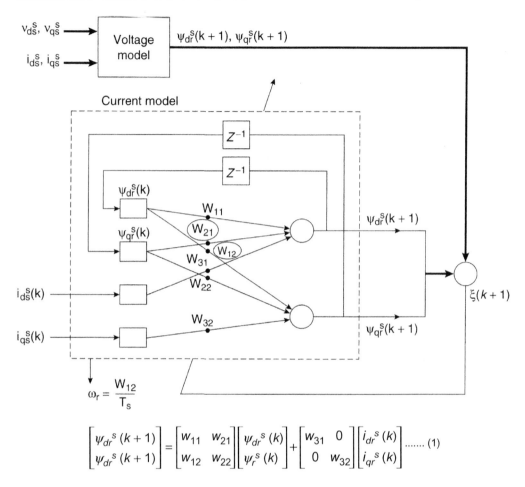

$$\begin{bmatrix} \psi_{dr}^{s}(k+1) \\ \psi_{dr}^{s}(k+1) \end{bmatrix} = \begin{bmatrix} w_{11} & w_{21} \\ w_{12} & w_{22} \end{bmatrix} \begin{bmatrix} \psi_{dr}^{s}(k) \\ \psi_{r}^{s}(k) \end{bmatrix} + \begin{bmatrix} w_{31} & 0 \\ 0 & w_{32} \end{bmatrix} \begin{bmatrix} i_{dr}^{s}(k) \\ i_{qr}^{s}(k) \end{bmatrix} \ \text{.......} \ (1)$$

The figure shows speed (ω_r) estimations for an induction motor by using a real-time RNN technique. Figure 7.59 discussed a similar estimation techniques that used the complex MRAC (model referencing adaptive control) method, where the reference voltage model flux vector estimation was compared with that of the speed adaptive current model to determine the speed. The current model equations can be expressed in finite difference form and expressed in the form of Eq. (1), where $W_{11} = 1 - T_s/T_r$, $W_{21} = -\omega_r T_s$, $W_{12} = \omega_r T_s$, $W_{22} = 1 - T_s/T_r$, $W_{31} = L_m \cdot T_s/T_r$, and $W_{32} = L_m \cdot T_s/T_r$, where T_s = sampling time. This equation can be represented by RNN shown in this figure using a bipolar linear transfer function for the neurons. With the machine parameters known, the weights (W_{21} and W_{12}) of the RNN are a function of speed, which can be tuned as shown in this figure. The solution of the voltage model generates the desired flux components as shown considering the fact that the parameters in this model remain invariant. These signals are compared with the RNN output signals and the weights are trained on-line so that the error $\xi(k + 1)$ tends to zero. Again, it is assumed that the current model parameters are known and remain invariant. The on-line training can be done by dynamic backpropagation method, and it should be fast enough so that actual speed variations can be tracked. Note that if ω_r is known, the rotor time constant T_r (or R_r) as an unknown can be tuned. In fact, if all weights are considered trainable, then ω_r as well as T_r can be tuned.

FIGURE 11.57 Two-stage programmable cascaded low-pass filter (PCLPF) for stator flux estimation.

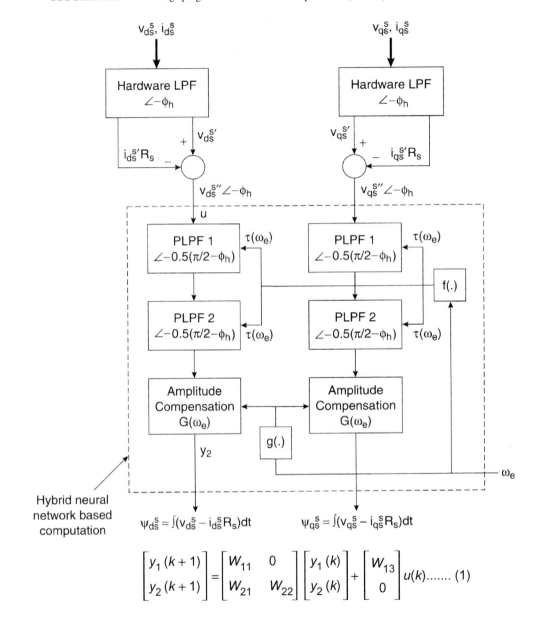

Stator flux vector estimation in a wide frequency range by means of the multistage PCLPF method was discussed in Figure 7.63. Instead of having a DSP solution, it is possible to train a neural network to solve the problem. Let us consider a simplified two-stage programable low-pass filters (PLPF) for each channel as shown in this figure, where the time constant τ and gain G are function of frequency (ω_e). Gain G can be split and merged as G equally with the identical series PLPF stages. The corresponding transfer functions are $Y_2(S)/Y_1(S) = Y_1(S)/U(S) = G/(1 + \tau S)$. The corresponding equations can be expressed in discrete time form and written as Eq. (1), where $W_{11} = W_{22} = 1 - T_s/\tau$, $W_{21} = W_{13} = \sqrt{G}T_s/\tau$, and T_s = sampling time. The functions to be implemented by the neural network are shown within the dashed box. Note that each PLPF stage is required to give a 45° phase shift if $\varphi_h = 0$. However, the φ_h angle increases with frequency and, therefore, the PLPF phase shift [0.5 (90° − φ_h)] decreases at higher frequency for integration.

FIGURE 11.58   Hybrid NNW topology for stator flux estimation at a very low frequency (0.01 Hz).

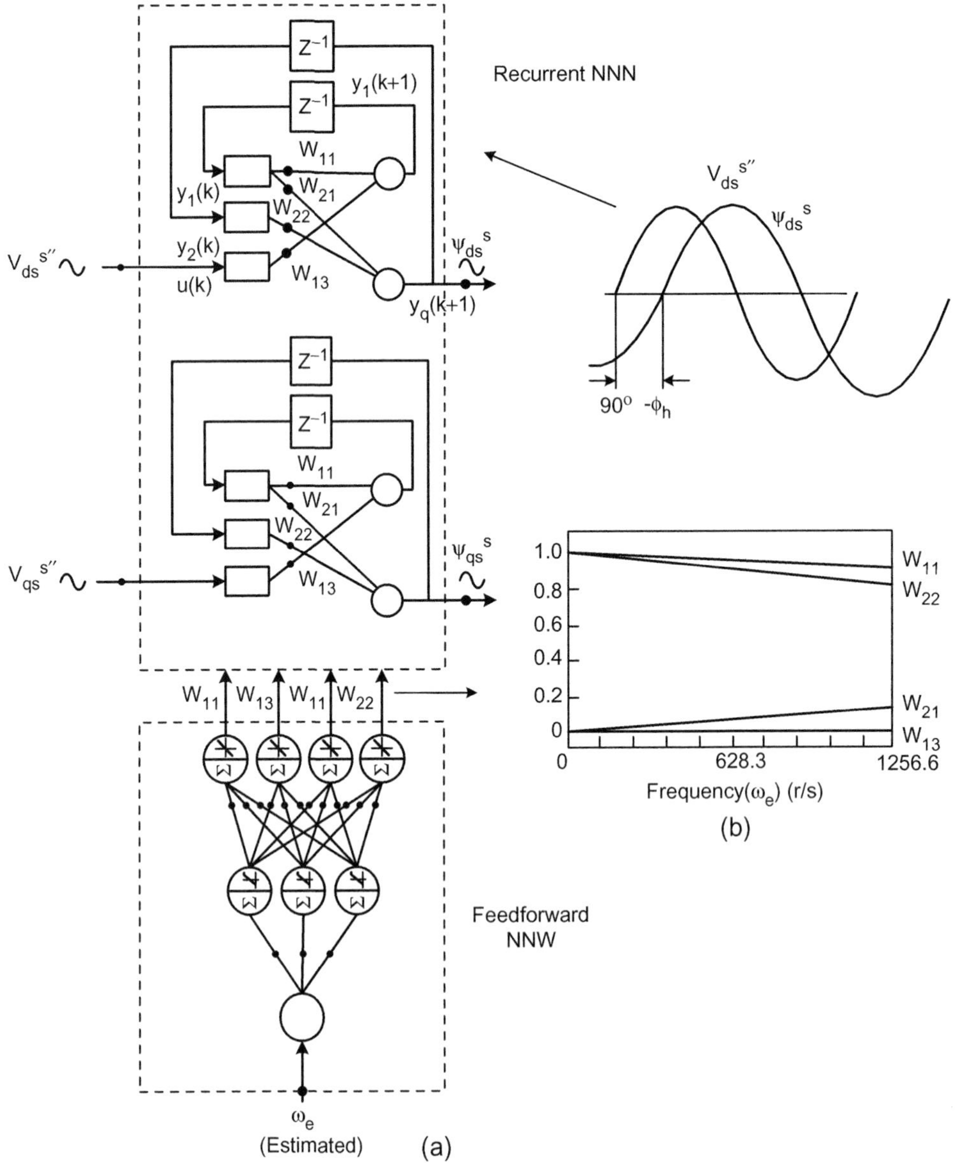

The upper part of this figure shows the real-time RNN-based flux estimation of each channel using Eq. (1) of the previous figure. The RNN neurons have bipolar linear TFs and all their weights, which are trainable, are functions of frequency ω_e. For each sinusoidal input voltage wave at a certain frequency, there is a corresponding output flux wave at $(90° - \varphi_h)$ lag angle, and the RNN weights are constant for that frequency. As the frequency varies, the weight matrix also varies. In this case, the RNN is trained online by using the EKF (extended Kalman filter) algorithm [18] so that the flux waves track the voltage waves. The data for the training were generated by simulation in the frequency range of 0.01 to 200 Hz with a step size of 1.0 Hz. At each frequency, 201 input–output data pairs were used for training. If the actual frequency falls within the steps, it is easily interpolated by the RNN. The trained weights as functions of frequency (slightly nonlinear) are shown at the right. An additional feedforward ANN shown in the lower part of the figure generates these weights as a function of frequency. This ANN uses a linear TF at the output but a sigmoidal TF in the hidden layer.

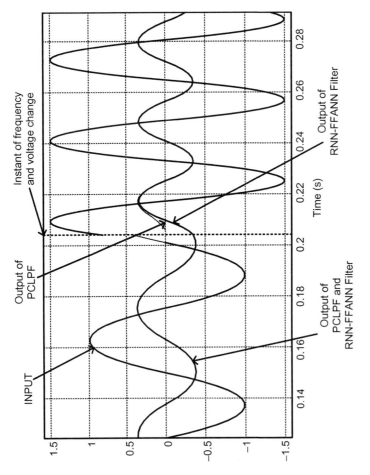

FIGURE 11.59   Comparison of response of DSP-based PCLPF with hybrid NNW-based PCLPF.

The figure shows the response of the hybrid NNW of Figure 11.58 when the frequency was stepped up from 20 to 30 Hz at 0.203 sec, keeping the volts/Hz ratio constant. The performance compares well with the corresponding DSP-based response except it shows superiority in smooth transitioning of the flux wave instead of a sudden change with that of the DSP-based response. Note that the flux wave lags the input voltage wave slightly less than 90° because of the finite φ_h angle, which was not corrected.

FIGURE 11.60 Stator flux-oriented vector drive with NNW-based SVM and RNN-based flux estimation.

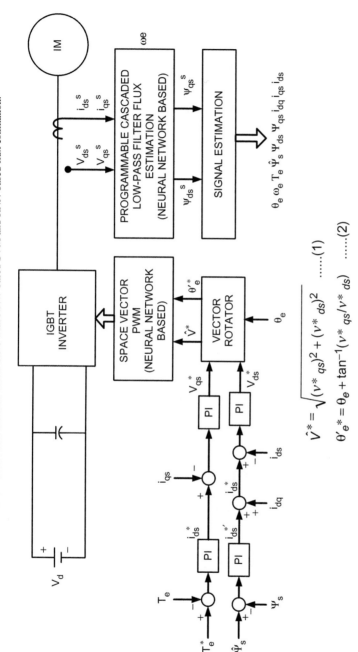

$$\hat{V}^* = \sqrt{(v^*{}_{qs})^2 + (v^*{}_{ds})^2} \quad \dots (1)$$

$$\theta'_e = \theta_e + \tan^{-1}(v^*{}_{qs}/v^*{}_{ds}) \quad \dots (2)$$

The figure shows the core EV-type drive system [19] using stator flux-oriented direct vector control incorporating the NNW-based SVM (Figure 11.44) and hybrid-NNW based flux estimator (Figure 11.58). Torque control is considered in the outer loop with closed-loop stator flux control. The flux loop requires the decoupling current injection (i_{dq}) as shown. The synchronous current control loops generate the voltage commands to the vector rotator, which generates the voltage vector for the SVM modulator using Eqs. (1) and (2), where θ_e = orientation angle between d^s and d^e axes. Note that the NNW implementation of the vector rotator is trivial. The PCLPF integrator generates the stator flux vector components from the machine terminal voltages and currents. The signal estimation block (described in next figure) then generates all appropriate feedback signals required for control.

FIGURE 11.61   Summary of equations for PCLPF integration and signal estimation block.

1. PCLPF-BASED INTEGRATION:

$$\psi_{ds}{}^{s} = \int (v_{ds}{}^{s} - i_{ds}{}^{s} R_{s}) dt \ \dots\dots(1)$$

$$\psi_{qs}{}^{s} = \int (v_{qs}{}^{s} - i_{qs}{}^{s} R_{s}) dt \ \dots\dots(2)$$

2. SIGNAL ESTIMATION BLOCK:

$$\hat{\psi}_{s} = \sqrt{(\psi_{ds})^{2} + (\psi_{qs})^{2}} \ \dots\dots(3)$$

$$\theta_{e} = \sin^{-1}\left(\frac{\psi_{qs}{}^{s}}{\hat{\psi}_{s}}\right) \ \dots\dots\dots(4)$$

$$i_{ds} = i_{qs}{}^{s} \cos\theta_{e} - i_{ds}{}^{s} \sin\theta_{e} \ ..(5)$$

$$i_{qs} = i_{qs}{}^{s} \sin\theta_{e} + i_{ds}{}^{s} \cos\theta_{e} \ \dots(6)$$

$$i_{dq} = \frac{\sigma L_{s} i_{qs}{}^{2}}{\psi_{ds} - \sigma L_{s} i_{ds}} \ \dots\dots(7)$$

$$\psi_{ds} = \psi_{qs}{}^{s} \cos\theta_{e} - \psi_{ds}{}^{s} \sin\theta_{e} \ ..(8)$$

$$\psi_{qs} = \psi_{qs}{}^{s} \sin\theta_{e} + \psi_{ds}{}^{s} \cos\theta_{e} \ \dots\dots(9)$$

$$T_{e} = \frac{3p}{4}\left[\psi_{ds} i_{qs} - \psi_{qs} i_{ds}\right] \ \dots\dots(10)$$

$$\omega_{e} = \frac{\left[(v_{qs}{}^{s} - i_{qs}{}^{s} R_{s})\psi_{ds}{}^{s} - (v_{ds}{}^{s} - i_{ds}{}^{s} R_{s})\psi_{qs}{}^{s}\right]}{\hat{\psi}_{s}{}^{2}} \ \dots(11)$$

All feedback signal estimation equations in the previous figure are summarized here using the standard symbols. The hardware filter effect has been included in Eqs. (1) and (2). All remaining equations are computed in the signal estimation block. This block also requires input from the $i_{ds}{}^{s}$, $i_{qs}{}^{s}$, $v_{ds}{}^{s}$, and $v_{qs}{}^{s}$ signals, but that is not shown for simplicity. Note that PCLPF requires the frequency (ω_{e}) signal, which is estimated in a circulatory manner by using its output flux vector signals. Also note that stator flux $\hat{\psi}_{s}$ is oriented to the d^{e} axis in stator flux-oriented vector control. All equations in the signal estimation block can be implemented by a single feedforward NNW chip similar to that shown in Figure 11.33(b).

FIGURE 11.62 Speed sensorless direct vector control of an EV drive with neuro-fuzzy-based stator flux estimation.

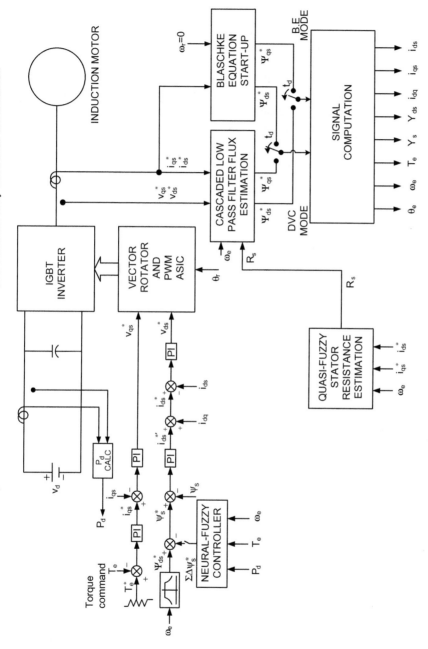

The figure shows the simplified block diagram for a complete stator flux-oriented vector drive for an electric vehicle [20]. The drive has torque and stator flux control in the outer loops, where the flux is programmed as a function of speed so that the drive can operate in a constant-torque as well as field-weakening regions. The flux from the function generator is subtracted by the output of the neuro-fuzzy controller (described later) to establish the command flux. The sinusoidal PWM control in the drive is implemented by an ASIC chip, the details of which are not described. The machine terminal voltages and currents are processed through DSP-based PCLPF and signal computation blocks as described before to generate the feedback signals. Of course, NNW-based PCLPF and SVM could also have been used in this case. The quasi-fuzzy stator resistance estimation block (Figure 10.51) estimates the correct R_s for the PCLPF. The signal computation block also receives the flux signals from the Blaschke equation (BE) mode during start-up (at zero speed but finite torque) (see Figure 7.65), but transitions to the direct vector control (DVC) mode with a time delay t_d, which was discussed in Figure 7.66. As the drive speed falls to zero in the PCLPF mode, it transitions back to the BE mode before starting again. A hysteresis band gap is maintained between the DVC and BE modes so that undue jerking does not occur during the transition.

FIGURE 11.63 Neuro-fuzzy-based efficiency optimization control.

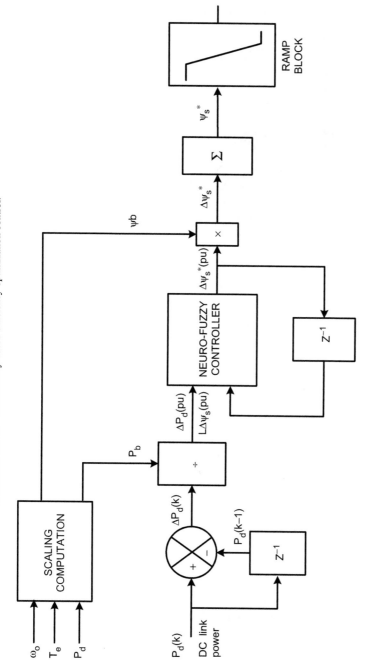

Fuzzy logic-based on-line efficiency improvement by rotor flux programming was discussed in Figures 10.35 and 10.36. This figure shows a fuzzy controller similar to that of Figure 10.36 for stator flux programming in the previous figure, where $\Delta\psi_s$ is the output signal instead of Δi_{ds} in Figure 10.36. In this case, the fuzzy controller is replaced by a feedforward NNW and the output ψ_s^* is ramped to prevent any abrupt change. Because both fuzzy logic and NNW basically implement input–output nonlinear mapping, it is logical that a fuzzy controller can be substituted by an equivalent neural controller. In this case, the input–output data table created by the designed and optimized fuzzy controller was used to train the NNW controller. Obvious advantages accrue with a NNW implementation in a controller.

FIGURE 11.64   Topology (2-10-1) of the neuro-fuzzy controller.

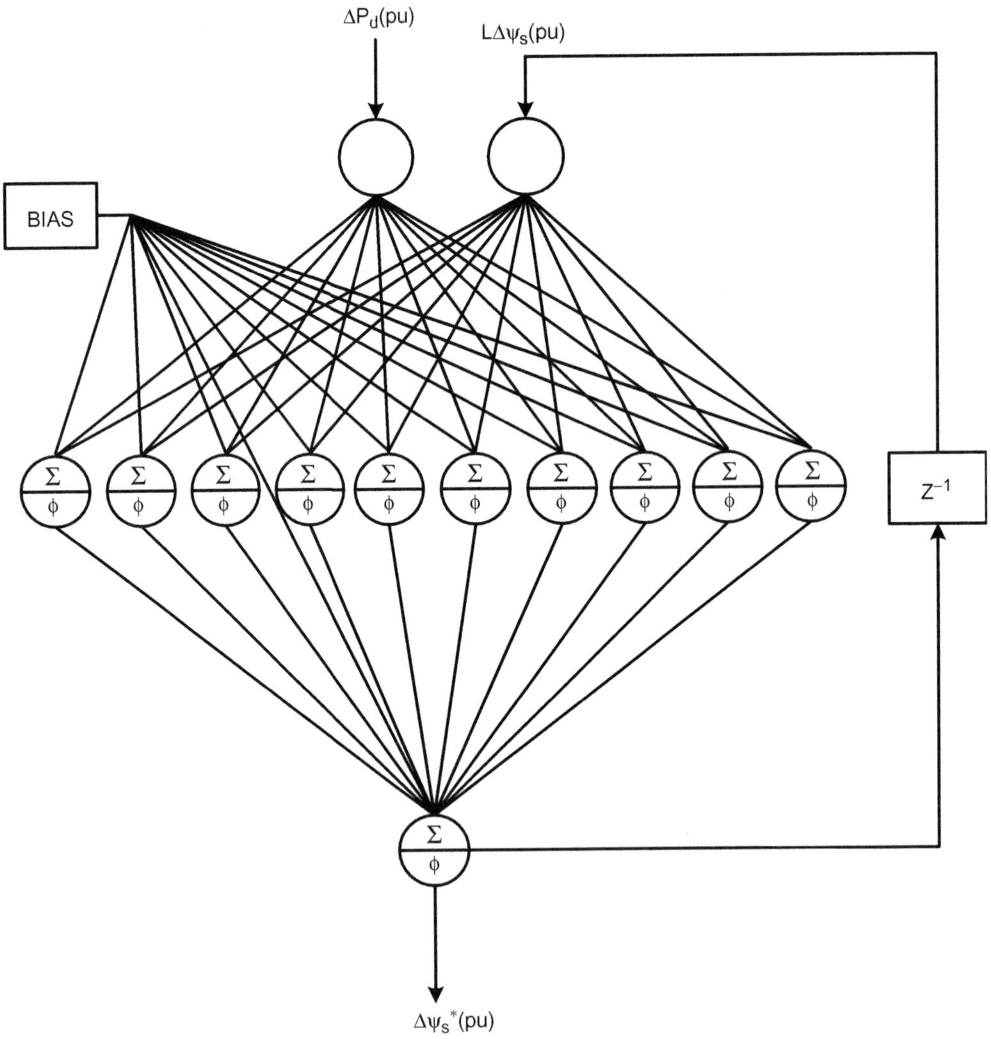

The figure shows the NNW topology of the neuro-fuzzy controller from the previous figure. The three-layer NNW has two inputs [$\Delta P_d(pu)$ and $L\Delta\psi_s^*(pu)$ or last-sampled $\Delta\psi_s^*(pu)$] and one output [$\Delta\psi_s^*(pu)$]. In addition, the bias input is shown. Both hidden and output layers use a hyperbolic-tan type TF. Because of the feedback signal with a sampling time delay ($T_s = 0.5$ sec), a dynamic backpropagation algorithm could be used for training.

FIGURE 11.65   DTC neuro-fuzzy controller: (a) DTC control block diagram and (b) neuro-fuzzy controller.

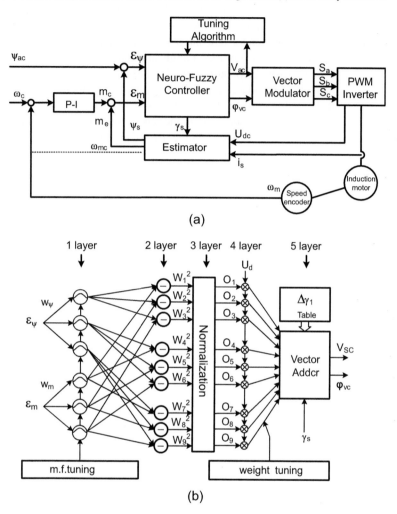

(a)

(b)

Conventional DTC control with HB PWM and its features was described in Figures 7.19 through 7.22. Performance of DTC control can be improved by the ANFIS technique described in Figure 11.21. The ANFIS-based DTC control block diagram [21] is given in (a) using somewhat different terminology, and the detailed neuro-fuzzy (NF) controller is shown in (b) of this figure. The 1 Layer receives the flux error (ε_ψ) and torque error (ε_m) signals multiplied by the respective weights w_ψ and w_m to the three bell-shaped MFs in both inputs. For simplicity of computation, triangular MFs can be chosen. The 2 Layer calculates a minimum of the input signals to generate W_1^2, W_2^2, etc., signals. The 3 Layer normalizes the input signals to calculate O_i (for ith line), which is then multiplied by dc link voltage U_d to generate V_{Sci}, where V_{Sci} is the ith component amplitude of the reference voltage vector. These vectors are added to deliver V_{sc} to the modulator. The angle of V_{sc} is calculated by the relation $\varphi_{Vc} = \gamma_s + \Delta\gamma$, where γ_s = actual stator flux vector angle and $\Delta\gamma_I$ = increment angle from $\Delta\gamma_I$ – Table as shown depending on the polarities (P, Z, N) of the ε_ψ and ε_m signals. The output vector magnitude (V_{sc}) and phase angle (φ_{Vc}) signals then generate the PWM signals (S_a, S_b, and S_c) through the SVM modulator. The ANFIS structure is tuned off-line by a least-squares error method for output MFs and a backpropagation method for output and input MFs. The NF controller adds complexity but smoothes the DTC performance. The estimator uses voltage model equations to estimate stator flux and torque.

FIGURE 11.66   NNW-based inverse dynamics (F^{-1}) MRAC control of a dc drive.

FIGURE 11.66   NNW-based inverse dynamics (F^{-1}) MRAC control of a dc drive.

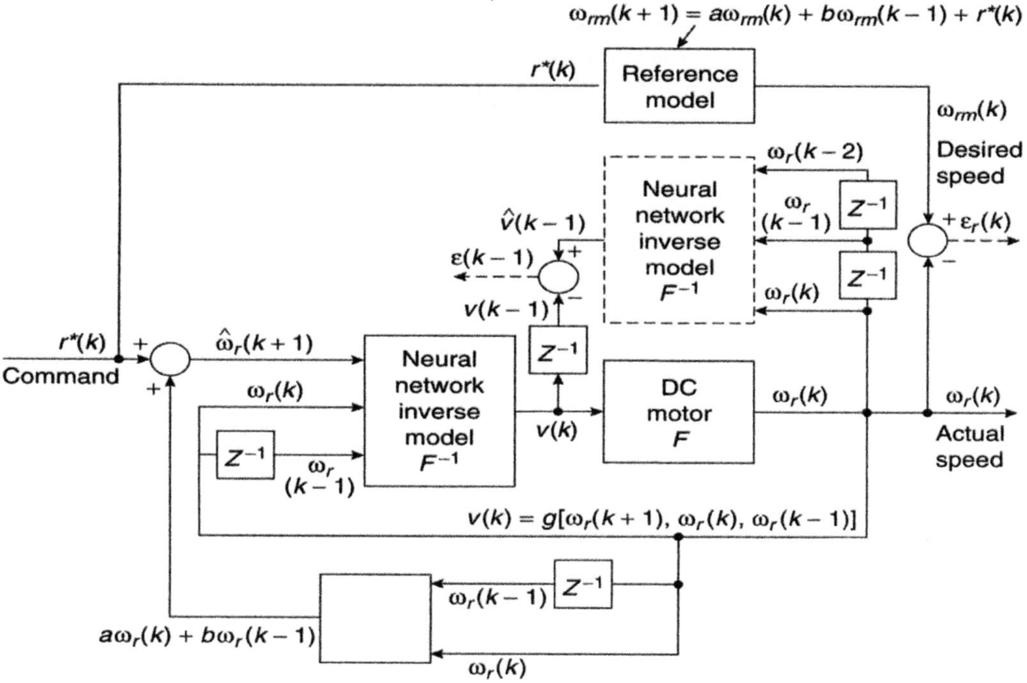

This NNW application describes an inverse dynamic model based on indirect model referencing adaptive control (MRAC) of a dc motor drive [22], where it is desirable for the motor speed to follow the desired speed trajectory from a reference model. The reference model equation (2nd order) in finite difference form is shown in the figure. The same concept is applicable to a vector drive that has a dc machine-like model. The dc motor to be controlled is shown by block F, which has a second-order nonlinear model shown below the block. It is assumed that the motor has a square-law load torque $[T_L(t) = \mu\omega_r^2(t)[\text{sign}\,(\omega_r\,(t))]]$ which contributes to nonlinearity. The two derivatives due to armature circuit di/dt and mechanical load $d\omega_r/dt$ contribute to the second-order system. The applied voltage is a function of speed, and in finite difference form, the variables are $\omega_r(k + 1)$, $\omega_r(k)$, and $\omega_r(k - 1)$. At first, a NNW-based inverse machine model (F^{-1}) was trained off-line by the input–output data generated by the machine. The F^{-1} network does not have any dynamics and is represented by feedforward three-layer NNW with five hidden layer neurons, where the input time-delayed speed signals are generated from actual machine speed by a delay line shown in the figure. The NNW essentially solves the equation, which follows the machine model except for an additional delay. After training the inverse model successfully, it is removed and placed in the forward path so as to cancel the machine forward model F entirely. With the command signal $r*(k)$ to the reference model, it generates the reference speed $\omega_{rm}(k)$, which the motor is required to follow. Considering that the tracking error $\varepsilon_r(k)$ tends to be zero, $\omega_r(k) = \omega_{rm}(k)$, and the speed at $(k + 1)$th time $\hat{\omega}_r(K+1)$ can be predicted from the model equation and the command signals for F^{-1} can be synthesized and impressed as shown in the figure so that it generates the estimated voltage $v(k)$ for the motor. If there are any parameter variation problems in the machine, the inverse model can be updated by training it on-line.

FIGURE 11.67 NNW-based inverse dynamics control of a multiple degree-of-freedom robotic manipulator.

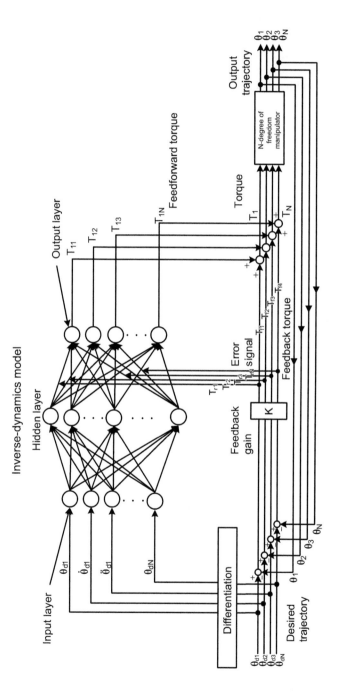

Multiple-axis robotic manipulators are used in manufacturing systems. This figure shows an N-axis manipulator with N degrees of freedom, where each axis of the robot has a drive that can be either dc or ac. These individual axis drives are nonlinear and they have a mutual coupling problem. How can we eliminate the nonlinearity and solve the coupling problem? The NNW-based MRAC drive discussed in the previous figure can be extended to the multiple-axis drive. In the figure, each desired position signal (θ_{d1}, θ_{d2}, etc.) establishes a desired trajectory, and the corresponding output signals (θ_1, θ_2, etc.) are the actual trajectories. The NNW-based inverse-dynamics model (F^{-1}), shown in the figure, represents the inverse model of the whole N-degree-of-freedom manipulator (F). It was trained off-line with the data of the actual manipulator and then connected in series as shown. Each axis model is a second-order dynamic system; therefore, each command signal and its two derivates (θ_{d1}, $d\,\theta_{d1}/dt$, and $d^2\theta_{d1}/dt^2$) are the input signals to the inverse model for that axis. The output of the inverse model is torque, which constitutes the input for the manipulator for that axis. If the F^{-1} model is accurate, it cancels the forward model F, and the desired trajectory reflects to the output trajectory. Because of the parameter variation of the manipulator, a feedback loop is added, and the feedback torque adds to the F^{-1}-generated feedforward torque. The error signals can be used for on-line training of the NNW as shown so that they tend to vanish. The system is very complex and has not yet been implemented in practice.

FIGURE 11.68 Some neural network IC chips.

- INTEL 80170NX ETANN (ELECTRICALLY TRAINABLE ANALOG ANN)
  - 64 NEURONS
  - 10,240 SYNAPTIC WEIGHTS
  - DEVELOP APPLICATION USING SIMULATORS AND DEVELOPMENT SYSTEM
  - COMBINE CHIPS FOR BIGGER SYSTEM
    (This chip has been withdrawn from the market.)
- MICRO DEVICE MD 1220NBS (Neural Bit Slice)
  - COMPLETELY DIGITAL
  - 8 NEURONS
  - 16-BIT WEIGHTS
- AMERICAN NEUROLOGIX NLX 420
  - MD1220 ARCHITECTURE
- NEURAL SEMICONDUCTOR NUSU32
  - 32 NEURONS
  - 1024 WEIGHTS IN SRAM

Once the NNW has been designed and tested successfully in a simulation, it requires practical implementation. If cost and speed are of no concern, the NNW can be executed directly from the PC-based simulator (discussed later). Currently, most NNWs are implemented by DSPs using sequential computation. Sequential computation is slow and does not have fault tolerance. However, DSP computation speeds and functionality are continuously improving. For higher speeds, multiprocessor systems (such as C40s, RISCs, or transputers) can be used with task sharing; this, however, tends to increase the cost. Of course, standard cell ICs and FPGAs can also be programmed for this [3] to a limited extent. However, a viable ASIC chip at an economical price does not currently exist commercially, although R&D in this area is very intensive. Several commercial IC chips are summarized briefly in this figure. Generally, they can be analog, hybrid, or digital, but a digital solution is the best. Intel's 80170NX ETANN was the first large analog ASIC chip, and it was introduced to the market in 1991. It was recently withdrawn from the market because it had an analog signal drift problem. Several small digital IC chips and their features have been summarized. A number of chips can be combined to implement a larger system, but this solution tends to be expensive. Following the bit-slice concept, MD1220 was possibly the first commercial NNW chip. Each chip has 8 neurons with a hard-limit threshold and eight 16-bit weights with 1-bit input. With bit-serial multipliers, the chip provides 9 MCPS. The NLX 420 chip is similar, but it has 16 neurons. The NUSU32 chip has a larger number of neurons. Digital chips are also available from Philips, Siemens, IBM, Hitachi, etc.

FIGURE 11.69   Architecture of 32-neuron NC3003 chip.

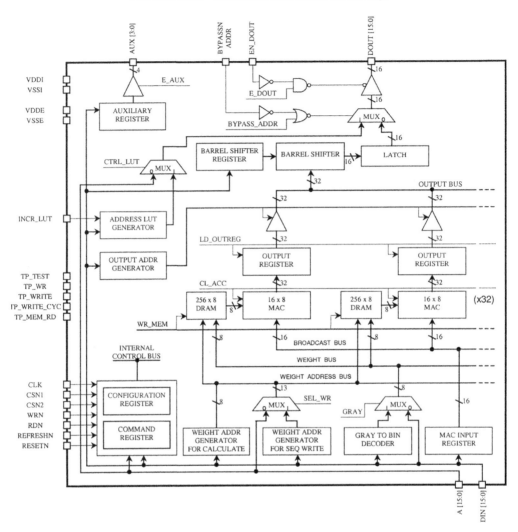

Neuricam of Italy recently introduced a fully digital VLSI NNW chip [25], the architecture of which is shown in this figure. The company has two more chips (NC3001 and NC3002) with slightly different architectures. The salient features of the NC3003 are summarized in the next figure. The chip contains 32 neurons that can be configured either as a single-layer or multilayer NWW. Each neuron is basically a 16×8-bit hardware multiply-and-accumulate (MAC) unit, which is coupled closely with 256×8-bit DRAM weight memory as shown in the figure. The memory bank can either be assigned to a single neuron or be partitioned among neurons of different layers. The pipelined digital data stream is operated in a single instruction multiple data (SIMD) architecture to enhance the speed of operation. The 16-bit data coming in the broadcast bus are multiplied by the 8-bit weight to constitute a 32-bit result, which is scaled by the barrel shifter to constitute the 16-bit output data. The chip supports four external 16-input, 16-output look-up tables (LUT) implemented in static RAM to calculate activation functions of the NWW. Unfortunately, it is not a standalone IC, but operates as a coprocessor in a host system. Up to four chips can be placed in parallel to increase the NWW's size. The vendor supplies an evaluation board including software drivers and graphical tools. Note that in many power electronics applications, 8-bit weight resolution may not be adequate to result in a training error that is sufficiently small.

FIGURE 11.70   Features of Neuricam NC3003 chip.

- DIGITAL VLSI CHIP– 32 PROCESSORS IN PARALLEL
- A PROCESSOR IS BASICALLY A HARDWARE MULTIPLY-
  AND-ACCUMULATE (MAC) UNIT (1.33 ns WITH 25-MHz CLOCK)
- CONFIGURABLE INTO SINGLE-LAYER OR MULTILAYER MLP
  (MULTILAYER PERCEPTRON) TYPE NETWORK
- 16-BIT DATA RESOLUTION
- 8-BIT WEIGHT RESOLUTION
- 64-KBIT DYNAMIC RAM STORAGE OF WEIGHTS
- FOUR (EXTERNAL) 16-BIT SRAM LOOK-UP TABLE
- ACTIVATION FUNCTIONS
- REACTIVE TABU SEARCH (RTS) TRAINING ALGORITHM
- 5-µs FEEDFORWARD COMPUTATION TIME
- MULTIPLE NC3003 CHIPS (UP TO FOUR) CAN BE CONNECTED
  IN PARALLEL WITH PROPORTIONATELY HIGHER COMPUTATION
  TIME
- CHIPS OPERATE IN HOST PROCESSOR ENVIRONMENT

The salient features of the Neuricam NC3003 IC chip discussed in the preceding figure are summarized here. Note that internally the chip consists of software programmable multiple DSPs, where the speed is enhanced by its pipelined architecture. Although individual MAC instruction takes only 1.33 ns, a typical two-unit NC3003 system takes about 10 µs for feedforward computation. This is far slower than the usually expected parallel processing ASIC chip. The training of the chip requires the usual input–output data table, but a special RTS algorithm is used. The advantage compared to the backpropagation algorithm is that it does not require derivatives of error signals, thus avoiding the need for high-precision computation. The fixed-point short-word digital implementation made possible by RTS leads to more economical VLSI architectures than its floating-point, long-word format counterparts, thus saving silicon estate, power dissipation, and speed. It appears that in spite of worldwide intensive research, the ultimate fully functional custom digital ASIC chip for NNW implementation is a far cry from being realized [27].

FIGURE 11.71 (a) Network/Data Manager window and (b) view of new perceptron Network window. (Courtesy of Mathworks, Inc.)

GRAPHICAL USER INTERFACE (GUI) OF MATLAB/NNW TOOLBOX

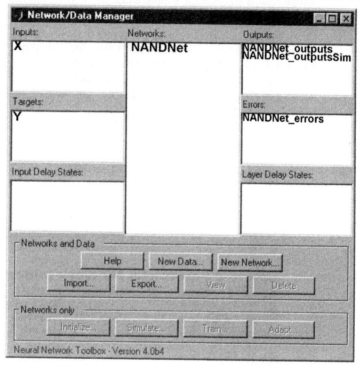

(a)

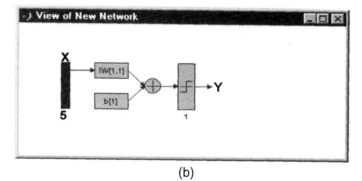

(b)

FIGURE 11.71 (a) Network/Data Manager window and (b) view of new perceptron Network window. (Courtesy of Mathworks, Inc.)

Let us now discuss NNW training by using MATLAB-based Neural Network Toolbox [5], which is in common use. A general flow diagram for NNW training was given in Figure 11.10. The graphical user interface (GUI) is very simple, and we will review the training steps by using a simple example of perceptron training, which is shown in symbolic form in (b). The example problem here is to train a 5-input single-neuron perceptron so that it behaves like a 5-input NAND gate. The perceptron neuron has five inputs each with a series weight, where the weight matrix is IW{1,1}, and there is a bias b{1}. The MATLAB command "nntool" brings up the Network/Data Manager window as shown. Once this window is running, we can create a network, view it, train it and export the results to the workspace with its help. Similarly, we can import data from the workspace to the GUI.

FIGURE 11.72   Perceptron training example of five-input NAND gate: (a) Create New Data window and (b) Create New Network window. (Courtesy of Mathworks, Inc.)

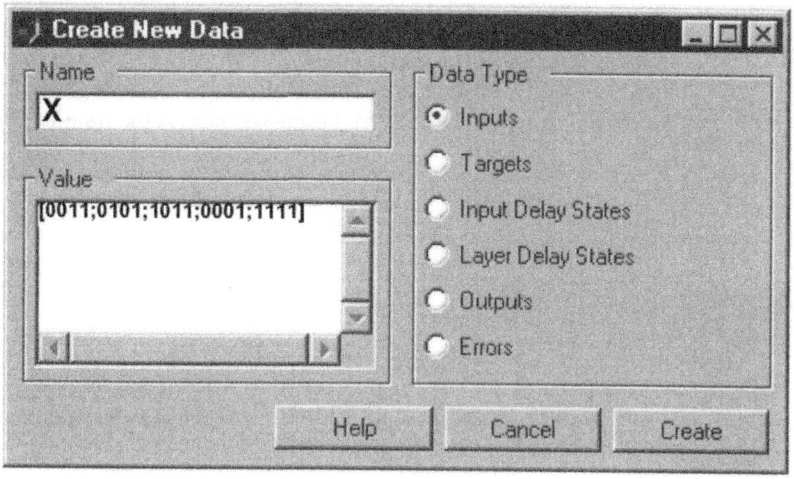

(a)

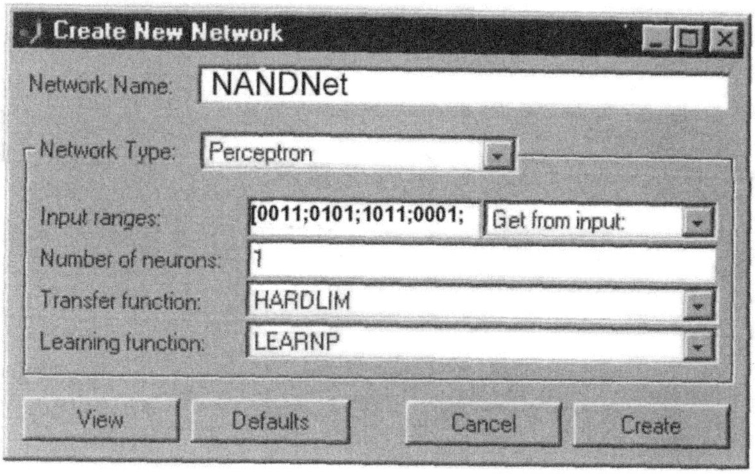

(b)

Let us name the network "NANDNet." We have four sets of five-element input data and the corresponding target data at the output for training. These are defined by the corresponding vectors $X = [0\ 0\ 1\ 1;0\ 1\ 0\ 1;1\ 0\ 1\ 1;0\ 0\ 0\ 1;1\ 1\ 1\ 1]$ and $Y = [1\ 1\ 1\ 0]$ that satisfy the NAND function. Clicking on New Data brings up the Create New Data window in (a) of this figure. The Name (X) and Values are entered and Data Type is marked as Inputs. Clicking Create creates the input file X. Then, the same steps are followed to create the Target data file Y. Both X and Y show up in the window of Network/Data Manager. Next the network is created in the Create New Network window shown in part (b). This window permits selection of various types of NNWs, transfer functions (TFs), and number of neurons. Input ranges are better selected from input X. Once the details are defined, clicking on Create generates the NNW, which can be viewed as shown in part (b). Discussed next is the training phase of the perceptron.

FIGURE 11.73   Perceptron training example of 5-input NAND gate: (a) Network NANDNet window and (b) training error curve. (Courtesy of Mathworks, Inc.)

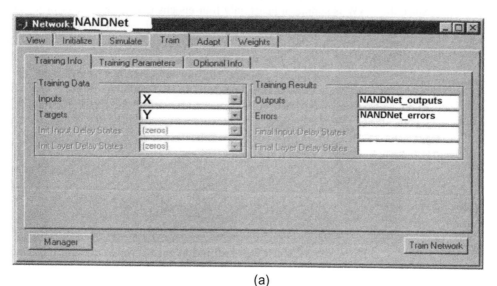

(a)

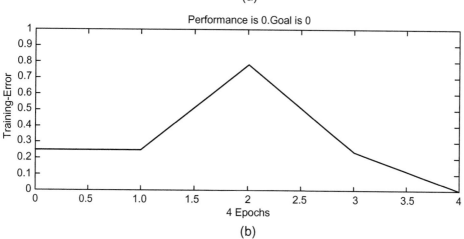

(b)

To train the NNW, click on NANDNet in the Network/Data Manager window to highlight it. Then, clicking Train shows the window Network: NANDNet, as shown in part (a). Click on the top tabs Initialize and then Train. Click Training Info and specify the Training Data as X and Y, respectively. The Training Results Outputs and Errors are specified by the names shown in this figure. Now clicking on Training Parameters shows the epochs and error goals, which can be changed in this window. At this point,

clicking Train Network at the bottom right corner starts the actual training. The training error vs. number of epochs appears in (b) of this figure, where the goal error is zero.
Since the NNW is simple, it takes only four epochs to converge the error to zero. In a
complex NNW, the final error remains finite but small, and the number of epochs may
be large. The network can be simulated to verify its performance. Click on the bottom
Network Only:Simulate on Network/Data Manager to bring up the Network:NANDNet
window. Click on Simulate. Use the pull-down menu X and label the output as
NANDNet_outputsSim to distinguish it from the training output. Now click on
Simulate Network in the lower right corner. In Network/Data Manager, double clicking
on NANDNet_outputsSim gives a small window. Data:NANDNet_outputs appears with
the value [1 1 1 0] that verifies the correct NNW operation. It can be tested with other
values also. The NANDNet, the results (NANDNet_outputs, NANDNet_outputsSim,
NANDNet_errors), and data (X, Y) can be exported to command line workspace. The
command NANDNet. Iw{1,1} shows the network weights as -2 ,-1, 0, -2,1 for X_1, X_2,
X_3, X_4, and X_5 , respectively, and NANDNet.b{1} shows the bias as 1. Simple hand
calculations will verify the NAND gate function. The GUI-trained NANDNet can be
converted to a Simulink model by the gensim(NANDNet) command and then linked
with the simulation of other system elements.

FIGURE 11.74   Backpropagation NNW training example for three-phase sine-wave generation for $0 > X > 2\pi$.

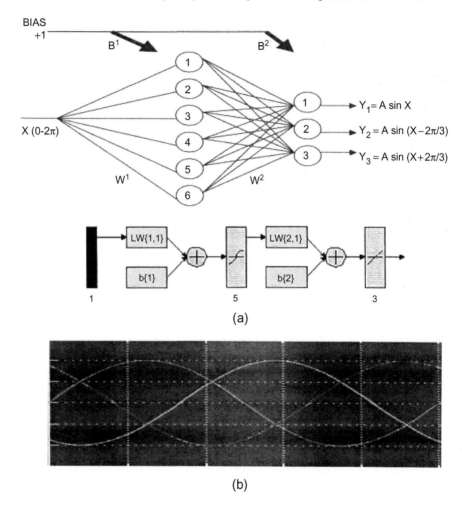

(a)

(b)

After discussing how to use the toolbox, we will now present a more complex demo program [26] to clarify our understanding about NNW training using the same toolbox. In the figure, the NNW receives the signal X in radians (0 to 2π) at the input and generates three-phase sine waves at the output as shown in (b) of this figure. Amplitude A is set to 1 for normalized data handling. A symbolic form of the NNW in the toolbox is shown below the network. It uses a three-layer (sometimes defined as a two-layer) backpropagation network with the topology 1-5-3, using a tan-sigmoid TF at the hidden layer and linear bipolar TF at the output layer. A bias of +1 couples to the hidden and output layer neurons. The bias signals may not be needed, but are presented for generality.

The network weights are not shown for simplicity. The input training data were generated with a step size of 0.1 rad, and the desired output waves were generated by MATLAB sine functions. The input–output training data file in MATLAB was imported to GUI for training. The Levenberg-Marquardt (L-M) backpropagation algorithm with a batch method of training was used.

FIGURE 11.75   Block diagram for testing the NNW (Sinet) in Simulink environment.

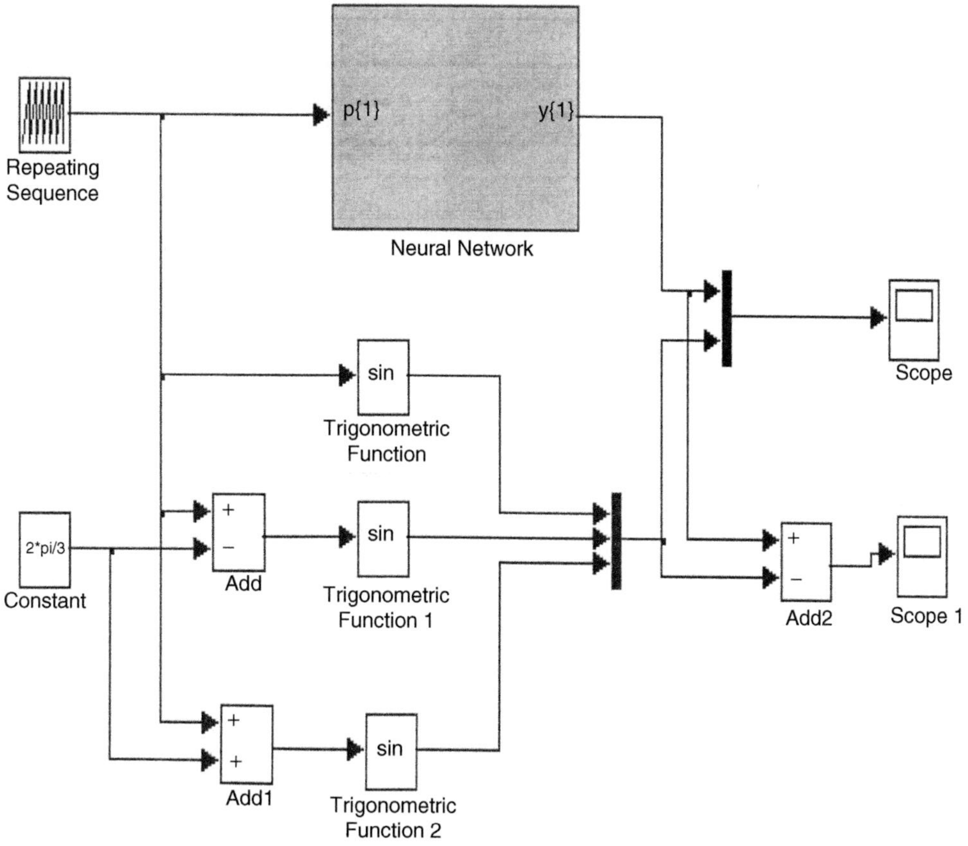

After the network was trained in GUI, it was converted to the Simulink program [by means of the gensim(Sinet) command, where Sinet is the network name] for testing. Testing of the trained NNW in the Simulink environment is shown in this figure. A sawtooth wave was generated for use as the NNW input and three-phase reference sine-wave generation as shown. The three-phase reference waves were compared with the NNW-generated three-phase waves in the scope.

FIGURE 11.76   Training of 1-2-3 network: (a) training error and (b) test results.

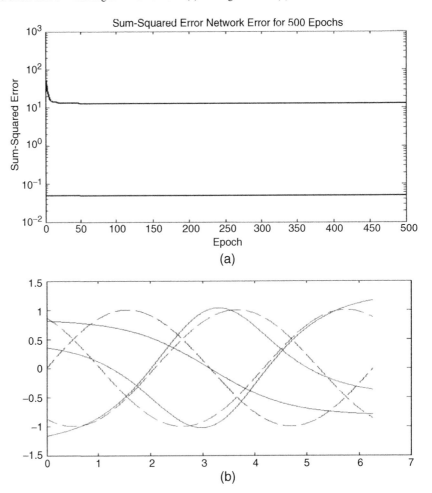

(a)

(b)

In the beginning, only two neurons were selected for the hidden layer. The specified goal of sum-squared error (SSE) was 0.04 and the number of epochs was 500. The SSE failed to converge to the desired value as shown in (a). It practically locked the SSE at the value of 15 at the end of 50 epochs. The Simulink test results in (b) indicate poor correlation of NNW output with the desired waves (dashed curves).

FIGURE 11.77   Training of 1-3-3 network: (a) training error and (b) test results.

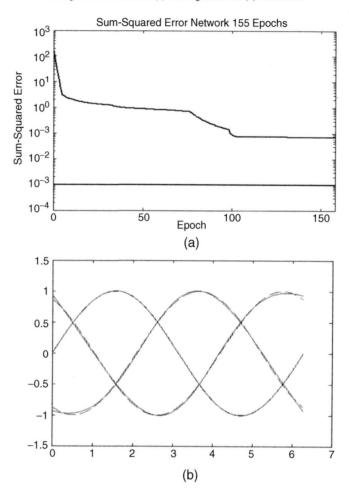

(a)

(b)

Next, the hidden layer was added with an extra neuron (1-3-3) and training was continued with the specified SSE = 1.0×10^{-1} and number of epochs = 155 as shown in (a) of this figure. In this case, the SSE locks as 0.07 at 100 epochs. The corresponding test results are shown in (b). Evidently, the results improved significantly, but it is not yet very satisfactory.

FIGURE 11.78   Training of 1-5-3 network: (a) training error and (b) test results.

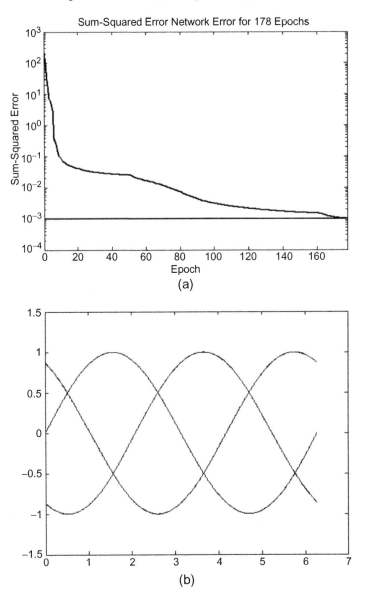

(a)

(b)

The network was next trained with five hidden layer neurons as shown in Figure 11.74. The corresponding results are shown in this figure, and they indicate excellent results. The number of epochs in this case was 178, and SSE converges to the desired goal of 1.0×10^{-3}.

FIGURE 11.79 Matrices of weights and biases for 1-5-3 network after training.

$$\mathbf{W^1} = \begin{bmatrix} W_{11}{}^1 \\ W_{12}{}^1 \\ W_{13}{}^1 \\ W_{14}{}^1 \\ W_{15}{}^1 \end{bmatrix} = \begin{bmatrix} 0.9746 \\ -1.0423 \\ 0.7937 \\ -0.9571 \\ 0.9504 \end{bmatrix} \qquad \mathbf{B^1} = \begin{bmatrix} B_{11}{}^1 \\ B_{12}{}^1 \\ B_{13}{}^1 \\ B_{14}{}^1 \\ B_{15}{}^1 \end{bmatrix} = \begin{bmatrix} -4.9313 \\ 6.5538 \\ -2.5281 \\ 1.1780 \\ -0.1443 \end{bmatrix}$$

$$\mathbf{W^2} = \begin{bmatrix} W_{11}{}^2 & W_{21}{}^2 & W_{31}{}^2 & W_{51}{}^2 \\ W_{12}{}^2 & W_{22}{}^2 & W_{32}{}^2 & W_{52}{}^2 \\ W_{13}{}^2 & W_{23}{}^2 & W_{33}{}^2 & W_{53}{}^2 \end{bmatrix} = \begin{bmatrix} 2.7096 & -4.1398 & -9.5330 & -1.4385 & -5.3972 \\ -8.2493 & -2.1779 & 4.4349 & 7.7484 & -8.3220 \\ 5.5397 & 6.3177 & 5.0981 & 9.1869 & 2.9248 \end{bmatrix}$$

$$\mathbf{B^2} = \begin{bmatrix} B_{11}{}^1 \\ B_{12}{}^1 \\ B_{13}{}^1 \end{bmatrix} = \begin{bmatrix} 3.4111 \\ 10.8206 \\ -14.2317 \end{bmatrix}$$

After successful training, the matrices of the weights and biases are summarized in this table. The matrix W^1 consists of the input weights of the hidden layer, W^2 consists of hidden layer output weights, B^1 consists of the hidden layer biases, and B^2 consists of the output layer biases. The values are given to four decimal places.

FIGURE 11.80   Advances and trends in neural networks for power electronics.

- AI IS EXPANDING THE FRONTIER OF POWER ELECTRONICS – A NEW CHALLENGE TO POWER ELECTRONICS ENGINEERS
- AMONG ALL AI DISCIPLINES, NEURAL NETWORKS WILL HAVE THEIR MAXIMUM IMPACT ON POWER ELECTRONICS
- CURRENTLY MOST APPLICATIONS USE BACKPROPAGATION- TYPE FEEDFORWARD NNWs
- MANY OTHER FEEDFORWARD AND RECURRENT NNW TOPOLOGIES REQUIRE EXPLORATION
- NONAVAILABILITY OF A LARGE NNW ASIC CHIP IS A PROBLEM – MOST APPLICATIONS USE DSP
- FAST ON-LINE TRAINING PROBLEM OF ADAPTIVE NNWs
- POWERFUL INTELLIGENT CONTROL AND ESTIMATION TECHNIQUES CAN BE DEVELOPED USING HYBRID AI (NEURO-FUZZY, NEURO-GENETIC, NEURO-FUZZY-GENETIC, FUZZY-GENETIC) SYSTEM
- SINGLE NEURO-FUZZY ASIC CHIP CAN UNDERTAKE SENSORLESS VECTOR CONTROL WITH ON-LINE FAULT DIAGNOSTICS AND FAULT-TOLERANT CONTROL IN FUTURE
- EXPECTED TO HAVE WIDESPREAD APPLICATIONS IN FUTURE

The emergence of AI technology has undoubtedly expanded the frontier of the power electronics and motor drives area, which is already a complex and multidisciplinary technology. Among all AI disciplines, the present trend indicates that neural networks will possibly have the maximum impact on power electronics. Most of our current applications are based on backpropagation (often called MLP or multilayer perceptron) networks, although a large number of feedforward and recurrent topologies are already available. Many novel applications can be explored with these networks. Unfortunately, large fully functional ASIC neural chips are not yet available [27], and most current applications use DSPs. For this reason, currently, industrial applications of NNWs are very few. Again, by far, the majority of NNW applications use off-line training. However, in a parameter varying or unfamiliar signal environment, adaptive NNW with on-line training is essential. Fortunately, NNW training time is decreasing progressively because of improvements in the training algorithms and rapid increases in computer speeds. Some of the off-line techniques can be used for on-line if the training time demand is not too stringent. Currently, much emphasis is being placed on R&D for hybrid-AI techniques, which will have a significant impact on power electronics.

Glossary

Activation or transfer function (TF)   A unipolar/bipolar linear or nonlinear function that shapes a neuron output.

Adaline   ADAptive LINear Element. An acronym for linear neuron.

ANN or NNW   A model intended to emulate biological nervous system.

ANN Toolbox   MATLAB-based software programs that help to design and test different types of neural networks.

Associative memory   A type of memory where an input pattern serves to retrieve the related pattern.

Backpropagation   A supervised learning rule in which weights are adjusted by error derivatives backpropagated through the network so as to minimize that error.

Backpropagation network   A feedforward network that a uses backpropagation training method.

Distributed intelligence (or memory)   A feature of neural works in which the intelligence or memory is spread throughout the network due to the synaptic weights.

Dynamic backpropagation   A backpropagation training algorithm that can be applied to a time-delayed feedback network.

Epoch   A computation cycle in training of a network that consists of a forward pass and a reverse pass.

Function approximation   The performance of a network so that the output approximates the desired function in response to the corresponding inputs.

Hyperbolic tan (or tan-sigmoid) TF   An S-shaped bipolar differentiable transfer function.

Learning rate   A training parameter (η) that determines the speed of convergence of error by fine-tuning the weights.

Mapping   An output corresponding to an input.

Momentum   A factor (μ) that is used in training a network to avoid a local minimum.

MSE   Mean square error is the averaged square error between the network outputs and desired outputs.

Pattern classification   A property of classifying different input signal patterns by neural networks.

Pattern recognition   A property of recognizing an input signal pattern and generating the related output pattern.

Perceptron   A single-layer network with hard-limit transfer function.

Recurrent ANN (or NNW)   A neural network with feedback.

Sigmoidal (or log-sigmoid) TF   A monotonic S-shaped unipolar and differentiable transfer function.

Supervised learning   A learning method that is supervised by a teacher or a learning algorithm

Unsupervised learning   A method of self-learning in which no teacher help is taken to correct network output.

Summary

This chapter describes the basic principles of neural networks and some of their applications in the power electronics and motor drive area. Neural networks are a vast and very important subject in artificial intelligence, and the technology has advanced tremendously in recent years with applications expanding into many different areas. The discussion is focused here on applications in the power electronics and meter drives area. The idea of NNWs comes from the biological neuron and biological neural network; in practice, however, a NNW's performance may be far inferior. However, NNWs help solve many problems efficiently that are difficult to solve by traditional methods. A biological network operates on an input–output mapping principle, in which the intelligence or memory (called *associative memory*) is distributed in the nervous system by the synaptic junctions of the cells. NNWs also have a similar mapping property. The mapping principle was discussed in detail for fuzzy systems in Chapter 10. Like the supervised training of a biological nervous system, NNWs are also trained by means of sample input–output data sets. This is different from programming a digital computer to solve a problem. A NNW has linear transfer characteristics with a linear activation or transfer function (TF), whereas a nonlinear TF can provide NNWs with a nonlinear mapping property. Most NNW-based projects require nonlinear TFs. Many NNW topologies have been proposed in the literature that can be classified as feedforward and feedback or recurrent NNWs (RNN). Feedforward networks have static nonlinear mapping property, whereas recurrent networks have dynamical or temporal mapping property. Therefore, feedforward NNWs are used for pattern recognition type applications, whereas the latter class is used to emulate dynamical systems. In this chapter, only perceptron and backpropagation feedforward NNWs, and real time recurrent NNWs are discussed. By far, the majority of power electronic applications use feedforward backpropagation-type NNWs. It can be shown that a three-layer topology can solve most problems, but occasionally, with more layers, the number of neurons and/or training time can be economized. Note that a three-layer NNW is often defined as a two-layer NNW because the input layer is essentially the distributor of signals. Both the hidden and output layer neurons use nonlinear TFs, but the output layer can also use linear TFs. The example data sets for training can be gathered from experiment or simulation provided that valid mathematical models exist for plant simulation. The training is done automatically with the help of a software program. Among a number of programs available commercially, the MATLAB-based Neural Network Toolbox is very popular. The graphical user interface (GUI) for using the toolbox has been described along with the training examples for perceptron-based NAND gate (NANDNet) and backpropagation NNW-based three-phase sinusoidal function generation (Sinet). The reader should be able to design and train NNWs easily with this user-friendly tool for solving his own problems. Once the NNW is designed successfully, it can be linked with a Simulink-based simulation of the entire power electronic system (if the model is available). A number of ANN applications in power electronics system have been described in the chapter: PWM of a voltage-fed inverter, waveform processing and delayless filtering, feedback signal estimation for vector drives, estimation of distorted waveforms, intelligent adaptive control of dc and ac drives, speed estimation, flux estimation by PCLPFs, neuro-fuzzy-based sensorless vector and DTC controls, etc. In particular, the space vector PWM technique has been emphasized because the nonlinear mapping property of NNWs suits this application very well, where complex DSP-based computations are usually used. All applications discussed here can be explored in detail by referring to the literature references given next. Unfortunately, NNW applications are very few in industry because of the current nonavailability of ASIC NNW chips. In a laboratory environment, a NNW has often been implemented in real time by high speed PC-based processors which interface the physical system under test. The usual PC or DSP-based serial implementation of NNWs not onlymakes the system expensive, but the speed and fault-tolerance properties are lost. Fortunately, the DSP speed is increasing dramatically in recent years. Of course, parallel DSPs can enhance computation speeds, but at the expense of higher cost. In the last several years, FPGAs have been applied to implement NNWs with somewhat reduced capabilities [3]. With the current R&D trends, it is expected that viable and inexpensive ASIC chips will be available in the future. Undoubtedly, NNW-based control and estimation will have a large impact on the power electronics and drives area in the future. With the current trend of R&D in powerful hybrid AI techniques and economical ASIC chips, it is expected that neural network-based control and estimation will have a large impact on power electronics and motor drives area in the future.

References

[1] S. Haykin, *Neural Networks*, Macmillan, New York, 1994.

[2] H. S. Tsoukalas and R. E. Uhrig, *Fuzzy and Neural Approaches in Engineering*, John Wiley, New York, 1997.

[3] M. N. Cirstea, A. Dinu, J. G. Khor, and M. McCormick, *Neural and Fuzzy Logic Control of Drives and Power Systems, Newnes*, Elsevier, 2002.

[4] B. K. Bose, *Modern Power Electronics and AC Drives*, Prentice Hall, Upper Saddle River, NJ, 2002.

[5] MathWorks, *Neural Network Toolbox User's Guide*, March 2001.

[6] MathWorks, *Fuzzy Logic Toolbox User's Guide,* 1998.

[7] B. K. Bose, "Expert system, fuzzy logic, and neural network applications in power electronics and motion control," *Proc. IEEE*, vol. 82, pp. 1303–1323, August 1994.

[8] B. K. Bose, "Artificial intelligence techniques—a new and advancing frontier in power electronics and motor drives," *Proc. Intl. Pow. Elec. Conf.,* Niigata, April 2005.

[9] A. M. Trzynadlowski and S. Legowski, "Application of neural networks to the optimal control of three-phase voltage-controlled power inverters," *IEEE IECON Conf. Rec.*, pp. 524–529, 1992.

[10] J. Zhao and B. K. Bose, "Neural network based waveform processing and delayless filtering in power electronics and ac drives," *IEEE Trans. Ind. Elect.*, vol. 51, pp. 981–991, October 2004.

[11] M. G. Simoes and B. K. Bose, "Neural network based estimation of feedback signals for a vector controlled induction motor drive," *IEEE Trans. Ind. Appl.*, vol. 31, pp. 620–629, May/June 1995.

[12] M. H. Kim, M. G. Simoes, and B. K. Bose, "Neural network based estimation of power electronic waveforms," *IEEE Trans. Power Electron.*, vol. 11, pp. 383–389, March 1996.

[13] J. O. P. Pinto, B. K. Bose, L. E. B. da Silva, and M. P. Kazmierkowski, "A neural network based space vector PWM controller for voltage-fed inverter induction motor drive," *IEEE Trans. Ind. Appl.*, vol. 36, pp. 1628–1636, November/December 2000.

[14] S. Mondal, J. O. P. Pinto, and B. K. Bose, "A neural network based space vector PWM controller for a three-level voltage-fed inverter induction motor drive," *IEEE Trans. Ind. Appl.*, vol. 38, pp. 660–669, May/June 2002.

[15] S. K. Mondal, B. K. Bose, V. Oleschuk, and J.O. P. Pinto, "Space vector pulse width modulation of three-level inverter extending operation into overmodulation region," *IEEE Trans. Power Electron.*, vol. 18, pp. 604–611, March 2003.

[16] C. Wang, B. K. Bose, V. Oleschuk, S. Mandal, and J. Pinto, "Neural network based SVM of a 3-level inverter covering overmodulation region and performance evaluation on induction motor drives," *IEEE IECON Conf. Rec.*, pp. 1–6, 2003.

[17] N. P. Filho, J. O. P. Pinto, B. K. Bose, and L.E.B. da Silva, "A neural network based space vector PWM of a five-level voltage-fed inverter," *IEEE IA Society Conf. Rec.*, 2004.

[18] L. E. B. da Silva, B. K. Bose, and J.O.P. Pinto, "Recurrent neural network based implementation of a programmable cascaded low pass filter used in stator flux synthesis of vector controlled induction motor drive," *IEEE Trans. Ind. Electron.*, vol. 46, pp. 662–665, June 1999.

[19] J. O. P. Pinto, B. K. Bose, and L.E.B. da Silva, "A stator flux oriented vector controlled induction motor drive with space vector PWM and flux vector synthesis by neural networks," *IEEE Trans. Ind. Appl.*, vol. 37, pp. 1308–1318, September/October 2001.

[20] B. K. Bose and N. R. Patel, "A sensorless stator flux oriented vector controlled induction motor drive with neuro-fuzzy based performance enhancement," *IEEE IAS Annu. Meet. Conf. Rec.*, pp. 393–400, 1997.

[21] P. Z. Grabowski, M. P. Kazmierkowski, B. K. Bose, and F. Blaabjerg, "A simple direct-torque neuro-fuzzy control of PWM-inverter-fed induction motor drive," *IEEE Trans. Ind. Electron.*, vol. 47, pp. 863–870, August 2000.

[22] S. Weersooriya and M.A.El-Sharkawi, "Identification and control of a dc motor using backpropagation neural networks," *IEEE Trans. Energy Conversion*, vol. 6, pp. 663–669, December 1991.

[23] K. J. Hunt et al., "Neural networks for control systems—survey," *Automatica*, vol. 28, pp. 1083–1112, 1992.

[24] K. S. Narendra and K. Parthasarathy, "Identification and control of dynamical system using neural networks," *IEEE Trans. Neural Networks*, vol. 1, pp. 4–27, March 1990.

[25] Neuricam , *NC3003 Totem Digital Processor for Neural Networks Data Sheet*, December 1999; http://www.neuricam.com.

[26] J. O. P. Pinto, "Artificial neural networks based three-phase sine function generation—a demo program," Internal Memorandum for The University of Tennessee, Knoxville, January 2005.

[27] L. M. Reyneri, "Neuro-fuzzy hardware: design, development and performance," *Proc. FEPPCON III*, Kruger National Park, South Africa, July 1998.

CHAPTER 12

Some Questions and Answers

[*Note:* This chapter supplements the main contents of the book, and often some amount of redundancy has been introduced intentionally. The answers consist of the author's own opinion and words based on his knowledge and experience in the subject.]

Q. What is the global energy generation scenario and the role of power electronics in that perspective?

A. Energy consumption in the world has been increasing by leaps and bounds in order to improve our standard of living. In fact, per-capita energy consumption has been the barometer of a nation's economic prosperity. The United States has the highest living standard in the world. It has around 5% of the world's population (6.5 billion), but consumes 25% of the total energy. Japan on the other hand, consumes 5% of total energy with 2% of world population. China and India together, with 38% of the world's population, consume only one-tenth the amount consumed by the United States. However, some developing countries (such as India and China) are now marching toward massive industrialization, which carries with it a corresponding increase in energy demand. As a result, we are experiencing a global energy shortage that is expected to have far-reaching consequences in the future. At present, around 28% of global total energy comes from coal, 21% from natural gas, 38% from oil, 6% from nuclear plants, and the remaining 7% from renewable sources (mainly hydro and wind). The U.S. energy generation scenario follows a similar pattern. It is interesting to note that 42% of U.S. energy comes from oil (mainly for automobile transportation), around 58% of which is imported from outside the United States. In the United States, approximately 37% of total consumed energy is in electrical form, of which roughly 55% comes from coal and 20% from nuclear plants. Unfortunately, fossil fuel (coal, oil, and natural gas) and nuclear resources in the world are limited. Surveys have indicated that with the current rate of consumption, the oil reserve may last around 100 years, the gas reserve around 150 years, and the coal reserve around 200 years, and natural uranium fuel for 50 years. Does this mean that the wheels of our civilization will come to a screeching halt beyond the 23rd century when all the energy resources are exhausted? Of course, there are possibilities of new resources in these areas. Ethanol and bio-diesel can supplement our oil needs. Breeder reactor can generate additional nuclear fuel. Unfortunately, nuclear energy has potential safety problems. In addition, nuclear waste remains radioactive for thousands of years, and we do not know yet how to dispose of it satisfactorily. Besides the limited reserve of fossil fuels, their environmental pollution effect is a serious concern in our society. The fusion energy scenario, on which we had placed so much hope, has not yet shown much promise thus far. Under the preceding scenario, energy costs will tend to rise dramatically in future, which will affect our quest to improve our standard of living. The current emphasis on "green" power (wind, photovoltaics, and fuel cells) will tend to ameliorate this condition to some extent, but will not solve the total energy problem. There is no doubt that power electronics will be asked to play an increasingly significant role in energy conservation by realizing more efficient utilization of electricity. According to an Electric Power Research Institute (EPRI) estimate, extensive (but viable) applications of power electronics can save approximately 15% of our total electricity consumption. Besides the economic benefit, the corresponding positive impact on the environmental pollution problem will be significant. Of course, the role of power electronics in renewable energy systems is evident [1].

Q. What is high-temperature superconductivity (HTS)? How is it important in power electronics?

A. A superconductor is defined as an element, intermetallic alloy, or compound that will conduct electricity without any resistance below a certain temperature. If there is no resistance, the power loss in the conductor is eliminated, and this becomes important, particularly in high-power applications. Ideally, once set in motion, electrical current in a superconducting electromagnet will flow forever in a closed loop. Practically, there will be finite loss and, therefore, the electromagnet requires occasional charging.

Superconductivity has a long and fascinating history of evolution. The traditional superconductor is based on liquid helium (boiling point ~4° K). This means that when the conductor is immersed in liquid helium, it loses its resistance entirely. High-temperature superconductivity (HTS) based on liquid nitrogen (boiling point ~77° K) was invented in 1986 by A. Muller and G. Bednorz, which made superconductors more practical for industrial applications. HTS wire is made of brittle ceramic material that consists of a very complex compound commercially known as BSCCO-2223. The current-carrying density is typically 1000 A/mm^2. With a lower cryogenic cooling temperature, the current density can be increased, whereas with a higher temperature, the density is decreased. Above a critical current density, the material loses its superconductivity property. HTS wire is extremely important to build powerful but lossless electromagnets that do not need any iron core. The size of the magnet becomes much smaller than that of a conventional electromagnet. HTS machines can be built that are completely ironless.

HTS has been applied in many applications that include SMES (superconducting magnet energy storage), large wound-field synchronous machines (motors and generators), fault current limiting transformers in utility systems, lossless power transmission through cables, linear synchronous motor MAGLEV transportation, and MRI machines. Power electronics is intimately involved in most of these applications, but note that power converters themselves do not operate at cryogenic temperature. In SMES, utility power is normally converted to dc (see Figure 5.32) to charge the superconductive coil, and the stored energy can be converted back to the ac line through the converter. In FACTS (flexible ac transmission system), SMES operating through the STATCOM can help improve the stability of the power grid. High-power multi-megawatt WFSM with HTS cooling of field winding has been considered for naval ship propulsion (Figure 8.62). The HTS machine efficiency is higher with lighter weight, and the ironless construction permits magnetic noise-free operation. In a MAGLEV system (Figure 4.31), HTS magnets permit magnetic levitation as well as cooling of the linear synchronous machine field winding losslessly on the carriage. Large dc link inductors in LCI drives also can use the HTS. In every application, the cost and reliability of cooling equipment should be critically considered against the benefits.

Q. What is the scenario on hybrid vehicles?

A. Although hybrid vehicles are still expensive as of this writing, as well as complex and technologically immature, suddenly they appeared in the limelight because of the rising price of gasoline fuel. Basically, a hybrid is an IC engine vehicle that is assisted by a storage battery. The battery has the important function of recovering the braking energy, which is wasted in the mechanical braking of a gasoline vehicle. This contributes to the improved mileage compared to gasoline vehicles. The battery, in addition, assists the engine to boost the vehicle power needed during acceleration and driving in upward slope. A typical input-split power system for a hybrid vehicle (Toyota Prius) [2] is shown in Figure 12.1. During starting and moving at low speed, it runs in pure EV mode,

FIGURE 12.1   Hybrid vehicle drive system.

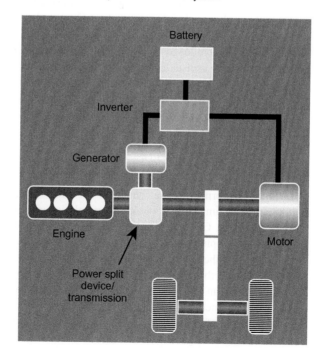

extracting power from the battery. In this mode, battery power is inverted by an IGBT PWM inverter to supply to the IPM synchronous motor that drives the axle. The battery is a Ni-MH (nickel–metal hydride) type, but it can also be a traditional lead acid, nickel-cadmium (Ni-Cd) or advanced lithium-ion (Li-ion) type. In normal driving, the vehicle runs in engine mode and the surplus engine power can be used to replenish the battery charge. This means that part of the engine power directly goes to the axle, and the remaining power is converted to electrical form by a PM generator and rectified through the rectifier (inverter acting in rectifier mode) for charging the battery. When the vehicle is in the garage, the battery can also be charged from the grid (although not in the Prius) through a rectifier (defined as plug-in HV). Of course, this involves extra cost. Besides, the distribution system capability with large number of HVs charging simultaneously should be carefully considered. During acceleration or running on a positive slope, the battery assists the engine power because the engine is not powerful enough. In this mode, the engine power goes to the shaft directly, whereas the battery power through the inverter is fed to the drive motor. However, during the braking mode, the stored energy in the vehicle is recovered, when the drive motor acts as a generator and the inverter acts as a rectifier to charge the battery. Of course, the battery has to be designed for a full-braking power rating. Recovering the braking energy in repeated start–stop cycles in urban driving improves the mileage/gallon rating of the car, but there is no such gain in highway driving. Unfortunately, the battery life becomes shorter with repeated charge–discharge cycles, and its replacement every few years becomes expensive. It remains a hidden cost to the hybrid buyer. In pure EV mode, when slow driving in the city, there is no urban pollution problem, but engine charging of the battery is inefficient and contributes to the usual emission problem. On the other hand, the grid charging the battery contributes to central power station pollution. In sustained highway driving in IC engine mode, there is no mileage improvement by regenerative braking. Besides, the pollution problem remains as usual. In a hybrid vehicle, possibly an ultracapacitor bank instead of a battery would be a good solution in the long run because its power rating is inherently high, has better efficiency, and there is no deterioration with cyclic power.

Note that a fuel cell vehicle (described in Figure 1.13) with an ultracapacitor or battery can also be defined as a hybrid vehicle.

Q. What really happened to the MCT, which was once claimed to be a very promising device?

A. Since the formal announcement of the MCT (MOS-controlled thyristor) by GE in 1988, it was claimed that it would be a revolutionary device of the future that would challenge thyristors, GTOs, and even IGBTs. Since then, enormous amounts of funds were spent on development of the device, although some power semiconductor experts were predicting that the device's performance as claimed was not feasible, particularly given the information that Siemens' previous project on MOS-GTO development did not succeed. Note that an MCT is a thyristor-like device that turns on and off by means of a MOS gate. The asymmetric blocking device has a thyristor-like low conduction drop (somewhat lower than that of an IGBT), but it has a tail time at turn-off that is worse than that of an

IGBT. Besides, the SOA (safe operating area) of the device is non-square (unlike IGBT), meaning a snubber is required in hard-switching converter applications. Apparently, the MCT is inferior to the IGBT in hard-switched converter (i.e., lower switching frequency, and snubberless operation is not possible) except that its lower conduction drop tends to offset the snubber losses. Apparently, MCT devices match well in soft-switched converters, where the tail current effect and poor SOA are not important. There were vigorous attempts to apply MCTs in soft-switched converters, but the attempt did not succeed because the soft-switched converter technology itself did not fly for motor drives and other power applications.

Q. Are IGBTs or IGCTs better for high-power multi-megawatt converters?

A. Both IGBT and IGCT devices are currently marching for higher power applications. Both devices are now available with forward blocking (asymmetric) and reverse voltage blocking (symmetric blocking) capabilities. However, high power IGBT is normally forward blocking. The IGBT PWM converters have been used in the 1- to 2-MW range, and have been the trend for higher power applications. The history of the IGCT is more recent (introduced in 1996), and it is closer to a GTO device. Because of high voltage and current ratings, IGCTs are inherently suitable for multi-megawatt power ratings. A general comparison of the two devices under high-power conditions is as follows:

- Both are competitive self-controlled devices for a multi-megawatt rating, although currently IGCT voltage and current ratings in the market are higher.
- The IGBT is a voltage-controlled device, whereas the IGCT is a current-controlled device.
- Both devices can be operated typically at 1.0 kHz PWM frequency at high power.
- Both devices can operate without snubbers.
- Both devices can be used in multilevel converters.
- Conduction drop in IGCTs is lower than that of IGBTs. Therefore, converter efficiency is higher. Trench gate technology lowers the conduction drop of IGBTs, but this is not available for higher voltage devices.
- Both devices can be used with a series-parallel connection for a high-power rating.
- Bypass diode is built monolithically with IGCTs, but is discrete with IGBTs.
- The SOAs of both devices are high.
- Reverse blocking IGCTs can be directly used in current-fed converters. However, IGBTs require additional series diodes.

In general, IGCTs are better for high-power multi-megawatt converters because of higher efficiency and fewer components. Note that traditionally, GTO converters were meant for such high power, but now they are on their way out.

[*Note:* The IGCT is essentially similar to SITH (static induction thyristor, which was introduced in Japan in the 1980s, but it was a normally on device. The life span of SITH was short because of a gate circuit fabrication problem with large turn-off current. ABB cleverly built the integrated gate driver with the device (see Figure 2.22) and commercialized it.]

Q. Can you explain overcurrent and short-circuit phenomena in an IGBT's intelligent power module (IPM) and how to protect the device?

A. An IGBT's specification sheet has two absolute maximum current ratings: One is dc or continuous current (I_C) and the other is peak but repetitive collector current (I_{CM}). The former rating is valid at $T_c = 25°C$ (infinite heat sink) and the later rating holds for $T_j < 150°C$. When the current tends to increase the rated value, the external current control circuit limits the gate drive to limit the load current within the safe value. If the overcurrent exceeds the specified limit value for more than a certain minimum time, it will be detected by current sense IGBTs (or desaturation of the device) and the gate drive will be disabled (overcurrent trip). The fault will be reported at the output.

There are two scenarios for short circuits in IGBT converters: (a) IGBT switching on a short circuit or (b) load or ground fault short across an already switched-on IGBT. The two cases are explained by the Figure 12.2 [3].

In case (a), when the IGBT turns on to the fault, the device current rises to a high value and the V_{CE} dips due to line leakage inductance (L_1) drop. Soon after this, V_{CE} rises

FIGURE 12.2 Short-circuit cases of IGBTs. (a) IGBTs switching into a short circuit. (b) Short across switched-on IGBTs.

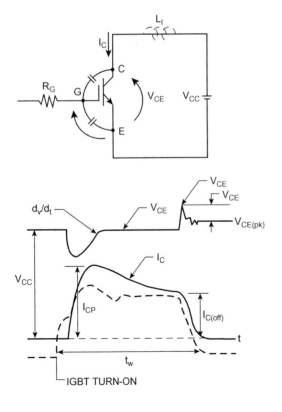
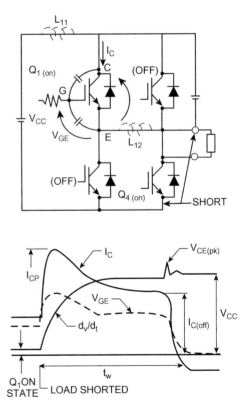

to V_{CC} value as shown. The dv/dt at this switch back is coupled to the gate through the transfer capacitance, causing a momentary rise in the gate voltage, thus causing higher peak collector current (I_{CP}) within 1 or 2 μs. The internal device temperature rises and reduces the short-circuit current to a lower value. To protect the device, the current is cut off by the gate circuit in time t_w (less than 10 μs). At turn-off, the sharp fall in the collector current increases the V_{CE} voltage. The whole excursion of transient voltage and current should be limited within the short-circuit SOA (SSSOA) for either fault condition.

In case (b), the increasing short-circuit device current desaturates the device and the initial $V_{CE(SAT)}$ goes to almost full V_{CC} value. The dv/dt during desaturation [higher than case (a)] is coupled back to the gate to increase the resulting short-circuit current, which is significantly higher than case (a). The device is turned off at time t_w and the resulting phenomena are similar to case (a). The typical limit time of t_w in either case is 10 μs. The non-repetitive fault current should typically be limited to 10 times I_C. The device is highly stressed in the fault condition and it can survive typically up to 100 short-circuit events over the life of the equipment.

Q. Can a power device be protected by a fuse?

A. Traditionally, thyristors have been protected by fast fuses. These fuses are expensive, and the recent trend is toward fuseless thyristor converters. One distinguishing feature of a thyristor (also a triac) is that once it is conducting, the gate current cannot turn it off. A thyristor has two types of current ratings: maximum rms current (I_{RMS}) and nonrepetitive peak current (I_{FM}) (half or multicycle). Besides for fuse protection, it has a subcycle I^2t rating for the time duration between 1.5 and 8.3 or 10 ms (corresponding to a half-cycle of 60 or 50 Hz). The fuse also has similar peak current and an I^2t short-time rating usually on a triangular current waveform basis. A fuse connected in series with a thyristor has to be properly coordinated so that the fuse blows before any damage occurs in the thyristor in case of a fault. A rule for protection is that the peak current and I^2t rating of the fuse should be smaller than the corresponding ratings of the device. A similar principle holds true for diodes and GTOs also. Note that a GTO's gate blocking capability is lost at high fault current. For an IGBT fault, fuse protection is not necessary because of its gate blocking capability as discussed before. The same principles hold true for power MOSFETs.

Q. What are the pros and cons of snubberless operation of an IGBT converter?

A. IGBT PWM converters may be available with a snubber or without a snubber. If a snubber is used, it is typically the R-C-D type [Figure 2.8(a)]. The snubber adds cost and additional loss during switching, although the switching loss is diverted from the device. The additional switching loss also means more burden on the cooling system. Therefore, the apparent advantages of snubberless operation are evident. The snubberless switching load lines are shown in Figure 12.3, where they are superimposed on a conventional SOA curve and nominal V_d-I_c envelope. At turn-off, the load line will exceed the dc supply voltage (V_d) because of series leakage inductance Ldi/dt (see Figure 2.19). At turn-on, the device current will exceed the steady-state current due to opposite-side diode recovery current.

FIGURE 12.3   Snubberless load lines in V_d-I_c area and SOA.

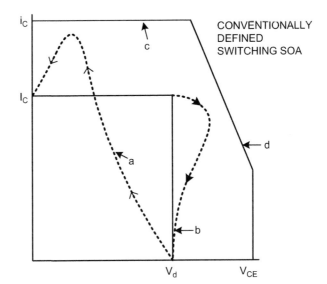

Evidently, the device turn-on and turn-off switching loss will be high. High switching loss in the device and active excursion in the V-I area will tend to stress the device and lower its life. The EMI, dv/dt, and di/dt effects will be high in the converter, resulting in the harmful consequences discussed before. Evidently, careful layout of the converter is essential to minimize leakage L. Varistor protection can be provided on devices to safeguard against voltage damage.

Q. Is a thyristor phase-controlled converter a soft-switched converter?

A. Ideally, thyristor converters (including cycloconverters) are soft-switched converters with zero current switching (ZCS). Therefore, unlike that of a hard-switched converter, the devices do not have any switching loss. The device turns on at zero current because at turn-on its voltage collapses, but the inductive current is transferred to it gradually from the outgoing device. The device turns off at zero current because the inductive current is diverted to the incoming device gradually before turn-off. If a snubber is used across the device to limit reapplied dv/dt and improve the recovery current transient, the device switching loss increases slightly and most of the stored snubber energy is dissipated in the snubber resistor. However, this loss is low due to low switching frequency. Therefore, the converter efficiency is very high (typically 98%).

Q. In a cycloconverter-fed synchronous motor drive (see Figure 8.59), what are the usual faults and the corresponding consequences?

A. Assuming that the equipment has been designed properly, one usual problem is thyristor commutation failure due to supply voltage fluctuation. The usual protections of the cycloconverter are gate circuit blocking and tripping of the line circuit breaker (on the transformer primary side). Both are normally commanded to operate simultaneously. However, there is a short delay in gate blocking (a few milliseconds), but longer delay (tens of milliseconds) in breaker tripping. Even if the cycloconverter operates at nominal unity DPF motoring load, there is a short angular interval in every fundamental half-cycle when it operates in the inverting mode. The commutation failure most likely occurs in the inverting mode. In this mode, there is current overlapping (μ angle) followed by reverse voltage switch-off of the outgoing device at turn-off angle γ. If there is a brown-out in the line voltage, the line current will increase for the same power output. The thyristor is vulnerable to commutation failure because of a longer overlap angle, lower turn-off angle, and reduced reverse voltage. If the load current tends to increase to high value, the trend will be the same, but gate firing control can regulate it within the safe limit. If the line voltage collapses to zero abruptly due to a line fault, the commutation failure is inevitable. If the line supply is open circuited, there will also be commutation failure. However, it is likely that counter emf (CEMF) due to parallel load or line stray capacitance will provide sufficient voltage and adequate time for successful gate blocking protection. Commutation failure may also occur if there is an undervoltage transient in the line.

In summary, commutation failure occurs if turn-off time is short and commutating reverse voltage is small. Even with gate blocking, the conducting thyristor may continue to conduct and build fault current if inverse device voltage is not adequate. The fault current will normally be fed by both machine CEMF and line voltage until the line breaker has tripped with the delay. The asymmetrical fault current has a braking effect on the machine and large oscillatory torque will be developed. This will tend to induce mechanical resonance unless there is a sufficient load braking effect. The whole phenomenon is extremely complex. The scenario with a phase-controlled converter is similar. Fault performance of a converter system is extremely complex. It should be studied by computer simulation and an appropriate protection system designed.

Q. Can you operate a converter with both sides inductive or capacitive?

A. With phase control in the thyristor converter, both sides can be inductive, because the current is commutated from the outgoing to the incoming device. If both sides are capacitive, neither thyristor or self-controlled devices can be operated because it will cause a current-hogging (in-rush) problem. Self-controlled devices cannot be operated if both sides are inductive. In conclusion, a PWM converter requires one side to be inductive and the other side capacitive.

Q. What is your prognosis on soft-switched power converters?

A. Although thyristor converters are essentially soft switched and have been known for a long time, the modern generation of soft-switched, self-controlled converters emerged in the middle of the 1980s. Since then, thousands of papers have been published proposing many different soft-switching techniques, but today they have hardly any industrial applications. Soft-switched converters were claimed to have a number of advantages (see Figure 4.65) and some penalties, but the advantages were often exaggerated in the literature. Unfortunately, the power circuit required auxiliary devices and passive components (such as a resonance circuit), and the technology was somewhat similar to forced-commutated thyristor converters. Although switching losses were minimal compared to hard-switched converters, the auxiliary components operating at high frequency contributed a considerable amount of loss, although the fundamental frequency was quite low. The added complexity of power circuits contributed to higher costs and lower reliability, besides the extra losses. Therefore, power electronics and drive manufacturers refused to accept soft-switched converters. Of course, soft-switched converters provide additional advantages for motor drives, such as lower dv/dt-induced insulation deterioration, acoustic noise reduction, reduction of bearing current, and soft motor terminal voltage at the end of a long cable. Such problems could be solved by alternate methods, such as by inserting a low-pass filter at the motor terminal (see Figure 7.23). The literature in soft-switched converters has recently dwindled down considerably due to lack of interest. In summary, it appears that soft-switched converters do not provide any future promise. However, note that soft-switched high-frequency link dc-dc converters are well accepted in SMPS. The additional advantage in this case is galvanic isolation between the primary and secondary circuits with a small high-frequency transformer.

Q. Can you define a thyristor force-commutated inverter as a soft-switched inverter?

A. Yes. Thyristor forced-commutated inverters, such as the McMurray inverter and McMurray-Bedford inverter, were introduced in the 1960s. In these converters, forced commutation of the devices was obtained at zero switching loss with the help of a resonance circuit. However, the converter efficiency was poor because of large losses in the resonant circuit occurring at the switching frequency. The phase control and forced commutation techniques indicate that the concept of soft switching is old.

Q. Can the kVA rating of an IGBT converter be increased by increasing the heat sink size?

A. The rating of a converter is based on two limiting parameters of power semiconductor devices: (1) peak current (I_p) and (2) peak junction temperature (T_{jmax}). The converter losses consist of device conduction loss and switching loss (neglect other losses). Assume that dc voltage and switching frequency are constant. With higher power, both the conduction and switching losses will increase. With a certain power loading, the losses that are pulsating in nature will flow through the heat sink to raise T_j. However, a thermal capacitance effect will tend to delay the T_j rise. As the power loading increases to the rated value at steady state, the T_j will rise to the limit value (T_{jmax}). If cooling is improved (i.e., heat sink thermal impedance is decreased), T_j will go down, and as a result, steady-state power can be increased within the limits of I_p and T_{jmax}. Again, with the same heat sink, a short time power rating will be higher (say, 50% overrating for 1.0 min) because of the delay in T_j rise. Higher rating for shorter time can be continued until peak power (with peak current) for a finite minimum time reaches with T_{jmax} limit. In conclusion, improved cooling increases the converter power rating within the limit of a peak device current.

Q. How do you compare SVM with SPWM for a three-phase load?

A. Space vector PWM (SVM) and sinusoidal PWM (SWPM) are the two most viable PWM techniques for a three-phase load. However, if the load neutral is connected for zero sequence current circulation (or if the load is single phase), only SPWM should be used. Note that both the PWM methods are open-loop carrier frequency-based, where the carrier frequency can be fixed and free-running (unsynchronized), or synchronized with the fundamental frequency. In free-running mode, unless the carrier-to-fundamental ratio is high, some amount of subharmonics is introduced in the load. Most of the three-phase loads including ac motors have isolated neutral. The general comparison between SVM and SPWM for an isolated neutral load can be given as follows:

- SPWM is simple to implement compared to SVM, which requires complex computations in real time. SPWM can be implemented by simple hardware or software.
- The linear undermodulation range of SPWM extends up to modulation index $m' = 0.785$ (where $m' = 1.0$ at square wave). In comparison, SVM has a higher undermodulation range, i.e., up to $m' = 0.907$. The bus voltage utilization is better with SVM in this region. However, sinusoidal modulating wave of SPWM can be mixed with an appropriate amount of triplen harmonics (zero sequence components) to achieve the same modulation index.
- The harmonic distortion in SPWM and SVM is comparable up to $m' = 0.4$. As m' increases, the distortion on SPWM increases nonlinearly as shown typically in Figure 12.4, where d^2 is the index for distortion.

FIGURE 12.4 Comparison of harmonic distortion in SPWM and SVM in the undermodulation region. (Note that triplen harmonics have been added in SPWM to increase m'.)

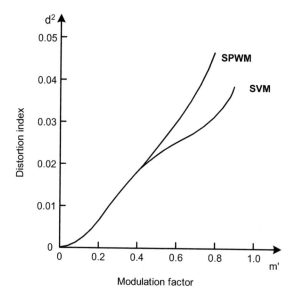

- In a PWM double-converter drive system, the line-side rectifier must operate in the undermodulation mode, whereas the load-side inverter can operate in either the undermodulation or overmodulation mode. Higher m′ in SVM permits line voltage synthesis with lower dc link voltage (V_d). This is a definite advantage.

Considering the overall merits, the SVM technique is invariably preferred in a modern three-phase system with isolated neutral load (including drives).

Q. Can you comment about the random PWM (RPWM) technique?

A. The objective of the RPWM technique is to attenuate the magnetic (or magnetostriction) noise in a machine supplied by a PWM inverter. In a traditional PWM inverter, the PWM or switching frequency is typically several kilohertz, which falls under the audio frequency range; therefore, the acoustic noise is generated by the machine iron (similar to transformer noise at 120 Hz with a 60-Hz power supply). In RPWM, the carrier frequency is varied randomly in a fundamental cycle so that the acoustic energy is spread over the frequency spectrum, creating a form of "white noise" that eliminates the acoustic noise. However, RPWM increases the machine harmonic losses reducing the drive efficiency. Modern IGBT PWM inverters within moderate power rating can operate at higher than audio frequency range and, therefore, this problem does not arise.

Q. In open-loop volts/Hz. control of induction motor drives, should you use nonlinear volts/Hz gain in the overmodulation region of SPWM?

A. The open-loop volts/Hz control block diagram with constant volts/Hz gain G is shown in Figure 7.10. In the undermodulation region, the transfer characteristics between the machine voltage (V_s) and modulating voltage (V_m) remain inherently linear as shown in Figure 12.5 (low-frequency boost voltage is neglected). Therefore, with constant gain G, V_s will rise linearly with ω_e maintaining the flux ψ_s constant at the rated value. As V_m exceeds A_1, the inverter enters the overmodulation region, where the gain between V_s and V_m decreases, which reduces the stator flux. With a linear increase in V_m, the inverter will go to the square-wave mode at frequency ω_1; i.e., the overmodulation region will extend from A_1 to C_1 values of V_m. The resulting flux reduction zone from A_2 to C_2 is shown in the lower figure. With the rated stator current, there will be torque loss in the overmodulation region because of flux loss. If the gain is made nonlinear (nonlinear function generator) from point A_1 to offset the inverter gain loss such that V_m traverses along the A_1–B_2' trajectory as a function of frequency, then V_s will increase linearly in the dotted line so that square-wave mode will be attained at point B. This will maintain the flux to the rated value and there will be no loss of torque. In this case, the overmodulation region will shrink from A–C to A–B. Note that with space vector PWM, the overmodulation nonlinearity is normally compensated within the algorithm and, therefore, the problem does not arise. If the drive has current control loops (such as synchronous current control), this nonlinearity is compensated within the loops.

FIGURE 12.5 Voltage and flux relations with frequency for linear and nonlinear gain factors.

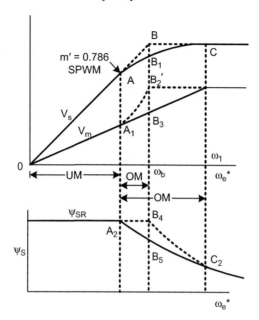

Q. Instead of linearizing the voltage transfer relation within the algorithm in the SVM overmodulation range, is it possible to attain the same by inserting nonlinear volts/Hz. gain as in SPWM?

A. Yes. SVM undermodulations and overmodulations for two-level and three-level inverters were discussed in Chapter 4. Operation in the undermodulation region remains linear as in SPWM. However, in the overmodulation (mode 1 and mode 2) range, the algorithm was implemented in such a way that the nonlinearity in voltage transfer characteristics was compensated within the algorithm. Instead, if the SVM is implemented by inserting nonlinear gain in volts/Hz control line (see, for example, Figure 11.45 for a three-level inverter), the operation can be explained by Figure 12.6. In the undermodulation range, the gain between the output and modulating voltage is constant. Therefore, as the modulating voltage signal increases, the output voltage vector V*, i.e., the radius of the circle increases proportionately until the circle is inscribed in the hexagon (limit of the undermodulation range). As it increases further, the circle crosses the hexagon and enters overmodulation mode 1 as shown in the figure. There will be a loss of output voltage during the portion of hexagon tracking (1-2), but this loss will be compensated by the nonlinear gain that will increase the radius of the command circle. In the limit of this mode, the locus of operation is completely on the hexagon. Beyond this, i.e., in overmodulation mode 2 that involves holding the command vector at the corner point, the operating principle is not very clear. One simple method might be to generate holding

FIGURE 12.6 SVM command voltage trajectory within hexagon covering overmodulation mode 1.

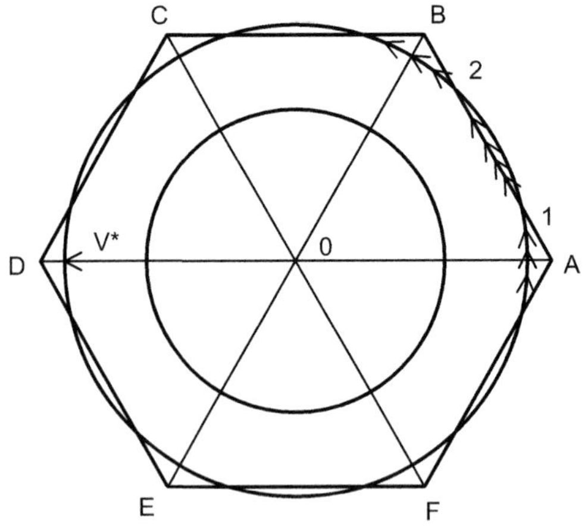

angle α_h [see Figure 4. 28(b)] linearly with modulation factor m′ in the range of 0 to 30°
until square wave is attained. This will give nonlinearity in the voltage transfer ratio that
can be compensated by nonlinear gain. The same principle is valid for SVM within current control loops.

Q. *How do you determine the switching frequency for a PWM converter in a drive system?*

A. The switching frequency of a PWM converter for a motor drive is determined on the
basis of total loss minimization of the converter and motor. The converter loss has two components: switching loss and conduction loss. The switching loss increases with frequency. It
also depends on the PWM algorithm, how fast the switching device is, whether operation
involves a snubber or no snubber, and whether the snubber is dissipative or regenerative.
Machine loss has two components: copper loss (which includes harmonic copper loss)
and iron loss (which includes harmonic iron less). At higher switching frequency, harmonic copper loss decreases, but the harmonic iron loss increases. If a series low-pass
filter is used, both the loss components will decrease. Of course, losses of both converter
and machine increase with loading. Rigorous loss curves can be plotted and optimum
switching frequency can be determined. It is assumed that both converter and machine
have adequate cooling so that temperatures are within safe limits at worst loading. On the
basis of the preceding observation, we conclude the following:

- Low-power drives use a faster device with low switching loss. Therefore, a high switching frequency is used. Minimization of acoustic noise may be another motivation for
  selecting higher frequency.
- For the same reason, high-power drives with slow switching devices have higher switching losses. Therefore, the switching frequency is low.

- The regenerative snubber used in high-power drives allows the switching frequency to be increased to a higher frequency than with a dissipative snubber.
- High-power IGBT and IGCT switching losses are lower than those of a GTO. Therefore, the switching frequency is higher in the former devices. Another reason is the snubberless operation capability of IGBTs and IGCTs. Typically in multi-megawatt applications, the GTO switching frequency may be 500 Hz, whereas IGBT or IGCT frequency may be 1000 Hz.

In a PWM rectifier application in the front end, a higher switching frequency will increase the converter loss but will improve the line current harmonics. For a certain waveform quality, the filter size can be decreased with a higher switching frequency (hence, a trade-off consideration).

Q. Is an energy recovery snubber a good idea for an IGBT converter?

A. Energy recovery (or regenerative) snubbers are commonly used in high-power GTO PWM converters. Instead of dissipating switching energy in the snubber, it is captured in the snubber capacitor, converted to high voltage by a dc-dc converter and then pumped back to the dc link. A GTO is a slow switching device with long tail current. Therefore, switching losses with lossy snubbers are so heavy that they tend to lower converter efficiency. Therefore, converter efficiency improvements and the reduced burden on the cooling system justify the extra cost, complexity, and even extra losses in regenerative snubbers. It also permits higher PWM frequency—an additional advantage. Note that 60- (or 50)-Hz square-wave operation in a multistepped GTO converter does not use a regenerative snubber. In comparison, IGBTs are much faster devices with much shorter tail times. Therefore, the additional cost and complexity of regenerative snubbers are not justified. The same reasoning holds true for IGCT converters.

Q. How do you select electrolytic capacitor size in the dc link of a voltage-fed inverter?

A. As mentioned before, the size of the electrolytic capacitor in the dc link is to be determined on the basis of the harmonic current it is required to sink, because otherwise the temperature rise due to ESR (equivalent series resistance) could exceed the safe limit. This may cause an explosion in the capacitor. For a three-phase PWM inverter load with battery supply, most of the harmonic currents are to be bypassed in the capacitor to prevent decrease of battery life. The analytical calculation of the ripple current is not easy. It is much more convenient to use a simple computer simulation to determine the ripple current. The instantaneous dc link current i_d can be determined on simulation from the balance of instantaneous power relation as follows:

$$i_d = \frac{v_a i_a + v_b i_b + v_c i_c}{V_d}$$

The rms ripple current should be calculated from i_d wave subtracting the dc component. Normally, ripple current is worst at the upper edge of the constant-torque region of a drive.

If there is a front-end converter, the estimation of ripple current is more complex. Again, simulation is the best method.

Q. How do you pre-charge the dc link capacitor in a double-sided PWM converter VFI drive?

A. The dc link capacitor is charged gradually at higher than peak ac line voltage so that the line-side converter can operate in the undermodulation mode. A simple method of charging the capacitor from the line side is to use the bypass diode rectifier with a series dc link resistance. The resistance is bypassed when Vd is the peak line voltage. The additional boost voltage can be created by temporarily running the rectifier in the overmodulation mode. The distorted line current will distort the line voltage. If this is not permitted, the inverter can charge the motor, which can then be regenerated to boost the capacitor voltage. Otherwise, a separate auxiliary charger is required for the capacitor (see Figure 4.57).

Q. Can multilevel converters fabricate multistepped waves?

A. Multistepped, high-voltage, high-power converters are extremely important for utility system shunt or series static compensator applications. This is particularly true with slow GTO devices, where switching frequency is the same as line frequency (50/60 Hz).

FIGURE 12.7   Three-level H-bridge converter and fabricated 12-step wave.

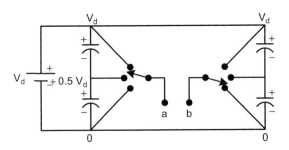

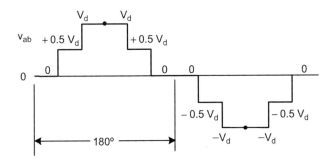

Multistepped waveform fabrication with phase-shifted two-level converters and transformers was discussed in Chapter 4. In such a configuration, transformer coupling permits not only the fabrication of multistepped voltage waves (for elimination of lower order harmonics) but also voltage boosts for the ac line. Similar fabrication of a stepped waveform is also possible in multilevel converters. However, more steps are possible with the same number of half-bridges in a multilevel converter. For example, an H-bridge in a two-level converter can fabricate 6-step, wave, whereas an H-bridge in a three-level converter can fabricate 12-step wave as explained in Figure 12.7. The fundamental voltage component and the harmonic content of the wave can be varied by varying the dwell time of each level. We can see from Figure 4.15 that six H-bridges of a two-level converter can fabricate 12-step three-phase waves. It can be shown that the same number of three-level H-bridges can fabricate 48-step three-phase waves [4].

Q. Is capacitor voltage balancing a problem in multilevel inverters? How does the unbalanced voltage affect converter performance?

A. In a multilevel inverter, more than one capacitor is connected in series in the dc link, and voltages in these capacitors should be kept equal. In addition, the harmonic currents flowing through them should be limited to prevent overheating. Consider, for example, the three-level inverter shown in Figure 4.33. If capacitor voltages are unequal, symmetrical PWM wave fabrication is not possible. If the inverter switching is such that current is drawn from the neutral point (0), the upper capacitor voltage will increase, whereas the lower capacitor voltage will decrease. If, on the other hand, current is fed at the neutral point, the lower capacitor will charge and the upper capacitor will discharge. This is a typical problem with split-capacitor power supplies. The voltage variation problem can be alleviated by oversizing the capacitors with a corresponding economic penalty.

Fortunately, the neutral current can be controlled to balance the neutral point voltage fluctuation without affecting inverter performance. Consider the hexagon switching states (Figure 4.34) of the inverter. Note that for the boundary space vectors, the switching states are unique, whereas for the inner vectors, there are multiple switching states. In the outer hexagon switching states, there is no neutral current flow and, therefore, no fluctuation problem arises. Similarly, there is no neutral current at zero states (000, PPP, NNN). For the inner hexagon corner states, consider, for example, the impressed inverter vector that corresponds to the inner states P00 and 0NN (i.e., either the P00 or 0NN state will satisfy the demand), and assume that the U phase is supplying load current, whereas the V and W phases are returning the currents. At state P00, the current from the P bus will flow to the motor through the U phase and then return to the neutral through the V and W phases, thus raising the neutral point voltage. On the other hand, for switching state 0NN, the neutral current will be extracted by the U phase and then returned to the N bus through the V and W phases. This will lower the neutral point voltage. Therefore, for balancing the neutral voltage, the P00 and 0NN states are to be distributed appropriately for the same time segment of the inverter vector. Either an open-loop or closed-loop method can be used to control the neutral voltage. The same principle is applicable to a multilevel converter with a higher number of levels. It should be mentioned here that for static VAR compensator applications, in which the loading is purely reactive ideally, this unbalanced voltage problem does not arise.

Q. Why is there a second harmonic problem in the dc link of a single-phase PWM inverter, whereas no such problem exists in a three-phase inverter?

A. In a single-phase PWM inverter that is supplying sinusoidal voltage and current waves to the load, the instantaneous power curve (neglecting PWM harmonics) has a dc component and a second harmonic component. Because a PWM inverter does not have any energy storage capabilities, the dc side will contain a dc current and second harmonic current along with the PWM related harmonics. Mathematically:

$$p_i = V_d i_d = v_0 i_0 = V_m \sin \omega t . I_m \sin(\omega t - \varphi)$$

Or

$$i_d = \frac{V_m I_m}{2 V_d} \left[\cos \varphi - \cos(2 \omega t - \varphi) \right]$$

assuming that V_d is an ideal voltage source. If the inverter has a dc link with finite capacitance, a second harmonic voltage fluctuation will occur. The voltage fluctuation will be less with a larger filter capacitor. This problem does not arise with a three-phase inverter because with sinusoidal voltage and current waves, the instantaneous power is always a constant irrespective of load DPF, i.e.,

$$p_i = v_a i_a + v_b i_b + v_c i_c = \text{constant}$$

Q. What is an ultracapacitor? Why is it so important for energy storage in a power electronic system?

A. An ultracapacitor (also called a supercapacitor or electrical double layer capacitor, EDLC) is an energy storage device like an electrolytic capacitor, but its energy storage density (Wh/kg) can be as much as 100 times higher than that of an electrolytic capacitor. In a commercial ultracapacitor, like that shown in the schematic of Figure 12.8, the electrodes are made of highly porous activated carbon. The porosity provides the large surface area, and the double layer formed by the ionic separation creates a very small distance between charge species, both factors contributing to the high-capacitance capability of an EDLC. Ultracapacitors are available with low voltage ratings (typically 2.5 V) and capacitor values that range up to 2700 farads. The units can be connected in series-parallel for higher voltage and higher capacitance values. However, the Wh/kg of an ultracapacitor are low compared to that of a battery. For example, the Wh/kg for an ultracapacitor is 6, whereas for a modern Li-ion battery, it is 120. Although its energy density is low, the power density (W/kg) of an ultracapacitor is very high, and large amounts of power can cycle through it millions of times without causing any damage or deterioration. Therefore, when compared to a Li-ion battery, the W/kg for an ultracapacitor is 10,000, whereas for the battery it is only 250. With large and frequent power cycling applications, the

FIGURE 12.8 Structure of ultracapacitor.

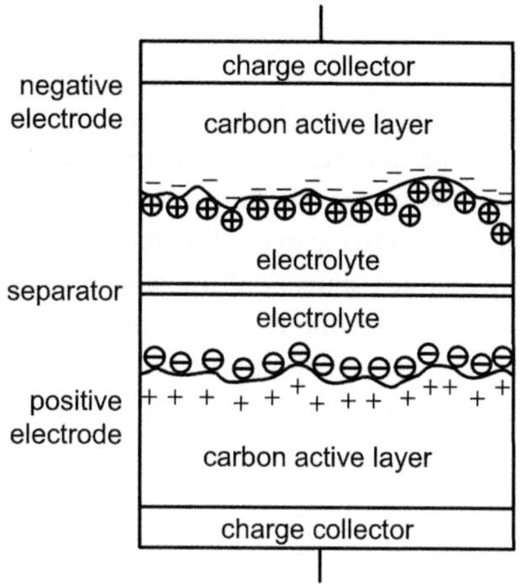

battery life will be short. Again, an ultracapacitor does not have the cold-weather functioning problems as in batteries.

Ultracapacitors are extremely important for next-generation fuel cell vehicles (see Figure 1.13). The devices, connected in parallel with FC (normally through a dc-dc converter), can supply large power transiently for fast acceleration of the vehicle, whereas for regenerative braking, it can absorb the braking power since FCs are nonregenerative. The battery solution in this case will be expensive because of its low cycle life and reduced power density.

Ultracapacitors have been evaluated as alternatives to Li-ion, Ni-MH, and lead-acid batteries for use in hybrid electric vehicle (HEV or HV) drives. Whereas these batteries can provide enough power to handle peak power cyclic loads, using them in this manner significantly reduces their cycle life. For example, at 60% depth of discharge, a typical lead-acid battery will last up to 1000 cycles before failure, and a Ni-MH battery will last up to 10,000 cycles before failure. Frequent replacement of batteries will increase the life-cycle cost of vehicles. There are many other applications in power electronic systems where ultracapacitor energy storage is extremely important.

Q. Is a double-sided PWM converter more viable than a phase-controlled cycloconverter?

A. Traditionally, thyristor phase-controlled cycloconverters have been used for multi-megawatt, four-quadrant motor drives in rolling mills, mining applications, etc. Recently, they are being replaced by double-sided three-level (i.e., neutral point clamped [NPC]) PWM converters. The pros and cons are as follows:

- A cycloconverter is prone to easy commutation failures as a result of line voltage dip, which makes the equipment unreliable. Self-controlled devices in an NPC converter do not have the same problem.

- A cycloconverter generates complex harmonic currents on the line and load sides that may include a subharmonic component. The harmonic problem is more complex with popularly used blocking mode operation (no intergroup reactor [IGR]). This increases machine heating, which in turn requires additional cooling. Excessive subharmonic current may cause machine magnetic saturation. Line-side complex harmonics are difficult to filter and may violate harmonic standards. In addition, excessive subharmonic current may saturate the line transformer. A large hybrid active harmonic filter (AHF) may be needed to clean the line harmonics. A double NPC converter with space vector PWM generates sinusoidal line and load currents, which do not need any AHF.

- A cycloconverter line DPF is poor, i.e., the reactive loading may be high on the line. This will cause additional heating on the transformer. The NPC rectifier can operate at unity DPF without such a problem.

- NPC converters have normally been used with GTO devices. Because of excessive switching losses, regenerative snubbers are normally used. On the other hand, IGBTs or IGCTs operates at higher frequency with or without a snubber. Replacement of GTOs with IGCTs has the additional advantage that conduction loss is lower, thus giving higher efficiency. Of course, a thyristor cycloconverter has better efficiency.

- The dc link capacitor bank will weaken the system reliability somewhat, but will provide energy storage and, in addition, line brown-out is not a problem. The cycloconverter (like the matrix converter) is a direct frequency changer without any energy storage. The overall cost of an NPC system will be lower than that of a cycloconverter with line AHF and SVC.

- When the drive is carrying a light load, the spare capacity of a PWM rectifier can be used as the SVC for parallel load with poor DPF.

In overall consideration, an NPC system is better than a cycloconverter. It is expected that thyristor cycloconverters will be obsolete in the near future.

Q. What is your prognosis concerning matrix converters?

A. The modern matrix converter (Figure 4.56; also known as a Venturini converter or PWM direct frequency changer) was proposed in 1980, and since then it has been discussed on and off many times in the literature, but there have been almost no product applications. Very recently, it has started to emerge again. A matrix converter (MC) is a direct competitor of the double-sided PWM converter (DSC) system (Figure 4.50), and its main pros and cons can be summarized as follows:

- A MC requires 18 forward-blocking IGBTs and 18 diodes compared to 12 IGBTs and 12 diodes in a DSC. The converter can also be realized in an anti-parallel configuration of devices by using 18 reverse (or symmetric) blocking IGBTs (no diodes), but reverse blocking devices have a somewhat lower switching frequency and higher conduction drop. (Note that when the IGBT was first introduced by GE Research Laboratories, it was reverse blocking and was known as an IGR or insulated-gate rectifier.)

- The dc link electrolytic capacitor in a DSC is replaced by a more expensive (but somewhat smaller in size) ac capacitor in the ac line of the MC. A small ac capacitor can also be used in the dc link, making a DSC more reliable, but the dc link voltage will not be smooth. Hysteresis-band current control will make ac currents (i.e., voltages) sinusoidal.

- A large electrolytic capacitor in the dc link provides energy storage for short-time supply interruption, making the drive more reliable, but this is not possible for MCs.

- Because the line-side converter in a DSC is a boost converter, sag in line voltage will not affect the dc link voltage.

- The maximum output voltage in a MC is typically limited to 86% of line voltage. This limitation does not exist in a DSC.

- Ease of current control and fault interruption due to the feedback diodes used in DSCs does not exist for matrix converters.

- In DSCs, a machine-side converter can be operated in the overmodulation mode while keeping the line-side converter undermodulated, but this flexibility does not exist in MCs.

- The line-side converter in a DSC can be operated as a static VAR compensator. This flexibility does not exist in MCs.

- DSCs can be integrated with the machine as in a MC, but it requires tapping the dc link to connect the capacitor.

- The line side of a MC may have a spurious harmonic resonance problem.

- There is some switching loss economization in MC because commutation voltage is less than dc link voltage in DSC.

Although the MC is a viable converter for ac drives, it cannot compete with DSCs when all of the pros and cons are considered. In the long run, it is expected to disappear from the market, even with the decreasing number and cost of power devices.

[*Further note:* In anti-parallel configurations, the converter topology is identical to that of a three-phase, phase-controlled thyristor cycloconverter (blocking mode) with isolated loads. The same topology with forced commutation or with reverse-blocking GTOs was proposed as a PWM frequency changer, but did not survive due to low frequency.]

[*Personal note:* Venturini visited General Electric's R&D Center in Schenectady, New York, in 1982 (when I was there) with the object of patenting and commercialization his converter through GE, but did not succeed.]

Q. What is FACTS? Why is it important?

A. The term FACTS stands for *flexible ac transmission system* and the idea was originated by the Electric Power Research Institute (EPRI) in the United States. Basically, it deals with active and reactive power flow control of a transmission system using power electronics [5]. The basic power electronic unit of FACTS is the multistepped static VAR generator (SVG or SVC) described in Figure 4.58, the principle of which was explained in Figure 4.57. An SVG is also known as a STATCOM or *static synchronous compensator.* It is a solid-state version of a rotating synchronous condenser. Of course, a thyristor-controlled reactor with parallel capacitor and thyristor-switched capacitor bank type SVCs have been traditionally used for real (P) and reactive (Q) power control in transmission systems. In recent years, high-power STATCOMs [4] have been developed using GTO-based multilevel (three-level) converters and applied in utility systems. Still, higher power

FIGURE 12.9 FACTS with shunt and series STATCOMs explaining unified power flow control.

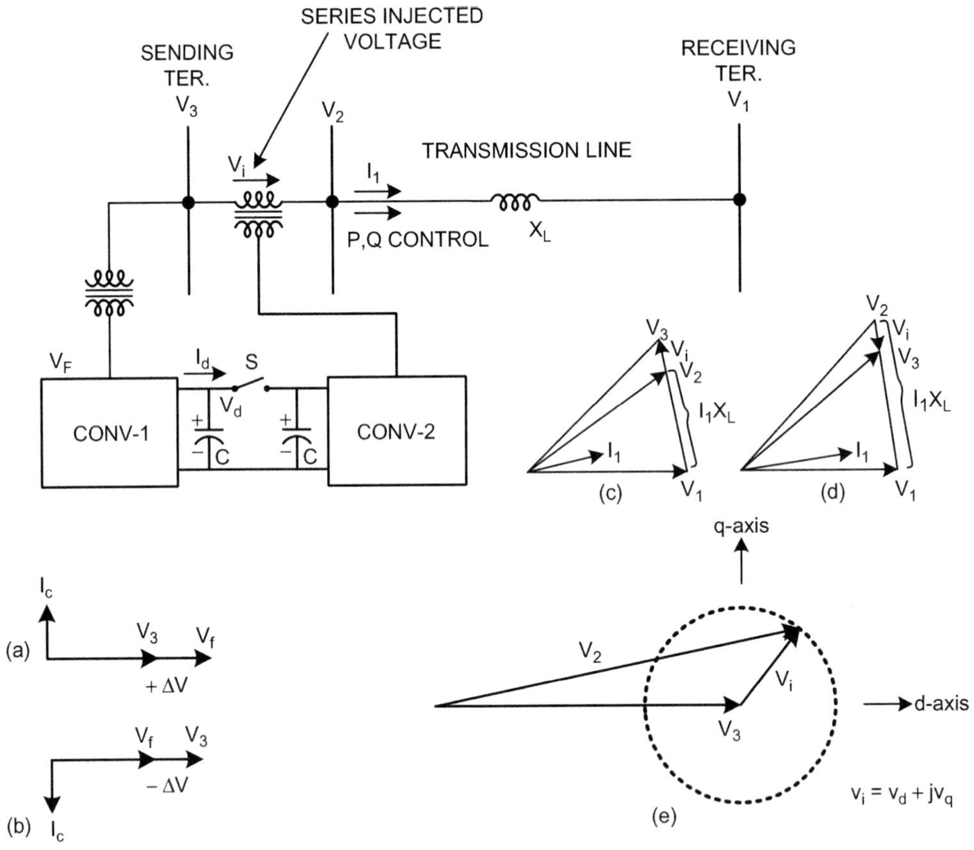

STATCOMS are yet to come with five-level (or higher) converters that use IGCTs. The principle of modern FACTS is explained in Figure 12.9. The main diagram shows a transmission line with receiving terminal and sending terminal, where V_1 and V_3 are the respective bus voltages, X_L = line reactance, I_1 = line current, P = active power flow, and Q = reactive power flow. It is desirable to control P and Q by the STATCOM units CONV-1 and CONV-2 shown in the figure. Each CONV is a multistepped (may be as high as 48 steps) voltage-fed converter with a capacitor on the dc side and coupled to the line by means of a transformer (see Figure 4. 58). Consider first the operation of CONV-1 with switch S open. It is connected to the sending terminal as a shunt element. The phasor diagram shown at the bottom left of Figure 12.9 indicates that it can be controlled to operate as a variable capacitive or inductive load by adding (Figure 12.9a) or subtracting (Figure 12.9b) series voltage ΔV, respectively as shown (V_f = counter emf of CONV-1). Therefore, CONV-1 alone can be used to control flow of Q or in bus voltage control mode at the sending terminal. Now, consider the operation of CONV-2 with switch S open. It injects series voltage V_i as a phasor so that $V_2 = V_3 + V_i$ (as phasors). Figures 12.9(c) and (d) show the phasor diagrams where V_i is aligned perpendicular to line current I_1 but subtracts and adds, respectively, from the sending terminal voltage V_3. In part (c), the injected voltage V_i is inductive (90° leading with current) and, therefore, reduces the line current I_1, whereas in part (d) it is capacitive and increases the line current. Note that due to the quadrature phase relation between voltage and current, no real power flows through the CONV-2 unit, but the series V_i injection controls P and Q of the line, which is evident from the phasor diagrams. In fact, the universal power flow control (UPFC) characteristics of the CONV-2 unit are explained in the phasor diagram of part (e), which indicates that the arbitrary d and q component of voltage of controllable amplitude (within limited values) can be injected in series to control arbitrary P and Q magnitudes. In such a case, real power has to flow in CONV-2 as shown by current I_d on the dc side with switch S closed. This means that CONV-1 supplies the real power ($V_d I_d$) which is circulated to the CONV-2. Of course, CONV-1 can control the flow of Q independently. This flexible P, Q and voltage control features are extremely important in utility transmission system.

The transient response of STATCOMs to supply and absorb energy pulses is very fast. Besides steady-state controls as discussed earlier, the units can control the transient stability and generator oscillation problems of the system. Successful example applications of FACTS include the New York Power Authority (NYPA) Mercy plant [4] and the TVA Sullivan plant [6].

Q. Can you clearly explain EMI phenomena in modern PWM inverter ac drives?

A. EMI noise problems and their remedial measures were discussed in Chapters 3 and 4. Modern power electronics equipment with increasing power ratings, fast switching IGBTs, and high PWM switching frequencies are being demanded with integration, where power and signal elements are in close proximity, thus creating difficult EMI problems. Systematic analysis of EMI problems and their solutions become quite involved. Often, the problems are solved by following established good practices of layout, shielding, grounding, and filtering of the equipment. Figure 12.10 [7] shows a three-phase PWM inverter drive with parasitic line-to-ground EMI noise currents. All of the inverter IGBTs (snubberless) generate common mode EMI on the three-phase output lines by hard switching with very high dv/dt (may be 5000–15,000 V/μs) depending on the system voltage and rise time of IGBTs. At each leading and trailing edge of a PWM pulse, a noise current I_{lg} is induced that is given by $C_{lg} \cdot dv/dt$, where C_{lg} is the parasitic line-to-ground capacitance. There are two components of C_{lg}: One is line-to-ground cable capacitance (C_{lg-c}), and the other is motor stator winding line-to-ground capacitance (C_{lg-m}). The common mode noise current flows through the grounding cable to the input transformer phases, and then to the inverter phases through the grounded neutral wire as shown. The ground cable resistivity shows several potential points due to I_{lg} current. It can also carry current due to other groundings of the equipment. The ground cable is connected to earth ground at Potential

FIGURE 12.10   EMI emissions in modern PWM inverter ac drives (with poor wiring practices).

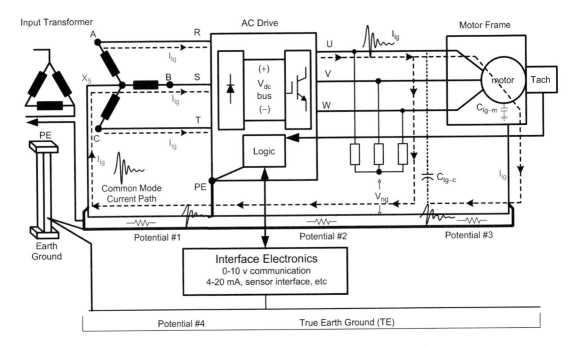

#4 shown in the figure. A virtual neutral reference node to ground with a common mode (zero sequence) voltage (V_{ng}) is considered for each line behind the I_{lg} current source. If a sensitive signal circuit is located nearby, voltage V_{ng} can couple to it capacitively and/or the current I_{lg} can couple to it, inductively polluting the signal. The noise is also coupled to the signal circuit through the ground wire potential differences. Several remedial measures, as discussed before, may be (1) physical distance between power and signal circuits; (2) shielding output cable and grounding the shield; (3) ground cable or plane with low resistivity; (4) grounded metal frame for transformer, converter, and machine; (5) common mode EMI filter; (6) lower switching frequency with slower devices; and (7) twisted-shielded signal wires with capacitor-decoupled signal power supply, etc.

Q. Can you use an active filter in a dc link to eliminate the need for an electrolytic capacitor?

A. The electrolytic capacitor in the dc link of a voltage-fed inverter has poor reliability, limited temperature rating, poor shelf life, and is the weakest link in a modern power electronics system. Yet, the voltage-fed converters are finding universal popularity nowadays. The electrolytic capacitor is more lossy than an ac capacitor due to its high ESR (equivalent series resistor), and its size is dictated by a harmonic heating effect rather than filtering bandwidth. Of course, an electrolytic capacitor can be replaced by an ac capacitor, which is more bulky and expensive. It is possible to eliminate the

FIGURE 12.11 DC-fed inverter drive with active harmonic filter in the dc link.

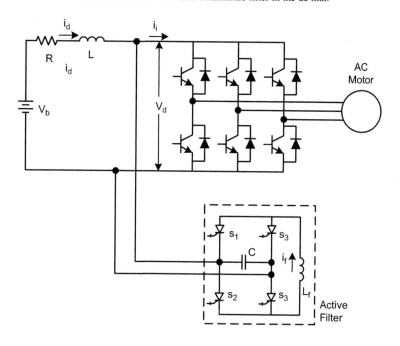

electrolytic capacitor in the dc link and replace it with an active harmonic filter as shown in Figure 12.11 [8]. The AHF has a single-phase current-fed topology, but unlike applications on the single-phase ac line, the filter has dc voltage at the input side. The inductance L_f traps dc current I_f, which can be switched in either polarity or free-wheeled as desired to fabricate the PWM current wave at input. The devices in AHF are self-controlled with symmetric blocking capability, and they usually operate at high frequency (typically power MOSFET with series diode). The small ac capacitor C at the input absorbs high-frequency currents and permits PWM self-commutation of the devices. Consider, for example, that the inverter is operating in square-wave mode. The input-side dc component is supplied by the dc source, whereas the ripple component (with six times fundamental frequency) is absorbed by the AHF. This means that the harmonic energy, instead of being absorbed by the electrolytic capacitor filter, is absorbed in the inductor L_f. Due to energy balance, current I_f will have a small fluctuation. For multistepped inverters, where the dc link harmonics are related to fundamental frequency, AHF operation is not difficult. However, for PWM inverters, the harmonic ripple frequency is much higher and will demand that the AHF switching frequency be much higher than the inverter switching frequency. In that case, the AHF will be more lossy. Note that the high-frequency current components of the spectrum will be filtered by C, and some amount of dc link voltage ripple can be tolerated. Of course, inverter switching frequency can be reduced by placing series inductor on the ac side.

Q. In a current-fed converter system, how to determine the dc link inductor size?

A. The dc link inductance (L_d) should have a large dc current (I_d) rating with reasonably low resistance so that the copper loss is low. In addition, it must absorb the ripple voltage impressed by the line-side and load-side converters without causing excessive ripple current. The relation is $\Delta i_d = (e_1 - e_2)\Delta t/L_d$, where Δi_d = permissible ripple current, e_1 = ripple voltage from the line side, e_2 = ripple voltage from the load side, and Δt = time increment. This means that for a permissible ripple current, the L_d value should be determined by knowing the e_1 and e_2 waves. Obviously, a small ripple current requires a more bulky inductor. Normally, it is an iron-cored inductor with an air gap to prevent saturation. The L_d value will be much higher in a load-commutated inverter (Figure 5.16) than in a double-sided PWM converter (Figure 5.29). It can be determined conveniently by simulating the system.

Q. Should you use a voltage-fed or current-fed inverter for high-frequency induction heating?

A. Either a voltage-fed or current-fed resonant inverter can be used for induction heating. The voltage-fed inverter requires a series resonant circuit, whereas the current-fed inverter requires a parallel resonant circuit. In both the cases, the inverter frequency tracks the load resonant frequency. In the former case, the load power factor is lagging, whereas in the latter case, the power factor is leading in order to have soft switching that eliminates the switching loss.

In a voltage-fed inverter scheme, as shown in Figure 4.71, a single- or three-phase diode rectifier with a boost chopper feeds a filter capacitor, which is then inverted to high frequency for an induction heating load. While the inverter controls the frequency, the load power is controlled by the dc voltage. The inverter devices are forward blocking. For lower frequencies or higher powers, IGBTs or GTO devices can be used. In a current-fed inverter (see Figure 5.10) scheme, a thyristor phase-controlled rectifier generates a variable current source through a dc link inductor, which is then inverted to ac for the parallel resonance load. The inverter controls the frequency, and the load power is controlled by the dc current source. The inverter devices are reverse (symmetric) blocking. Therefore, for high frequencies, IGBTs with a series diode (or reverse blocking IGBTs) can be used. For higher power at lower frequency, thyristors, IGCTs (symmetric), or GTOs (symmetric) can be used.

In overall comparison, the voltage-fed converter scheme is superior because of its lower costs, higher efficiency, and higher frequency capability.

Q. Should you use a voltage-fed or current-fed converter for an active harmonic filter?

A. Either a voltage-fed PWM rectifier with capacitive load or a current-fed PWM rectifier with an inductive load can be used as an active harmonic filter (AHF) or static VAR compensator (SVC). In the former case (see Figure 4.57), a three- or single-phase PWM rectifier at high PWM frequency feeds a capacitor load. The converter is controlled to sink the harmonic current on the ac side and regulate the dc voltage on the output side. A small line inductance is needed to filter out the PWM frequency voltages. The PWM rectifier uses forward blocking devices (such as IGBTs or power MOSFETs).

In the current-fed scheme (see Figure 5.32), a three- or single-phase PWM rectifier feeds a regulated dc current to an inductive load, while sinking harmonic current on the ac side. A filter capacitor bank is needed on the ac side to absorb the PWM frequency current ripple and permit self-commutation of the devices. The devices are reverse blocking (such as IGBTs or power MOSFETs with a series diode or reverse blocking IGBTs).

In overall comparison, the voltage-fed converter scheme is superior because of its lower cost, higher efficiency, and higher frequency capability. The same statement is applicable for SVCs. However, SVCs can operate at low PWM frequencies. Often, a multistepped phase-shift voltage-fed rectifier is satisfactory (see Figure 4.58).

Q. AC machine dynamic models can be described either in Cartesian d-q form or in complex polar form. Which one is better?

A. The representation of dynamic ac machine models and their analysis can be done either in Cartesian d-q form or in complex space vector form. Although the complex form is convenient for compact representation, the author prefers the d-q form because of its intuitive simplicity and because it's easy to understand. The beginners in this area, particularly the students, tend to get scared by the complex space vector form of representation.

Q. Which type of machine is best for EV/HEV drives?

A. Apparently, an interior permanent magnet (IPM) synchronous motor with NdFeB magnet is the best choice for EV/HEV propulsion systems. The reasons are as follows:
- The motor has higher efficiency compared to induction or switched reluctance motors.
- It has a large field-weakening range compared to surface PM machines, which is essential for EV/HEV operation. However, when the motor is running at high speed, the leading DPF reactive current due to large machine CEMF is excessive, which causes excessive losses in the machine and converter.
- Further, efficiency improvements in the low–speed, low-torque region are possible by means of flux programming (see Figure 8.26).
- With higher energy NdFeB magnets, the motor power density is high with lower inertia. The machine is more expensive compared to IMs, but the price differential is falling recently.

One problem of the machine is that at high speed, if there is a converter fault or control failure, the fault current will be high due to large counter emf. This will cause large pulsating braking torque. Of course, the inherently large motor inductance will tend to limit the fault current. If the (healthy) converter is tripped suddenly at high speed, the free-wheeling bypass diode rectifier will regenerate, causing a large braking effect. Again, this braking effect will be limited by large machine inductance.

FIGURE 12.12 Comparison of reluctance and magnet torques in recent Toyota hybrid vehicles [2].

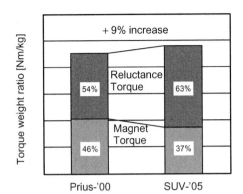

In modern IPM-HEV drives, the machine is designed such that the reluctance component of torque is significantly high to create the following advantages:

- Reduction of flux (ψ_f) will reduce CEMF at high speed so that fault current and pulsating torque are reduced.
- Reduction of flux also reduces iron loss, which becomes particularly important at high-speed operation.

A typical comparison of reluctance torque and magnet torque for a Toyota Prius I (2000) and Toyota SUV (2005) at maximum torque condition is shown in Figure 12.12. This indicates that the machine runs more as an SyRM motor rather than PMSM.

[*Personal note:* At General Electric's R&D Center, when the EV drive (ETX-II) project with an IPM machine was started around 1985, the induction motor was considered to be the supreme candidate for EV applications. The use of an IPM machine was thought very unwise at that time.]

Q. Can you provide a perspective on the use of variable-speed drives in subway transportation?

A. Railway traction applications reflect to some extent the modern advances in power electronics and motor drives. In urban metro or limited distance applications, the power supply is invariably dc with the help of trolly wire or a third rail. Traditionally, the supply voltage has been 600 V or 750 V, but recently, it has been increased to 1500 V, and even to 3000 V in the latest installation. On the other hand, for long-distance railways, such as the Japanese Shinkansen line, a 25-kV, 50/60-Hz supply is usually used. In either application, dc motors have traditionally been used, but have been replaced for the most part by induction motors since the early 1980s. In traction applications, the drive should have four-quadrant capability, i.e., it must regenerate in both forward and reverse rotations. In the old metro drive system (still used now), simple resistance switching was used to control the motor speed. In such a system, the efficiency was very poor due to the dissipative resistors used for motoring and dynamic braking. At the advent of thyristors, thyristor-based force-commutated choppers were used in the 1960s for speed control, which significantly improved the drive efficiency. However, from the early 1980s, induction motor drives were used with force-commutated thyristor inverters. The commutation circuits of thyristor-based converters were complex, dissipative, and contributed to poor reliability. The advent of GTOs quickly replaced the thyristor inverter drives. From the mid-1980s, commercial IGBTs were introduced and gradually found wide application in traction drives. The use of IGBTs improved the system cost and efficiency, and it has been the preferred device until today, and possibly will remain so in the future. With the higher voltage rating of modern IGBTs, the supply voltage could also be raised. For a 750-V system, 1200-V IGBTs required series connection in a conventional two-level inverter because of the wide fluctuation of supply voltage. A three-level inverter (or two-level inverter with input chopper in some cases) solved this problem. The advent of 1700-V IGBTs permitted the converter to fall back to the simple two-level topology again. However, the advent of 3.3-kV IGBTs permitted the use of simple two-level inverter in a higher voltage (1500-V dc) system. The highest voltage IGBTs with 6.5-kV ratings have been available since 2001. This permitted a 3000-V dc supply with a two-level inverter to be used in the latest traction system.

[*Personal note:* In my old home city of Kolkata, India, netro drives use rheostatic controlled dc motor drives. In my recent trip to Santiago, Chile, I found that the same drives are used in some sections where the stations are far apart.]

Q. What is the prospect of a variable-frequency starter for induction motors?

A. Induction motors require some type of starter for constant rated speed operation. Cage-type induction motors can be started directly from the power supply, but the in-rush current will be high. Traditionally, rheostat, wye-delta, or auto-transformer starters have been used. Solid-state anti-parallel thyristor starters with phase control have also been used widely, but high-distortion current does not satisfy line harmonic standards. In addition, because starting torque is low in all the preceding methods, they may not be satisfactory for starting with a large connected load. With WRIM, starting with a rotor rheostat is somewhat easy, but primitive and wasteful of energy. The three-phase rheostat can be replaced by a diode rectifier and dc-to-dc PWM converter with fixed resistance load.

A cage-type induction motor can be started conveniently with a front-end diode rectifier–dc link filter–variable-frequency voltage-fed inverter. The scheme is somewhat expensive compared to that of a solid-state phase controller, but has the following advantages:

• The starting torque can be the same as the rated torque.
• The motor can be bypassed to the line supply after starting to eliminate the converter loss.
• The same starter can be used on a time-sharing basis for other motors.
• At light-load running condition, the flux can be programmed by voltage control to improve drive efficiency.

- The line current can be made sinusoidal and DPF can be controlled to unity with a PWM rectifier.

   With the recent trend toward decreasing costs for power electronics and stringent line harmonics regulation, a variable-frequency starter looks very attractive and will tend to replace phase-controlled starters.

Q. How can you use a single and centralized variable-frequency starter for multiple drive units?

A. A thyristor phase-controlled (solid-state) starter for induction motors was discussed in Figure 3.26(a). The disadvantages of this starter are excessive line and machine harmonic currents and poor starting torque. On the other hand, use of a variable-frequency drive (VFD) as the starter solves the problems discussed above, although it is more expensive. Considering the VFD's many merits and the significant decrease in power electronics costs in the future, the VFD starter looks promising. If an installation has multiple drives, a single centralized starter can be used on a time-sharing basis. Consider, for example, the multiple induction motor installation in Figure 12.13 with machines of reasonably high power ratings, and motors used as pumps. Each motor can be started from zero speed with the help of a VFD operating in the open-loop volts/Hz control mode. As the motor frequency and voltage approach the line condition, it can be synchronized with the line and transferred to it. All three machines can be started in sequence. The machines can also have different power ratings. In addition to start-up operation, one or more machines (in parallel) can be operated continuously at variable frequency with the starter to control the speed, i.e., the pumping need (gallons/m) within the power rating of the VFD.

FIGURE 12.13   Multiple machine operation with a central VFD.

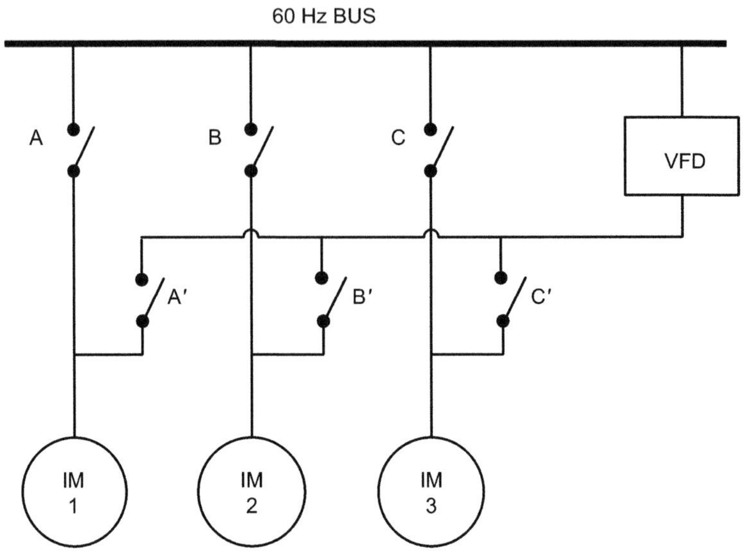

The frequency and voltage of multiple machines in parallel can be raised and then transferred to the line. Similarly, if machine(s) are already on the line, they can be transferred to the VFD and speed can be controlled. Instead of induction motors, PM machines can also be used. Programming a central VFD for multiple machine start-up and speed control can be very economical. Large-power WFSMs can also be started with a centralized VFD, which can be a voltage-fed or current-fed converter system.

Q. In a speed sensorless vector drive of an induction motor, how is the drive started in the very low frequency region?

A. The speed signal in a vector drive may be needed for two purposes: (1) for a feedback signal for the speed control loop and (2) for synthesis of the unit vector. Consider first the indirect vector control with a speed encoder. Because of the encoder, there are no problems with vector control in the entire region from zero frequency to field-weakening region. Assume now that the speed sensor is replaced by a speed adaptive flux observer (Figure 7.60). In spite of the inaccuracies induced by the parameter variations, the drive will operate well down to, say, 50 rpm. However, currently, no speed observer works down to zero frequency. In this low-frequency range, the drive can be started in open-loop volts/Hz control, and then switched into vector control. For a direct vector drive above a minimum speed (say, 50 rpm), the unit vector can be calculated from machine terminal voltages and currents. If a speed signal is needed for the speed loop, it has to be estimated by an observer. However, in the very low frequency range (including zero frequency), the machine can be started in the open-loop volts/Hz control mode and then switched over to vector control.

[*Note:* Speed estimation strategy at zero frequency by auxiliary signal injection is still under development.]

Q. Is the DTC control self-starting without any speed encoder?

A. No. Also, DTC control does not work at very low frequency. The operation of DTC control depends on correct estimation of the stator flux vector ($\psi_{ds}{}^s$ and $\psi_{qs}{}^s$) from the machine terminal variables based on voltage model equations. Normally, the voltage model does not work at very low frequency. The problem is similar to that of stator flux-oriented vector control. Therefore, the machine has to be started with open-loop volts/Hz control, and then switched over to DTC control. Of course, there are complex methods of flux vector estimation from zero speed. With the speed sensor, the flux vector can be estimated from zero speed with current model equations and, therefore, starting and DTC operation from zero speed are not problems.

Q. Why DTC control generates more harmonics?

A. DTC control normally uses a hysteresis-band PWM-like control-type technique; therefore, the usual features of HB-PWM hold true here, i.e., variable switching frequency and large ripple in the current, flux, and torque (particularly when switching frequency is low).

During a switching cycle of DTC control, only one inverter state is selected for the period. In comparison, for example, with space vector PWM, the sample time (Ts) is constant and the three nearest inverter states are selected with correct time segments so that the ripple is minimum. If HB control in DTC is replaced by fuzzy or neuro-fuzzy control [9] along with the SVM, the ripple can be substantially minimum. Thus, the simplicity of DTC control is lost at the cost of high complexity.

Q. Why is the name "vector control" more appropriate than "field-oriented control"?

A. The term "field-oriented control" originated because of its first application in machine drives. For example, in induction motor drives, the stator field or excitation component of current is oriented in the direction of rotor flux (rotor flux orientation), whereas the torque component of current is oriented perpendicular to it. Similarly, in stator flux-oriented vector control, the excitation current is oriented in the direction of stator flux. However, later, the same control technique was applied in non-machine applications. For example, in a PWM rectifier, the active or in-phase component of current (I_P) is controlled in the direction of the voltage vector, whereas the reactive current (I_Q) is controlled perpendicular to it. In such a case, the term "field-oriented control" has no meaning, but the term "vector control" is appropriate. Therefore, in general, the term "vector control" is most appropriate.

[*Personal note:* In 1988 at the Kyoto IEEE Power Electronics Specialists Conference (PESC), the author appealed to attendees to use the term "vector control" instead of "field-oriented control," and the trend has been clearly in that direction. In all of his books and published literature, the author has used the term "vector control" only.]

Q. Can you operate a static Kramer drive at synchronous speed?

A. No. At synchronous speed (slip or S = 0), the rotor-induced emf is zero. This cannot establish dc link current I_d. Therefore, developed torque remains zero at synchronous speed. The machine speed should be slightly lower than the synchronous speed for proper operation.

Q. Is the transformer essential in slip power recovery drives?

A. Yes. If the application is such that it permits reduced speed range from the synchronous speed, then the converter rating can be economized, but the price is a bulky 60-Hz transformer (and, of course, the more expensive wound rotor machine). Due to recent decreases in converter costs, the importance of this drive is diminishing.

Assume, for simplicity, the unity turns ratio machine and consider the scenario of three classes of converters:

1. *Current-fed converter system in static Kramer drive* (Figure 7. 43) – Both the diode rectifier and thyristor inverter always handle the same rated current (dc current is proportional to torque), but the rectifier voltage rating depends on the slip, and the inverter voltage rating is determined by its supply ac voltage. If the lowest speed of the machine is 50% of synchronous speed (S = 0.5), the rectifier is at maximum voltage

and its power rating is maximum (50% compared to standstill condition). Without a transformer, the inverter power rating is 100%. Since $V_d = V_I$, the inverter ac voltage can be cut down to 50% by a transformer. This results in a converter system power rating of 50%. With further reductions in the speed range, the converter rating will be reduced proportionally. The line transformer increases the inverter firing angle toward 180° at maximum V_I, thus improving its line DPF near unity. Because of machine lagging excitation current, the total DPF will be less than one. With a 100% converter system rating (no transformer), the speed can be controlled from zero to synchronous speed. In that case, the converter rating is the same as that for stator-side speed control.

2. *Cycloconverter in static Scherbius drive* (Figure 7. 45) – The cycloconverter acts as a step-down frequency and voltage changer, where its output voltage and frequency match with those of the machine. With a line frequency of 60 Hz, the typical maximum output frequency range is restricted to 0–20 Hz, i.e., the speed range is limited to 33% (S = 1/3) from synchronous speed. This also means that the machine output voltage range is 0–33%. The cycloconverter current rating is dictated by the load torque, and its voltage rating is determined by its line voltage. Therefore, a transformer will reduce the power rating of the cycloconverter, and also the line DPF will improve because of its reduced voltage ratio.

3. *Voltage-fed PWM converter system in static Scherbius drive* (Figure 7.50) – The voltage rating of either converter is the same and is dictated by the dc link voltage (V_d). Again, the dc link voltage is higher than the peak line voltage or peak machine voltage (whichever is higher). This means that with a transformer, the voltage rating of the converter units, i.e., their power rating can be decreased. Note that at rated torque, machine converter current loading is the same at any slip, i.e., the power handled by it is always the rated power although slip power flow varies with the slip. However, the current loading of the line converter increases at higher slip because of higher slip power flow. This means that both converters handle maximum power at the lowest speed. If the speed range is limited to 50% from the synchronous speed, the transformer can reduce V_d by 50%, thus reducing converter power rating to 50%. This implies that a 100% converter system rating without a transformer will permit 100% speed range.

Q. Can you operate the static Scherbius drive at unity or leading DPF?

A. Yes. The basic fundamental is that if a synchronous motor is overexcited, it operates at leading power factor. Consider a Scherbius drive (Figures 7.45 or 7.50) and assume that it is running at ideal synchronous speed, i.e., operating as a synchronous motor. At this condition, the excitation current is dc. Higher dc excitation current will make the terminal DPF leading. At any sub- or supersynchronous speed, the excitation current is ac at low frequency. An increase in the ac excitation current will make the machine DPF leading. The leading machine DPF can not only cancel the lagging DPF of a cycloconverter to make the line DPF unity, but the line DPF can also be controlled to be leading. With double-sided PWM converters, the conditions are more flexible. Here, a line-side converter can also be operated at leading DPF to control the total line DPF.

Q. How do you start the static Kramer drive and static Scherbius drive from zero speed?

A. Neither machine can be started directly from zero speed for two reasons: (1) Rotor current must be established to develop the starting torque and (2) the machine acts as a stationary transformer (S = 1) with full voltage at rated frequency (60 Hz) appearing at the secondary. The standard converter ratings are exceeded and sometimes operation becomes abnormal. A separate resistive starter can be used that will establish the rotor current and limit the converter voltage and frequency.

Q. Which type of wind generation system is best?

A. A variable-speed wind generation system operating on the utility grid is basically a regenerative ac drive, in which a motoring propeller is replaced by a generating turbine. In principle, any type of machine, such as induction (cage or doubly fed with slip power recovery), PMSM or SRM, can be used. In fact, both induction and synchronous machines of different types have been widely used. Slip power recovery static Scherbius drives with a double-sided phase-controlled converter or cycloconverters operating in the regenerative mode have been traditionally favored for such applications. However, the usual disadvantages of a doubly fed machine, phase control, and a 60-Hz transformer remain. Recently, static Scherbius drives with double-sided PWM IGBTs converters and PMSM drives have found acceptance. In the author's opinion, a cage-type induction generator with double-sided PWM converters (see Figure 10.42) is the best selection for this application. Several reasons can be summarized as follows:
- A cage-type machine is cheaper than a doubly fed induction machine.
- The speed range for energy capture is high.
- Does not need a 60-Hz power transformer.
- Line and machine currents are sinusoidal with line at unity or programmable DPF.
- With no wind, the line-side converter can be operated as a static VAR compensator.
- Advanced control for performance optimization can be easily applied.

   (With the recent reductions in the cost of NdFeB magnets, the PMSM drive also looks attractive because of efficiency improvement.)

   Although the converter cost is somewhat high, the current trend toward decreasing costs of power devices and elimination of the 60-Hz transformer makes the system more justified.

Q. Premature insulation failure has been observed in IGBT inverter-fed induction motor drives. What are the possible reasons for this type of failure?

A. Stator winding insulation of an induction motor is normally designed for sine-wave voltage operation. With a modern PWM inverter that uses fast switching IGBTs, additional stresses are imposed on the insulation, causing accelerated aging and premature dielectric failure [10]. The primary reason for this aging is high *dv/dt* stress on the insulation due to the steep wavefront of the PWM wave. Displacement current and some dissipative current flow through the insulation. The turn-to-turn, phase-to-ground, and phase-to-phase insulation of the stator winding is therefore subjected to increased dielectric stress over a period of time. Again, at the leading edge of the PWM wave, there can be a ringing effect that causes an overvoltage transient that adversely affects the life of the insulation. With a long cable between the inverter and the machine, overvoltage may also result from the wave reflection. The common mode stray current that flows through the bearing also has a deteriorating effect. There is, of course, the usual deterioration of machine insulation over a period of time due to temperature, mechanical vibration, humidity, and chemical gases. A soft-switched inverter may eliminate inverter-induced insulation deterioration. Otherwise, a low-pass filter can also be used at the machine terminal.

Q. Can you tell us something about the bullet train drives used in Japan?

A. The bullet train (or Shinkansen system) in Japan is a very successful high-speed transportation project originally started in the 1960s. In its early phase, the trains used dc drives from a single-phase ac line supply through tap-changing transformer and diode rectifier. This was replaced by thyristor phase-controlled rectifier dc drives. Since the early 1990s, GTO-based PWM rectifiers and two-level inverters with induction motor drives had been used. However, since the end of the 1990s, the three-level IGBT PWM inverter system shown in Figure 12.14 [11, 12] has been used in several lines.

FIGURE 12.14   Schematic diagram of modern Shinkansen system.

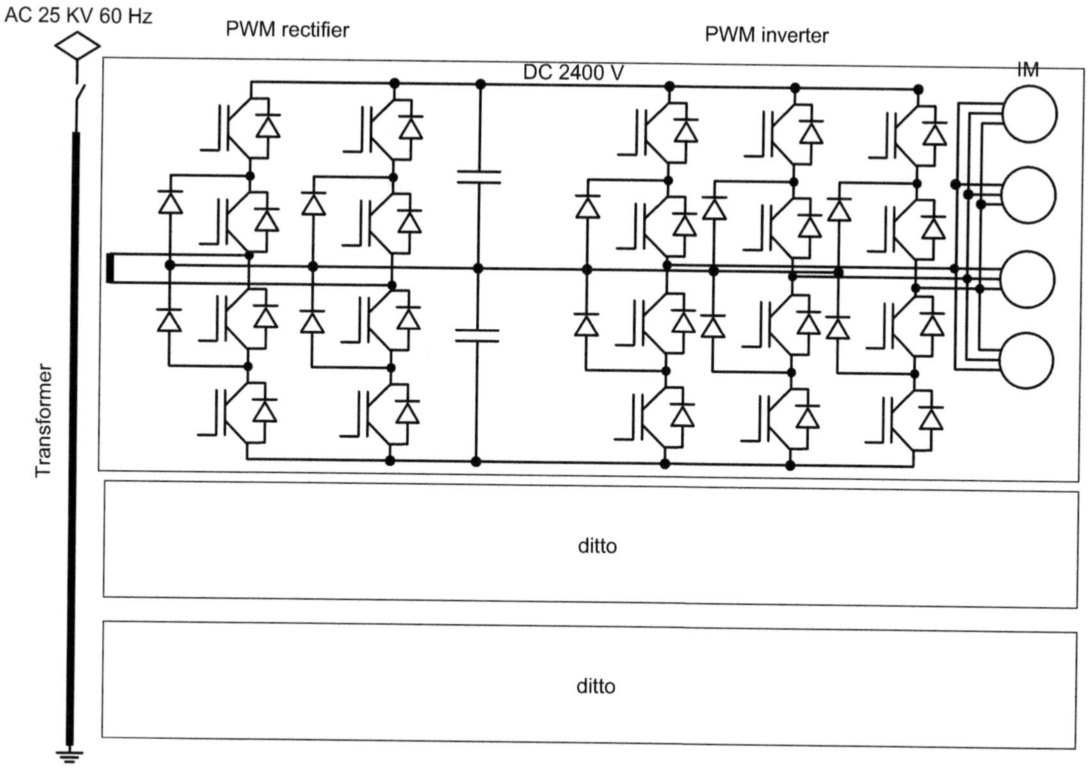

The single-phase line power supply is at 25 kV and 60 Hz. It is stepped down by a transformer and feeds three identical drive units as shown. The single-phase PWM rectifier and three-phase PWM inverter with IGBTs use three-level half-bridge units as indicated and permit four-quadrant operation with supply DPF at unity. Each inverter has four identical induction motors (275 kW each) with closely matched wheel diameters so that parallel motor speeds are nearly equal. Vector control is applied to the parallel machines. The maximum speed of the train is around 180 miles/hr.

Q. An induction motor is running with a variable-frequency inverter. What will happen if one phase becomes isolated due to an inverter fault?

A. A power device in an inverter may fail as a result of a short or open circuit, although a failing by short is more common. Whatever the cause, assume that the protection system isolates the faulty inverter leg, injecting only single-phase supply to the machine. In the running condition, injecting only a single-phase supply will allow the machine to continue operating, but the pulsating torque will be high. If the frequency is low, the pulsating torque and corresponding speed pulsation will be very damaging, but at high frequency, an inertia filtering effect smoothes the speed. This scenario is somewhat identical in single-phase machines with a 60-Hz supply starting with a two-phase supply. At 60-Hz supply, the pulsating torque at 120 Hz is tolerable because of inertia filtering.

Q. Can you explain clearly how you can get ideal integration of machine stator voltage with dc offset minimization using the programmable cascaded low-pass filter (PCLPF)?

A. Consider the three-stage identical PCLPF shown in Figure 7.63, and assume that the hardware low-pass filter phase shift angle $\varphi_h = 0$. This means that each identical LPF is required to shift $-30°$ angle for ideal $-90°$ phase shift in the integration irrespective of fundamental frequency variation. An equivalent op amp-type circuit is shown in Figure 12.15.

First, consider the difficulty of integration with a one-stage op amp only, and assume that the desired dc offset attenuation factor is 0.1. Because

$$\tau\omega_e = \tan\varphi \ ... \ (1)$$

FIGURE 12.15   Op amp representation of PCLPF for integration.

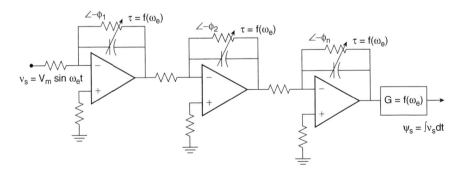

$$\tau = R_f C_f \ \dots (2)$$

$$\text{i.e.,} \ C_f = \frac{\tan \varphi}{\omega_e R_f} \ \dots (3)$$

and assume $R_f = 10K$ for $R_i = 100K$ for 0.1 attenuation of dc offset. Ideal integration ($\varphi = -90°$) is possible with $R_f = \infty$, but dc offset will build up in this case. With finite τ, select $\varphi = -89.9°$ at 60 Hz. Calculation shows that C = 151.9 μF for 60 Hz, but for the same φ at 0.1 Hz, C = 91, 140 μF (very high).

Now select a three-stage PCLPF such that $\varphi = -30°$ for each stage to get ideal integration. Calculation gives $C_f = 91.8 \ \mu F$ at 0.1 Hz for 0.1 dc attenuation per stage (10^{-3} for there stages). Obviously, for the same phase shift at variable frequency, τ has to vary inversely with ω_e ($\tau \omega_e$ = constant). With hardware LPF, $-\varphi_h$ varies with ω_e. The PCLPF has to compensate for this variation. The PCLPF can be computed by a DSP in real time.

Q. What is a hybrid elevator?

A. Elevators normally use variable-speed drives for speed control and regenerative braking. Traditionally, thyristor-fed dc motor drives have been used, but recently ac motors with

FIGURE 12.16 Hybrid elevator system.

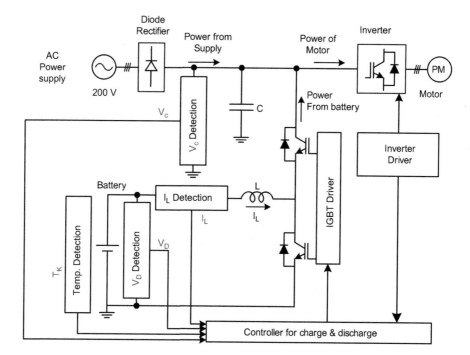

variable-voltage, variable-frequency control have invariably been used. Again, an induction motor had been the usual choice, but recently, PM motors have gained preference because of their higher efficiency. For variable-frequency ac drives, double-sided PWM converters with IGBTs and vector control are the most common. Recently, hybrid elevator drives have been used in Japan [11]. Figure 12.16 shows the drive system. The PM drive motor receives the usual power supply through the diode rectifier and IGBT PWM inverter. The regenerative braking power of the motor in the dc link is absorbed by a Ni-MH battery through a dc-dc converter. During power failure in the utility system, the battery operates as a UPS system and supplies power to the inverter through a dc-dc boost converter. Hybrid elevators have been used in machine roomless (MRL) systems for low-rise buildings.

Q. What do you understand by a hybrid train?

A. The concept of a hybrid train is similar to hybrid electric vehicle (HEV), and it is being developed in Japan [11]. The train is supposed to have frequent start–stop operations like a subway drive, but there is no electric power supply as in the usual subway train. Such a train has a diesel engine as its power source and a battery as an energy storage device. The engine runs at constant speed to give optimum efficiency. While accelerating the train, the battery boosts the engine power, and while braking, the battery absorbs the regenerated energy. A depleted battery can be charged either by the engine or from utility supply.

Q. What can you say about the variable-frequency starter for synchronous motors?

A. The application of a variable-frequency converter for starting a constant-speed synchronous motor is not always evident, although in principle it can be used. For sinusoidal line-start PMSM, a cage winding is used on the rotor so that the machine can be started as an induction motor, and then run as a synchronous motor. At synchronous speed, the cage winding has no effect. A high-power WFSM at no load can be started with a pony motor mounted on the shaft. A variable-frequency converter has often been used for starting WFSMs. The motor is usually started at no load to reduce the converter size, and then the load is picked up after the bypass operation to the main supply. Either voltage-fed or current-fed converters can be used. However, a current-fed load-commutated converter system is preferred because thyristor converters are simple. An excellent example is the propulsion system for the QE2 cruise ship described in Figure 8.49. If there are a number of machines, one single converter system can be used as the starter on a time-sharing basis. Note that SRMs and BLDMs (brushless dc motor with trapezoidal machine) are electronic motors with integral converters in the front end that can be used as starters as well as continuous-speed controllers.

Q. In a speed sensorless vector drive for a PMSM, how is the drive started in the very low frequency region?

A. Most of the explanations given before for sensorless induction motor drives are also valid here. The sinusoidal PMSM drive can have only direct vector control (because slip $s = 0$), where the unit vector above a minimum speed (or frequency) can be estimated from the machine terminal voltages and currents. In the lower range (including zero frequency), the motor can be started in the open-loop volts/Hz control mode and then switched to the vector control mode. For a BLDM, following the same principle, the machine can be started in open-loop control and then switched to the self-control mode (see Figure 8.70).

Q. Is it a good idea to put a conducting girder on the PM machine rotor to protect the magnets against centrifugal force?

A. Electrically, the conducting girder acts like a cage (an *amortisseur* or damper) winding on the rotor, which is normally used in a line-start PM machine. In this case, after starting function with cage winding, it has no function electrically at synchronous speed with a sinusoidal 50- or 60-Hz line power supply. However, with a variable-frequency PWM inverter-fed drive, harmonic currents are induced in the girder that may heat it up to high temperature; the additional loss decreases the machine's efficiency. The skin effect will increase the resistance, thus worsening the heating effect. Therefore, cage winding (or use of a metal girder) is not a good idea in variable-frequency drives. (It can, however, provide some damping against oscillatory torque.)

Q. Is the vector control of synchronous machine drives somewhat sluggish compared to that of induction motors?

A. Yes. In an induction motor drive, when torque is commanded by the i_{qs}^* command, the flux is not affected. The flux is kept constant by the independent current i_{ds}^*. Consider a wound-field synchronous machine drive (Figure 8.54). When torque is commanded by I_T^*, in order for the flux magnitude to remain constant, it requires angular alteration by change of field current I_f. This response is very sluggish. Of course, injecting magnetizing current from the stator side in the vector drive alleviates this condition, but the response remains somewhat sluggish. For an interior magnet PM machine, a similar flux vector angle alteration is needed by the stator-supplied magnetizing current, making the response somewhat sluggish. The sluggishness is minimal for a surface magnet machine. Note that vector control is only applicable for a sinusoidal machine.

Q. What are the trends of variable-speed air conditioner drives?

A. Traditional air conditioners use a single-phase constant speed induction motor in which the switching is controlled by a thermostat. The efficiency of such a system is poor. On the other hand, a variable–speed, load-proportional drive is more efficient. Japan is the leader in variable-speed home air conditioners because energy costs in Japan are quite high. About 94% of residential air conditioners in Japan use variable-speed drives. Originally, induction motor drives were used, but because of the higher efficiency of PMSMs, these machines are invariably used nowadays. Figure 8.69 described a typical drive. But a more modern drive is shown in Figure 12.17 [11].

FIGURE 12.17 Configuration of modern variable-speed air conditioning system using PMSM.

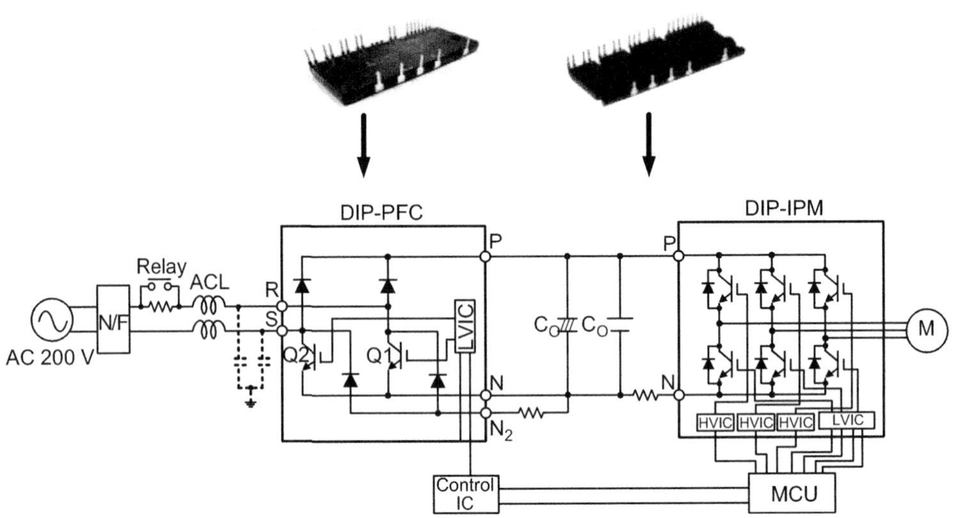

The PWM rectifier uses DIP-PFC with power factor control (PFC), and the PWM inverter uses an IGBT DIP-IPM module as shown in the figure. The motor is vector controlled with the help of a microprocessor, whereas the rectifier is controlled by an IC. Refrigerators and washing machines also use similar types of drives.

Q. Why is the bearing current problem less serious in PWM inverter-fed PM machine drives then that of induction motor drives?

> *A.* Chapter 4 revealed that common mode dv/dt-induced bearing current flows through the stray capacitances of motor air gaps and bearings. The machine air gap capacitance will decrease as the machine air gap increases. The PM machines (SPM [both sinusoidal and trapezoidal] and IPM) have larger effective air gaps (PM permeability $\mu \sim 1$) compared to induction machines. Therefore, other conditions remaining the same, bearing current will be smaller in PM machine drives compared to induction motor drives.

Q. Why are variable-frequency drives gaining more acceptance for ship propulsion?

> *A.* Traditionally, large diesel engines have been used for ship propulsion. The disadvantage of this type of propulsion is that the engine shaft is directly connected to the propeller, and engine speed is dictated by propeller speed. In such an installation, the flexibility of engine location is lost (also the problem of long shaft), and engine efficiency is poor because of the need for variable speed. In variable-speed, diesel-electric propulsion, the engine is coupled to a synchronous generator to generate constant voltage at 60/50 Hz. Its constant-speed operation makes the engine efficiency higher, thus saving a considerable

amount of fuel—quite a good advantage considering the recent increases in the price of diesel fuel. The electrical power from the generator is distributed (drive by wire) to the variable-frequency drive that couples to the propeller. The installation site of engine-generators is very flexible. Synchronous motor drives with load-commutated inverters or cycloconverters are normally used in ship propulsion. In both cases, the line DPF is poor. There will be additional harmonic load on the generator. Regeneration is not essential because the engine cannot absorb regenerated power. However, the ship auxiliary load connected to the 60/50-Hz bus can utilize regenerated power. The braking energy is usually absorbed by the load.

[*Note:* A double-sided PWM VFI, although somewhat expensive and more lossy, is beneficial in the propulsion system. The PWM rectifier can regenerate power to be used as auxiliary power, if needed. It can maintain unity DPF sinusoidal line current that will make the generator size economical and more efficient.]

Q. What is your prognosis on switched reluctance motor drives?

A. Peter Lawrenson, a pioneer in SRM, had a dream (if you have ever seen his video) that SRM would revolutionize electric drives in the future—unfortunately, that dream has not materialized so far, and will not ever materialize. Thousands of university papers have been published and dozens of doctoral degrees have been awarded in this area, but the prognosis of the technology remains essentially the same as before. Some of the features of SRM drives were discussed in Figure 8.66. Of course, like other machines, in principle, SRM can also be used in any application. However, considering all of the aspects, the SRM drive is quite inferior to PMSM (or BLDM) drives, inferior to induction motor drives, and somewhat inferior to SyRM drives. In fact, SRM is the closest relative to SyRM, but in the literature, SRM is mostly compared with PMSM and IM drives. Recently, SRM drives have been proposed for EV/HEV drives [13], but in reality, the more popularly used IPM drives for such applications are moving closer to SyRM drives [2].

Simple and economical machine construction is possibly SRM's only favorable point, but these criteria only—especially in light of its other drawbacks—cannot favor SRM over PMSM, IM, and SyRM drives.

[*Personal note:* In 1996, at the IEEE-APEC Rap Session on SRM drives, I was invited to talk against SRM drives. After my presentation, I got long, loud applause from the audience. After all of these years, my view essentially remains the same. Of course, SRMs can be justified for very specialized applications.]

Q. After simulating a power electronic system with Simulink, can you use the controller part of the simulation program for real-time control with DSP?

A. Yes. The MATLAB-based Real-Time Workshop [14] automatically generates standalone C code from the Simulink source code either for the entire system or for an individual subsystem. The resulting code can be used for a variety of target platforms (the hardware or operating system environment in which the generated code will run). The C code is then compiled to generate the object code for a particular DSP/microprocessor. However, the code size may not always be optimum for real-time DSP implementation of power electronic systems. This means that the handwritten code may run in shorter time and occupy less storage space.

Q. Can you accelerate the execution speed of the Simulink simulation program?

A. The MATLAB-based Simulink simulation program for power electronic systems is slow but very user-friendly. Recently, dramatic improvements in the speed of personal computers have helped to make program execution faster. In addition, the execution of a Simulink program that uses built-in blocks can typically be enhanced two to six times with the help of Simulink Accelerator. The Accelerator uses portions of the Real-Time Workshop to automatically generate C code from Simulink models, and then the C compiler creates the fast, executable code. Note that although the Simulink Accelerator takes advantage of Real-Time Workshop, separate Real-Time-Workshop is not required to run it. Also, if a C compiler is not installed in the Windows PC, the Icc Compiler provided by MathWorks can be used. However, the Simulink Performance Tools option must be installed in the system to use the Accelerator program.

Q. In a multitasking system of DSP control, how are the task sampling times selected?

A. A power electronic system under the control of a DSP normally has to perform multiple tasks (multitasking system) in a periodic manner (see Figure 9.41), such as current control and speed control. The current control loop has higher bandwidth and, therefore, it requires faster sampling. A rule of thumb may be loop response time $> 10\times$ sampling time. Therefore, the speed loop will require a longer sampling time. A preliminary estimate can be made for all sampling times (T_1, T_2, T_3, and T_4). Normally, the shortest sampling time (T_1) is generated by a hardware timer, and the other intervals are generated by software timers. For example, a down-counter to generate T_2 can be clocked by T_1 pulses. Similarly, the T_3 counter can be clocked by T_2, and T_4 can be clocked by T_3. This makes all the count ratios integral. The task execution times (t_1, t_2, t_3, and t_4) should be estimated by the respective code lengths. Alternately, operation of the simulator program can provide this estimation. The loading of DSP should never exceed the limit. In fact, some time margin should be provided for last-minute code additions. The estimation can be made as follows from Figure 9.41:

$$\text{Loading factor} = \frac{\sum \text{Computation time}}{T_4} \quad \dots(1)$$

But, $n_1 = \dfrac{T_2}{T_1}$, $n_2 = \dfrac{T_3}{T_2}$, $n_3 = \dfrac{T_4}{T_3}$, where $n = $ an integer. Therefore,

$$\text{Loading factor} = \frac{1}{T_4}\left[\frac{T_4}{T_1}t_1 + \frac{T_4}{T_2}t_2 + \frac{T_4}{T_3}t_3 + t_4\right]$$

$$= \frac{t_1}{T_1} + \frac{t_2}{T_2} + \frac{t_3}{T_3} + \frac{t_4}{T_4} \quad \dots (2)$$

Equation (2) can be written as

$$\text{Loading factor} = \frac{1}{T_1}\left[t_1 + \frac{T_1}{T_2}t_2 + \frac{T_1}{T_3}t_3 + \frac{T_1}{T_4}t_4\right]$$

$$= \frac{1}{T_1}\left[t_1 + \frac{t_2}{n_1} + \frac{t_3}{n_1 n_2} + \frac{t_4}{n_1 n_2 n_3}\right] \dots (3)$$

The loading factor should always be less than unity and (1– loading factor) is the per-unit idle time. i.e.,

$$t_1 + \frac{t_2}{n_1} + \frac{t_3}{n_1 n_2} + \frac{t_4}{n_1 n_2 n_3} < T_1 \dots (4)$$

Again, individually,

$$t_1 < T_1$$

$$t_2 < \left(T_2 - \frac{T_2}{T_1}t_1\right)$$

$$t_3 < \left(T_3 - \frac{T_3}{T_1}t_1 - \frac{T_3}{T_2}t_2\right)$$

$$t_4 < \left(T_4 - \frac{T_4}{T_1}t_1 - \frac{T_4}{T_2}t_2 - \frac{T_4}{T_3}t_3\right) \dots (5)$$

These equations should be satisfied for the selection of time intervals. Signal sampling and feedback signal estimation should be done in respective sampling times. For example, feedback current should be sensed every T_1 interval, whereas speed sensing and flux estimation can be done in T_2 intervals.

Q. Can you implement fuzzy control with a look-up table in DSP?

A. Yes. A fuzzy system basically implements nonlinear mapping from input space to output space. Therefore, once a fuzzy system has been designed, a look-up table can be generated between input variables and output variables and implemented by DSP. For example, in Figure 10.24, a three-dimensional look-up table between the input variables e and ce and the output variable du can be defined to implement the fuzzy controller. This table permits fast implementation, but requires large amounts of memory to obtain precise data. To the author's knowledge, the first application of FL in power electronics used the look-up table method [15], where the first table is coarse and the second table is fine to reach the desired precision. Note that a fuzzy controller in a Simulink simulation program can generate C code that can be compiled to generate object code for a DSP.

Q. Which type of adaptive control works best for parameter variation and load disturbance condition in the drive?

A. Consider the speed control loop of a vector drive with traditional *P-I* control. If there is a sudden increase in torque load (T_L), the speed will fall but will recover because of the closed loop. With higher loop gain, the dip will be smaller, but the loop gain should be limited to avoid stability problems. The effect is similar if moment of inertia (J) increases. However, an inertia change directly affects the loop gain, i.e., the stability is affected. In open-loop volts/Hz control, increase of T_L will cause a dip in speed that will not recover. The inertia change ideally has no effect at steady-state operation.

A number of adaptive control techniques are available, such as self-tuning control, MRAC, sliding mode control, and fuzzy control. Among all of them, fuzzy control appears to be the best. Fuzzy *P-I* control is essentially *P-I* control with adaptive proportional (*P*) and integral (*I*) gains. Advantages include these: This type of control is easy to implement, transient response is excellent with deadbeat, the speed dip is small, the steady-state error is zero, and a mathematical plant model is not really essential.

Q. Is the bias signal essential in a neural network?

A. Without a bias signal, if the input signals are zero, the output will be zero. The bias generates offsets at the output from the zero point. Therefore, a network can be trained with or without a bias signal as desired. However, if one is not sure if a bias signal is needed, it is always better to train the network with the bias signal. The bias circuit weights are automatically adjusted during training for zero or nonzero offset.

Q. Can you clarify the linearly separable pattern classification property of a perceptron?

A. A perceptron (see Figure 11.6) is the simplest form of NNW used for classification of signal patterns that are linearly separable. The term *linearly separable* generally means that the patterns lie on the opposite sides of a hyperplane. This requires that the patterns to be classified must be sufficiently separated from each other to ensure that the decision surfaces consist of hyperplanes. If only one neuron is used with two-dimensional input (p_1 and p_2), then the decision boundary is a straight line in p_1-p_2 plane as shown in Figure 11.6. A pair of linearly separable patterns for a two-dimensional, single-layer perceptron is shown in Figure 12.18(a) [16]. If the two patterns ζ_1 and ζ_2 are allowed to move too close to each other, as shown in part (b), they become nonlinearly separable and the elementary perceptron fails to classify them. If the input is three dimensional (p_1, p_2, p_3), then the decision surface is a plane separating the patterns. An example synthesis of a 5-input NAND gate with perceptron training by the Neural Network Toolbox was given in Figure 11.73. For the same input signal vector, a different perceptron can be trained for a different decision boundary. The perceptron network in Figure 11.6 can be used, for example, for mortgage approvals in banks with one profile of a person and employment approval with another profile. Its application for power electronics is left to the imagination of the reader. An example application might be identification of response

FIGURE 12.18 (a) A pair of linearly separable patterns and (b) a pair of nonlinearly separable patterns.

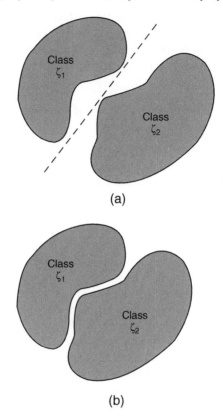

(a)

(b)

pattern (overdamped, oscillatory, critically damped, etc.) [17] of a drive system and correspondingly adjusting the P-I gains.

Note that the computational power of a perceptron is very limited. If more unique input–output mapping is required, a multilayer perceptron (MLP) type backpropagation NNW is required, which has more computational power.

Q. Will a perceptron work with other type activation functions?

A. The decision-making characteristics of a perceptron basically remain the same regardless of whether a hard-limiting or soft-limiting activation function is used as the source of nonlinearity for the neuron. We may, therefore, state that as long as we limit to the model of a neuron that consists of a linear combiner followed by a nonlinear element, then regardless of the form of nonlinearity used, a single-layer perceptron can perform pattern classification only on linearly separable patterns.

Q. Can a three-layer feedforward NNW also be defined as a two-layer NNW?

A. Yes. In a conventional three-layer NNW (Figure 11.8), the input layer is basically the distributor of the input signals. It can be shown as a layer of linear neurons (linear activation function), but does not perform any computations. Therefore, the NNW is also defined as a two-layer NNW.

Q. How do you explain the learning capability of a neural network?

A. A neural network has a distributed intelligence (or memory) property that is contributed by the synaptic weights distributed in the network. The biological neural network in a human brain also has similar distributed intelligence. This distributed intelligence gives the associating memory property of NNWs. If a signal pattern is impressed at the input, an associative pattern will appear at the output by selecting the weight pattern. Supervised training with example input signal patterns fixes this weight pattern. If the input pattern is then changed, the output pattern will be different for the same weights. Note that N number of weights can give N degrees of freedom for input–output mapping. If there are a number of input–output signal patterns to be matched by the same NNW, the weights are trained back-and-forth for the different patterns by means of a supervised training algorithm. After successful training, every input pattern will generate the corresponding output pattern even if the weights remain the same.

This concept is somewhat difficult to visualize. The idea comes from the way a biological neural network behaves. It is explained in Figure 11.11 for the optical character recognition (OCR) problem. Consider the expertise of a man in power electronics. How does he learn his expertise? He has to train his brain (a biological neural network) in this area by attending courses and listening to presentations in conferences. Without this learning (or training) he cannot solve any problem in this area. By training, the synaptic junctions of brain cells are appropriately programmed. Training of an artificial network is similar to the training of a human brain in a certain domain. If a NNW is impressed with an input signal pattern that it does not recognize (i.e., it is not trained for this particular pattern), then the output will be erratic. In the same way, a power electronics engineer cannot solve a problem unless he is trained to solve similar problems.

If a power supply is cut off and then resumed, NNW intelligence or memory remains unaltered. In the same way, if a power electronics engineer goes through anesthesia for surgery, his expertise remains unaltered, indicating that human intelligence or memory is distributed in the synaptic junctions of brain cells, but not in a computer-like central memory. We are then mimicking the natural intelligence property in NNWs (artificial intelligence).

Q. How do you define a three-layer backpropagation NNW as a universal function approximator?

A. As mentioned earlier, a neural network has a multidimensional input–output nonlinear mapping property. Consider, for simplicity, one-dimensional input (X) and the corresponding one-dimensional output (Y). For a set of X input values and the correspondingly desired Y values, the NNW can be trained. The network has an interpolation property.

This means that the NNW describes a continuous approximate function between Y and X although it was trained at discrete values. It has been proven theoretically that a three-layer NNW (with one hidden layer) is sufficient to have a multidimensional function approximation. A universal function approximation generalizes an arbitrary p-input q-output continuous functional relation. However, the three-layer NNW may not be optimum in terms of number of total neurons, training time, or ease of implementation.

Q. How can you make backpropagation training perform better for the NNW?

A. The backpropagation training algorithm (in MATLAB/Neural Network Toolbox) helps to train the NNW automatically. However, the following steps help to train the NNW better:

- The training of NNW becomes faster with hyperbolic-tan type TFs of neurons. This type of TF is applicable for both unipolar or bipolar signals. However, for bipolar signals, hyperbolic-tan TFs must be used.
- The target or desired values should be chosen within the range of sigmoidal or hyperbolic-tan TFs (with some amount of offset from the limiting values).
- The initial synaptic weights and biases of the NNW should be generated randomly but with uniform distribution in a small range. The reason for making the range small is to reduce the likelihood of saturation of the neurons.
- It is good practice to randomize the order of presentation of training examples from one epoch to the next. This improves the speed of convergence and avoids the possibility of limit cycles in the evolution of the synaptic weights. The training process is maintained on an epoch-by-epoch basis until the synaptic weights and biases of the NNW stabilize and the MSE (mean square error) over the entire training set converges to some minimum desired value.
- For on-line training, pattern-by-pattern example data updating rather than batch updating should be used for weight adjustments. This method of training is much faster.

Q. Can the backpropagation NNW perform linearly separable pattern classification like the perceptron? Can it also perform nonlinearly separable pattern classification?

A. Yes. The backpropagation network has a much higher computation capability than a perceptron. It can be trained for arbitrary input–output nonlinear mapping that includes linearly and nonlinearly separable pattern classifications.

Q. A feedforward neural network filters out the fundamental component in a delayless manner from a distorted wave. Will the NNW work properly for any arbitrary distortion of the input wave?

A. No. The distortion of the wave should be recognizable by the NNW. For a predetermined range of distortion, the NNW can be trained off-line thoroughly to extract the desired fundamental wave. If the distortion exceeds a predetermined range, the NNW will not perform properly. Consider an active harmonic filter (AHF), in which the filter is supposed to sink the ripple current. For predetermined distortion, a NNW will extract the fundamental current in phase, which can be extracted from the input current to generate the current command for the AHF. For any distortion beyond the predetermined range, the NNW should be trained on-line to function properly. A small amount of extrapolation is satisfactory for NNWs. In summary, NNWs can only interpolate a signal, but their extrapolation property is very limited. The NNW will not function if the waveform is arbitrary.

Q. Can you implement a neural network with a look-up table in a DSP?

A. Yes. A feedforward NNW is basically an input–output nonlinear mapping. For example, three-phase sine-wave generation, as shown in Figure 11.74, is usually implemented by means of a DSP look-up table. It is fast, but requires large memory space if highly precise data are required. The NNW can also be computed by a DSP at the expense of computation time. However, the advantage here is that a NNW has inherent interpolation capabilities. This means that coarse data training will permit precision data generation. With more input and output variables, look-up table implementation becomes complex. In that case, DSP computation is desirable in the absence of a NNW ASIC chip.

Q. Can you implement a fuzzy controller by means of a feedforward NNW?

A. Yes. Because fuzzy systems and neural networks basically implement nonlinear input–output mapping, a fuzzy controller can be implemented by a NNW.

Q. Does the power electronics and machine drives technology appear to be almost saturated?

A. Generally, in the author's opinion, the answer is yes. But it is always difficult to predict the future course of a technology. Our past experience can only be extrapolated to predict the future. Recently, there has been an explosion in the global power electronics literature. Carefully looking into these publications, it should be evident that the majority of these contributions are incremental in nature. Note that every power electronics application is different, and some incremental contribution is needed in each case, thus giving opportunity for new papers. Large numbers of papers are also generated from the results of analytical, simulation, and sometimes experimental studies on systems with some perturbation of structure or performance criteria that may not have direct application relevance. True "new" contributions such as vector control, DTC control, sliding mode control, fuzzy or neural control, space vector PWM, or novel converter topologies are very rare nowadays. In the early days of power electronics, major contributions that advanced the technology used to come from large corporations such as GE, Westinghouse, Siemens, and Brown-Boveri, etc. But, nowadays, at the international conferences, these are being reported by university professors and graduate students. The company papers are rare.

In a technology evolution, any new invention alters its direction and brings new momentum to R&D efforts. For example, in the history of power electronics evolution, the invention of the transistor and then the thyristor ushered in new eras. But, at the end of the thyristor era (in the 1970s), it was felt that the technology was coming to a halt.

Fortunately, the advent of new power devices opened new frontiers in power electronics. The advent of microprocessors and modern personal computers gave new momentum to the technology. Fortunately, decreases in cost and size and performance improvements accelerated the number of power electronics applications and promoted a tremendous surge in application-oriented design and development work. But truly they contribute only incremental technology advancements. Currently, AI-based intelligent control and estimation, particularly artificial neural networks, has brought a new dimension to the technology. The advent of large band-gap power semiconductors will again bring new application frontiers to the technology. There is a tremendous potential for R&D in power devices and VLSI technologies that will affect the power electronics area, but truly speaking, they do not fall into the mainstream of power electronics.

In summary, and in the author's opinion, power electronics and motor drives area are nearing saturation as shown by the curve in Figure 12.19.

The technology will continue to evolve at a gentle rate for a prolonged period of time provided another new invention does not alter its course.

FIGURE 12.19   Saturation trends of power electronics and motor drives.

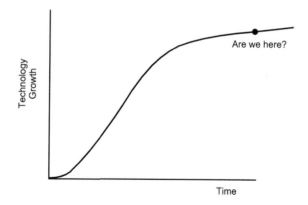

References

[1] B. K. Bose, "Energy, environment, and advances in power electronics," *IEEE Trans. Power Electronics,* vol. 15, pp. 688–701, July 2000.

[2] M. Kamiya, "Development of traction drive motors for the Toyota hybrid system," *Intl. Power Electronic Conf. Rec.,* Niigata, Japan, Paper no. S43-2, April 2005.

[3] *POWEREX IGBT Intelligent Power Modules,* Application and Data Book, 1998 (see pwtx.com).

[4] S. Bhattacharya, B. Fardenesh, B. Shperling, and S. Zelingher, "Convertible static compensator: Voltage source converter based FACTS application in the New York 345 kV transmission system," *Intl. Power Elec. Conf.(IPEC) Rec.,* Niigata, Japan, April 2005.

[5] L. Gyugyi, C. D. Schauder, S. L. Williams, T. R. Rietman, D. R. Torgenson, and A. Edris, "The unified power flow controller: A new approach to power transmission control," *IEEE Trans. Power Delivery,* vol. 10, pp. 1085–1097, April 1995.

[6] C. Schauder, M. Gernhardt, E. Stacey, T. Lemak, L. Gyugyi, T. W. Cease, and A. Edris, "Operation of ±100 MVAR TVA STATCOM," *IEEE Trans. Power Delivery,* vol. 12, pp. 1805–1811, October 1997.

[7] G. L. Skibinski, R. J. Kerkman, and D. Schlegel, "EMI emissions of modern PWM ac drives," *IEEE Industry Applications Magazine,* pp. 47–81, November/December 1999.

[8] B. K. Bose and D. Kastha, "Electrolytic capacitor elimination in power electronic system by high-frequency active filters," *IEEE IAS Annu. Meet. Conf. Rec.,* pp. 869–878, 1991.

[9] P. Z. Grabowski, M. P. Kazmierkowski, B. K. Bose, and F. Blaabjerg, "A simple direct-torque neuro-fuzzy control of PWM-inverter-fed induction motor drive," *IEEE Trans. Ind. Elec.,* vol. 47, pp. 863–870, August 2000.

[10] M. Fenger, S. R. Campbell, and J. Pedersen, "Motor winding problems caused by inverter drives," *IAS Applications Magazine,* pp. 22–31, July/August 2003.

[11] M. Yano, S. Abe, and E. Ohno, "Historical review of power electronics for motor drives in Japan," *Proc. IPEC, Niigata, Japan,* pp. 81–89, 2005.

[12] M. M. Bakran and H. G. Eckel, "The use of IGBTs in traction converters," *Intl. Power Elec. Conf.,* Niigata, Japan, April 2005.

[13] R. W. De Doncker, J. O. Fiedler, N. H. Fuengwarodsakul, S. E. Bauer, and C. E. Carstensen, "State-of-the-art of switched reluctance drives for hybrid and electric vehicles," *IPEC Rec.,* Paper no. S43-1, April 2005.

[14] *Real-Time Workshop for Use with Simulink,* The MathWorks, Natick, MA, 2004.

[15] Y. F. Li and C. C. Lau, "Development of fuzzy algorithms for servo systems," *IEEE Control Systems Magazine,* April 1989.

[16] S. Haykin, *Neural Networks,* Macmillan, New York, 1994.

[17] M. T. Kyaw, "Modular adaptive neuro-fuzzy PID controller based on online process response estimation," Doctoral thesis, University of Hyderabad, India, May 2005.

INDEX

A

Absolute position encoder
  for self-controlled synchronous machine,
    489–490
  slotted-disk-type, 491–492
ac-ac converter, Simulink simulation, 594–596
ac-HFAC-ac converters, 266–267
ac motor drives
  dc-HFAC-ac converter, 277
  three-phase ac controllers, 117
  thyristor ac voltage controllers, 112–113
  use proliferation, 395
  variable-frequency ac drives, 395
  variable-speed, for induction motor
    control, 394
  voltage-fed converters for, 232–233
ACS1000, function and features, 416–417
ACSL Sim, for computer simulations, 583
ac switches, and ac-HFAC-ac power conversion,
    266–267
Active harmonic filter, PWM rectifier
    application, 318
ADC, as microcomputer component, 608–609
AHF, *see* Active harmonic filter
AI, *see* Artificial intelligence
Airgap flux-oriented vector control, induction
    motors, 430–431
Algebraic loops, in ac-ac converter model
    equations, 596
Analog resolver, in position encoder, 493
ANN, *see* Artificial neural network
Application specific integrated circuits, as digital
    controller type, 606
ARCP converter, *see* Auxiliary resonant
    commutated pole converter
Artificial intelligence
  classification, 653
  definition, 652
  intelligent control definition, 654
  in power electronics technology, 17

Artificial neural network, as artificial
    intelligence, 653
ASCI inverter, *see* Autosequential current-fed
    inverter
ASIC, *see* Application specific integrated circuits
Assembly language instructions
  for MCS-51 microcontroller family, 613
  for Texas Instruments TMS320C30, 618
Asymmetric six-phase synchronous machine, 303
Autosequential current-fed inverter
  features, 293
  for induction motor control, 291–292
  performance comparison, 320–321
Auxiliary resonant commutated pole converter, 261
Auxiliary signal injection method, for induction
    motor speed estimation, 458

B

Band-pass filter, in induction motor speed
    estimation, 458
BE, *see* Blaschke equation
Bipolar junction transistors
  basic features, 48
  overview, 44–45
  properties, 47
Bipolar power transistors, device evolution, 27–28
BJT, *see* Bipolar junction transistors
Blaschke equation, and induction motors, 463–465
BLDM, *see* Brushless dc motor
BPF, *see* Band-pass filter
BPT, *see* Bipolar power transistors
Brushless dc motor
  closed-loop current control, 532
  closed-loop speed control, 531
  definition, 383
  as self-controlled machine, 488, 526
  torque-speed characteristics, 530
Buck-boost converter, speed control, 161–162
Buck converter, computer simulation example, 586
Buck dc-dc converter, 159–160

Printed and bound by CPI Group (UK) Ltd, Croydon, CR0 4YY

08/05/2025

01864862-0002